CANADIAN SPECIAL PUBLICATION OF FISHERIES AND AQUATIC SCIENCES 56

Oceanography

of the

British Columbia Coast

RICHARD E. THOMSON

Department of Fisheries and Oceans
Ocean Physics Division
Institute of Ocean Sciences
Sidney, British Columbia

DEPARTMENT OF FISHERIES AND OCEANS

Ottawa 1981

Available in Canada through
Associated Bookstores
and other booksellers
or by mail from
Canada Communication Group — Publishing
Ottawa, Canada K1A 0S9

Catalogue No. FS41-31/56E
ISBN 0-660-10978-6

Reprinted 1983, 1984, 1991

Correct citation for this publication:

THOMSON, R. E. 1981. Oceanography of the British Columbia coast.
Can. Spec. Publ. Fish. Aquat. Sci. 56: 291 p.

for Justine and Karen

45-

Contents

Contents *(concluded)*

FOREWORD

Oceanographers are a cult unto themselves. To outsiders they appear as people who, for reasons apparently known only to themselves, have a passion for analyzing water. Their god is the sea, their creed to understand, to interpret, and ultimately to respect its seemingly endless mysteries. The language they use in their chapels is mathematics and their priests are computers. No wonder then that the members of the cult find themselves locked away in remote institutions, removed from view by the public.

This book represents an attempt by one of them to explain the workings of the cult and to describe the nature of their religion in language that all can understand. This work is not so much a text book as it is a biography, a biography of the sea. In an easy conversational style and with numerous illustrations the author describes the origin, physical characteristics, and nature of the forces that act within and upon the eastern Pacific Ocean. He describes the origin, shape, and structure of the sea floor as well as the development of coastlines and beaches. Without being turgid or condescending, the author carefully explains the complex movements of water masses, first in general terms and then specifically with regard to the various coastal regions of British Columbia. He describes the seasonal variations in salinity and temperature and, in a remarkably lucid manner, the complex interaction of planetary body forces which result in the variability of tides. Many of his own experiences as well as those of his colleagues are used to illustrate some of the more spectacular effects of water mass motions, such as the tidal rips in the northern Strait of Georgia, internal gravity waves in Knight Inlet, and the unique tidal whirlpools in Gilford Passage. These and many other aspects of coastal oceanography are presented in such a way that the excitement of the author for the nature of his work is transferred to the reader.

This book should have wide appeal. Not only is it suitable for scientists and their students, but should become a standard reference to be found in the libraries of commercial vessels as well as in the collections of yachtsmen. Although the author has no pretensions as to its evangelical significance, he has, with this book, made an outstanding contribution to the enjoyment of the coastal waters of British Columbia.

C. J. YORATH
Geological Survey of Canada
Sidney, B.C.

BACKGROUND INFORMATION

Introduction

The origins of this book date back more than six years when I first began a series of oceanographic articles for an innovative Vancouver-based yachting magazine. Then, as now, numerous books had been written about the sea but only rarely did one of them refer to the fascinating oceanic processes that characterize the coastal waters of British Columbia and Washington State. This work has evolved from the same information void that inspired the original yachting articles. As in any metamorphosis, however, its contents mark a substantial advance in quantity, diversity, and sophistication.

In writing the book, I have endeavored to present a text that would be readily comprehensible to the intelligent layman without compromising the scientific content. To attain this sometimes elusive goal, the book has been divided into three major sections: the first (Part I) covers the origin and present physiographic makeup of the oceans; the second (Parts II and III) deals with many principal features and concepts of physical oceanography and provides the necessary fundamentals to an understanding of oceanic phenomena discussed in later chapters; the third (Parts IV and V) describes in detail the physiography, meteorology, and oceanography of the coastal and deep-sea regions of the Northeast Pacific Ocean. All material, including the appendices and more than two hundred and seventy diagrams, has been subjected to careful scrutiny by experts in the various fields of marine science to ensure its relevance and authenticity. It is hoped that the end product will provide the reader with an enjoyable, as well as practical, guide to the nature of the world oceans and instill a new degree of respect for the physical beauty of the Pacific west coast.

Acknowledgments

I gratefully thank the many colleagues and friends who have contributed to the contents of this book. I am especially indebted to Professor George Pickard, recently retired Director of the Department of Oceanography at the University of British Columbia, and to Dr Christopher Yorath, Head of the Marine Geology Section of the Pacific Geoscience Centre, who not only provided valuable advice and criticism of the manuscript but also were a source of personal encouragement and inspiration. Considerable thanks are due Ms Sharon Thomson and Mr Robert Sandilands for carefully reviewing an early draft of the text and Dr Rob Macdonald for expertly reviewing the penultimate version. My sea-going colleagues Mr Stan Huggett, Dr Patrick Crean and Mr James Holbrook are acknowledged for so generously providing me with technical information while Mr Allan Douglas is acknowledged for his excellent work on the many computer graphics in the figures. My appreciation also to Ms Doris van Aanhout, who skillfully transformed my handwriting into the first draft of the manuscript; to Ms Susan McKenzie, who assisted with the many subsequent revisions; and to Ms Billie Mathias, who helped complete the manuscript. Not least, I gratefully thank those colleagues in the Department of Fisheries and Oceans who helped get the book published and my wife Irma who, in addition to reviewing the earlier drafts, patiently endured the countless nights and weekends I spent cloistered in my study.

Lastly, I thank the following companies, societies, government departments, and associations for granting permission to reproduce material from their publications: Aanderaa Instruments Ltd. (Victoria); Academic Press, Inc.; American Association of Petroleum Geologists; American Geophysical Union; American Meteorological Society; Canadian Hydrographic Service, Pacific Region; Dodd, Mead and Company; Doubleday and Company, Inc.; Dowden, Hutchinson and Ross, Inc.; Edward Arnold Ltd.; Elsevier Scientific Publishing Company; Geological Society of America, Inc.; International Hydrographic Review; International North Pacific Fisheries Commission; John Wiley and Sons, Inc.; Department of Supply and Services; Pergamon Press Ltd.; Prentice-Hall, Inc.; Seismology Society of America; U.S. Department of Commerce; U.S. Naval Oceanographic Office; Webb Institute of Naval Architecture.

Abstract

THOMSON, R.E. 1981. Oceanography of the British Columbia coast. Can. Spec. Publ. Fish. Aquat. Sci. 56: 291 p.

This book deals with the physical aspects of the sea as exemplified by the Pacific Ocean and the contiguous waters of the British Columbia coast. Although principally devoted to waves, currents, and tides, the book spans a broad spectrum of topics ranging from meteorology and marine biology to past and present marine geology. It attempts to elucidate the nature of oceanic motions and to relate them to everyday experience for the general interest of the casual reader and for the practical benefit of the professional mariner, scientist, or engineer.

Key words: oceanography, British Columbia, tides, currents, upwelling, waves

Résumé

THOMSON, R.E. 1981. Oceanography of the British Columbia coast. Can. Spec. Publ. Fish. Aquat. Sci. 56: 291 p.

Le présent ouvrage traite des aspects physiques de la mer avec, comme exemple, l'océan Pacifique et les eaux continguës de la côte de la Colombie-Britannique. Bien qu'il porte surtout les vagues, les courants et les marées, le volume aborde une variété de sujets, de la météorologie et de la biologie marine à la géologie marine passée et présente. L'auteur tente d'élucider la nature des mouvements de la mer en rapport avec l'expérience de tous les jours, tant à l'adresse du profane que du professionnel, marin, scientifique ou ingénieur.

PART 1
HISTORY AND NATURE OF THE COAST

Chapter 1. Historical Setting

Origin of the Oceans

Water is abundant within the solar system. It is locked into the icy cores of the larger planets from Saturn to Pluto, over 6000 million km from the sun. It resides in the frosty polar regions of Mars, and forms the swirling clouds of Jupiter's lower atmosphere, and the slowly drifting oceans that cover 71% of the earth's surface. Without such an abundance of water it is conceivable that life could not have evolved on earth, for it was in the vast protective seas that the synthesis of organic compounds first began. It is not surprising, therefore, that the need to explain the origin of the oceans occupies a special place in man's endeavor to understand the nature of the universe.

The widely accepted theory in science today is that the solar system was born over 4½ billion yr ago from an immense rotating cloud of gas and dust (Fig. 1.1a). Contained in this primordial cloud of uncertain origin were all the ingredients that now constitute the planets: hydrogen, helium, and other gases; dust of diverse composition, metals, radioactive elements, and ice. In addition, a large amount of water was locked within the chemical bonds of the dust. As gravitational forces pulled this material together it began to spin with ever-increasing rotation until it deformed into a disk (Fig. 1.1b). At the center of the disk a massive body was created that eventually became so dense and hot that nuclear fusion of hydrogen and helium began and transformed it into a star known as the sun. During its early development, the sun jettisoned huge quantities of mass that drifted through the infant solar system as an intense solar wind, and left behind the much reduced nuclear furnace that warms the earth today. About the same time the sun was evolving, but away from the central core of the spinning disk, other portions of the primordial cloud were accreting into planets and their accompanying moons (Fig. 1.1c). Much of the remaining dust that lay exposed between the planets was then swept out of the solar system by the early solar winds.

For the first few million years after the consolidation of the sun, the earth coalesced by sweeping up enormous amounts of minute particles from the spinning disk. During these formative years, the earth was a cold and solid planet with an inhospitable atmosphere rich in hydrogen and the noble gases, but devoid of water. As the earth's size increased, temperature and pressure also increased due to the combined effects of enhanced internal gravitational pull, collisions with the multitude of meteorites within the cluttered solar system, and the heat given off by the radioactive decay of elements like uranium, thorium, and an isotope of potassium. Over tens of millions of years so much heat was accumulated that most of the interior of the earth became molten. Liquid iron, nickel, and other heavy elements then sank toward the center to form the core which, except for a small solid central portion, remains liquid to the present time.

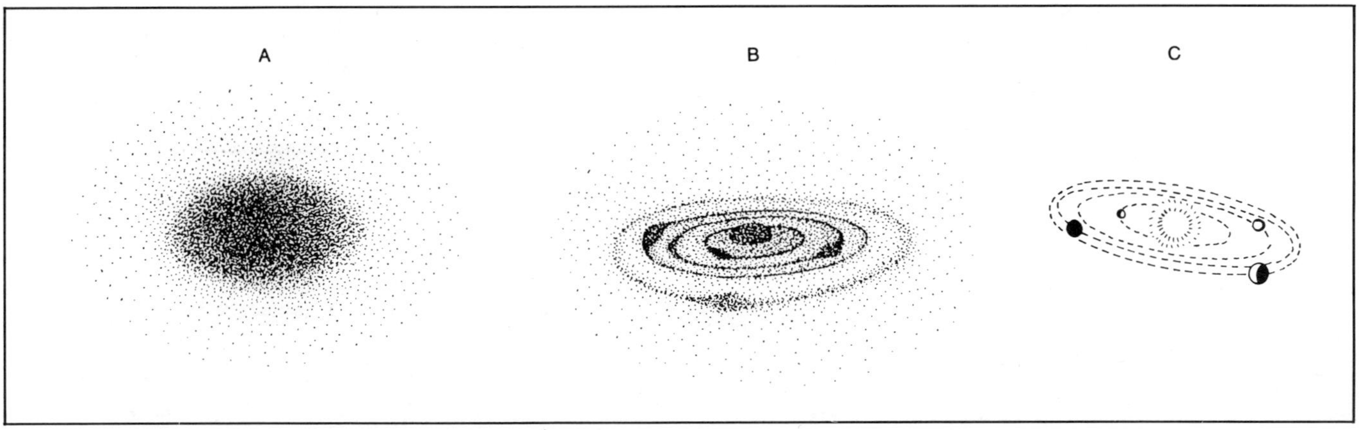

FIG. 1.1. Three stages in evolution of the solar system. Over 4.5 billion yr ago a great cloud of dust and gas (a) far out along an arm of the spiral galaxy began to collapse into a rotating disk. Concentration of this material at center of disk (b) led to formation of a massive body whose increasing density and temperature became enough to ignite its nuclear fuel and turn it into a star, the sun (c). At the same time, material away from core was accreting to form planets and their satellites.

On the other hand, lighter materials, such as aluminum, sodium, silicon, and potassium, together with large quantities of water, floated upward to form the plasticlike upper mantle and rigid crust of the planet. The solar wind by this time had long swept away the accretion atmosphere and in its place was an atmosphere rich in methane, ammonia, and water vapor. Much of this water vapor had been squeezed out of the solids by heat and pressure and had escaped to the surface. It is most likely that the liberated water formed an uninterrupted ocean over the face of the earth, although a considerable portion may also have resided within the atmosphere as steam.

Regardless of whether the water was in liquid or vapor form, it is thought that sometime between 3.7 and 2.2 billion yr ago the earth's surface had cooled sufficiently to allow condensation of most water vapor, to ensure an early presence of oceans on the evolving earth. Moreover, these were salty oceans only slightly less saline than those of today. The salt was extruded to the surface along with the water that resulted from the outgassing process. At present, more than 97% of the world's water is in the ocean basins and only 0.001% is in the atmosphere.

Within the protected, nutrient-rich environment of these primordial seas the synthesis of complex organic compounds first took place and the earliest living organisms came into existence. The evolution of life had begun. The earth itself has continued to evolve in a geological sense and at present has a layered structure like an onion (Fig. 1.2). At its surface, large crustal plates have formed

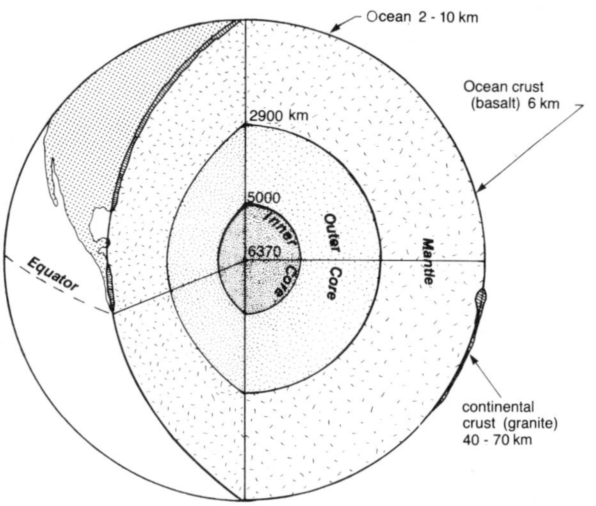

FIG. 1.2. Internal structure of earth. Mantle is magnesium-iron silicates, outer core molten iron and nickel, and inner core solid iron and nickel.

which, akin to thin paving stones that float on a viscous layer of molten tar, migrate slowly over the surface of the globe, and continually reshape the boundaries and dimensions of the oceans contained between them.

A fortunate combination of circumstances has permitted a continued existence of the oceans. Earth is close enough to the sun that sea water doesn't freeze in great quantities, yet distant and massive enough that gravitational pull prevents any significant amount of water vapor from escaping into outer space. Consequently, water in the oceans is old stuff indeed, dating back billions of years when it first emerged onto the surface of the primeval earth. If new water is still being squeezed out somewhere from the interior of the earth it has yet to be detected. We truly sail upon ancient seas!

Drifting Continents

The slaglike crust that formed on the cooling exterior of the earth eventually became distributed as continental land masses, with thicknesses of around 50 km, and thinner oceanic basins, with thicknesses of around 10 km. In the mantle beneath the crust, the partially melted rocks of the asthenosphere have been kept in continual motion by the heat from radioactively decaying material. Resembling vertically circulating motions in a pot of heated broth (Fig. 1.3), the slow upward convective currents associated with internal heating have resulted in the mas-

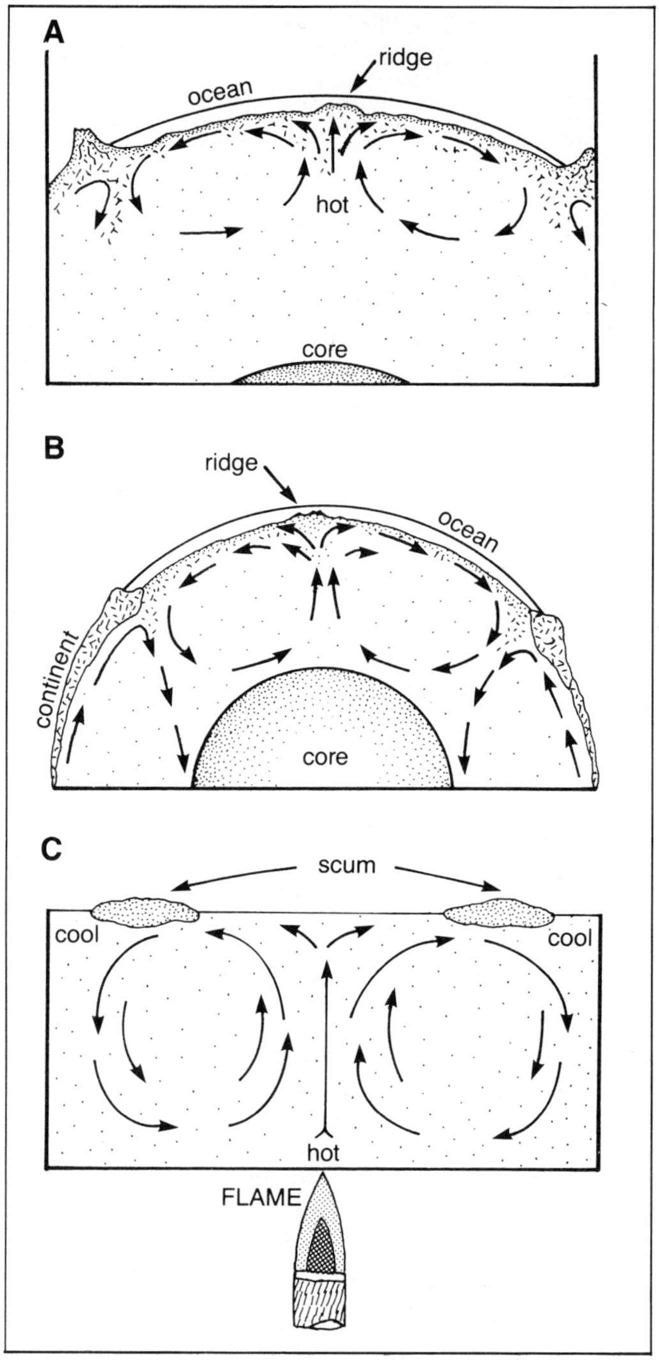

FIG. 1.3. Two possible convection-cell models of earth's mantle (A, B) and convection in a pot of heated broth (C).

sive extrusion of asthenospheric magmas through the seafloor along oceanic mountain chains called mid-ocean ridges. Solidification of these magmas to form basaltic rocks has added new material to the crust and underlying lithosphere, so modern ocean floors are among the youngest tectonic features of the earth's surface. The ridges then become centers for the divergent processes of seafloor spreading and newly created lithosphere is transported laterally away from the ridges toward regions of convergence (Fig. 1.4). Continental land masses drift away from the divergent regions, and are passively carried along at the speed of seafloor spreading.

At zones of plate convergence, subduction of the oceanic lithosphere occurs and the lithospheric plate is bent downward, often accompanied by the formation of trenches, the deepest regions of the world oceans. Frictional heat generated at the surface of the sinking plate causes it to undergo partial melting as it descends into the earth's interior. Molten materials may subsequently reemerge at the surface as continental volcanoes landward of the trench, or as an arc of volcanic islands in proximity to the trench. The volcanic island chains of Indonesia exemplify island arcs that originated in the vicinity of a subduction zone. [There are also volcanic island arcs that have no connection with converging plate boundaries. The Hawaiian Islands for instance, originate where thermal plumes of magma a few hundred kilometres in diameter rise to the lithosphere from the outer core, to create a hot spot of volcanic activity. Some 20 hot spots exist. The most recent volcanic activity occurs at one end of the chain (the big island of Hawaii, for example) while dormant or extinct volcanoes are at the opposite end of the chain (Kauai). The chain is aligned in the direction of motion of the plate on which the volcanoes reside, with the most active volcanoes at the leading edge of the advancing plate.] Examples of continental volcanoes associated with crustal subduction are Mount Baker, Mount Rainier, and the recently active Mount St. Helens in Washington State, and Mount Garibaldi in British Columbia. Other examples within the Pacific "Rim of Fire" include the Aleutian Islands with spectacular volcanic peaks and world-famous Mount Fuji in Japan. The Aleutian trench just south of the Aleutian Island chain reaches depths over 8000 m in a region where typical ocean depths are in the neighborhood of 5000 m. Less than 200 km to the north, the trenches give way to volcanic islands, whose conical peaks rise abruptly as much as 2800 m from the surface of the bordering seas.

It is believed that the processes of seafloor spreading have been going on continuously throughout most of geological time. The present distribution of continental and oceanic plates, however, derived from a stage of seafloor spreading that began some 200 million yr ago with the breakup of the supercontinent Pangea, meaning "all lands." The breakup resulted in the formation of the six major global "paving stones" or lithospheric plates, which totally encompass the surface of the earth (Fig. 1.5). In Fig. 1.6, boundaries of the plates are delineated by regions of convergence, as found at trenches or certain mountain ranges, by regions of divergence such as oceanic ridges, and by faulted zones where plates are sliding past one another. The divergent boundaries, or spreading ridges, stretch the lengths of the Atlantic, India, and South Pacific oceans, whereas in the North Pacific, spreading regions

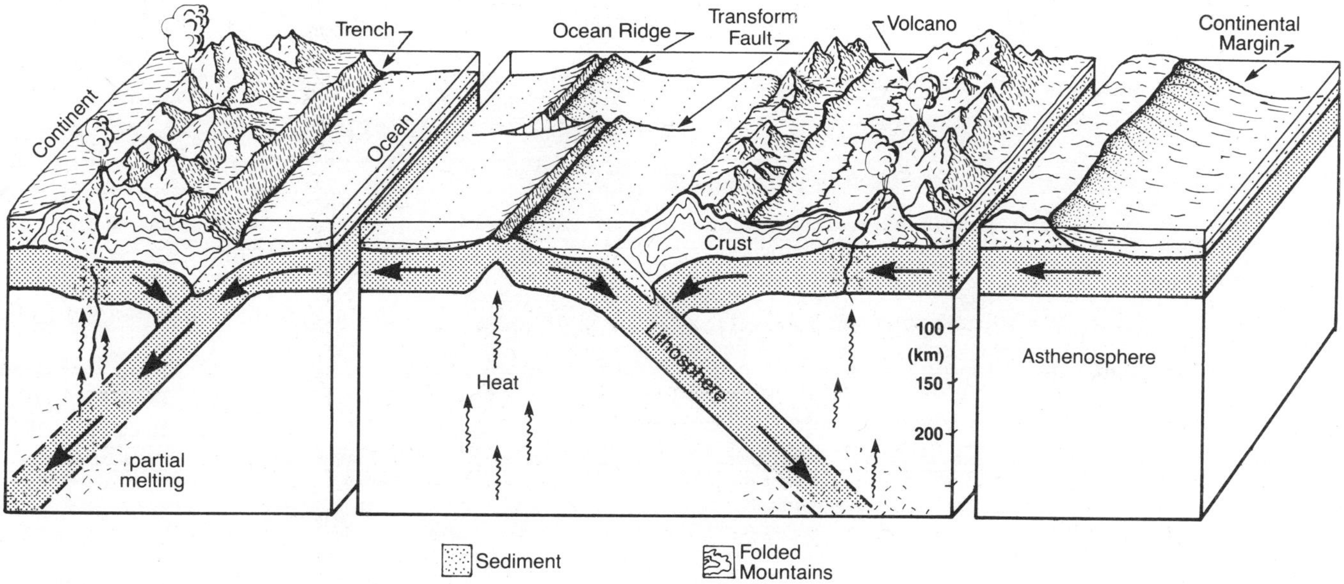

FIG. 1.4. Schematization of seafloor spreading and continental drift (Plate Tectonics). Remelting of oceanic lithospheric basalt follows subduction to depths of around 200 km within the asthenosphere. High rugged mountains and deep oceanic trenches are formed by the crust buckling at contact zone between oceanic and continental plates. Volcanoes occur where magma from remelt zone penetrates fractured crustal rocks.

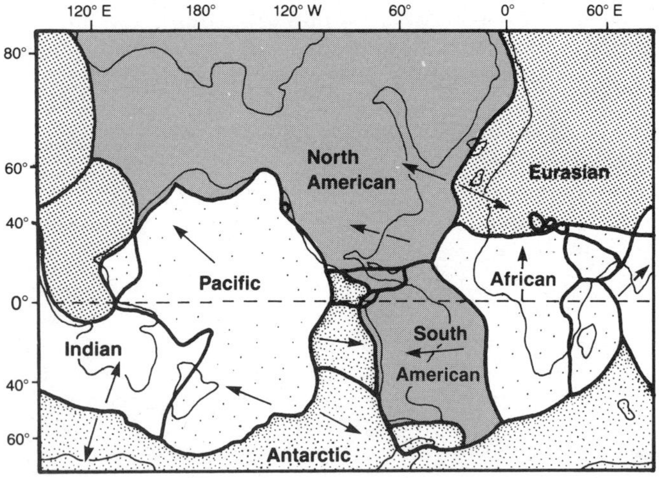

FIG. 1.5. Extent of the six major lithospheric plates and associated plate boundaries on earth's surface and present coastline. Arrows indicate plate motions relative to adjoining boundaries. (From Wyllie 1976)

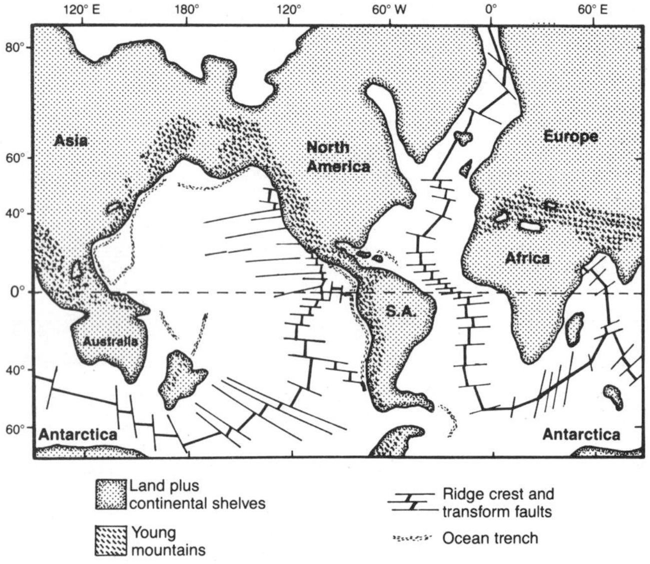

FIG. 1.6. Active geological features associated with the major lithospheric plates. (From Wyllie 1976)

appear to be confined to the eastern side of the ocean. A particularly active system of oceanic ridges is off the coasts of British Columbia and Washington. Convergent boundaries around the Pacific rim are found beneath the continental slopes (Fig. 1.5). Earthquake activity tends to go hand in hand with lithospheric motions at the convergence zones, and make the coastal belt of the Pacific Ocean one of the most earthquake-plagued regions of the world (Fig. 1.7).

The rock injected into the ocean floor at the ridges requires over tens of millions of years to reach a subduction zone, and accumulates a thin covering of sedimentary "dust" from the organic and inorganic matter that settles out of the overlying ocean. A portion of these sedimentary rocks descends into the mantle with the subducting lithosphere while the remainder is compressed, folded, and converted to metamorphic rocks by the powerful contact forces at the plate boundaries. Tectonic mechanisms such as these account for the extensive system of young mountain chains that almost encircle the entire rim of the Pacific Ocean, from the Andes of South America, to the coastal ranges of North America, along the rugged Aleutian and Kamchatka Island chains, and then through to the mountainous backbones of Japan, the Philippines, and New Zealand (Fig. 1.6). In a sense, these mountain belts resemble a crumpled "bow" generated as the continental plate plows its way into the oceanic plate.

Early evidence

The notion that there are continental and oceanic plates that move independently of one another over the surface of the earth is embodied in the Theory of Plate Tectonics; a new theory which has only gained worldwide acceptance within the last decade. However, similar ideas were proposed as early as 1620 by Sir Francis Bacon, who suggested that the Americas were once joined to Europe and Africa, and by a Frenchman named Placet who in 1668 wrote an article in which he attempted to show that America was not separated from the rest of the world before the "Deluge." Near the turn of the 19th century, important independent geological studies by Eduard Seuss of Austria and Alfred Wegener of Germany presented strong evidence for the possibility of continental drift. Like working with pieces of a jigsaw puzzle, Seuss found that the land masses of the southern hemisphere fitted into a single continent he called Gondwanaland, whereas Wegener proposed that all continents were joined over 200 million yr ago as the vast supercontinent of Pangea.

Pangea began to break up roughly 200 million yr ago by rupturing along what is now the Mid-Atlantic Ridge. This led to the formation of the great continent of Gondwanaland in the southern hemisphere, comprised of South America, Africa, India, Antarctica, and Australia, and the equally vast continent of Laurasia in the northern hemisphere, comprised of North America and Eurasia. Further rupturing within the two great continents eventually allowed the smaller continental areas of today to drift free. Figure 1.8 attempts to trace the history of continental drift over the last 200 million yr. As an example of the kind of motions involved, note that India drifted to the northeast and eventually collided with Eurasia; the Himalayas were formed where the rocks of the two converging continental plates were uplifted by compressive forces, whereas the Tibetan Plateau was formed as the northern edge of the Indian subcontinent slid beneath, then uplifted, the southern margin of Eurasia.

Because geophysical studies of the ocean floor off the southwestern corner of British Columbia have played an important role in the development of the Theory of Plate Tectonics, it is worthwhile to delve a little deeper into the nature of the geological processes involved. This also will serve as a prelude to understanding the evolution of the coastline. There are two basic facts that must be grasped from the onset. First, the iron and nickel compounds within the molten rock that pushes through the oceanic ridges are influenced by the magnetic field, which origi-

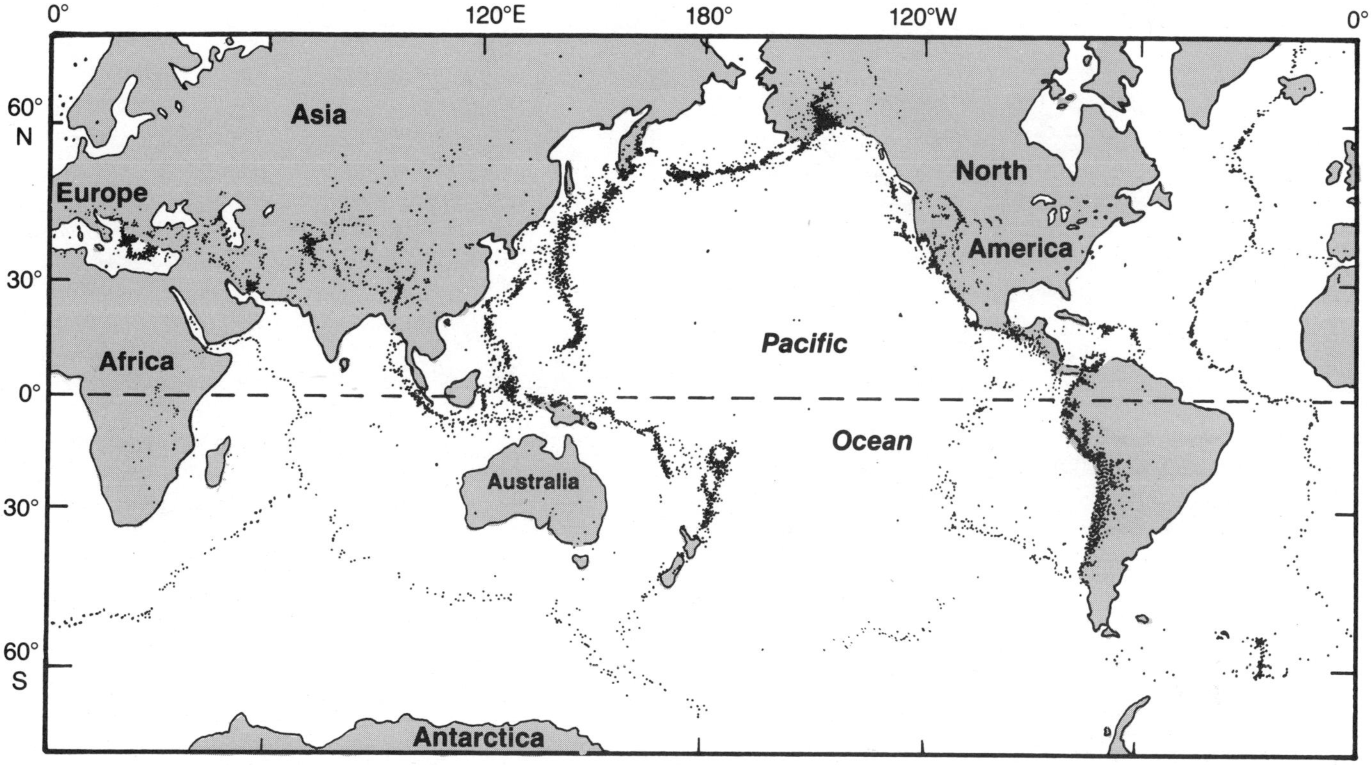

FIG. 1.7. Distribution of earthquake epicenters recorded between 1961–67. Note close relationship between earthquake zones and plate boundaries of Fig. 1.5. (Adapted from Barazangi and Dorman 1969)

nates with the liquid outer core of the earth. As the rock solidifies, the magnetic domains within these compounds become "frozen," and thereby provide a historical record of the ambient magnetic field of the earth. Once these small magnets are frozen in this manner, the magnetic recording of the rock cannot be altered by a change in the earth's magnetism. Second, the earth's magnetic field periodically undergoes abrupt reversals in direction of polarity accompanied by changes in intensity; the magnetic north pole migrates toward the geographic south pole and the magnetic south pole migrates toward the geographic north pole, with a few thousand years in between when there is no magnetic field at all. In the last 10 million yr reversals have occurred approximately once every million years. (The geographic poles, of course, remain fixed at the ends of the axis about which the earth spins.) Therefore, as the seafloor spreads away from both sides of a ridge at the rate of a few centimetres per year, the rocks that contain iron–nickel compounds will carry with them a natural record of the earth's past magnetism. In regions where these crustal rocks carry the same magnetic polarity as that of the earth's present magnetic field, the total measured magnetic intensity will be anomalously high (the two intensities reinforce one another); where the polarity is opposite to that of the earth's, the measured intensity will be anomalously low (the two intensities oppose one another). The difference between the actual measured magnetic intensity and an average value based on more widespread regional measurements is called the magnetic anomaly (Fig. 1.9). Positive anomalies occur in regions where the rocks solidified when the earth's magnetic field had the same polarity it has today; negative anomalies occur in regions where the rocks were formed during periods of reversed polarity.

Maps of magnetic anomalies obtained from air and shipborne magnetometer surveys over seafloor spreading regions consist of parallel bands of alternating polarity (Fig. 1.10). (Magnetometers are precision instruments capable of measuring the intensity of the earth's magnetic field to an accuracy better than 1 gamma in 1,000,000. The intensity of the earth's field today ranges from 25,000 to 60,000 gammas, depending on geographical location.)

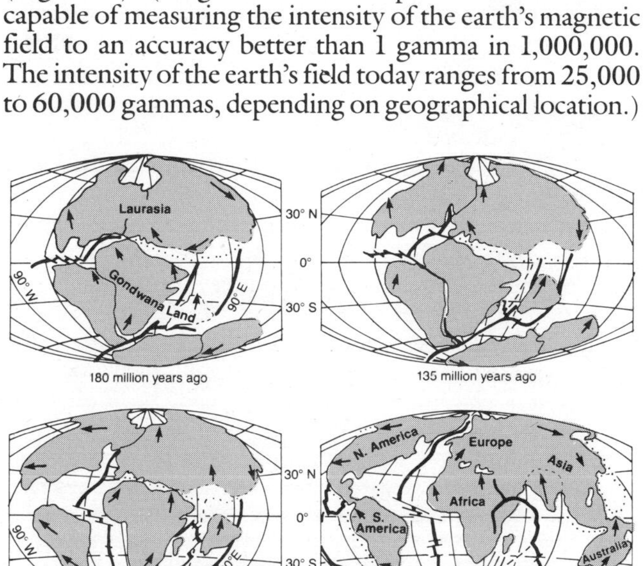

FIG. 1.8. Continental drift during past 200 million yr. Arrows show movement of continents beginning with breakup of the super continent of Pangaea. Solid lines, divergent plate boundaries; dotted lines, convergent plate boundaries. (Adapted from Dietz and Holden 1970)

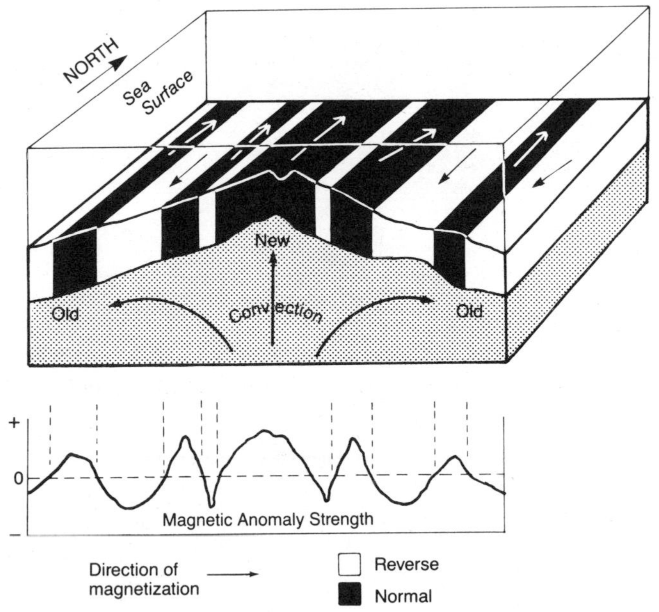

FIG. 1.9. Magnetic anomaly pattern produced by magnetized rocks within the oceanic crust. The ambient magnetic field is indelibly recorded within solidifying magma that spreads away from oceanic ridge; striped pattern of altering direction of magnetization results from combined effect of seafloor spreading and periodic reversals in the polarity of earth's magnetic field.

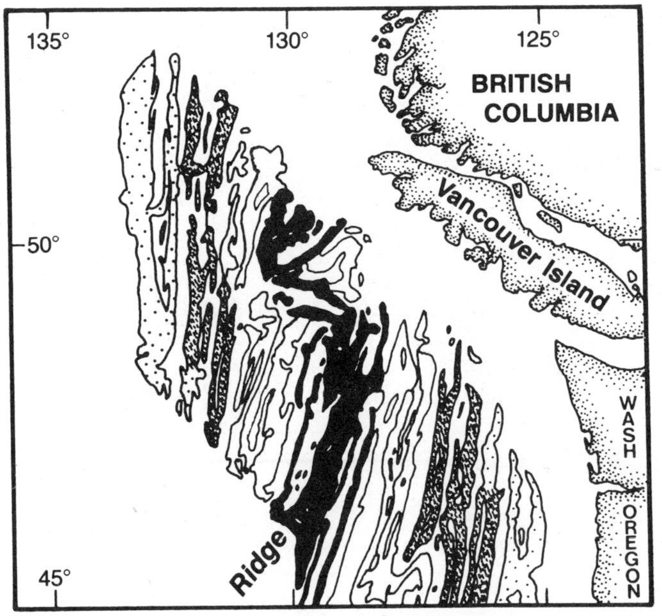

FIG. 1.10. Portion of a magnetic anomaly map of the Northeast Pacific Ocean. Rocks are youngest along darkest stripes, a region of active seafloor spreading, and increase in age away from ridge. Dotted stripes are approximately 8–10 million yr old. (Modified after Raff and Mason 1961)

In conjunction with other geological dating techniques such as radiometric dating, these bands can then be used to establish the rate and direction of spreading over approximately the last 200 million yr. These results also may be compared to paleomagnetic studies, which give the relative positions of the continents to as far back as 600 million yr. One of the earliest magnetic anomaly maps was off the British Columbia–Washington coast by Raff and Mason (1961) (Fig. 1.10). The dramatic striped pattern in this map led to a hypothesis by Vine and Mathews (1963) that linked seafloor spreading with reversals in the earth's magnetic field and paved the way for firm establishment of the Theory of Plate Tectonics.

Finally, changes in the earth's magnetic field may produce some adverse effects. It has been estimated for example that 10,000 yr prior to a reversal the magnetic field strength decreases by 60–80%, reverses direction over the next 2000 yr, and then increases again but in the opposite direction during the next 10,000 yr. Over this period, magnetic compasses would first be rendered useless. Eventually new compass cards would be needed! On a more serious vein, some scientists believe that magnetic field reversals in the past were responsible for the extinction of many species of animals. This is because a much greater number of charged particles and cosmic rays from the outer reaches of the atmosphere would have reached the earth's surface by penetrating the low-intensity magnetic field. This increased radiation might then have caused an alteration in climate together with harmful mutations in genes which, over thousands of generations, would have made the animals less fit for survival. Although there are a multitude of objections to this notion, future scientists will have an opportunity to test its validity. There has not been a reversal for 800,000 yr, much longer than at any time in the last 9 million yr, and a change is decidedly overdue.

Evolution of the Coast

The inside waterways of British Columbia and Washington State occupy a relatively young coastal trough on the surface of the earth known as the Georgia–Hecate Depression (Fig. 1.11). Its formation began around 150 million yr ago as part of a general downfolding of the crust along the Pacific coast that followed commencement of the latest era of continental drift. The depression extends some 4000 km from Alaska to the Gulf of California. The forces necessary to buckle the crust originated with the underthrust of the North American continent by the Pacific Ocean floor. Accompanying the formation of the Georgia Depression was an uplift and folding of the adjoining land form into the Vancouver Island Range and Olympic Mountains to the west, and the Coast Mountains and Cascade Mountains to the east. Since its inception, the Georgia Depression has undergone a number of uplifts and downwarpings. These have resulted in retreats and advances of the sea in this area and in variations in the type of overlying sediment layers. One such uplift between 5 and 10 million yr ago led to a full retreat of the ocean with subsequent formation of an expansive valley to the east of Vancouver Island. The mountain building processes apparently ceased around 2 million yr ago, but subsidence, glacial scouring, erosion, and deposition subsequently modified the Georgia Depression since it took its present form about 1 million yr ago. In addition, other subsidences have occurred during the evolution of the coast, most notably the Juan de Fuca depression that runs nearly east–west between the Olympic and Vancouver Island mountain ranges.

Recent research off the British Columbia–Washington coast has revealed two small crustal features known as the Juan de Fuca and Explorer plates, which have added to the detailed movement of the offshore region. Confined between the two massive crustal plates formed by the North America continent and the Pacific Ocean basin, the two smaller plates have relatively independent movements. As shown schematically in Fig. 1.12, the western boundary of the two small plates consists of a series of oceanic ridges and submarine fracture zones; the ridges mark local regions of seafloor spreading and the fracture zones represent faults along which sliding of adjoining sides takes place. The broken line in Fig. 1.12 marks the eastern extent of the Explorer and Juan de Fuca plates and the geographical position where they begin to dip or subduct under the outer edge of the continent. Figure 1.13 is a three-dimensional cartoon of the geometry of seafloor spreading along the same section of coastline.

The numerous volcanic peaks within the Cascade Mountains and Garibaldi volcanic belt suggest that extensive volcanic activity has accompanied this subduction of the oceanic crust. Although British Columbia volcanoes have been dormant for the past 2500 yr, the spectacular eruption of Mount St. Helens northeast of Portland, Ore., on May 18, 1980, provided firsthand evidence that

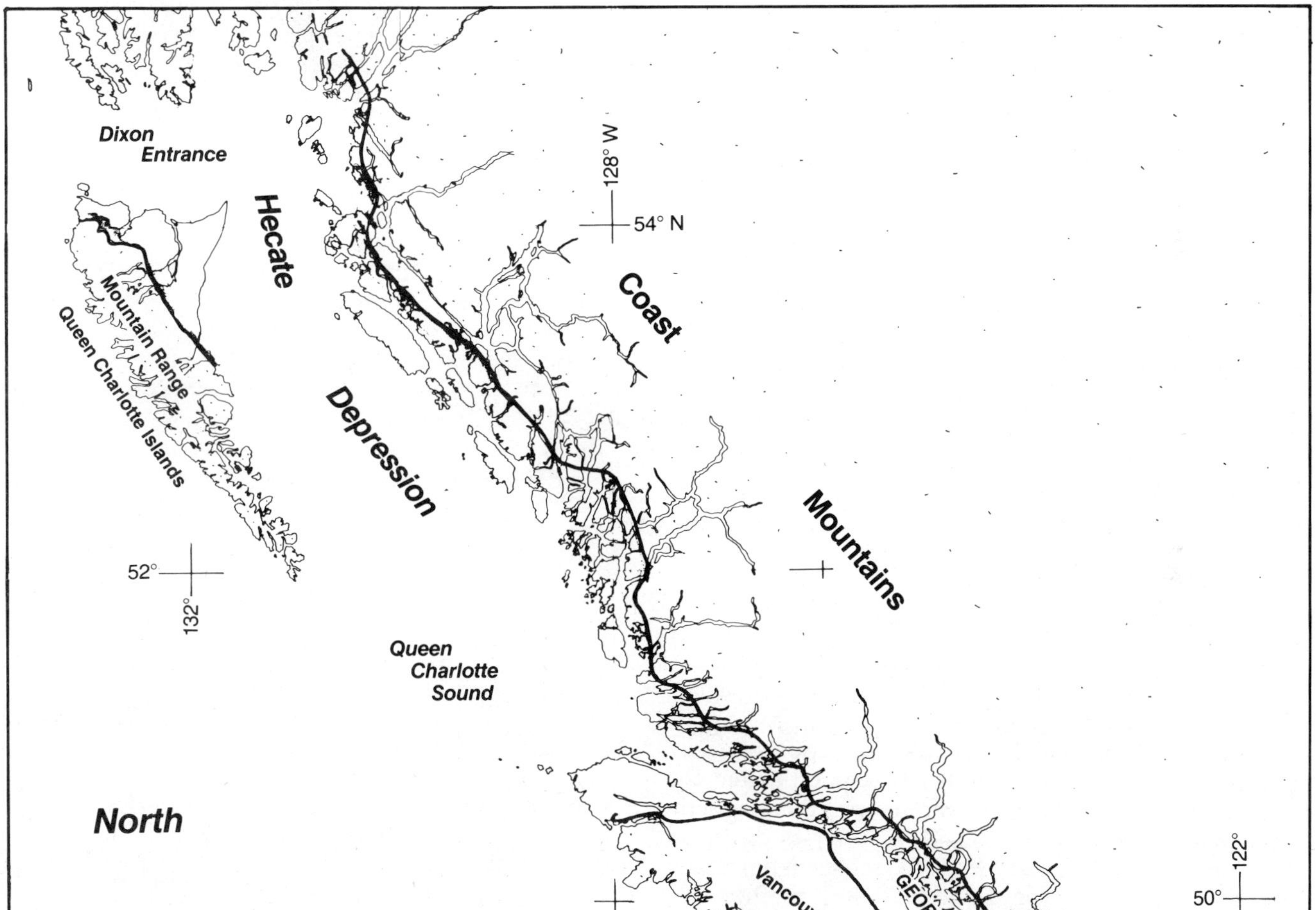

FIG. 1.11. Major geographical features of coastal British Columbia and Washington. (Adapted from Clauge and Bornhold 1980)

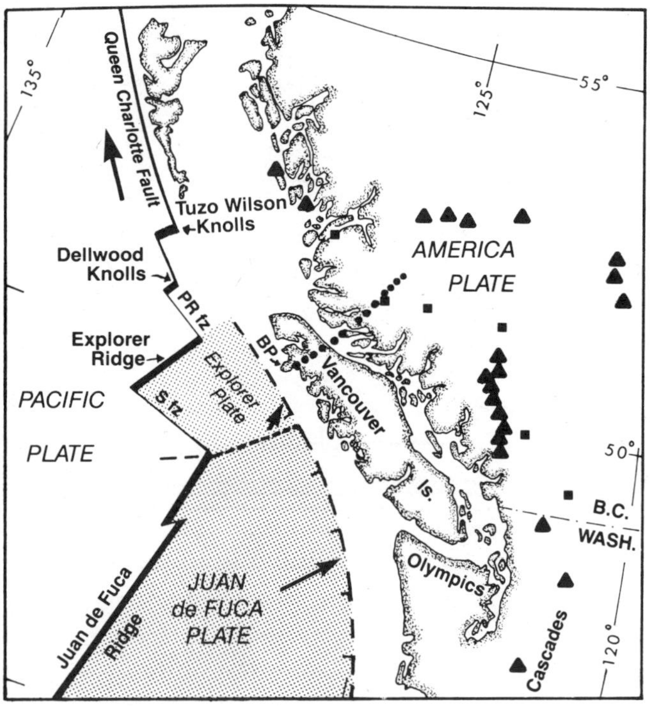

FIG. 1.12. Tectonic features off British Columbia and Washington coasts. BP, Brook's Peninsula; Sfz, Sovanco fracture zone. Broken line is a suggested northern edge of the subducted Juan de Fuca plate. Triangles denote Quaternary volcanoes (1 million yr ago to present), squares Miocene volcanic centers (about 5–22.5 million yr ago). Arrows show probable plate movements relative to American Plate (Modified after Riddihough and Hyndman 1976)

the tetonic forces that have shaped the coast are still extremely active. Accompanying earthquake activity has been intense. Since seismic recording devices were first installed in British Columbia early in this century, hundreds of weak to moderately strong earthquakes have originated within a belt that extends the length of the offshore fracture zones. A secondary region of earthquake epicenters also exists between the Gulf Islands and southern Puget Sound. The fact that the latter earthquakes often originate about 60 km deep in the earth's crust, compared to about 30 km for the offshore earthquakes, is considered proof that the ocean floor is dipping sharply at an angle of 20° under the North American continent.

Such underthrusting is not characteristic of the entire coast. Evidence indicates that in recent times subduction of the ocean floor has not taken place north of Brooks Peninsula on the west coast of Vancouver Island (Fig. 1.12). Instead, the oceanic crust appears to be shifting northward relative to the continental crust along the Queen Charlotte Fault, which meets the Aleutian trench in southeastern Alaska. Subduction of the Pacific plate again occurs south of the Alaska Peninsula and Aleutian Island chain.

Magnetic anomaly maps off the British Columbia coast suggest that the Juan de Fuca plate began to thrust under the continental margin of British Columbia over 50 million yr ago. As a consequence, the leading edge of the plate is now some 320 km inland of the continental margin. Between 10 and 5 million yr ago the plate appears to have moved northeast toward the coast at around 5 cm/yr (1.9 in./yr), but then decreased its speed to about 4 cm/yr over the last 5 million yr. (Although this is not an excessive speed by most standards it can, over millions of years, add up to many hundred kilometres.) The reduced volcanism in the area in recent times has been considered to reflect the decreased rate of subduction to around 3 cm/yr. This compares with subduction rates of around 5–10 cm/yr in the volcanically active regions of Japan and the Aleutian Islands.

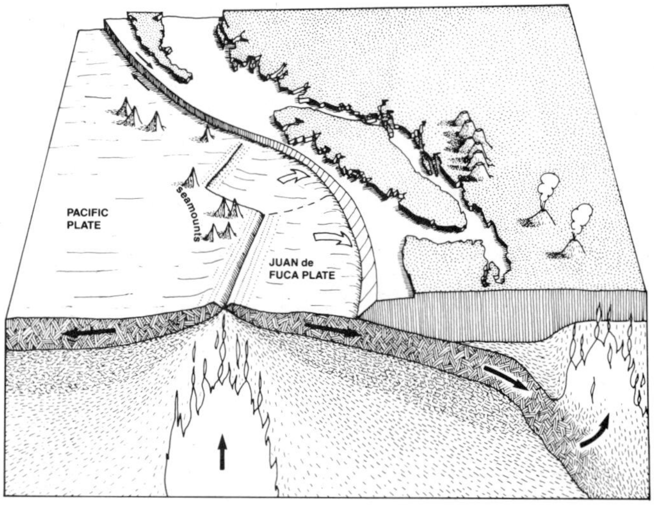

FIG. 1.13. Artistic interpretation of larger structures in Fig. 1.12 with underwater volcanoes (seamounts) included. (Courtesy of R. Riddihough and L. Carnes)

Quite clearly, plate tectonics and related processes have substantially influenced the structure and evolution of the Pacific west coast.

Glaciation

During the four glacial periods in the last million years, the British Columbia coast has been considerably altered by the advance and retreat of massive glacial systems. The deep U-shaped valleys and inlets and deposits of unconsolidated sediments in this area are reminders of the immense power of the glaciers that once gouged their way seaward from the interior of the province. The last ice sheet, for example, is estimated to have been about 1200 m thick at what is now the Fraser River delta. When it retreated around 10,000 yr ago, the decreased load on the earth beneath allowed the land to rebound upward as much as 140 m in parts of the northern Strait of Georgia and 80 m near Victoria. Moreover, a postglacial uplift of 15 m in the lower Fraser Valley seems to have diverted the Fraser River from its previous course into Boundary Bay to its present course into the Strait of Georgia. Therefore, Lulu Island and Sea Island are relatively new features of the delta that were formed from river deposits. Point Roberts, on the other hand, seems to be a more permanent structure of this area and may at one time have been an island off the mouth of the Fraser River. Present-day extension of the front of the Fraser River delta varies with location; the region around Sand Heads at the end of the Steveston Jetty, for instance, is growing into the Strait at 6 m/yr, whereas regions like Sturgeon Bank appear to be changing more slowly (see Chapter 10).

To the south, glaciers extended as far as the Seattle area. Ice from the last glacial period retreated from Puget Sound around 14,000 yr ago and the present shoreline is faced with bluffs as high as 150 m that consist of glacial till deposited during the last ice age.

In addition to suppressing the elevation of the land directly beneath them, extensive continental glaciers removed vast quantities of water from the world oceans. Estimates based on different kinds of geological data indicate that world sea level was over 140 m below its present level at the height of a glacial period more than 35,000 yr ago and about 100 m below its present level at the height of the last ice age 15,000–20,000 yr ago. However, the actual height of sea level relative to the coast in a given region during maximum glaciation depends on the comparative importance of crustal subsidence, which effectively raises sea level, versus ocean water removal, which effectively lowers it; during postglaciation, local sea level depends on the comparative magnitude of crustal rebound, which tends to lower sea level, and worldwide glacial melt, which tends to raise it (Fig. 1.14). Vertical movements of the earth's crust caused by changes in crustal loading are called isostatic effects whereas glacier-induced changes in worldwide sea level are called eustatic effects. Since the last glaciation, isostatic effects have generally dominated eustatic effects within the inland coastal areas of British Columbia and Washington and sea level has fallen, whereas the opposite is true of the outer coasts of Vancouver Island and the Queen Charlotte Islands. In the last few thousand years, however, trends have reversed along most of the coast, as world sea levels rise and local uplift ceases. Extensive melting of glacial ice would eventually lead to a sea level about 60 m higher than at present, accompanied by a slow but continual resettlement of a large portion of the world's population. The shoreline of the Georgia Depression would move many kilometres inland and cities like Victoria, Vancouver, and Seattle would no longer exist in their present locations. On the other hand, the reverse would occur if another glacial epoch is approaching, as some scientists suggest.

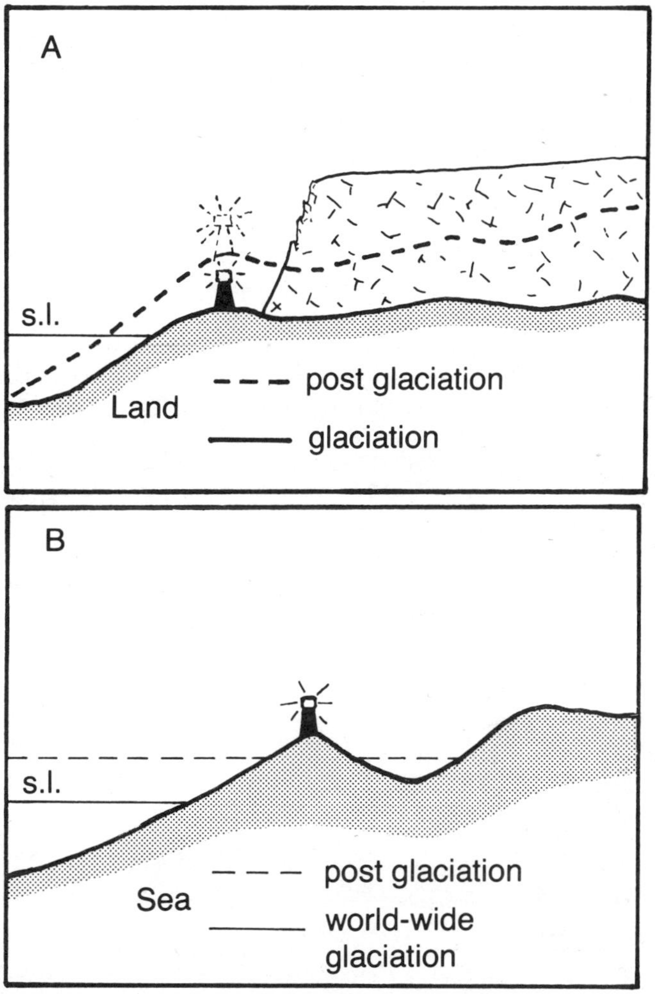

Fig. 1.14. (A) Isostatic changes in sea level. As the land subsides under weight of continental ice sheet, mean sea level rises. Removal of ice allows land to rebound, and causes mean sea level to fall. (B) Eustatic changes in sea level. Worldwide glaciation removes water from the oceans, lowering mean sea level. Melting causes mean sea level to rise.

Early Exploration

Human beings are recent arrivals in the evolutionary time scale of the earth. Anthropologists state that true man and woman first appeared in the open savannahs of southeast Africa some time between 1 and 2 million yr ago. From there, migration took place to the distant corners of the earth. Over 20,000 yr ago ancestors of the modern Eskimo and Indians first began to move eastward across an ancient land bridge linking Asia to the Americas, a bridge facilitated by a 100-m lowering of sea level during the height of the last glacial period. By the time the white man arrived, the coastal tribes of British Columbia and Washington had attained a sophisticated, albeit not always peaceful, civilization adapted to use of the sea and inland waterways. Ethnic groups such as the Coast Salish, Nootka, Bella Coola, Kwakiutl, and Haida had in fact become dependent on the sea early in their history. The Haida of the Queen Charlotte Islands were particularly skillful seamen who made regular, and generally unwelcome, forays to the Strait of Georgia and Puget Sound.

Following the merciless conquests of the Aztec and Inca civilizations of Central and South America in the early 16th century, Spanish authorities in Mexico began to turn their interest to the lands to the north. In 1592, a Greek pilot with the Spanish Navy named Apostolos Valerianos of Cephalonia, better known as Juan de Fuca, set sail northward on a voyage of discovery and is credited with being the first European to enter the Strait which bears his name. It is possible that de Fuca was also the first to enter the Strait of Georgia as his log mentions he discovered many islands and a broad inland sea after sailing 20 days along Juan de Fuca Strait. The concept of an inland channel from the sea in this region, however, proved too bizarre for his contemporaries and the discovery was subsequently rejected. It was not until nearly 200 yr later, in July 1787, that the entrance to the Strait was rediscovered by Captain Charles Barkley sailing a British ship under Austrian colors (which he did to circumvent the necessity of a licence from the East India Company). (Captain James Cook, who 9 yr earlier had been set-off the southern coast in bad weather, also denied the existence of de Fuca's Strait.) Barkley's rediscovery was verified in 1788 by Captain John Meares of the British Merchant Service. His statement in his log that ". . . we shall call [the Strait] by the name of its original discoverer, John de Fuca . . ." (Walbran 1971) helped establish the name in use today. It is interesting to speculate on the course of history had the Spanish quickly followed up on the voyages of de Fuca.

During the latter half of the 18th century, knowledge of the outside coast of British Columbia had been obtained by the Spaniards, Juan Perez, Heceta, and Quadra. Moreover, during his last voyage of discovery on the coast in 1778, Cook landed at Nootka where he bartered for sea otter furs. Traders arrived by land and sea soon after this profitable venture. It was not until 1791 that the Strait of Georgia was officially discovered by Lieutenant Eliza of the Spanish Navy who named it the "Gran Canal de Neustra Senora del Rosario la Marinera." Undoubtedly ignorant of Eliza's voyage, Captain George Vancouver entered the Strait the following year and called it the "Gulph of Georgia" after King George III of England, the monarch credited with losing the American colonies. The term "Gulf" is still incorrectly used to this day, although it was officially recognized as a strait by British navy hydrographer Captain Richards in 1865.

Vancouver, who had been a midshipman on Cook's earlier voyage to the coast on the *Resolution*, surveyed most of the coast from California to Alaska from May 7, 1792, to Aug. 19, 1794 (Fig. 1.15). It was part of his commission directed to "The acquiring [of] accurate information with respect to the nature and extent of any water communication which may tend . . . to facilitate an intercourse for the purposes of commerce between the north-west coast, and the country upon the opposite side of the continent" (Walbran 1971). In addition, Vancouver was instructed to negotiate with Quadra at Nootka for the peaceful division of the land claimed by the English and Spanish in this corner of the New World. Despite the hostilities between the two nations, their mariners often cooperated in charting the coast. When Vancouver and Broughton, commanders of HMS *Discovery* and *Chatham*, returned to the Strait of Georgia from Puget Sound in 1792, they met the Spanish vessels *Sutil* and *Mexicana* under Galiano and Valdes off Spanish Bank near Vancouver. The four small ships then proceeded northward and for 3 wk exchanged information eventually incorporated into the first chart of the southern B.C. coast. Commanders Valdes and Galiano were only a few days behind Vancouver when he completed his passage along the east coast of Vancouver Island. Valdes' name for Johnstone Strait, Canel de Descubierta (Discovery Strait), was given in honor of Vancouver's ship.

Following Vancouver's departure there was a lull in government hydrographic charting on the coast. Consequently, most soundings and surveying in the early 1800s were conducted by trading companies like the Hudson's Bay Company. In 1846 the British Admiralty began surveys of Esquimalt and Victoria harbors. Juan de Fuca Strait also was charted around this time and soundings published as one of the first Admiralty charts of the coast. The need for more extensive hydrographic knowledge became readily apparent during the conflict over the boundary between British Columbia and Washington. According to the Treaty of Oregon of 1846 "The line of boundary . . . shall be continued westward along the 49th parallel of north latitude to the middle of the channel which separates the Continent from Vancouver's Island and thence through the middle of the said Channel and the Fuca Straits to the Pacific Ocean" (Sandilands 1971/72). But which channel? The British argued that Rosario Strait was the main channel while the Americans argued for Haro Strait. An "impartial" German, Kaiser Wilhelm I, was then called in to arbitrate the dispute and ruled in favor of the U.S. position, a decision that may have been motivated more by politics than by fact.

The first Canadian lights on the west coast were installed on Race Rocks and Fisgard Island (Victoria Harbour) in 1860, to complement the lights on Cape Flattery and New Dungeness set earlier by the U.S. government. A buoyage system was implemented on the lower Fraser River in 1859 followed by a lightship at the river entrance in 1866. The first tidal current measurements were made from the HMS *Nymphe* in Seymour Narrows during the summer of 1895.

Vancouver's original track along the east coast of Vancouver Island was resurveyed between 1898 and 1905 from HMS *Egeria*, a steam screw sloop with the distinction of being the last hydrographic ship of the British Royal Navy to be stationed on the coast. Her commanding officers, Smyth, Simpson, and Parry are remembered in B.C. place names along with such earlier hydrographers as Pender, Richards and Prevost.

In 1890 the CPR Canada–Orient steamship *Parthia* struck a shoal in Vancouver Harbour. This prompted the first Canadian saltwater survey under the direction of Canadian Hydrographer William J. Stewart and led to the Admiralty Chart of 1893 for Burrard Inlet. The distinction for the first truly Canadian chart goes to Prince Rupert Harbour, which was then the terminus of the Grand Trunk Pacific Railway that helped link Canada to the far east.

Because soundings alone are not enough to establish water levels in tidal waters, tide recording stations were set up in the 1890s and the first tide tables for the southwest coast appeared in 1901. Analysis and prediction of tides from these records were performed at the Liverpool Tidal Institute in England until the 1950s, but have since been done by the Canadian Hydrographic Service in Ottawa, Canada. Slack water predictions for the most dangerous passes were published in the Pacific Tide Tables for 1908. At present, daily tide height predictions are published for 14 reference stations along the B.C. coast; with current predictions made for 15 specific localities, a number that will continue to grow as more studies are completed.

Modern hydrographers take soundings by highly sophisticated instrumentation and techniques. They also are involved with tidal current predictions for all coastal waters, pollution studies, deep-sea tides, Arctic Ocean tides and currents, and cooperate with oceanographers in a

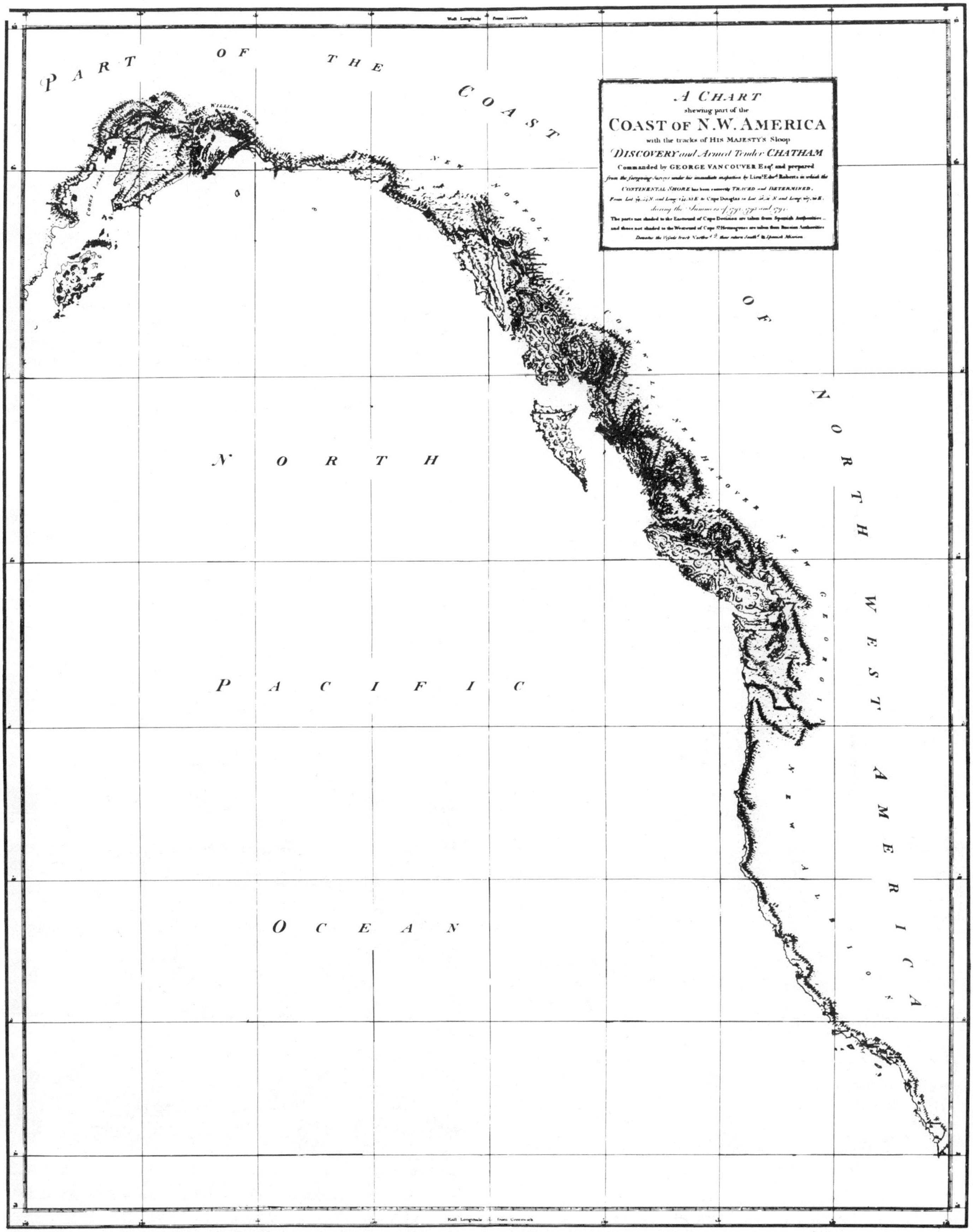

FIG. 1.15. Copy of Captain George Vancouver's original chart of west coast of North America, surveyed from 1792 to 1794. Vancouver included the work of both Spanish and Russian navigators in preparing his chart. Ship tracks and most place names have been edited from this reproduction due to illegibility. (Courtesy Canadian Hydrographic Service, Pacific Region)

variety of research projects. Since the early 1970s, most B.C. charts have been printed on the west coast rather than Ottawa. The conversion of charts to metric units is expected to be completed some time in the 1990s.

Ocean Science

Oceanographic research on the British Columbia coast began around 1908 with establishment of the Fisheries Research Board, Marine Biological Station, at Departure Bay near Nanaimo. For the most part, initial studies were fish-oriented with particular emphasis on salmon, although the need to measure water properties soon became apparent. In 1932, a program began that employed lighthouse keepers to obtain local water temperatures and salinities. Known as the Lighthouse Data Series, it has proved useful to fisheries investigators who have the monumental task of investigating the nature of fish stocks within the coastal environment.

In the 1930s, joint oceanographic–hydrographic cruises were undertaken in an effort to understand the physical and biological processes in the Strait of Georgia, Juan de Fuca Strait, and the offshore region. Following World War II, when oceanic research was linked to antisubmarine warfare, the Pacific Oceanographic Group was set up in Nanaimo and began pioneer research into the physical oceanography of the B.C. coast. Two years later, in 1948, the Pacific Naval Laboratory was established. Now known as the Defence Research Establishment Pacific, this group continues to study ocean acoustics and turbulence. The early measurements of turbulence in Seymour Narrows are still widely quoted in modern scientific literature.

The Institute of Oceanography at the University of British Columbia was founded in 1949, and reorganized into the Department of Oceanography in 1979. Under the direction of the professors, physical, chemical, geological, and biological studies have been conducted in most inlets and inside waterways in the southwest corner of the province. Moreover, the institute is considered to be a world leader in studies of the generation of ocean waves, ocean turbulence, and physical processes at the air—sea boundary. Many of Canada's present oceanographers are graduates of degree programs at the institute.

Within the last decades, numerous other public and private oceanographic establishments have been founded on the west coast. These include the large and prestigious Department of Oceanography at the University of Washington in Seattle, and the Institute of Ocean Sciences, Patricia Bay, B.C., where a wide spectrum of oceanographic research is conducted from the polar seas to the equatorial oceans by ships, aircraft, satellites, and submersibles (Fig. 1.16). A partial list of other publicly financed research institutes is in Appendix B; consulting and engineering firms that specialize in oceanographic research may be found in local telephone directories.

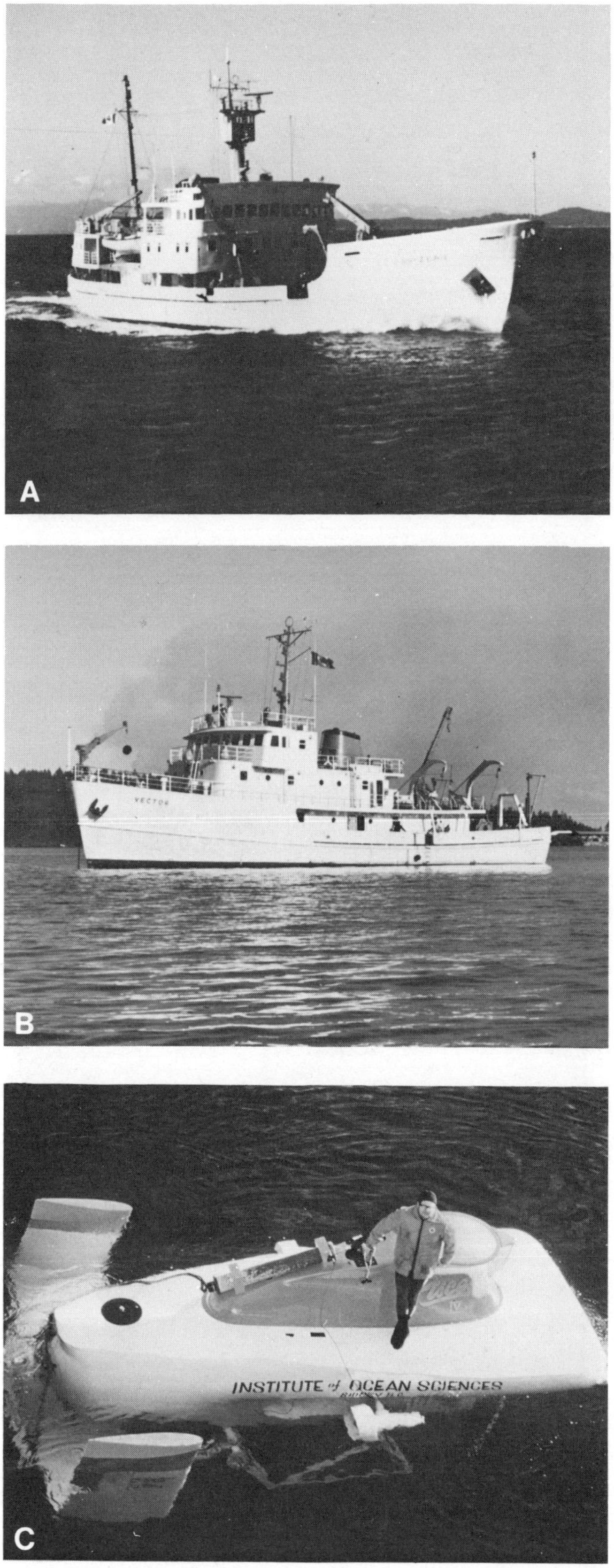

FIG. 1.16. Research vessels of the Institute of Ocean Sciences, Patricia Bay, B.C. (A) Canadian Scientific Ship (CSS) *Parizeau*, displacement 1929 t, length 64.3 m; (B) CSS *Vector*, displacement 505 t, length 39.6 m; (C) *Pisces IV* submersible, length 6.1 m, maximum depth 1800 m. (Submersible photo courtesy C. MacKay; ship photographs by author)

Chapter 2. The Coast Today

Seafloor Topography

The Pacific is the largest world ocean. With an area roughly equal to the Atlantic and Indian oceans combined, it contains almost half the earth's water in a basin with the largest average depth (3940 m, 12,925 ft). It also has the deepest regions in the world, where depths exceed 11,000 m in several trenches along the western shores. In the vicinity of Ocean Weather Station "P" (Fig. 2.1), 1500 km west of Vancouver Island, the floor of the Pacific basin is relatively flat and featureless with depths around 4200 m. Such topography typifies the abyssal plains that stretch over a wide area to the south of this location. To the north and west on the other hand, the bottom is more irregular.

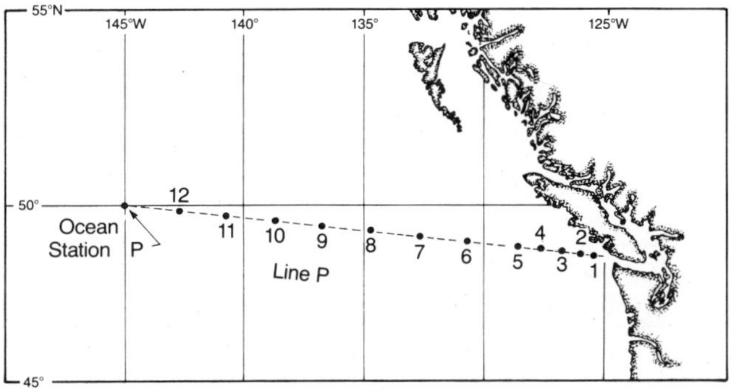

Fig. 2.1. Locations of Line P stations.

The first half of the sounding line in Fig. 2.2 illustrates a gradual shoaling of the bottom eastward of the weather station to a decrease in depth of 380 m over 650 km; yet the gently sloping region is far from featureless because of numerous inactive or extinct submarine volcanoes called seamounts. Projecting thousands of metres above the bottom, some undersea mountain peaks are high enough to penetrate into the sunlit surface layer of the Northeast Pacific Ocean and have become the habitat for diverse and abundant varieties of fish.

Roughly halfway to the coast, the water depth begins to decrease more rapidly in an almost steplike fashion. Within this section of the ocean basin, the bottom is characterized by broad underwater ridges, with peaks and valleys reminiscent of rift mountains on the continents. As discussed in the previous chapter, these submarine ridges mark regions of active seafloor spreading off the coast of British Columbia. Toward the entrance of Juan de Fuca Strait, the sounding line again crosses a relatively flat oceanic region (the Nitinat Deep Sea Fan) before it encounters the outer edge of the continental margin. The more gentle slopes of the seaward edge of the continental margin form the continental rise, an apron of thick sediments thought to have been deposited by sediment-laden water known as turbidity currents that flowed downward along the seafloor from the adjacent continental slope.

Covered with land-derived silt and sand, continental slopes are relatively steep oceanic regions that approximately mark the seaward extent of continents. Off the coast of British Columbia, the outer edge of the continental slope is delineated remarkably well by the 1800-m (1000 fa) contour, whose distance from land varies from about 90 km off the southeast coast of Vancouver Island to less than 45 km off the northern tip of the Island. The slope has a highly rugged terrain of bumps, knobs, and steep-faced canyons (Fig. 2.2).

At around the 180-m (100 fa) contour, the sounding line begins to cross the shallow, gently ascending area of the continental shelf. Essentially submerged portions of the coast, continental shelves cover roughly 5% of the earth's surface and are by far the most highly productive fisheries areas in the world. Much of man's future supply of oil and gas is thought to lie beneath the layers of sediments that cover the shelves, although exploratory drilling by Shell Canada Ltd., in 1967–69 off the southwest coast of British Columbia failed to find hydrocarbons of any substantial quantity despite the presence of promising geological structures. Compared to its Atlantic counterpart, the continental shelf along the Pacific side of North America is narrow. With the exception of the shallow basins of Hecate Strait and Queen Charlotte Sound, the width of the shelf off British Columbia rarely reaches 95 km and is typically less than 45 km, whereas off the Nova Scotia coast it generally exceeds 185 km. Along the west coast of the Queen Charlotte Islands the shelf is almost nonexistent as water depths plummet more than 2500 m to the ocean floor in less than 35 km. Although the sounding record in Fig. 2.2 makes the shelf appear as an almost featureless bank, other portions of the shelf along the coast are much more rugged and more reminiscent of the topography of the adjacent land. Other sounding routes, of course, would show different specific structures, but the general nature of the bottom is typical of ocean basins throughout the world.

Landward of the continental shelf, the coast of British Columbia gives way to a complex network of inlets, straits, passes, sounds, and narrows. Including islands, the coastline stretches almost 27,300 km from the border of Washington State to the Alaska Panhandle. Towering over this network of waterways are the Coast Mountains with the highest peak, Mount Waddington (3994 m), near the head of Bute Inlet. Numerous rivers empty into this oceanic system via glacier-formed valleys; larger rivers such as the Skeena, Nass, and Fraser, and their tributaries constitute an essential link in the life cycle of Pacific salmon. Moreover, these rivers have a profound effect on the structure and circulation of coastal waters. The Fraser River, in particular, is known to influence the currents in the entire inside passage from the entrance of Juan de Fuca Strait to the entrance of Queen Charlotte Sound, and also may affect physical variations in surface waters on the outer coast of Vancouver Island. Water depths inside this

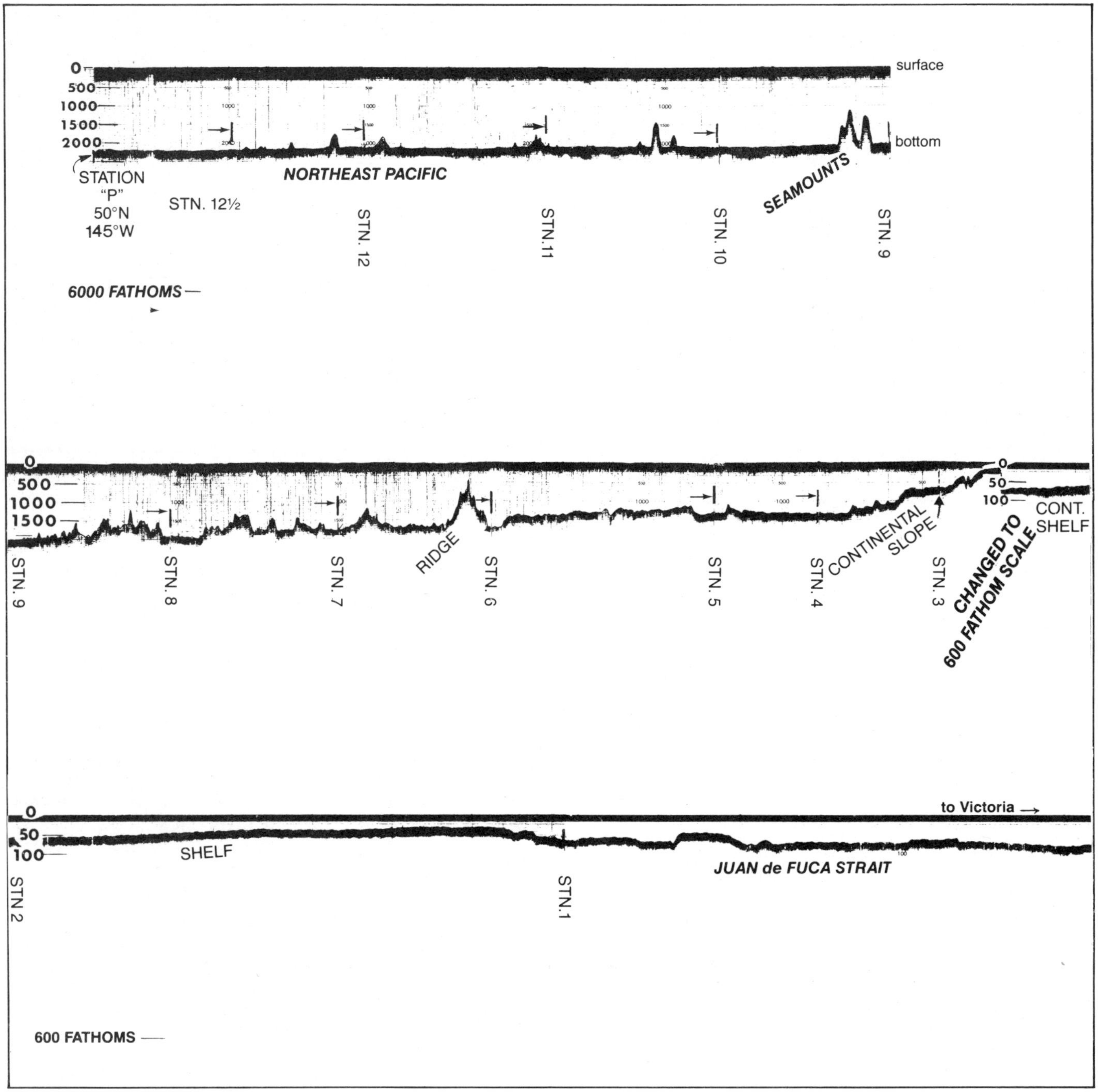

Fig. 2.2. Depth profile taken along Line P from Ocean Weather Station P to Juan de Fuca Strait. Depths in fathoms. There is a 10-fold increase in the vertical scale on the echo-sounder at the continental shelf from 6000 fa full-scale to 600 fa full-scale, with a corresponding change in the horizontal scale. (Courtesy Canadian Weathership CCGS *Quadra*)

protected coastal domain vary considerably over short distances. Within the comparatively expansive basins of the Strait of Georgia, Puget Sound, and Juan de Fuca Strait, depths are generally shallower than 400 m, whereas in certain inland regions such as Bute and Jervis inlets they can reach 650 m or more.

Seamounts

Seamounts are underwater volcanic peaks that rise more than 1000 m above the neighboring ocean floor. Estimates based on echo soundings indicate there could be as many as 14,000 in the Pacific basin alone, most of them clustered into groups or stretched out into chains similar to the continental volcanic areas along the Pacific Rim. The flanks of some smaller seamounts attain average slopes of 55° although the larger ones rarely have slopes exceeding 15°. Their size may range from relatively small conical bumps on the bottom (Fig. 2.2) to massive submarine mountains such as the Great Meteor Seamount, 1500 km off the northeast coast of Africa. First discovered in 1937 by the German research vessel *Meteor*, this large oceanic feature has a base diameter of 110 km and rises abruptly 4200 m to within 269 m of the surface of the

Atlantic Ocean. Its flat-topped summit covers an impressive area of 1450 km^2 (560 $miles^2$).

Oceanic Islands, like those of Hawaii, are seamounts that have penetrated the ocean surface. Of the 5000 or so within the world ocean, the vast majority are in the Pacific between the Philippines and Easter Island.

Seamounts are linked to the formation of other oceanic features. Atolls, for example, are ringlike islands in tropical waters constructed from coral reefs built on the submerged outer cone of an extinct volcano. As Charles Darwin explained in 1842, the reefs survived by continuing to grow upward as the original oceanic island subsided below the sea surface. Guyots or "tablemounts" are flat-topped seamounts generally found below depths of 200 m. They appear to be formed by subsidence of an atoll or oceanic island whose surface topography has been leveled by wave erosion. More than 200 guyots exist in the world oceans, the tops of many are deeper than 2000 m.

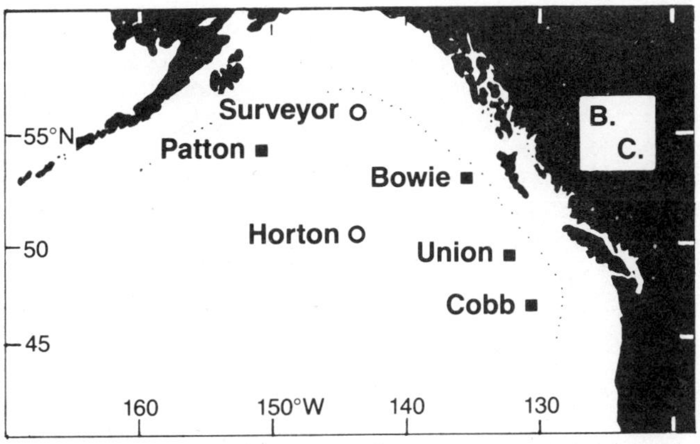

FIG. 2.3. Prominent seamounts (squares) and guyots (circles) in the Northeast Pacific Ocean. Dotted line marks northern limit of northeast Pacific seamount province.

Wave Erosion

Most seamounts investigated to date show clear evidence of ancient wave erosion. Beach terraces (relict beaches) at various depths mark elevations that a particular volcanic peak once had above sea level. Some wave-washed features can be related to the lowering of sea level by about 100 m during past glaciation. However, this explanation cannot account for the eroded tops of some deeper guyots nor the deeper wave-cut terraces. An alternative theory is that many seamounts were formed along the comparatively shallow ridges over seafloor spreading regions where they were subject to wave action and were then conveyed into deeper water by the crustal motions. This account is consistent with the fact that seamounts are younger than 60 million yr and that older ones are in deeper water. Lastly, the reestablishment of isostasy under the volcanic mass could account for a certain amount of subsidence. In this case, a newly formed volcanic intrusion slowly settles until its weight is just balanced by the buoyant effect of the magma below, a process that is analogous to what happens to any object floating in a highly viscous fluid when it is raised above its equilibrium level and then released.

Coastal Seamounts

Cobb, Bowie, and Union seamounts are among the most prominent undersea volcanoes off the British Columbia coast and are part of a cluster of about 100 large seamounts and guyots that extend from the Gulf of Alaska to the Oregon coast (Fig. 2.3). The pinnacles of Cobb and Bowie seamounts are unique in that they are accessible to scuba divers. Recently, they have been considered possible sites for studying ocean circulation, deep-sea tides, ocean winds, and marine life, and as testing grounds for ocean technology such as manned underwater habitats.

Cobb Seamount, about 500 km southwest of the entrance to Juan de Fuca Strait, is a terraced mountain with a base width of 32 km that rises about 2750 m from the ocean floor at an average slope of 12°. Discovered in August 1959 by the U.S. Fish and Wildlife Service, its summit is a 47.6-m central spire that rises out of the top of the mountain to within 34 m of the sea surface (Fig. 2.4). The spire forms an oval 10.5-ha platform and appears to

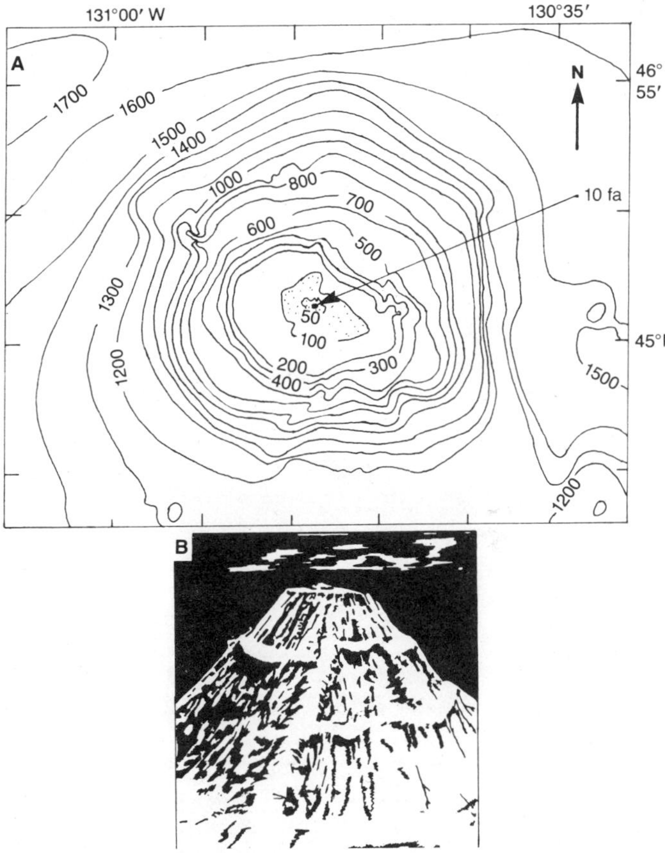

FIG. 2.4. (A) Bathymetry of Cobb Seamount (in fa). (B) Pictorial rendering of depth soundings shows relict beach terraces. (From Budinger 1967)

be the neck of an ancient volcano that submerged below the rising seas for the last time at the end of the last ice age about 10,000 yr ago. Beach terraces at 83, 146, 183, and 195 m are proof that the seamount once stood well above sea level and were probably formed during periods of standstill followed by submergence associated with seafloor spreading and regional or local subsidence. Bottom samples from the 83- and 183-m terraces include well-rounded pebbles and cobbles, plus an abundance of intertidal mussel shells that one might expect to find near coastal shores. The deepest terrace also appears to have been formed by wave action, and suggests the seamount was once an oceanic island at least 195 m above sea level some 27 million yr ago.

In May 1976, a survey of Cobb Seamount by Canadian Hydrographic Service researchers in the *Pisces IV* submersible showed the following features: considerable grass on the pinnacle with appreciable swell-induced motion; steep-sided cliffs terminated by fine-shell beaches that vary in width from 3 to 20 m with numerous current-induced ripples (Pl. 1); a visibility of 40–50 m at a depth of 200 m, and numerous red snapper and sea perch.

Bowie Seamount is approximately 220 km west of the Queen Charlotte Islands and rises to within 37 m of the surface from depths of over 3000 m. As with Cobb Seamount, fauna exist in abundance in the clear sunlit water over the three shallow pinnacles that protrude from the upper portion of the mountain peak (Fig. 2.5). All

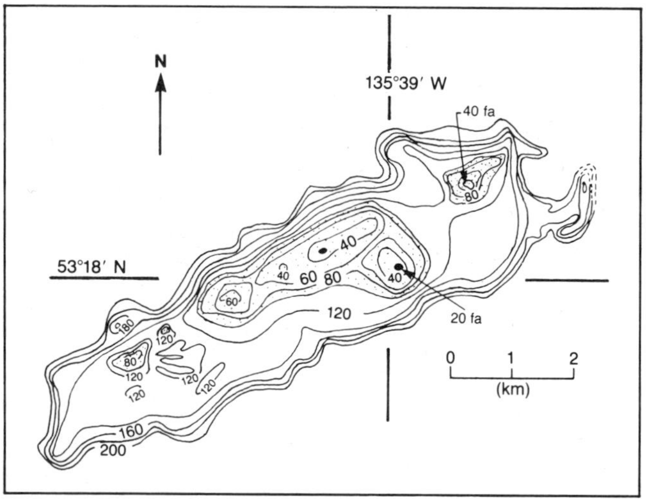

FIG. 2.5. Bathymetry of Bowie Seamount above 200-fa contour. Stippled area is shallower than 100 fa. (From Scringer and Halliday 1971)

told, the tops of these pinnacles cover 4 ha and are the gathering place for numerous rockfish, red snapper, and perch. Observations from the *Pisces IV* submersible a few days prior to those of Cobb Seamount showed a smooth-sided terrain of black sandy material with relatively little surface growth. Visibility at 200 m was around 50 m. A terrace at 238 m on Bowie Seamount, similar to the 183-m terrace on Cobb Seamount, is evidence of past wave erosion during a period of lower sea level. Presumably the 55-m difference between the two is related to differences in subsidence during recent geological time.

Union Seamount, about 108 km west of Estevan Point on Vancouver Island, is the third most shallow peak off the British Columbia coast. It is an almost conical volcanic mountain (Fig. 2.6) that rises to within 293 m of the sea surface from depths of 3300 m. Secondary summits are at 402 m and 512 m. The absence of terraces on the flanks of the seamount indicates it never rose far enough above sea level to be exposed to wave action. Its only distinctive feature appears to be a small crater on its eastern flank with a diameter of 148 m and a relief of 18–27 m. Although shallow enough, Union Seamount has yet to be explored by submersible.

Inlets, Estuaries, and Sills

Inlets are perhaps the most distinctive oceanographic features of the coastal zone of British Columbia. Although this term applies to any small arm of the sea that cuts into a coast, many of the province's 79 inlets consist of long, narrow channels bounded by steep mountainous terrain reminiscent of Norwegian fiords. Towered over by snow-capped peaks as high as 3300 m, Knight Inlet is generally regarded as the most spectacular fiord in the southwestern sector of the province, though other main-

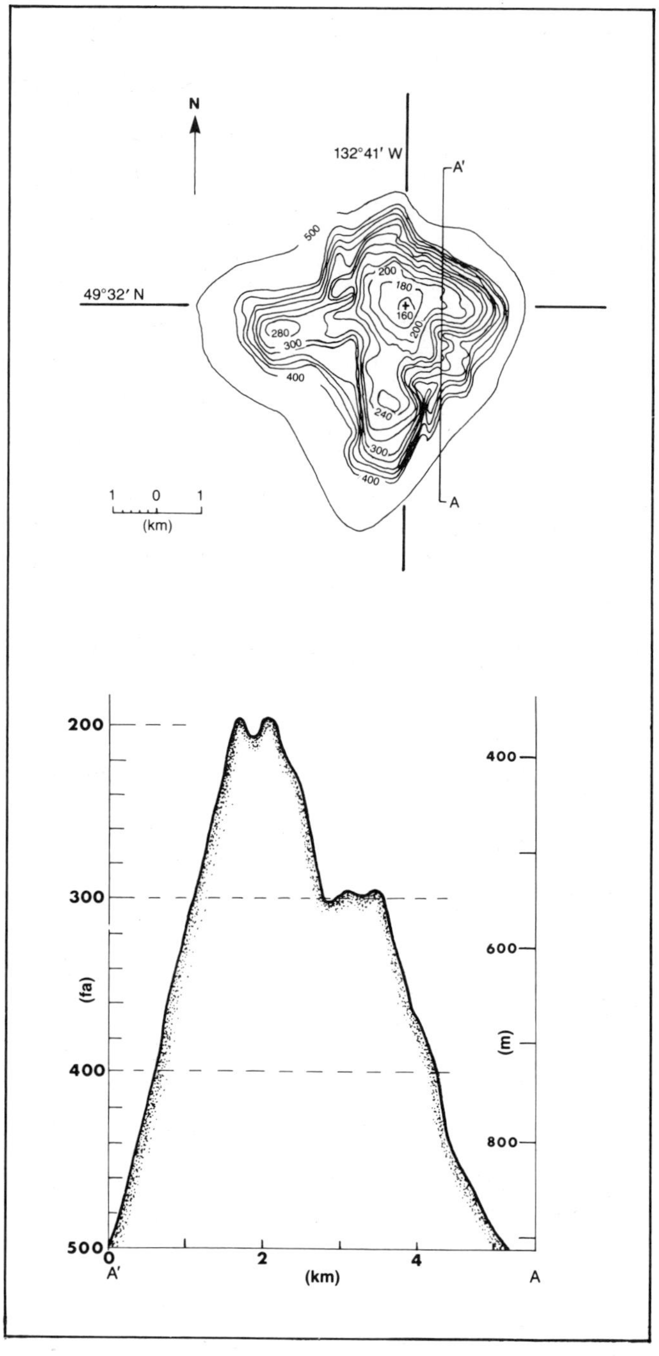

FIG. 2.6. (Top) Bathymetry of Union Seamount above 500-fa contour. Peak (+) has a depth of 160 fa. (Bottom) Profile was taken north to south along sounding track A′ to A. (From Scringer and Halliday 1971)

land fiords, such as Bute and Jervis inlets, come close to equaling it for sheer rugged beauty. The inlets of Vancouver Island are less spectacular and less extensive than those of the mainland and, with the exception of Saanich Inlet near Victoria, are only on the west coast of the Island. Remnants of the large continental glaciers that once gouged their way seaward to create the drowned valleys that inlets now occupy presently form the icefields at the heads of Knight and Bute inlets. (Glaciers still terminate directly in many of Alaska's inlets where they break up into icebergs.)

A typical west coast inlet is a deep U-shaped basin with a glacial-mud bottom, a river at the head, and an underwater ridge (a sill) across the mouth (Fig. 2.7). The amount of fresh water that flows into an inlet from the river depends on the drainage area, the time of year, and whether the river is fed primarily by rainfall or snow-melt. As runoff into most Vancouver Island inlets results from rainfall, the freshwater input is greatest during the winter–spring rainy season and least during the summer–fall dry season. Freshwater input into the larger mainland inlets, on the other hand, reaches its peak during the freshet or snow-melt period, which begins in May. During the warm summer months, the volume of water discharged by the river per unit time can exceed by a factor of 10 or more the corresponding discharge rate for the period from late fall to early spring, when the river is fed by rainfall. In summer, surface waters of inlets are often rendered a milky-green color by the large amount of glacial silt carried downstream.

Large quantities of fresh water within a coastal basin create an oceanographic region called an estuary, a partially enclosed body of water where seawater is measurably diluted by mixing with river runoff. Traditionally, the term has applied to lower deltaic portions of a river but has now been extended to include inlets, bays, sounds, and arms that receive measurable volumes of fresh water. It is possible to extend the definition to even more extensive regions. For example, the entire southwestern passage from the entrance to Juan de Fuca Strait to the entrance to Queen Charlotte Strait might be considered part of the Fraser River estuary because up to 60% of the dilution of this region can arise from this one river. But such a broad definition is of little practical use so it will be limited to confined geographical regions.

There are basically two types of estuaries on the west coast: salt-wedge estuaries, like the Fraser River downstream of Deas Island, where river runoff is large and little mixing takes place between the fresh water above and salt water below; and partially mixed estuaries, typical of most inlets and sounds where there is enhanced mixing between the two layers because of greater tidal action and lower runoff. A third category, well-mixed estuaries, where strong tidal currents combine with low runoff to produce water that is nearly homogeneous from top to bottom, are uncommon in British Columbia. They are limited to small bays close to turbulent tidal passes such as Menzies Bay near Seymour Narrows (see Fig. 3.27).

One of the most important aspects of estuaries is that they act as nutrient traps where river-borne organic and inorganic materials collect in concentrated amounts. This makes them biologically active areas that support large populations of mammals, birds, and marine life, particularly the area in the immediate vicinity of the river

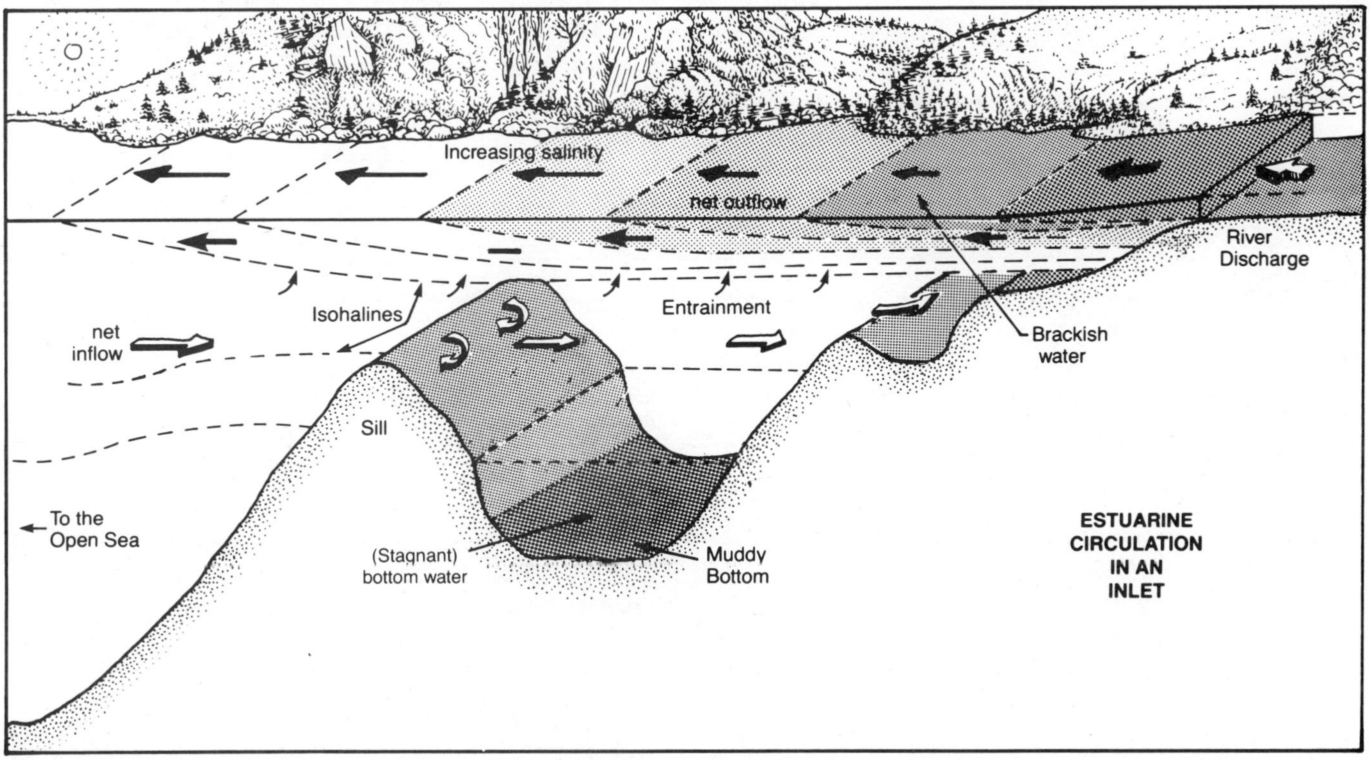

FIG. 2.7. Estuarine circulation in a typical British Columbia inlet. Salt water entrained and carried seaward by river outflow is replenished by a net inflow at depth. Sloping isohalines (lines of equal salinity) indicate a down-inlet increase in salinity in surface brackish layer. Turbulent mixing occurs in vicinity of sill.

mouth. Deltas associated with estuaries have long formed the nuclei for human settlements and industrialization, as well as providing fertile agricultural land and convenient transportation routes. In many places in the world, estuaries are highly endangered environments, threatened by man's overcrowding, wastes, and dams that alter the nature of rivers. The delta of the Fraser River is a prime example of an estuarine environment that would appear to be doomed to eventual destruction.

Besides having a profound effect on the biology, river runoff alters the current patterns in a basin and regions beyond. Consider an inlet. Because fresh water is appreciably lighter than salt water, the freshwater runoff makes its way down-inlet as a shallow surface layer, which absorbs or entrains some of the saltier water from below as it progresses (Fig. 2.7). Therefore, the saltiness of the surface layer increases down-inlet while that of the lower layer decreases up-inlet. To compensate for this loss of salt to the shallow but rapid (50 cm/s or so) outflow at the surface, there must be a deep but slower net inflow at depth. In this manner, there is no loss of salt within the basin of the inlet as salt entrained and carried seaward by the top layer is replaced by an increased inflow of salt in the bottom layer. (Looked at separately, the volume of inward flowing water in the deeper layer may be 10–20 times greater than the volume initially discharged by the river. No net increase in volume takes place in the inlet, however, as the river eventually entrains a volume equal to the deep inflow and carries it seaward.) Known as an estuarine-type circulation, this two-way flow is particularly well established during periods of large river discharge or periods of strong down-inlet winds when the entrainment process is especially vigorous. A boat proceeding up inlet would be bucking an opposing current while a submarine would have a favoring current.

Often the "jet" of river water that enters an inlet remains partly intact, and moves down inlet as a thin ribbon of silty surface flow. For example, the Squamish River can be seen from aerial photographs to bounce from one side of Howe Sound to the other as it makes its way toward the Strait of Georgia (Pl. 2). Depending on winds and tide, a boat moving in clear water may experience completely different currents than a neighboring boat in the silty river water. Fresh water can, of course, enter an inlet in other ways than smoothly flowing rivers. Among the most impressive are the plummeting streams that free-fall tens of metres to the inlet from glacier formed hanging valleys (Fig. 2.8).

As will be shown later, estuarine-type circulation is not confined to partially enclosed basins such as inlets, but is found throughout the inside waterways of the west coast where land drainage makes its way seaward. It even exists in the open North Pacific where a relatively dilute surface layer overlies the more salty bulk of the ocean. In other areas of the world, a kind of reversed estuarine circulation can take place. The Mediterranean Sea has a strong two-way circulation even when it receives relatively little fresh water either as direct rainfall or runoff. This is because extensive evaporation causes the surface waters in the basin to become saltier and, therefore, heavier. These waters then sink and eventually flow out as a bottom layer

FIG. 2.8. A hanging valley near Cascade Point in Knight Inlet. (Compare with Fig. 6.19) (Courtesy H. Freeland)

through the Straits of Gibraltar where they are compensated by a surface inflow of less salty Atlantic Ocean water. During World War II, German submarines "trapped" in the Mediterranean by the Allied blockade of the Straits of Gibraltar were able to escape detection by silently drifting into the Atlantic with these bottom currents (Fig. 2.9).

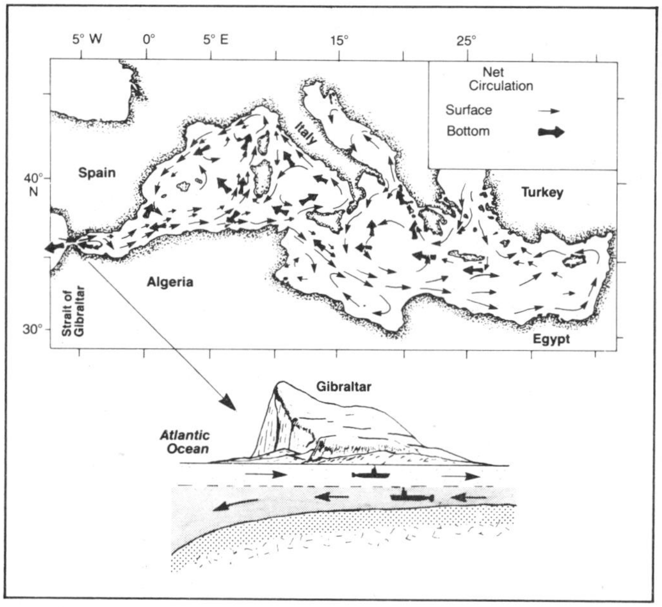

FIG. 2.9. Net circulation in Mediterranean Sea in summer. Evaporation of surface water from the sea leads to a net inflow of Atlantic water in upper portion of Straits of Gibraltar and a net outflow at depth. (Top figure adapted from Lacombe and Tchernia 1972)

Sills are another important feature of most coastal waterways in British Columbia and Washington. Typically, these underwater ridges exist where moving glaciers once deposited crushed rock and silt, similar to the manner rivers deposit sediments at a delta. When the glaciers began their final retreat some 10,000 yr ago, the piles of rubble, or moraines, left behind formed ridges which account for the relatively shallow depths at the entrances to most inlets. Similar circumstances led to the formation of sills in the Strait of Georgia, Juan de Fuca Strait, and other coastal areas. As shown in nautical charts, the change in water depth from one side of the sill to the other can be quite pronounced. At the entrance to Jervis Inlet, for example, the depth increases rapidly from 300 m over the sill to more than 600 m on the inside. In addition, some basins have more than one sill; each marks a former terminus or interstadial deposit of some glacier. Howe Sound has two main sills, one at the western entrance and another north of Anvil Island (see Fig. 10.38). The latter sill in effect marks the transition of the sound to an inlet. On occasions the combination of glacial dumping and upward rebound of the earth, which accompanies the removal of the weight of ice, was enough to create a rocky sill that now protrudes above the level of the inlet, thereby separating it from the sea by an isthmus of land. One of the best illustrations of this is Powell Lake north of the town of Powell River. Despite its name, the lake is actually a deep (300 m) land-locked inlet, which became isolated from the Strait of Georgia 7000–11,000 yr ago. Although the top half of the lake has long since been sweetened by land drainage, the salty bottom half provides conclusive evidence of an ancient connection with the sea. Nitinat Lake on the west coast of Vancouver Island is illustrative of an inlet that was almost completely land-locked. Its present connection with the ocean is via a relatively long, 2.5-km channel with four very shallow sills; an outer bar is only 2 m deep and three inner sills are about 3 m deep at lowest normal tides. Because some seawater enters on each tide, however, the fresh land drainage is kept confined to about the upper 10 m of the 180-m deep lake. Like so many inlets with shallow confined openings, tidal currents through the entrance to Nitinat Lake can be quite strong, though the actual volume of water exchanged is rather small. An even better example is Nakwakto Rapids that connects the nearly land-bound basins of Seymour and Belize inlets with Queen Charlotte Sound where rates of over 8 m/s (16 kn) makes it the swiftest tidal channel in the world.

To oceanographers, the presence of sills is of more than passing interest. One reason is that the proximity of these ridges to the surface tends to prevent oxygen-rich ocean water from penetrating into the deeper regions behind them. As a consequence, water in these sheltered depths may become low in lifesupporting oxygen and, under prolonged stagnation, may in fact become anoxic (lacking in oxygen) and contain large concentrations of hydrogen sulphide or "rotten egg gas," a lethal situation for all but primitive anaerobic bacteria (those capable of "living in the absence of oxygen"). Bottom waters in Powell Lake have already reached this stage; those in Saanich Inlet, Nitinat Lake, and Hood Canal in Puget Sound suffer from such conditions part of the year. If there is a comparatively rapid inflow of dense water into such regions, the poorly aerated water can be forced upward where it can cause mass suffocation of fish and other marine organisms. This could be detrimental to local fishing industries if it were ever to occur on a large enough scale.

On a somewhat less dramatic note, the 60-m deep sill off Victoria impedes the passage of cold, salty Pacific Ocean water into the inner portion of Juan de Fuca Strait. Similarly, the sill across Admiralty Inlet hinders the flow of deep Juan de Fuca Strait water into Puget Sound.

Temperature Distribution

Water temperatures in the coastal areas of British Columbia vary widely with location and season. Off the west coast there is a persistent northward decrease in temperature at a given depth in the Northeast Pacific, and isotherms (lines that join points of equal temperature) tend to run east–west (Fig. 2.10). Typical sea surface temperatures during winter vary from about 8°C off Vancouver Island to around 6°C off the Alaska Panhandle. With the advent of spring, the greater elevation of the sun and longer hours of daylight, together with decreased storm activity, lead to a substantial warming of the upper layer of the ocean. From then until late August there is northward "march" of the isotherms; sea surface temperatures off the coast of Vancouver Island reach typical maximum values of 18°C and at times may exceed 20°C in a thin (less than 1 m) top layer. Towards fall, solar radiation reaching the ocean surface begins to decline rapidly and storm tracks again move southward from the Gulf of Alaska. Heat that was stored in the upper 100 m or so of the ocean surface in summer is quickly lost to the atmosphere, and the isotherms begin the annual southward march to repeat the cycle. Below about 200 m there is little seasonal variation in water temperature.

As expected, some years are warmer, others colder, than normal. In the summer of 1957 there were excep-

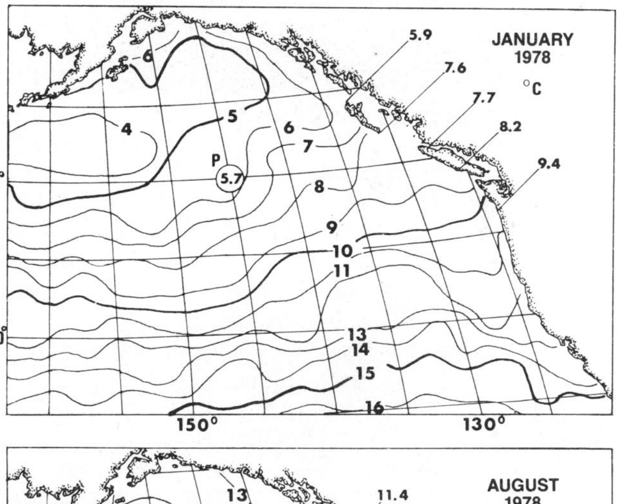

FIG. 2.10. Mean monthly sea surface temperatures over Northeast Pacific Ocean in winter and summer 1978 (°C). Circled values are from Weather Station P. Coastal values are from lighthouses (N to S): Langara Is., Cape St. James, Kains Is., Amphitrite Pt., Columbia River, Blunts Reef (lightships). (From Natl. Mar. Fish. Serv. Bull. 1978)

tionally high sea-surface temperatures of 20°C and more along the coast. Some scientists credit these anomalously large amounts of heat in the upper layer of the ocean for both the mild winter of that year and the hot, dry summer in 1958. In contrast, the abnormally low sea-surface temperatures off a large area of the coast in 1971 may have led to the bitter winter of that year and the rather mediocre summer of 1972.

Within the protected areas of the coast the variability of water temperatures depends on the intensity of solar heating as well as the degree of vertical mixing by wind and tide. In turbulent tidal channels like Juan de Fuca Strait and Johnstone Strait the water is always cold. Only in isolated nooks and crannies, such as Sooke Basin, does any appreciable summer warming take place. Off Victoria, sea-surface temperatures range from around 7°C in winter to 10°C in summer, the greater summer values apparently related to warm surface water in the Strait of Georgia that eventually makes its way seaward. In Puget Sound, water temperatures vary little throughout the year. Tidal current mixing at the various passes that lead into the Sound keep the temperature around 10°C, although extremes of 4.5°C in winter and 15.5°C in summer occur on occasions. Near some river mouths the fresh surface water may cool below freezing and turn to ice.

The rather strong vertical mixing in the southern tidal passes like Haro Strait and Active Pass keeps the surface temperatures in the southern Strait of Georgia a chilly 10°C year round. Beaches like those in Departure Bay near Nanaimo are regularly cooled by upwelled water during periods of northwest winds in summer (see Fig. 5.8). Only in the northern Strait and in sheltered areas of the Gulf Islands are tidal currents and winds sufficiently weak to allow long periods of swimmable conditions — by adult standards, that is! (An indication of the ability to survive in cold water is presented in Appendix C.) One of the most notable areas for good swimming on the B.C. coast is Desolation Sound where slow currents and complex topography create almost ideal conditions. One particularly warm spot is Pendrell Sound near the entrance to Toba Inlet, where, in early August, temperatures can rise to around 21°C to a depth of 5 m, with water visibility easily greater than 30 m. (Anchorages unfortunately are few and far between.) Pendrell Sound is also one of the major oyster spawning areas of British Columbia, and produces commercial quantities of oyster larvae that are cultivated then shipped to various locations in the Strait of Georgia, where they are grown to edible size. Oysters, it should be pointed out, don't reproduce just anywhere. They require warm water with little variance in temperature as well as a confined circulation pattern that won't carry the freely floating larvae out of the protection of the embayment — conditions fulfilled by few areas in the world.

Generally speaking, the surface temperatures in inlets with rivers at their heads remain cool throughout the summer; the snow-melt that swells the rivers has little opportunity to warm during its course seaward. Surface temperatures off the mouth of the Fraser River are also colder than the Strait of Georgia for similar reasons. Only in some longer inlets, such as Jervis and Bute, will summer heat produce an appreciable down-inlet warming of the upper layer. Saanich Inlet, on the other hand, with weak tidal currents and low river runoff, can warm in some sections to over 20°C on a hot, calm day in midsummer.

In many areas of the inner coast, the upper half metre or so may warm to "bathtub" temperatures when the afternoon tide floods slowly over large sandy beaches that have been heated by the summer sun (low tides in summer in the Strait of Georgia system occur in the early afternoon). Miracle Beach north of Courtenay, and Boundary Bay south of Vancouver, are like this, although anyone who has been swimming there will have noticed how much colder the water is an arms-length below the surface. In extreme cases in fact, the temperature drop may be as much as 8 Celsius degrees.

Within Burrard Inlet, the warmth of surface water in the recreational areas of Spanish Bank and English Bay is due in part to the heating effect of the sandy beaches. But, of greater importance is the top layer of silty Fraser River water that flows around Point Grey during much of any given tidal cycle (see Fig. 10.36). Not only does this layer absorb solar radiation but it also prevents any mixing of the cold salty water from below, a direct consequence of the lower density of fresh water compared to salt water. The fact that the Fraser River water seldom reaches the northern shore of Burrard Inlet partly accounts for this area's comparatively cold surface temperatures in summer; local winds and the flow of cold, well-mixed water from the vicinity of First Narrows also help keep temperatures cool.

Finally, it is interesting to note that oceanographers have found that, although surface waters in the Strait of Georgia and Juan de Fuca Strait are warmest in summer and coldest in winter, the deep waters are often coldest in late summer and warmest in early winter. This effect is a consequence of the complex interaction between the Pacific Ocean and inland waters, and undoubtedly has important ramifications for the overall biological and physical characteristics of the marine environment of southwestern British Columbia and northwestern Washington.

Salinity Distribution

Seawater is essentially a salt solution with more than 99% of the sea salt comprised of only six elements: sodium, chlorine, calcium, potassium, magnesium, and sulphur; sodium and chlorine comprise 86% of the total. Because they are dissolved in water, these salts don't exist as compounds (e.g. sodium chloride, ordinary table salt) but are broken down into ions, individual atoms or groups of atoms that carry an electrical charge. Moreover, although seawater is chemically renewed over tens of millions of years due to river runoff and recycling through the earth's crust, the comparatively rapid mixing processes are such that the proportion of these major elements, or ions, is kept constant throughout the world's oceans. This allows oceanographers to determine the overall salt content in a specified volume of seawater by simply measuring the concentration of the easiest measurable element, chlorine. (In the chemical analysis, bromine and iodine are unavoidably included but they are many thousands times less abundant than chlorine and, therefore, unimportant.)

To be specific, oceanographers define the salinity of seawater as the total weight of salts in grams dissolved in one kilogram of water at a temperature of 15°C. The salinity can then be calculated by determining the weight in grams of chlorine (plus bromine and iodine) in a kilogram of seawater, known as the chlorinity, and by the formula

$$\text{salinity (‰)} = 1.80655 \times \text{chlorinity (‰)}$$

(As a rough indication, taste buds are able to detect salt in water when the salinity approaches 1‰.) Modern oceanographic instruments now use seawater's ability to conduct electricity to obtain rapid measurements of salinity; the principle of operation is that electrical conductivity increases as saltiness increases.

The presence of salt is of consequence to numerous processes within the ocean. An increase of dissolved salts increases the density of the water, and thereby enables differences in the salt content to drive ocean currents. It also acts like antifreeze to lower the freezing point of

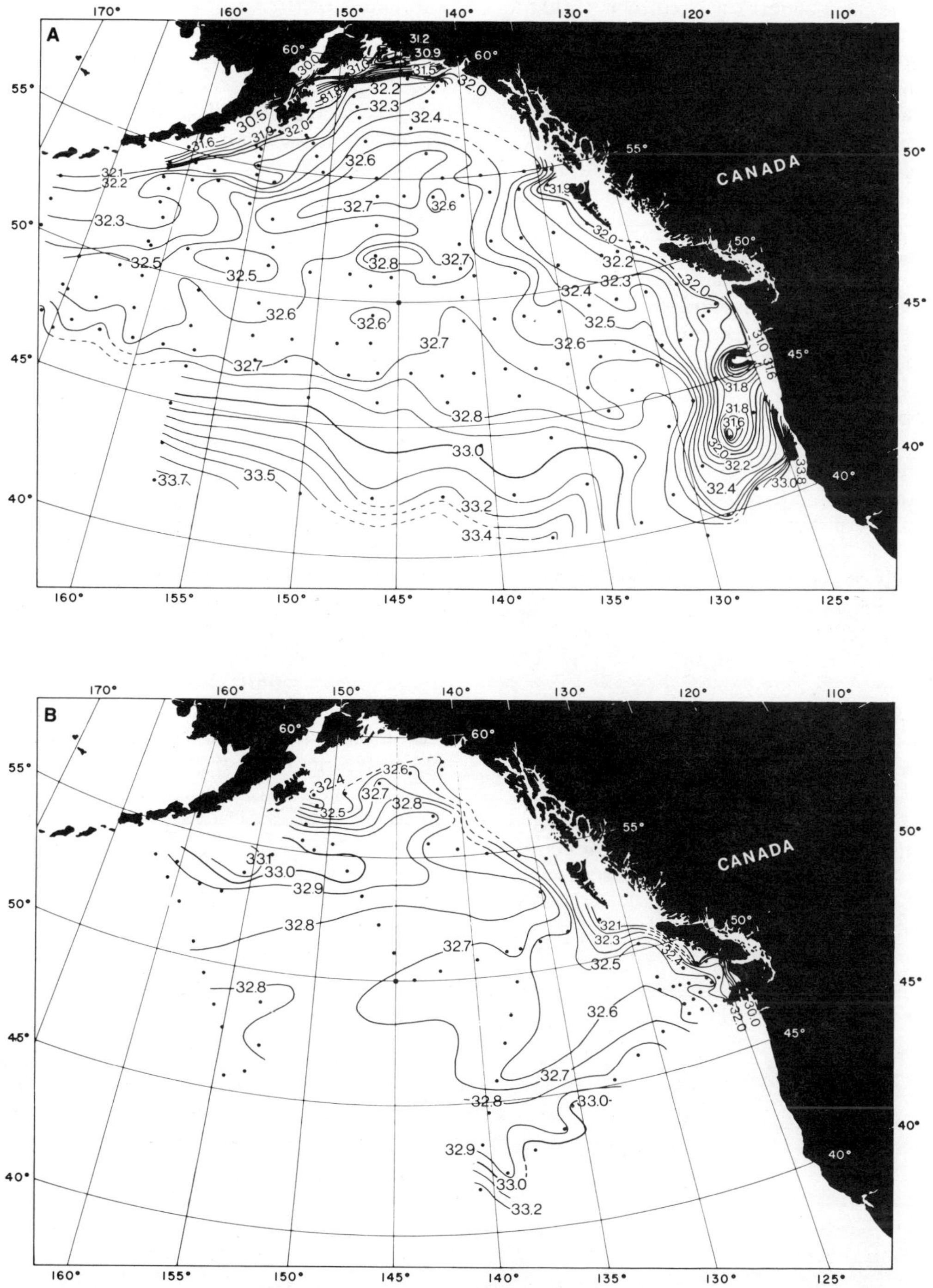

Fig. 2.11. Surface salinity (‰) from two oceanographic surveys: (A) August 1955, (B) February 1957. Dots show locations of measurements. Low values near coast are due to dilution by land drainage; high values off California presumably result from upwelling (see Chapter 5). (Courtesy S. Tabata)

seawater, helps retard the formation of ice in polar seas, and creates osmotic pressures, which affect both physical and biological mechanisms, and lastly causes flocculation (joining together into larger chunks) of fine sediments carried into oceanic regions by rivers. Off the mouth of the Fraser River, for instance, the mixing of fresh river water with salty ocean water causes the finer clay particles to settle out of river water as it overflows the Strait of Georgia.

In the vicinity of Weathership Station P, salinities vary from about 32.6‰ (32.6 g salt to 1000 g seawater) at the surface to around 34.4‰ near the bottom at 4200 m. Within the upper 100 m or so, there is also a small seasonal variation in salinity which, at the surface, ranges from about 32.3‰ in summer to about 33.0‰ in winter (Fig. 2.11a, b). Near the coast, patches of surface water often have salinities below 29.5‰ in summer due to land drainage and river runoff. Between depths of 100–200 m in the North Pacific there is a permanent feature called the main halocline (halo = salt; cline = slope) within which salinity, and, therefore, density, increases rapidly with depth (Fig. 2.12). The top of the halocline marks the deepest penetration of wind-induced mixing and convective overturning while the bottom of the halocline marks the beginning of a gradual increase in salinity to the ocean floor.

Within the protected inside waters the salinity decreases appreciably due to the diluting effects of river runoff. In fact, if salinity was continuously measured up Juan de Fuca Strait from the Pacific Ocean a gradual sweetening of seawater at all depths would be observed. Typical surface values in summer, for example, might show a decrease from 31.5‰ near Swiftsure Bank to about 29‰ off Victoria; corresponding values in winter would range from 31.5 to 30‰. These are average values, however, and large patches of less or more salty water are commonly carried back and forth in the Strait, especially in summer.

Sills act as barriers to the penetration of subsurface, highly saline Pacific Ocean water into the inside passage. Across the shallow sill south of Victoria the salinity at 100 m might change from 33.5‰ on the seaward side to 31.5‰ on the landward side. Similar jumps occur across the sill at the entrance to Puget Sound, across either end of the passes (Haro Strait, Rosario Strait) leading into the Strait of Georgia, and across every major sill throughout the inside waterway and at the mouths of inlets. In Johnstone Strait there is a rapid change in salinity below 50 m from one end of Race Passage to the other; salinities decrease from 31.5‰ on the western side of the sill to less than 28‰ a few kilometres to the east.

During the summer freshet, the Fraser River discharge forms a thin, relatively low-salinity layer over much of the Strait of Georgia. This layer is said to be brackish when salinities range between 0.5 and 17‰. Within a few kilometres of the front of the delta the thickness of the brackish layer varies between 1 and 10 m, depending on proximity to the main outflow channels, the state of the tide, and the strength of the wind. Beneath this depth, salinities increase rapidly as the thin silty layer gives way to the clear salty oceanic water. Salinity of the deeper water ranges from 29 to 31‰ throughout the year.

Spanish Bank in Vancouver is commonly inundated with brackish Fraser River runoff in summer as many swimmers can testify, while, in contrast, the water adjacent to the north shore is generally much more salty. The exception to this general rule occurs during strong flood currents or moderate-to-fresh southerly winds, when brackish water may be carried toward the northern shore of Burrard Inlet. At times, silty water reaches First Narrows but is then mixed by the turbulent tidal motions. Consequently, the salinity of the surface water in Vancouver Harbour is typically greater than that in the outer portion of Burrard Inlet.

Winter conditions in the Strait of Georgia differ appreciably from those of summer. Because of a factor of 10 decrease in freshwater discharge from the Fraser River and an intensified mixing by winter winds, maximum annual salinities occur at the sea surface in February. For reasons that will be explained in Chapter 10, salinities of the deeper water are also at their greatest in late winter.

The salinity structure within the inlets of British Columbia and Washington is also tied to the amount of freshwater discharge by adjoining rivers. Surface salinities will be low during periods of comparatively large runoff and high during periods of relatively low runoff. As a

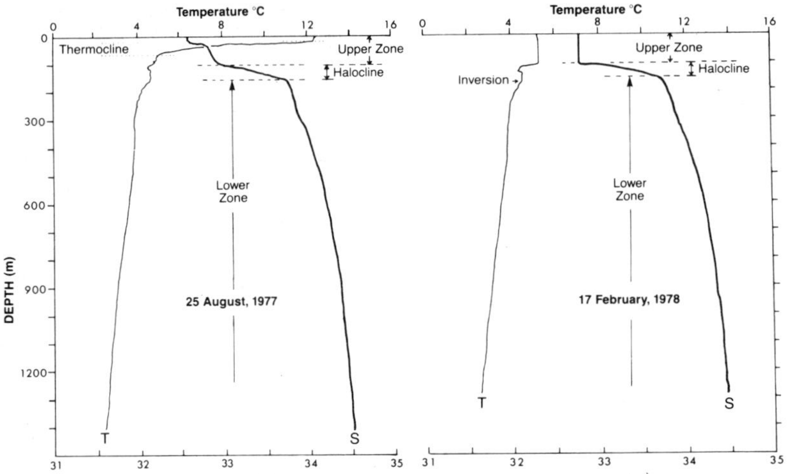

FIG. 2.12. Vertical profiles of salinity (S) and temperature (T) at Ocean Station P in summer and winter.

result of differences in the type of drainage patterns, therefore, inlets on Vancouver Island experience minimum saltiness in winter, whereas those on the mainland experience minimum salinities in summer; and vice versa.

Wind Patterns

Air, like water, is a fluid that can be set in motion through a variety of mechanisms, and each exerts a particular influence on the spatial and temporal variability of the flow. In many respects, therefore, atmospheric circulation can be considered a speeded up version of the oceanic circulation, with some important differences. An obvious difference is that current patterns are confined by the limited extent of the oceanic basins whereas large-scale wind patterns are only partially distorted by coasts and other topography. Winds, moreover, are generally many times faster than even the swiftest currents. On the other hand, winds are considerably less persistent. Whereas the water, once it is set into motion by some mechanism, will tend to maintain its flow properties for some time after the mechanism has faded away, the air will not. In other words, the atmosphere quickly "forgets" past influences on its motion and structure. This short memory, related to the much lower inertia and heat capacity of the atmosphere compared to the ocean, accounts in part for the greater variability of winds compared to currents, and for the near impossibility of making accurate weather forecasts more than a week ahead.

Another important distinction is that winds are driven primarily by pressure differences associated with horizontal gradients in air temperature whereas ocean currents are driven by the mechanical drag of the wind and by horizontal gradients in salinity; temperature differences are often of secondary importance. Once set into motion, however, both air and water are subject to similar kinds of effects including the earth's rotation, friction, topography, centripetal forces, and tidal forces. Tidal forces induce motions in the atmosphere but these are insignificant compared to tidal currents in the ocean and have a negligible effect on surface winds. High altitude observations, moreover, show that, unlike the ocean where tides are mainly produced by the gravitational attraction of the sun and moon, atmospheric tides are generated by the daily heating of the sun's rays (see Chapter 3). Finally, for historical reasons, there is a terminological distinction in direction between the wind and the current that often can be confusing. The direction given to an ocean current is that *in which* it is moving whereas the direction given to the wind is that *from which* it is blowing. Thus, a westerly current is moving *to* the west and a westerly wind is blowing *from* the west.

Wind patterns exist on a wide variety of time and distance scales over the surface of the earth. At the lower end of the scale are localized phenomena such as "dust devils" a few metres across that last for a matter of seconds; at intermediate scales are the highs and lows of weather systems with dimensions of hundreds to thousands of kilometres that persist for days; and at the far end of the scale are "planetary" pressure systems with dimensions of the earth's diameter that last for many months and control the general circulation of the atmosphere. For the intermediate and planetary scale pressure systems, whose spatial extent exceeds 100 km, high altitude winds blow nearly parallel to lines of constant pressure (isobars). In the northern hemisphere these so-called geostrophic winds constitute a balance between the horizontal pressure gradients and the rightward turning Coriolis effect of the earth's rotation (Fig. 2.13a), so that the air moves counterclockwise around a low-pressure system (cyclonic winds) and clockwise around a high-pressure system (anticyclonic winds). In the southern hemisphere, the Coriolis effect turns motions to the left and directions are reversed: clockwise (cyclonic) around a low-pressure cell and counterclockwise (anticyclonic) around a high-pressure cell. The strength of the wind increases as the distance between isobars decreases.

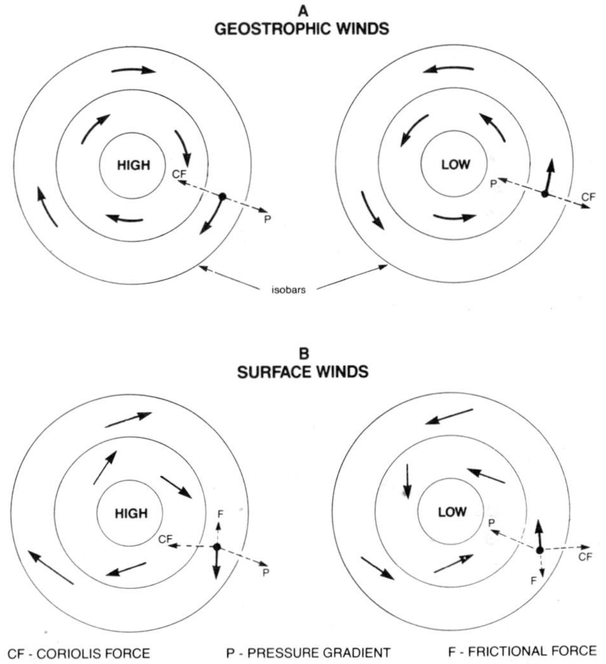

FIG. 2.13 (A) Geostrophic winds in northern hemisphere. Winds blow parallel to isobars (lines of equal pressure) with pressure gradient balanced by Coriolis force. (B) Near surface winds in northern hemisphere. Frictional forces upset geostrophic balance and cause winds to blow away from high-pressure regions and toward low-pressure regions. Coriolis force balances combined frictional force and pressure gradient.

Frictional drag causes a modification of these wind fields near the earth's surface. Weather maps show surface winds over the ocean blow slightly toward areas of low pressure at angles of roughly 15° to the isobars, but blow slightly away from high-pressure areas at approximately 15° to the isobars (Fig. 2.13b). This angle is generally greater over land where topographic features both deflect the winds and increase the net frictional drag.

Prevailing wind patterns along the west coast of Canada and the U.S. are controlled by the locations and intensities of two major semipermanent atmospheric pressure cells: the Aleutian Low and the North Pacific High (Fig. 2.14). The former gradually increases in intensity from August to December as its center shifts southeastward from the northern Bering Sea to the Gulf of Alaska.

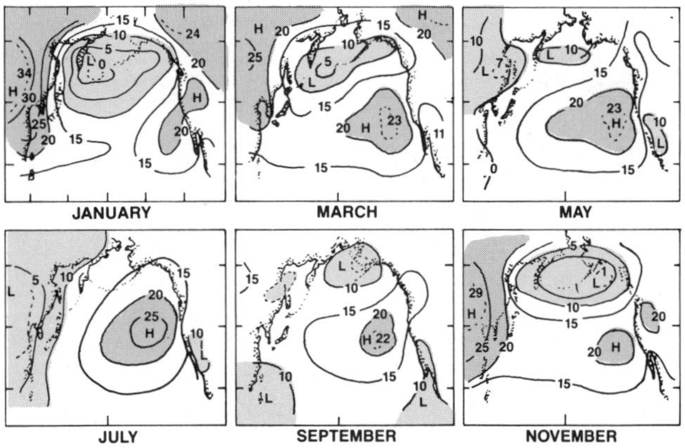

FIG. 2.14. Monthly mean air pressure at sea level from 1951 to 1970. Value on diagram + 10,000 divided by 10 gives pressure in millibars. (Adapted from Favorite et al. 1976)

Maximum intensity of the low-pressure cell and its associated counterclockwise winds occurs in January, with an accompanying abrupt shift of its center to the western Aleutian Islands. This is then followed by a progressive weakening until July when the system is no longer evident. Generally light and variable winds occur in the Gulf of Alaska during this month. The North Pacific High, which centers around 30–40° latitude off the California coast, is present year-round and reaches maximum intensity from June to August, when it almost encompasses the entire Northeast Pacific Ocean. The combined pressure pattern produced by these two systems means that from late fall to early spring winds will be predominantly from southeast to southwest along the British Columbia–Washington coast as the air circles anticlockwise around the dominant Aleutian Low. Although not evident in Fig. 2.14, there is often a "bunching-up" of the isobars against the coastal mountain ranges with a subsequent acceleration of the associated winds. From May through September, the combined effect of a greatly weakened Aleutian Low and intensified North Pacific high results in a clockwise flow of air over the ocean. Coastal winds at this time are predominantly from the northwest, again with some strengthening due to orographic effects. Hence, winds from directions other than northwest and southwest to southeast are only incidental to transitional weather systems.

Depressions and Fronts

Although departures of the wind from the large-scale prevailing directions can be affected by orographic features and by sea–land breezes (see Chapter 10), greatest departures generally originate with passing cyclonic depressions or lows. With characteristic extents of hundreds of kilometres and durations of days, these features often intensify into storms that invade the coast between early fall and late spring. Strong and rapidly changing winds are inherent to the fronts associated with developing storms.

The modern concept of depression formation originated with the now famous Frontal Theory of Cyclones proposed by a brilliant group of meteorologists at Bergen, Norway, during World War I. According to this theory, lows first begin when a warm, moist air mass from midlatitudes moves side by side with a cold, comparatively dry air mass from polar regions (Fig. 2.15a). At first, the interface or front between the two adjacent air masses is shaped like a long, straight wedge where colder air intrudes beneath warmer air at a slope of about 1 in 100 along a surface of separation called the polar front. From its inception, this arrangement is unstable to wavelike perturbations in the atmospheric flow. The flow causes the initially straight-line front to warp, so warm air bulges into cold air at the crest of the perturbation and cold air into warm air at the contiguous trough (Fig. 2.15b, c). The size of the bulge amplifies over a period of a few days as the entire developing system translates eastward at typical speeds of 250–500 km/day. Passage of the leading warm front and trailing cold front at a particular location is marked by an abrupt backing of wind direction. Eventually, the more rapidly moving cold front overtakes the warm front and warmer air begins to be lifted aloft by the cooler air beneath. Within the depression, the leading edge of warm air climbs over the cold air, condensing water vapor into rain or snow as it ascends. At the rear of the bulge, cold air pushes underneath the warm air, resulting in additional cloud and precipitation. Finally, the entire area of warm air is lifted over the cold air mass and the two fronts coalesce to form an occluded front (Fig. 2.15d). The low-pressure system then weakens as it evolves into a large, slowly counterclockwise turning mass of nearly uniform temperature air.

During the formation of a depression, winds intensify as the potential energy stored in the atmospheric thermal field is converted into kinetic energy of motion. More specifically, warm air rising and cold air sinking in the storm results in a lowered center of gravity that leads to an increase in the spin of the storm about its axis. This process ceases when the air becomes mixed to a uniform temperature so there is no longer a source of potential energy. Although the depression might be expected to produce a rise in pressure, this is more than compensated for by a divergence of the air aloft that accounts for the rapid drop in surface pressure associated with the developing storm.

More recent theories on the formation of depressions no longer rely on the existence of fronts. Instead, they show that lows (or highs) develop whenever there is a sufficient south-to-north decrease in air temperature and a strong enough increase of westerly wind speed with altitude. During the formation of lows, strong vertical motions driven by heat released by the condensation of water vapor draws colder air toward the center of the storm. Like an an ice skater who draws her arms toward her body to spin faster, the counterclockwise winds of the storm intensify as a greater mass of air is drawn closer to its axis to feed the updraughts. Falling pressures are produced by the ascending air currents. In contrast, high-pressure cells are formed in regions of sinking air currents when the mass of air tends to be pushed outward from the center of the cell near its base. When this is intense enough, the effect of the earth's rotation is to cause the air to circulate in a clockwise direction.

Off the British Columbia coast, the speed of the wind

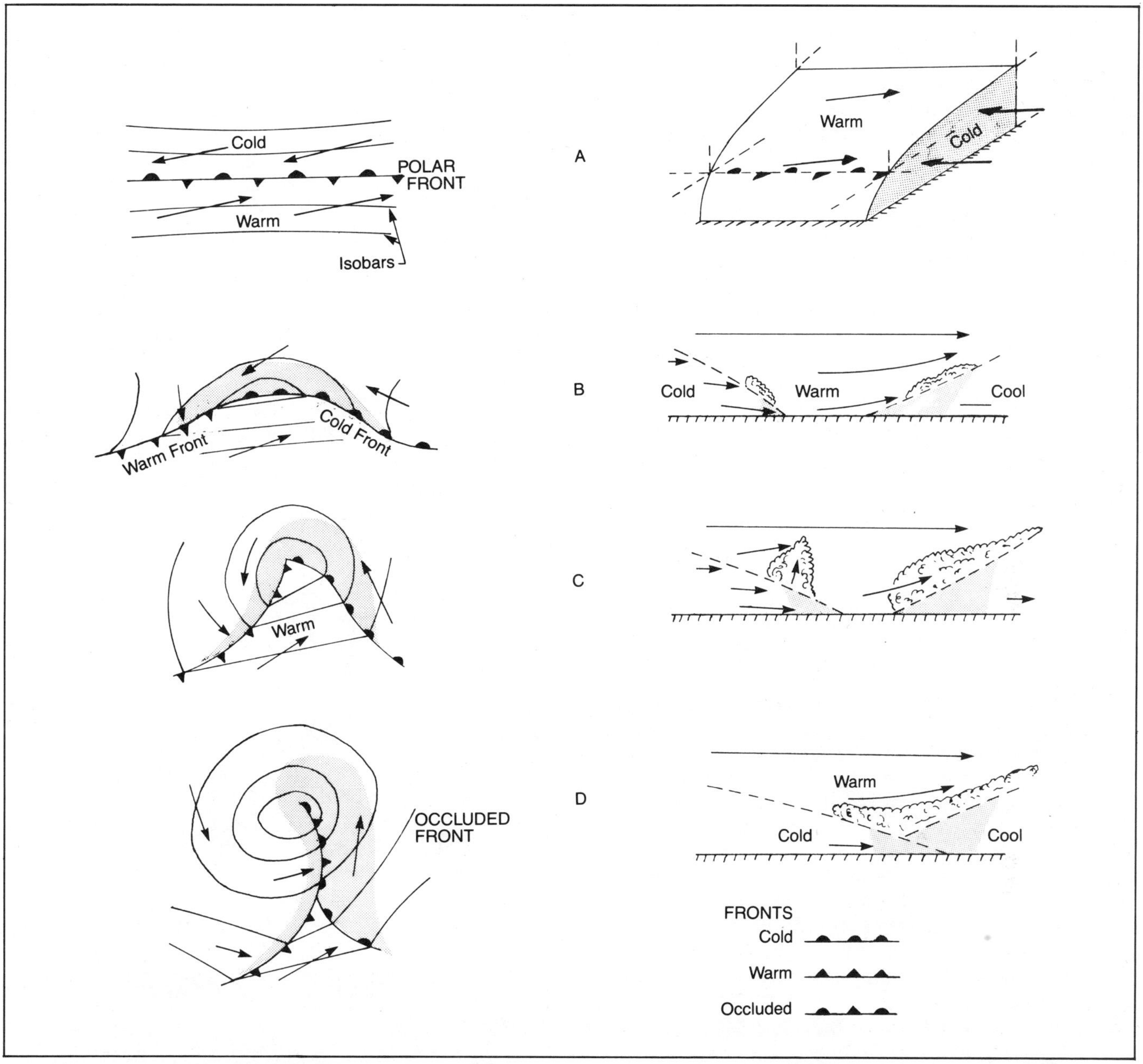

FIG. 2.15. Birth of a depression and cold front occlusion. *Left side*, plan view; *right side*, sectional view. A polar (stationary) front (A) develops a kink (B) that amplifies into a wavelike bulge (C) delineated by warm and cold fronts. Bulge grows and an occluded front (D) is formed as advancing cold front overtakes warm front and lifts it upward. Shaded area denotes precipitation; arrows give direction of air flow.

favors a cyclic variation in strength roughly every 3 days due to the eastward passage of highs, lows, and associated frontal systems. On the average, therefore, a ship in the Northeast Pacific Ocean could expect comparatively strong winds about every 3 days as the wind strengthens, weakens, and strengthens again over this period. Maximum wind speeds vary with season and are greatest in fall and winter, somewhat less in spring, and least in summer.

Cloud cover immediately adjacent to the B.C.–Washington coast is typically between 75 and 90% both winter and summer. In the vicinity of the weathership station the range is slightly larger, increasing from 75% in January to overcast conditions of nearly 100% in July when sea-fog is prevalent.

Shores and Beaches

The "rapid" sea-level changes that accompanied the retreat of the continental glaciers from 15,000 to 7000 yr ago left British Columbia and Washington with highly irregular and unstable coastlines. Since then, oceanic processes have continually modified the shoreline, attempting at all times to maintain a state of equilibrium at these coasts with regard to the supply of sediment, the action of waves and currents, the range of the tide, and the configuration of the shoreline. In the process, headlands have been cut back, cliffs eroded or formed, beaches established, spits formed, and drowned valleys partially filled in with sand and mud from present-day rivers. The net effect has been to simplify the geometry of the coastline. Though these features may appear more or less permanent to the casual observer, they continue to be subject to modification. Erosional coastal forms, such as headlands for example, accompany a local landward retreat of the shoreline whereas depositional coastal forms, such as sandspits, are associated with a local seaward advance of the shoreline. In only a few instances, as certain beaches, has the coastline managed over thousands of years to establish a quasi-permanent form through the various forces that attempt to restructure it.

Among the numerous factors that have affected B.C. coastal forms the more important are: the degree of glacial scouring and deposition; the internal structure of shore rocks and their resistance to erosion; the energy and direction of the prevailing wind, waves, and currents; the range of the tide; the availability of unconsolidated sediments; and the shoreline configuration. Regions of highly resistant igneous or volcanic rocks that border on low-wave environments, such as inlets for instance, have experienced little perceptible erosion. By contrast, weakly cemented sedimentary rocks exposed to the high-energy waves of the open Pacific over a tidal range of many metres have undergone considerable erosion. Shorelines are eroded during storms, when powerful breakers crash against the shore aided by abnormally high tides caused by strong onshore winds. Such combinations have been responsible for a wide diversity of coastal features in the area, ranging from sea cliffs to rocky wave-cut platforms, to beaches, to tidal flats. Aside from a desire to delve into the workings of nature, man has an obligation to understand the mechanisms involved in the formation and modification of the shoreline so that the effects of jetties, breakwaters, marinas, and other structures can be predicted or, alternatively, that undesirable natural erosion or deposition can be properly controlled. Too often in the past, ill-conceived interference with these processes has wrought more harm than good, with detrimental consequences to both the marine environment and to property along the shore.

Of the many features of the coast, beaches are the most popular. They are promoted as marketable assets by the tourist industry and many coastal communities have developed economies dependent on their recreational uses. Beach sands have sometimes been used for construction purposes or transported, with limited success, to sandless shores to make them more attractive to the public. Beaches further serve as natural buffers to sea cliffs and coastal property by causing the waves to dissipate their energy.

Though the term "beach" is often used synonymously with "shore" it applies to a specific type of shore feature with distinct characteristics that distinguish it from other coastal forms. To be precise, a beach is an accumulation of unconsolidated sediments (sand, shingle, cobbles, other fragmented rock) that extends from a physiographic change such as a sea cliff or series of dunes, or from the seaward limit of permanent vegetation, to the furthest point offshore where sediments can be carried by the waves. Unlike common usage that places the seaward edge near the low-tide line, this definition extends the beach (derived from an Anglo-Saxon word for shingle) into water depths of tens of metres. There is one obvious conclusion from the definition: for a beach to exist in the first place there must be a source of sediment somewhere along the coast and mechanisms vigorous enough to extract and deposit the loosened sediment in appreciable quantities. Materials that constitute a beach may be derived from local shore deposits carried seaward or from offshore deposits moved landward; alternatively they may consist of sand and mud carried to the coast by rivers and streams, or sands of distant origin transported parallel to the coast by longshore currents. Whenever the volume of unconsolidated material added to the beach by the various redistributing mechanisms (waves, currents, rivers, winds) exceeds that removed, a net deposition occurs and the extent of the beaches increases. If there is a net loss of material, due for example to a change in the wave or current conditions or to a depletion in the sediment supply, beach erosion occurs. A state of equilibrium exists only when the supplies and losses are in balance. Except in the vicinity of river mouths and near comparatively few large deposits of unconsolidated glacial deposits, sediments are scarce on the west coast and few extensive beaches have formed. The larger beaches on the outer coast, such as Long Beach on Vancouver Island (Pl. 3) and Second and Third Beaches on the Olympic Peninsula, as well as those in Puget Sound and along the north-western shore of the Strait of Georgia, are derived from extensive glacial deposits. The extensive beaches in the Boundary Bay area and on the eastern shores of Juan de Fuca Strait can be linked to river-borne sediments. For the most part, however, the trend along the characteristically resistant coastlines of British Columbia and Washington has been to form small pocket beaches of relatively coarse sand, cobbles, and boulders derived from the erosion of cliffs and bluffs. Finer sediments are apparently removed from these areas by bathymetric focusing of wave energy and the steep shores usually associated with them (Pl. 4). Many areas of the Strait of Georgia and Juan de Fuca Strait provide excellent examples of well-developed pocket beaches.

A cross-section of a typical beach would show a well-defined profile with a number of distinguishable features. The basic ones are presented schematically in Fig. 2.16 in conjunction with the near-shore features of the wave field. A partial list of relevant definitions also is presented.

Backshore The region of a beach that extends seaward from the limit of vegetation or distinct change in the physiography, such as a cliff or sand dune, to the sloping foreshore. In many areas along the coast the backshore is absent and the sea terminates abruptly at a vertical bluff. On river deltas, such as the Fraser River delta, the landward limit is delineated by a salt marsh.

Beach face The sloping portion of a beach normally exposed to the swash of waves.

Berm A nearly flat portion of a beach formed by the deposition of sediments by receding waves. This is the

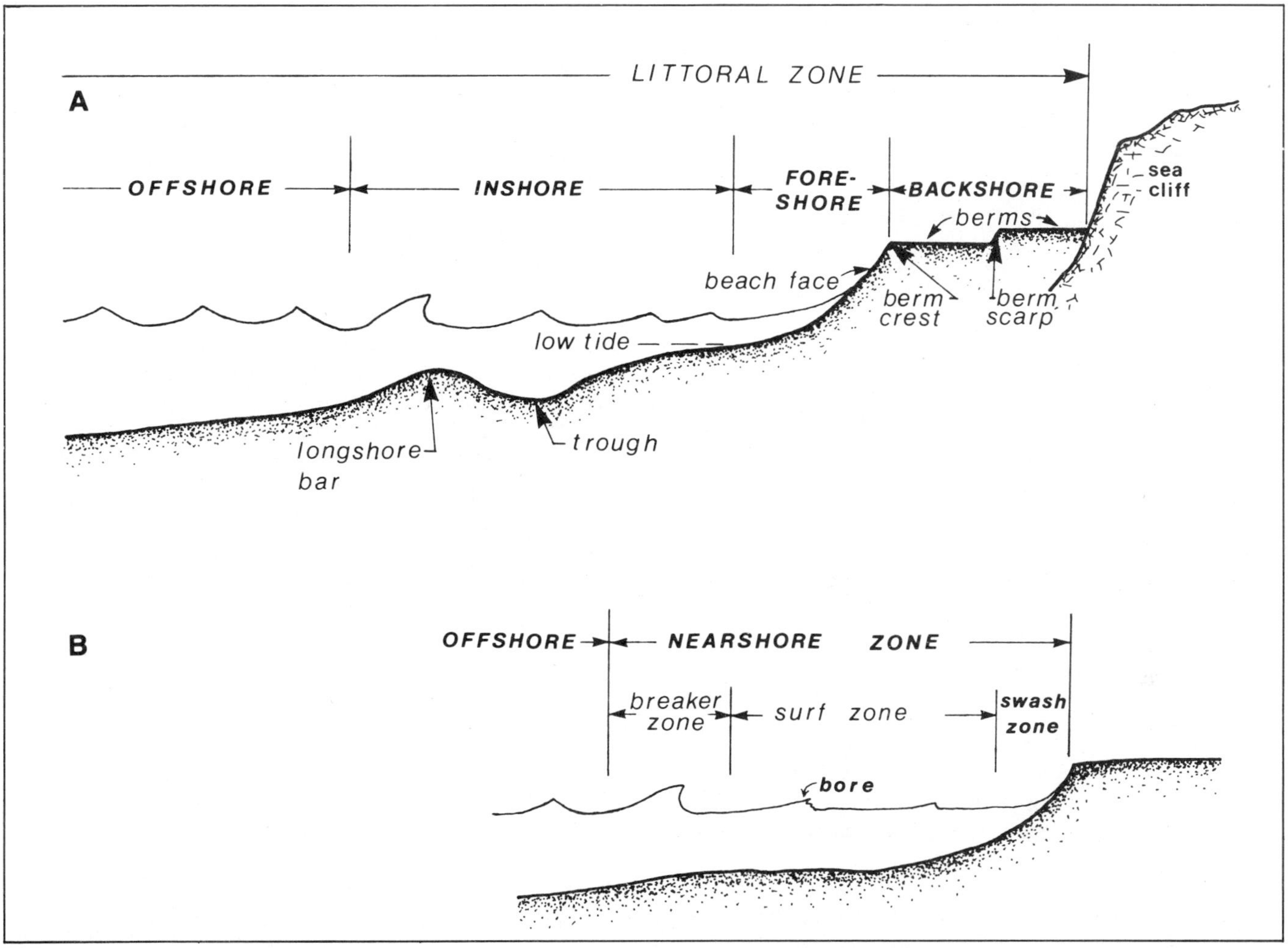

Fig. 2.16. (A) Typical beach profile and beach terminology; (B) terminology to describe wave action and currents in nearshore region. (From Komar 1976)

region sunbathers call the "beach." The berm crest is the seaward edge of a berm.

Bore A steep-faced wave that advances forward like a wall of water prior to breaking in shallow water or on an opposing current.

Breaker A wave that has become unstable because of steepness or because the crest has overtaken the trough in shallow water. (Various types of breakers are discussed in Chapter 8.)

Breaker zone The region of breaking waves. On wide, flat beaches such as Long Beach there may be more than one breaker zone as the waves reform with reduced height after each successive breaking.

Foreshore The sloping portion of a beach located between the berm crest (or upper limit of wave swash at high tide) and the seaward limit of the backrush at low tide.

Inshore The region of a beach from the foreshore to the area immediately seaward of the breaker zone.

Littoral zone Synonymous here with beach although marine biologists often limit the definition to the intertidal zone between mean high- and mean low-water levels.

Longshore bar A ridge of sand nearly parallel to the shoreline. In near tideless areas (as in Hawaii) a series of longshore bars separated by longshore troughs may form, whereas areas of large tidal range like the British Columbia–Washington coast are likely to have at most one large bar exposed at low tide.

Offshore The relatively flat area that extends from the breaker zone to the edge of the continental shelf or, in protected waters, to the edge of a sharp increase in depth.

Shore The strip of land that borders a body of water. A shore may be solid rock or unconsolidated sediments such as sand or cobbles (as in the case of a beach).

Shoreline The line of demarcation between the water and the exposed beach.

Surf zone The region that extends landward from the first line of breakers to the swash zone. (The surf zone, and the breaker and swash zones, comprise the nearshore zone of the wave environment.)

Swash zone The landward limit of wave action alternately covered by the uprush of the wave swash and exposed by the backrush.

Essentially four main factors determine the above features of a typical beach profile: type of beach material, energy of the wave environment, strength and direction of prevailing winds, and range of the tide. Steep-faced beaches consist of coarse sand, shingles, pebbles, or boulders whereas flat beaches consist of fine sand. This particular aspect of beach profiles can be related to two of the above factors and is worthy of explanation. First, the backrush of water created by a wave is weaker than the shoreward uprush in the swash zone, primarily due to percolation of water into the beach face and secondarily to frictional drag on the swash. The energy source of the uprush is the incoming wave whereas the backrush must begin from rest and accelerate down-slope. As the backrush is less capable of moving the sediment than the uprush, the beach face must steepen until the increasing downslope pull of gravity is sufficient to increase the velocity of the backrush to bring about a dynamic balance to the onshore–offshore transport of sediment. The type of beach material is important as water percolates much more readily into a gravel or cobble beach than into a fine sand beach. Hence, the slope of the beach face can be considerably greater for coarse grain beaches as a consequence of the much weaker return backrush associated with them. This is not the full story, however, as beaches of similar composition can have widely different slopes. It appears that the energy of incoming waves also is important; the more energetic the waves the lower the beach slope for a given grain size. The wave steepness, stage of the tide, and level of the water table within the beach further contribute to its inclination. (The cohesive surface tension of percolated water accounts for the fact that wet sand is more compact and firm than dry sand; try building a sandcastle out of dry sand! Berm faces of saturated sand can be vertical or even overhanging structures many centimetres high, whereas those composed of dry sand are necessarily rounded.)

In addition to differences in sediment grain size from one beach to another, a given beach may exhibit a gradation in grain size across its profile. On the numerous pocket beaches on the B.C.–Washington coast, beach material often changes from cobbles or pebbles on the steeply sloping backshore to finer pebbles or sand on the more gently sloping portions toward deeper water, as on China or French Beach. On flat, sandy beaches a gradation also occurs, though it is sometimes impossible to distinguish visibly owing to the small size range of the sand grains; typically the grain size is greatest near the plunge point of the breakers and decreases both toward deeper water and landward through the surf and swash zones.

Wave-induced influence on the beach profile is most pronounced within the surf zone, for it is here that sediments are actively moved both on and offshore as well as parallel to the shoreline. As will be explained in Chapter 8 waves in the surf zone are responsible for the formation of longshore currents (or littoral currents) and their seaward flowing counterparts known as rip currents. The width of the surf zone depends on the beach slope and, to a certain degree, the stage of the tide. Broad, flat beaches have wide surf zones, and steeply sloped beaches have narrow surf zones where waves break close to shore causing a rush of water up the beach face. Moderately sloping beaches have a distinct surf zone at low tide when the waves break over the outer portion of the beach; at high tide the surf zone may be lacking. Because of the large tidal range on the British Columbia–Washington coast, it is possible to distinguish between a "high-tide beach" and a "low-tide beach." Because waves sort sediment grain sizes, the high-tide portion of the beach is generally composed of coarser material than the sandy low-tide beach.

Perhaps the most dramatic effect waves can have on exposed sandy beaches is the ability to change the entire profile from a comparatively steep swell (or summer) profile to a more gentle but complex storm (or winter) profile (Fig. 2.17). The summer profile is characterized by a wide berm and the absence of a longshore bar and trough. In the winter profile, most sand in the berm has shifted offshore to form a bar (or bars on tideless beaches) parallel to the shoreline. Therefore, waves produce an annual shifting of the sands, offshore in winter to the bar and then onshore in summer to the berm. Sand is removed from the berm in winter during storm-wave conditions, accompanied by a landward transport of sand from seaward of the breaker zone, leading to a convergence of sand to form a large bar within the breaker zone. This process can be quite rapid, only a few hours of large seas cause considerable erosion of the beach face.

During the summer when flatter swell conditions prevail, the wave action causes a landward transport of sand over the entire beach profile with a subsequent erosion of the offshore bar and building of the berm.

Coastal winds also appear to play an important role in the structure of a beach profile. The onshore winds that generally prevail at a coast can be instrumental in removing sand from the beach by blowing it inland to form dunes. Offshore winds on the other hand will carry sand seaward to the backshore. In addition, winds at right angles to the coast set up currents that will augment or counter the transport of beach material by waves in the surf zone. Onshore winds generate a landward flowing current at the surface and a compensating seaward flowing current near the bottom that moves the beach material seaward; by the same token offshore winds generate a landward bottom current and a subsequent onshore transport of beach material. The amount of sediment moved in this manner increases with the strength and duration of the winds.

Minor Beach Features

Most beaches have a variety of interesting surface features. As such features are indicative of the beach environment, they are useful to geologists who study ancient beach deposits.

The irregular, thin wavy lines or ridges of fine sand,

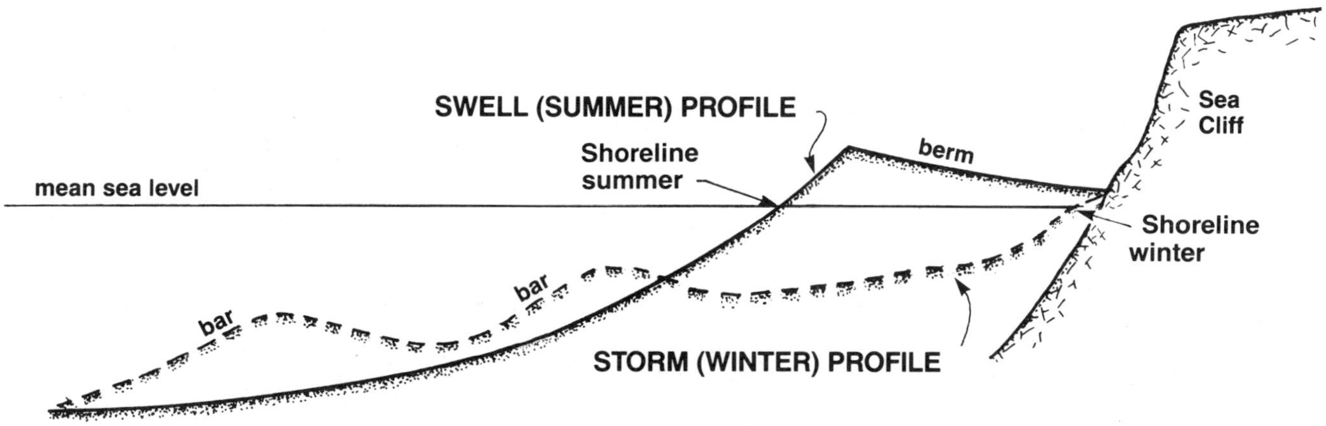

FIG. 2.17. Swell profile of a beach. Sands are moved onshore during low swell conditions (summer) and offshore during steep wave conditions (winter).

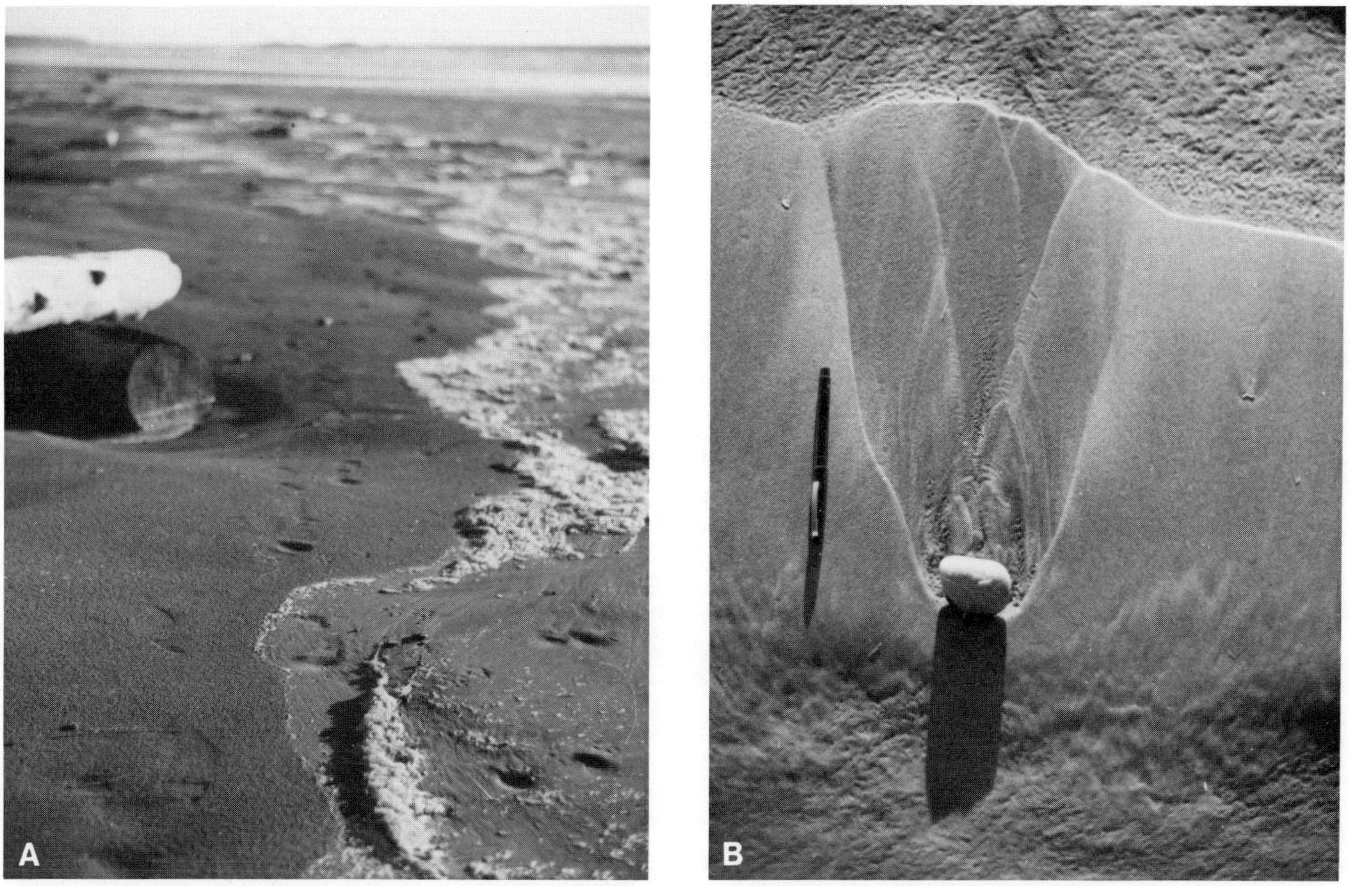

FIG. 2.18. (A) Swash marks. (B) V-swash marks in lee of a beach pebble. Pen and shadow = 20 cm (8 in.).

foam, and bits of debris that form at the leading edge of the wave swash on sandy beaches are known as swash marks (Fig. 2.18a). The irregularity of these patterns is due to the irregularity of the swash produced by each successive wave, and to the fact that the swash mark by one wave is partially obliterated by the next. Because the material is gathered together through surface tension at the forward edge of the swash, swash marks tend to form on the drier upper portion of the beach and not on the wetter lower portions.

V-swash marks (Fig. 2.18b) are formed by the deflection of the backwash around pebbles, shells, or other small objects on the beach. Figure 2.19 shows schematically how the sand is hollowed out by turbulent motions in front of the obstacle and by the enhanced water speed in a V-pattern away from its sides, and then deposited in the obstacle's wake. Diamond-shaped rhomboid ripple marks (Fig. 2.20) are another common feature of the swash zone on sandy beaches. As with V-swash marks, one pointed tip of each diamond is oriented in the direction of the back-

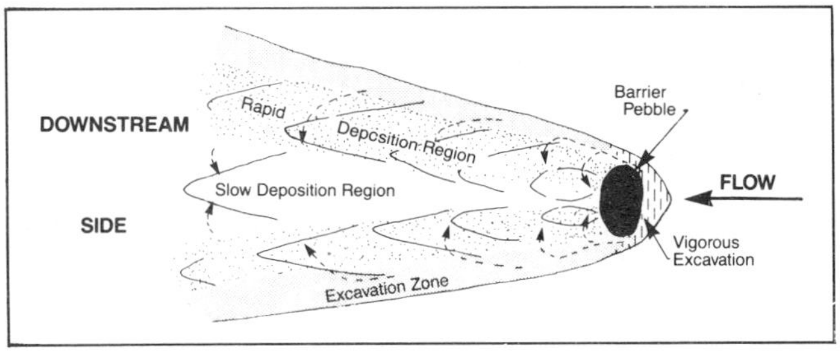

FIG. 2.19. Zones of excavation and deposition associated with a V-swash mark. (From Sengupta 1966)

FIG. 2.20. Rhomboid ripple marks. Pen length = 13.5 cm (5.3 in.).

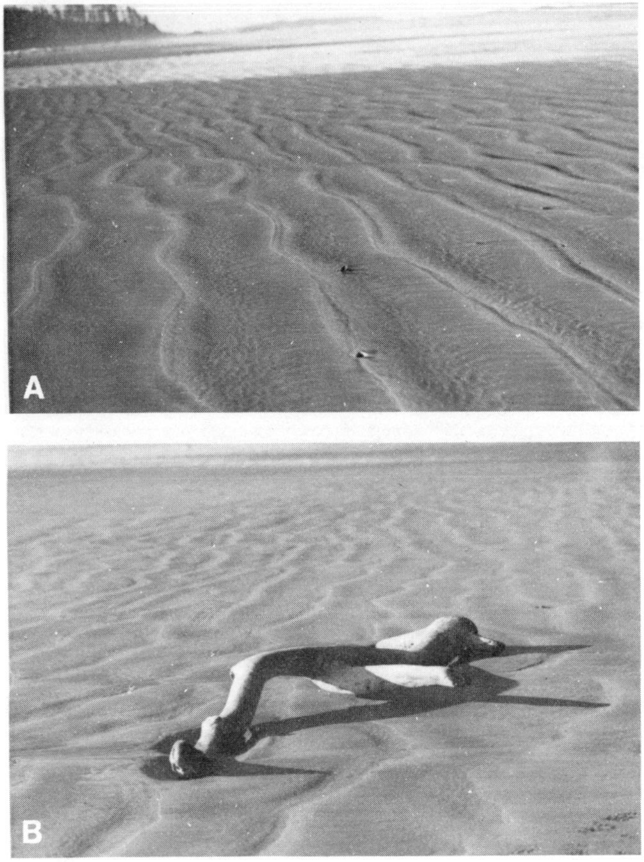

FIG. 2.21. Backwash ripples at Long Beach, west coast Vancouver Island, (A) distance between troughs about ½ m; (B) log roughly 3 m.

FIG. 2.22. (A) Details of rill marks, (B) series of rill marks.

wash. Because the current sorts grains by size, these patterns often have a distinct color differentiation, with darker heavy minerals in the gaps between lighter quartz grains. Formation of rhomboid marks apparently requires a rapidly flowing backwash less than 2 cm deep, but their origin is still uncertain.

The series of low sand waves and troughs produced across the swash zone on flat sandy beaches are known as backwash ripples (Fig. 2.21). Crest-to-crest separations are typically of the order of 50 cm or less and heights are usually less than a few centimetres. On a falling tide, water becomes trapped in the troughs, which often have a bordering of darker materials to give the ripples a bandlike appearance.

Rill marks (Fig. 2.22) are small erosional features that closely resemble the dendritic drainage patterns produced by streams or rainfall draining down a slope. Sand carried seaward along the rill fans out as it reaches a flat portion of the beach much the same as stream-borne sediments fan out into a valley. Rill marks are apparently produced by accumulations of percolated water that seep out of the beach at low tide. Sand domes are small mounds in an otherwise level beach raised up by air trapped within the sand. The air in this case is forced out as the wave swash percolates into initially dry sand. Small holes resembling those made by sandhoppers are produced when air escapes through wet sand in the swash zone.

Other types of ripples commonly seen on beaches include oscillatory ripples, formed by the back and forth motions under waves, and current ripples, produced by undirectional flow such as longshore currents, rip currents, or tidal currents. The former have symmetrical crests and rounded troughs oriented parallel to the shoreline (that is, at right angles to the direction of the waves), whereas the latter are asymmetrical with gentle up-current slopes that give way to steeper slopes on the downstream side of the flow (Fig. 2.23). Current ripples are typically oriented at an angle to the shoreline as currents tend to align themselves parallel to the coast; where rip currents lead out to sea, however, ripples tend in the direction of the coastline. Both types of ripples are often

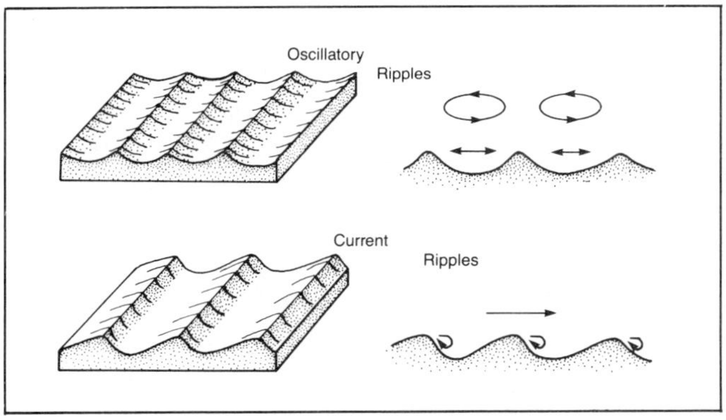

FIG. 2.23. Oscillatory ripples versus current ripples.

exposed at low tide on expansive tidal flats such as the Fraser River delta where, because of their large size (100 m or more from crest to crest and up to 1 m high), they have become known as megaripples. These sand waves are not fixed, but migrate slowly in the direction of the prevailing flow, both up and down current.

On a recent underwater geological survey on the continental shelf off the northwest coast of Vancouver Island on the *Pisces IV* submersible, geologists further observed extensive oscillatory ripples in depths of 100 m that were oriented parallel to the coast. These ripples of carbonate shell hash and volcanic sand and pebbles had spacings of about 100 cm and heights of over 15 cm, with rounded crests and near-flat troughs (Fig. 2.24). Calculations have shown that these deep ripples are most likely caused by long, storm-generated swells originating to the southwest with heights of 10–15 m. Smaller ripples crossways on top of the ripples probably are produced by strong tidal currents that flow parallel to Vancouver Island. Similar kinds of ripples have been observed at depths to 125 m on the Oregon continental shelf.

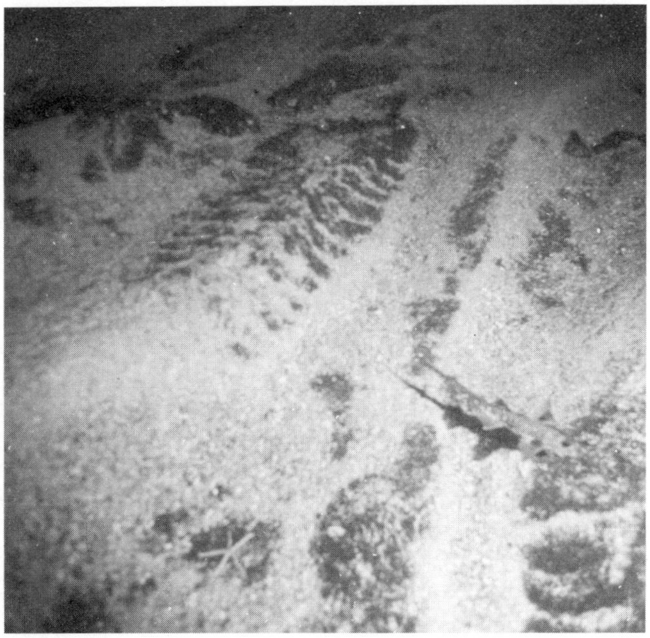

FIG. 2.24. Oscillatory ripple marks at 90 m depth off west coast Vancouver Island, photographed from *Pisces IV* submersible, June 1977. Ripples 30 cm high, about 30–60 cm apart, and oriented parallel to coast. Coarse carbonate shell hash is concentrated in troughs, and finer carbonates extend up the flanks and across crests. Small current ripples on flanks at right angles to the host ripples. Ratfish is about 30 cm long. (From Yorath et al. 1979)

Sand ripples above the high-water mark are generated by coastal winds. The orientation of these patterns depends on the prevailing wind direction so ripple crests may run at various angles to the shoreline. Wind-generated ripples are usually asymmetric with gently rising windward sides and steep-sided leeward sides. Length scales range from a few centimetres to many metres (as for dunes). Frequently there are ripples on low isolated mounds that are separated by relatively flat areas of sand (Fig. 2.25). Miniature wind-shadow ridges a few centimetres long are often formed in the lee of driftwood, pebbles, shells, or other objects with the ridge axis aligned parallel to the prevailing wind direction.

FIG. 2.25. Wind-induced sand ripples above higher high-water level at Cox Bay, south of Tofino, west coast Vancouver Island. Piece of wood at lower right is about 15 cm (6 in.) long. (Photo by the author)

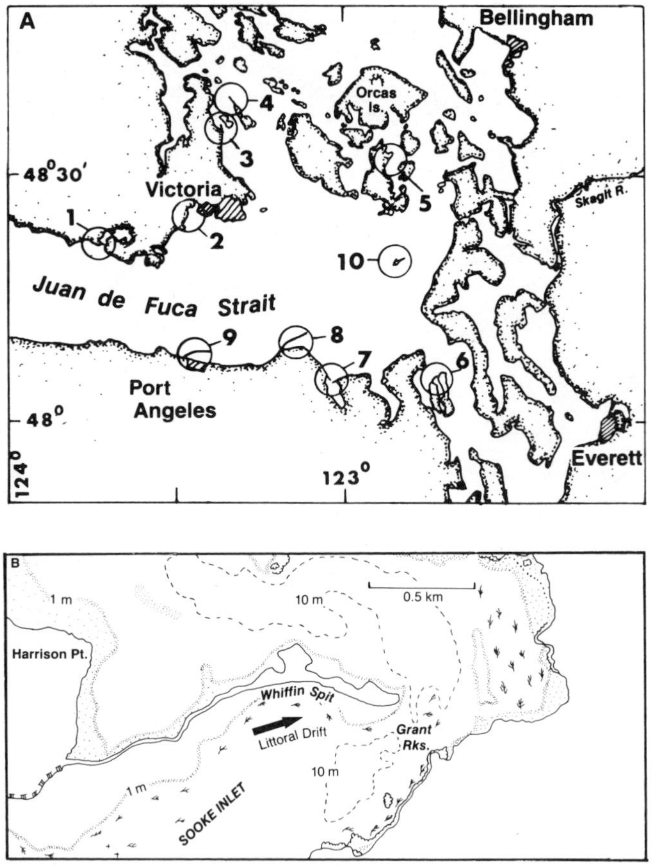

FIG. 2.26. Prominent spits in the eastern sector of Juan de Fuca Strait (A) 1. Whiffin Spit; 2. Esquimalt Lagoon (Cobourg) Spit; 3. Cordova Spit; 4. Sidney Spit; 5. Spencer Spit; 6. Spit (Marrowstone Is.); 7. Gibson Spit; 8. Dungeness Spit; 9. Ediz Hook; 10. Smith Island Spit; with expanded view of Whiffin Spit (B).

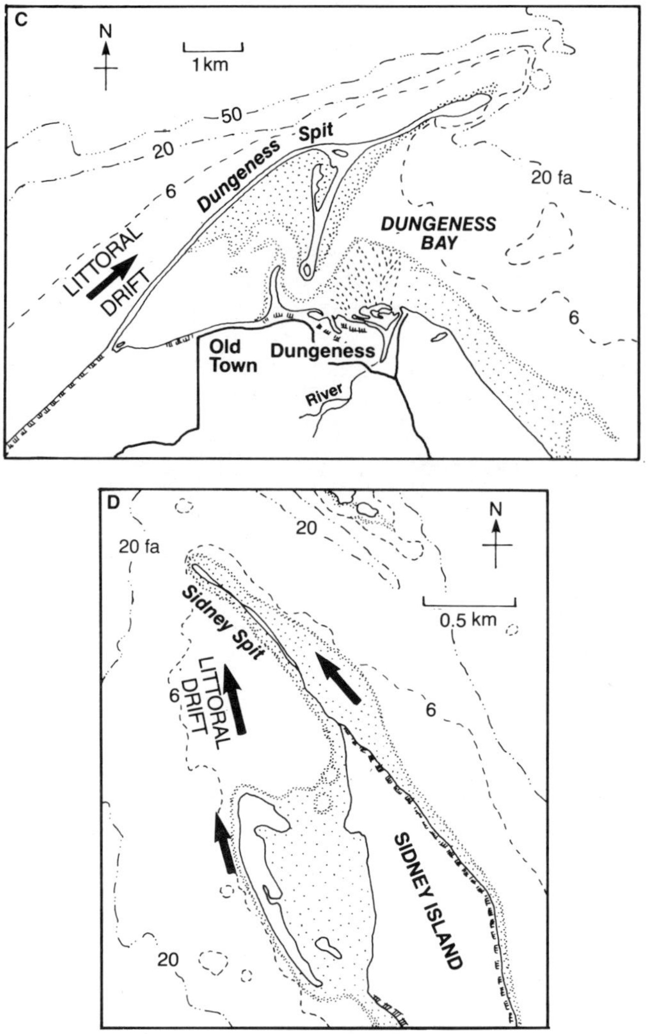

FIG. 2.26. Expanded view of Dungeness Spit (C), and Sidney Spit (D).

Spits and Cusps

A spit (or hook) is a beach with one end joined to the shore and the other end free where it terminates in a hook or recurve (Fig. 2.26). The spit elongates in the direction of longshore sediment drift and can be an alongshore extension of an existing beach or may be aligned across the direction of the prevailing waves. Spits are most common on irregular coasts where they often grow across bay mouths and the entrances to rivers and extend them in the direction of the littoral drift. In this way, spits provide an effective mechanism to straighten out existing bumps in the coastline. Some of the more striking examples of spits (as well as those in Fig. 2.26) include Rose Spit and Sand Spit in the Queen Charlotte Islands, Ediz Hook and the Smith Island Spit (Fig. 2.27) in the eastern sector of Juan de Fuca Strait, Goose Spit (Pl. 5) near Comox, and Rebecca Spit on Quadra Island. Sequim Bay southeast of Dungeness illustrates an embayment that has been almost totally cut off from the sea by the growth of two spits at its mouth.

Ediz Hook (Fig. 2.28) is a textbook example of how spits evolve and what effects man can inadvertently have on their stability. Formation of this spit started about 14,000 yr ago when glaciers began retreating from Juan de Fuca Strait and local sea level was roughly 30 m lower than it is today. As the ice melted and sea level rose, the Elwha River, 13 km west of Port Angeles, cut through the glacial deposits and carried sand and gravel to the Strait. These sediments were then transported eastward by the prevailing littoral current set up by the action of up-strait winds and waves, and the river delta grew to the east (Fig. 2.28a). Sea level continued to rise and additional sediments were added to the alongshore drift through the erosion of adjacent sea cliffs. This led to the creation of small spits to the east of the delta, the forerunners of the present day hook (Fig. 2.28b). Growth of Ediz Hook began in earnest once sea level had approximately reached its present level a few thousand years ago. With the continued deposition of material from the Elwha River and the erosion of sea cliffs, the spit grew eastward as a natural extension of the shoreline, which itself turns abruptly southeastward at the base of the spit (Fig. 2.28c). At the same time the sea cliffs were eroding southward, the westerly base of the spit was migrating southward also. Periodically, moreover, waves would breach the spit and carry sand to its inner side by "overtopping," a process

FIG. 2.27. Smith Island Spit, eastern Juan de Fuca Strait, looking eastward.

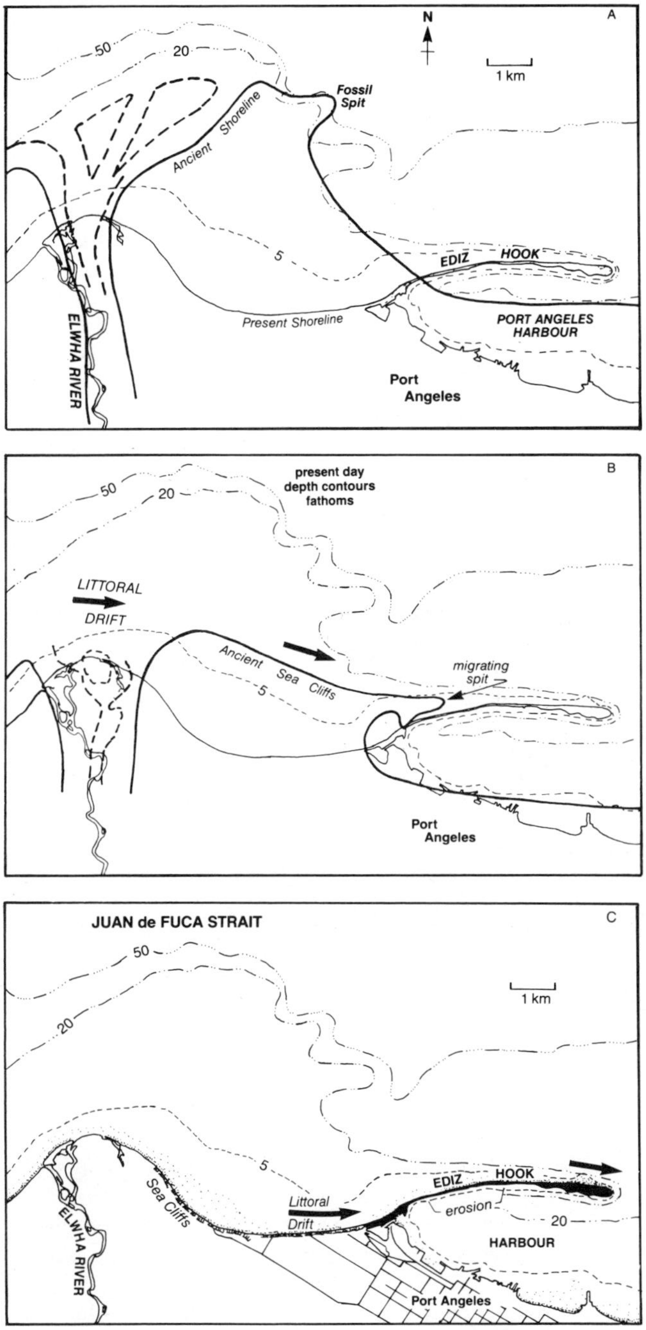

FIG. 2.28 Evolution of Ediz Hook (see Fig. 2.26). Approximately 14,000 yr ago, when local sea level was 30 m lower than present (A) eastward littoral drift began to build spits down-current of Elwha River; (B) sea cliffs east of river eroded southward as the spit that was to become Ediz Hook migrated to the east and south; (C) present-day spit forms natural harbor for city of Port Angeles. (Depths in fa.) (Adapted from U.S. Army Corps of Engineers 1976 and Canadian Hydrographic Chart 3448)

that broadened the hook. Until recently, new material from the west was available to heal the breaches and to provide for the hook's eastward growth of roughly 100 m per century.

Today, Ediz Hook extends about 6.5 km into Juan de Fuca Strait and has an average height of 4.2 m; width varies from 90 m in the west to 27 m at midsection, then widens to 225 m at the eastern end. It forms a natural breakwater for Port Angeles Harbour and is occupied by a U.S. Coast Guard Air–Sea Rescue Station and a paper mill. But all is not well with the spit. Recent events show that its foundation has become susceptible to considerable permanent damage by storm driven waves in Juan de Fuca Strait and, unless steps are taken to protect it, the spit will eventually wash away completely.

The problems all began in 1911 with the completion of a water supply dam for Port Angeles on the Elwha River, which virtually cut off the estimated 38,000 m^3/yr bed load sediment the river had been feeding into the Strait. Next, in 1930, a water supply pipeline was laid along the base of the sea cliffs between the mouth of the river and Ediz Hook. To prevent breaks in the line, rocks and metal sheet-piling were added for protection, and effectively reduced by a factor of 3 the 215,000 m^3 of sediment that had been eroded annually from the sea cliffs. Additional protective steps in 1961 further reduced the amount of erosion. This, combined with river dams, resulted in a 75% drop in materials that nourish the spit. Unfortunately, the wave-induced littoral current retained its capacity to transport large quantities of sediment to the east and, because it could no longer derive these materials from the river deposits or the sea cliffs, it began to remove them from the spit. According to the U.S. Army Corps of Engineers, "Ediz Hook is an active state of erosion due to lack of adequate feed material and is in imminent danger of breaching permanently . . . . Breaching of the Hook would deny land access to the U.S. Coast Guard Base, forcing its relocation. The breakwater protection to Port Angeles would also be destroyed, rendering it unusable as a deep-water port." The suggested solution? First, build a rock revetment (retaining wall) 5 m above the low-tide level and 2 m thick on top along the northern side of the spit. Then nourish the seaward edge of the spit with sufficient beach sand (about 83,000 m^3) to raise the beach profile enough to prevent initial undermining or scour of the revetment toe. And finally, truck in an additional 10,000 m^3/yr to maintain a stable beach profile. The costs for this effort runs into the millions and there would be no guarantee of success. Only time would tell if man can rectify his interference in the natural processes of beach formation.

Beach cusps are crescent-shaped shoreline features of seaward protruding ridges or mounds of sand or gravel separated by small rounded embayments. These may be isolated formations, but more commonly occur in a series with fairly uniform spacing along the shore (Fig. 2.29).

FIG. 2.29. Beach cusps. French Beach, near Point No Point, western Juan de Fuca Strait, February 1978. Ridges of coarse gravels and pebbles; adjoining troughs mostly fine gravels and sand. (For aerial view of cusps see Fig. 2.32.)

Cusps are more likely to form on coarse sand and gravel beaches than on fine sand beaches, where they tend to be sporadic. They also form more readily where waves strike the shore head on and are more common in coastal embayments than on long extensive beaches, where they tend to be obliterated by longshore currents and diagonally approaching breakers. The spacing between beach cusps is directly proportional to the height of the waves that produced them (doubling the wave height doubles the spacing) and ranges from about a metre or less on beaches with low, short waves to as much as 60 m on beaches exposed to large pounding breakers. Sometimes widely spaced cusp remnants are left by storm waves on the higher levels of a beach at the same time that more closely spaced cusps at the lower levels are generated by smaller waves.

Ridges of the cusps generally are of coarser material than the intervening troughs (Fig. 2.29) and there is usually a decrease in sediment grain size down the length of the ridge. This is due primarily to the greater permeability of the coarse material. Once the swash has deposited the larger sediment grains into comparatively steep ridges, rapid percolation diminishes the ability of the backrush to carry them seaward or return them to the troughs. Cusps attain maximum development during the transition from winter to summer beach profiles, and are often destroyed when the transition is reversed. The reason for their remarkably even spacing is only now becoming understood. According to the latest notions, beach cusps are formed through the action of an alongshore pattern of persistent nearshore circulation cells. These current cells are in turn generated by the interaction of incoming surface waves with another type of wave known as an edge wave, essentially a low, visibly undetectable oscillation of the sea surface with crests and troughs perpendicular to the shoreline rather than parallel as with ordinary waves. Simply put, the uniformly spaced troughs and crests of the edge wave modify the incoming breakers along the beach in such a way that the wave swash converges to form cusp ridges and diverges to form cusp troughs at regular intervals. Edge waves will be included in the discussion of rip currents in Chapter 8.

Sea Cliffs to Tidal Flats

Many localities along the west coast are bordered by vertical or near vertical sea cliffs that often plummet abruptly beneath the level of wave action without the hint of a backshore. Within the fiords and interconnecting channels of the sheltered inshore water are precipitous bluffs tens of metres high, where massive glaciers once gouged their way seaward through existing valleys. Erosion of these cliffs has generally been minimal because of the resistant nature of the rock and the low energies of the waves. In many places, large vessels can approach the cliff edge without fear of grounding. Stream valleys, which were truncated during the formation of sea cliffs, sometimes now discharge their flow via water falls from hanging valleys (Fig. 2.8). Sea cliffs in the low-resistance sedimentary rocks that underly the Vancouver Island side of the Strait of Georgia and Gulf Islands are commonly undercut below the high-tide mark through the action of waves and currents. In Active Pass, cliffs that border Galiano and Mayne islands have been undercut a metre or so, the process probably accelerated in recent years by ship and ferry wakes. Wave erosion in such cases is chiefly due to the hydraulic pressure exerted during impact, which can be immense for large waves striking the cliff head-on.

Sea cliffs on the western shore of the Strait of Georgia, in the Gulf and San Juan Islands, and on the outer coast are often bordered by gently sloping wave-cut platforms, partially exposed at low tide (Fig. 2.30). Erosion of these particular cliffs is augmented by the abrasive action of sand and rock fragments thrust at the cliffs by waves breaking over the adjacent platforms. Spectacularly high sea cliffs are found in the San Juan Islands and eastern Juan de Fuca Strait, e.g. Smith Island, Whidbey Island, and the coast immediately west of Port Angeles.

Waves exploit weaknesses in sea cliffs to excavate sea caves and wave-cut galleries. (Galleries carved in sandstone on Gabriola Island were used by Spanish explorers to cache goods.) Excavations of wave-cut platforms may also form tidal pools like those at Botanical Beach near Jordan River on the Canadian side of Juan de Fuca Strait. They contain a colorful display of marine flora equal to

FIG. 2.30. Wave-cut platform, west coast of Graham Island, Queen Charlotte Islands. (Courtesy C. Yorath)

any in the world. Fracture zones that cut at right angles to the rocky shorelines of Juan de Fuca Strait and the outer coast have also allowed the surf to penetrate inland along narrow steep crevices. The proliferation of these crevices, together with the low sea cliffs and dense nearshore plant growth, make hiking the West Coast Trail both interesting and challenging. Another common erosional feature along the west coast is the sea stack formed where more resistant portions of a cliff remain standing within the surf but separated from the retreating cliff (Fig. 2.31). Examples of sea stacks are the pinnacle of rock offshore from Long Beach, Duncan Rock seaward off Cape Flattery, and Siwash Rock off Stanley Park in Vancouver. A much rarer erosional form is the sea arch, where waves have hollowed out a section of an otherwise resistant promontory (Pl. 6). Such features are quite common along the rugged Oregon coast. Under certain conditions, offshore rocks or islands become linked to the mainland by a spit or cusp-shaped deposit called a tombolo, a term that originated in Italy where such features are especially well developed. One of Canada's most famous landmarks, Percé Rock on the Atlantic coast of Quebec, is joined to the mainland by a tombolo. There are numerous tombolos on the west coast of Vancouver Island and the Olympic Peninsula as well as in the protected inside waters. Fisgard Island off Esquimalt, Whyte Islet adjacent to Whytecliff Park, West Vancouver, and Isabella Island south of Saltspring Island are examples of tombolo-linked islands. Figure 2.32 is an excellent example of a tombolo.

With the exception of the high sea cliffs, extensive sandspits, and small lagoons in the eastern end of Juan de Fuca Strait, most of the Georgia Depression is characterized by a low coastal relief and a highly resistant shoreline. Shores of the comparatively low-wave environment from Queen Charlotte Strait through the Strait of Georgia into Puget Sound generally consist of rocky intertidal platforms up to 300 m wide, with heavily wooded backshores that extend down to the high-water mark. Large kelp beds frequently grow in the shallower more protected areas of the platforms and great accumulations of stranded logs are common along the backshore. As stated earlier, eroded sediments are relatively scarce except in the vicinity of river mouths, and thereby limit beaches to the pocket variety formed between resistant headlands adjacent to unconsolidated glacial deposits and erodable rocks. Where streams or rivers enter marine waters, deltaic tidal flats of mud and sand are formed; the extent of the deltas is partially limited by the degree of shelter from waves and currents.

In more exposed waters, wave action modifies the backshore of the intertidal platforms more readily and bluffs are more pronounced. The western portion of Juan

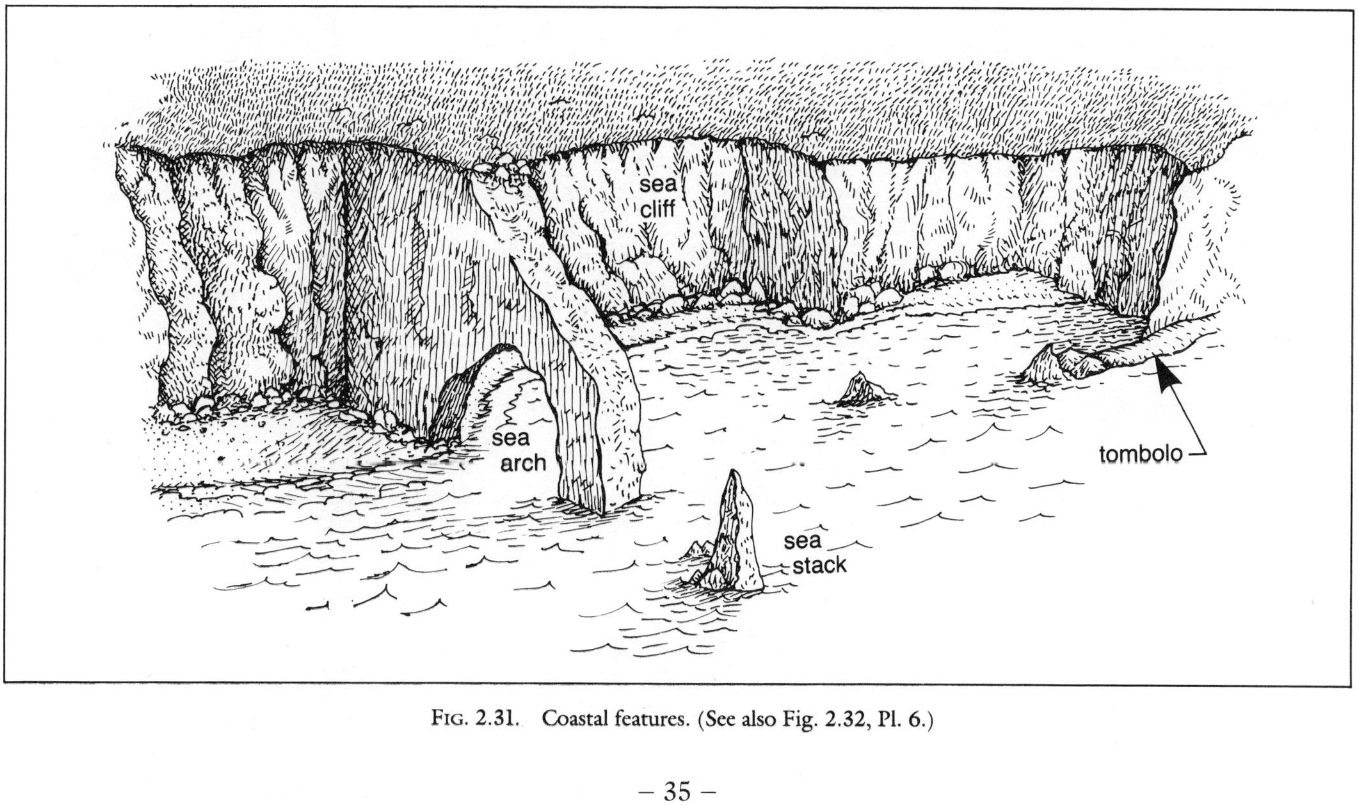

FIG. 2.31. Coastal features. (See also Fig. 2.32, Pl. 6.)

FIG. 2.32. Tombolo joining Frank Island to Esowista Peninsula just south of Tofino on outer coast of Vancouver Island. Breaking ocean waves refract around the barrier island and set up a longshore current regime that transports sediments toward lee of island. Resulting spit has grown seaward until it reached island. Notice also beach cusps along northern (top) shore of tombolo. (B.C. Government Air Photo 1969)

FIG. 2.33. Rocky intertidal platform and cliffs near Point No Point, western Juan de Fuca Strait. View toward west. Pocket beach right of center of photograph. (Photo by author)

de Fuca Strait, for example, is characterized by low rocky platforms to 200 m wide with wave-cut bluffs to 20 m high (Fig. 2.33). Limited amounts of sediment have accumulated in small pocket beaches with large accumulations of logs at the high-water mark. Storm waves in these regions have forced the retreat of the tree line to higher elevations on more gently sloping shores and kelp beds are confined to the sheltered areas. Small rivers that enter the Strait are truncated by bars across their mouths and deltas are of limited extent.

Aside from the narrow, 20-km wide coastal plain in the region of Barkley Sound in central west Vancouver Island and a few isolated beaches, the outer coasts of British Columbia and Washington consist of resistant volcanic rock. These are extensions of the coastal mountain ranges whose elevations reach 1000 m within 10 km of the coast. Extensive glaciation followed by a drowning of the coast at the end of the last ice age has formed a rugged and complex shoreline of steep-sided fiords, broken island groups, and wide intertidal platforms inundated with numerous offshore shoals and isolated rocks. The portions of the coastline that face seaward are exposed to the high-wave energies of the North Pacific Ocean and to the brunt of huge storm-generated surf. Though high cliffs are rare (except on the west coast of Moresby and Graham islands in the Queen Charlottes where they may reach 150 m), steep banks and low undercut bluffs are common, with narrow sand or pebble–cobble beaches at their bases near the high-tide level. The accumulation of wood, fragments of coral, and other debris on these beaches is a beachcomber's delight. In the lee of the many offshore islands and in sheltered areas of Barkley Sound and the numerous fiords, wave effects are considerably diminished and trees can extend to the high-water level. Where glaciers dumped large accumulations of poorly consolidated sediments at the coast, wave erosions and other natural processes created flat, broad, expansive beaches of fine sand between the more resistant headlands. The larger beaches typically have alongshore lengths of around 10 km and offshore extents of 200–300 m. At Long Beach, the shore is backed by logs and low vegetated dunes or cliffs. In the more quiescent areas of nearby Orice Bay, wide mud flats have evolved.

With the exception of the region north and east of Graham Island, the coastlines of Queen Charlotte Sound, Hecate Strait, and Dixon Entrance are similar to those to the south. That is, they consist of either low-lying rocky shorelines adjacent to high wave-energy environments or precipitous cliffs bordering low wave-energy environments, as in fiords and connecting inland channels. Beaches are mostly small and made up of pebble–cobble size material. The north and eastern shores of Graham Island, however, are a gentle, lowland plain, less than 200 m in elevation, of well- to poorly-consolidated muds, sands, and gravels deposited by ancient streams that once flowed from glaciers that covered the Queen Charlotte Mountains. Sediment presently scoured from the adjacent cliffs by waves is transported within the littoral zone toward Rose Point where it has produced wide sandy beaches on the northern shore and narrower coarse gravel beaches on the east shore. Hydrographic charts of this area reveal a broad shallow area where depths less than 50 m extend almost to the mainland from Graham Island, nearly 50 km. Swell heights in this region during a southeast gale can be "mountainous" so operators of small craft are advised to check the marine forecast prior to sailing these waters.

Most unconsolidated sediment within the nearshore zone of the coast is in deltas formed at the seaward terminus of streams and rivers. Small but well-defined deltas are located at the heads of inlets and certain sounds and along sections of coastal embayments. Extensive deltas exist at the mouths of larger rivers that drain vast areas of the Pacific hinterland. The extent of a particular delta is determined by the amount of sediment transported seaward by the river, by the degree of sediment removal through the action of local winds, waves, and currents, and by the topography of the basin into which the river empties. River-borne sediments of silt and sand are carried downstream as suspended load within the water or are dragged along the bottom of the main channels as traction load. Within the larger navigable rivers of the coast, (Skeena, Fraser, and Columbia) the traction load is often in the form of seaward migrating sand waves, which may disrupt routine dredging operations and lead to rapid changes in channel depth. Bed waves with heights of 4.5 m and crest-to-crest lengths of 150 m have been observed downstream of the George Massey (Deas Island) Tunnel in the Fraser River. The downstream migration rate of these sand waves was as great as 75 m/day. For the larger melt-fed rivers of the coast, the annual sediment load is mostly dispersed via a deltaic distributary system during the peak runoff period from May through July. In the Fraser River, immense amounts of sediment (approximately 20,000,000 t) are discharged annually, though this is still small compared to the 275,000,000 t discharged annually by one of the world's largest rivers, the Mississippi, or the 200,000,000 t by the Mackenzie River.

A typical delta consists of very low gradient tidal flats of mud and sand between the seaward edge of the delta front and the seaward limit of the salt marshes and sand

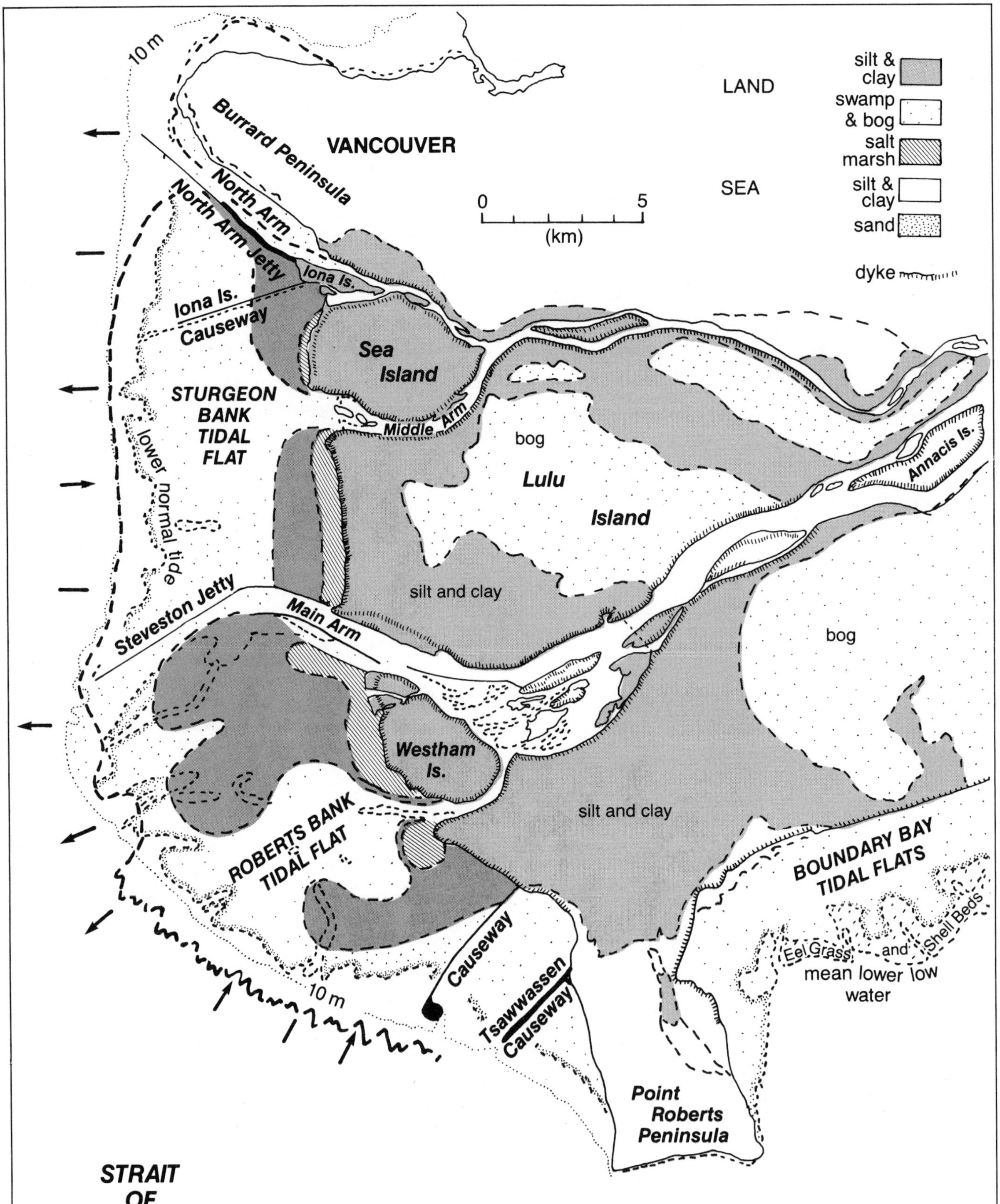

FIG. 2.34. Geologic map of the Fraser River delta. Heavy broken line marks seaward limit of delta foreslope. Arrows indicate direction of advance or retreat of delta front; solid bars denote no change. (Adapted from Hoos and Packman 1974)

dunes of the backshore (Fig. 2.34). Logs and other debris commonly collect at the leading edge of the high-tide line. In many instances man has used dikes and fill to reclaim the marshland (as in the municipalities of Richmond and Delta at the mouth of the Fraser River) and has altered the natural restructuring of the delta by building jetties and training walls.

Despite their extremely small slopes, tidal flats are far from featureless. In addition to the constantly shifting dendritic drainage channels and distributaries that cut across the flats, the redistribution of surface sediments by waves, tidal currents, and river flow creates a variety of migrating sand ripples. These may have symmetrical crests (wave formed) or asymmetrical crests (current formed). Sand-based megaripples 50–100 m long and up to 2 m high are common on the Fraser River tidal flats, and are especially well formed between Iona and North Arm jetties and along the northern side of Steveston Jetty (Fig. 2.35).

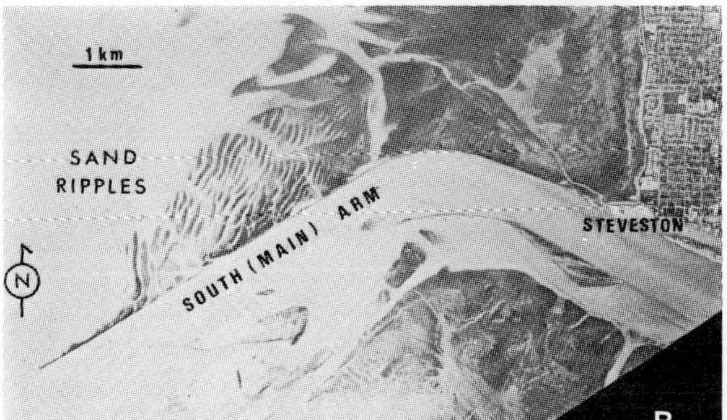

FIG. 2.35. Megaripples on Fraser River tidal flats. (A) vicinity of North Arm Jetty; (B) Steveston Jetty (see Fig. 2.34). Ripple spacings about 100 m; heights 1 m and less. (Courtesy J. Luternauer)

Man-Made Structures

The natural mechanisms of erosion and deposition, combined with the accompanying processes of redistribution, provide a gradual modification of the coastline. Often these changes are in conflict with the requirements of man, who endeavors to modify them by building, what he considers, appropriate structures in the nearshore zone. Such coastal engineering structures serve three main purposes: jetties, training walls, and breakwaters aid navigation and provide protected moorage for boats; seawalls and groynes reduce erosion of the coast.

Jetties are seaward projecting walls constructed at the mouths of rivers or at the entrances to lagoons or tidal estuaries, to stabilize the positions of navigable channels and to prevent shoaling through littoral drift. As with the Steveston, Iona, and North Arm jetties at the mouth of the Fraser River (Fig. 2.34), jetties tend also to direct and confine the flow, and thereby enhance the ability of a channel to self-scour. (Training walls like those on the Squamish River serve a similar function.)

Jetties help also to protect channel entrances from waves and allow land access to offshore docking facilities (e.g. the Tsawwassen Ferry Causeway). To avoid sedimentation problems associated with the longshore drift of sand, most jetties extend beyond the breaker zone and act as barriers to the natural littoral drift induced by the action of waves. This results in a seaward advance of the shoreline. On the downdrift side, on the other hand, the shoreline retreats as the longshore current removes shore material to compensate for the loss of sediment load trapped upstream of the barrier (Fig. 2.36). Although

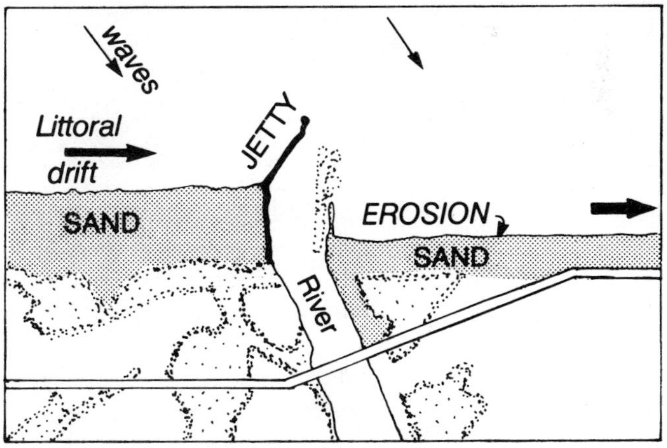

FIG. 2.36. Effect of jetty on alongshore transport of sediment. Deposition occurs upstream and erosion downstream of jetty.

such problems are not severe in the Strait of Georgia or other protected waters because of the comparatively low-energy waves, this is not the case on exposed coasts, for example California, where considerable erosion of beaches and property adjacent to jetties has occurred.

Breakwaters (Fig. 2.37) are artificial harbors built to shield boats from the waves. They are usually attached to the shore at one or both ends, with a narrow opening for boat traffic, and extend beyond the breaker zone. In regions of persistent longshore drift, the entrances commonly require periodic dredging to remove accumulated sands. It was once thought that detached breakwaters, built parallel to the shore and not attached to it, would allow the littoral drift to continue unimpeded and avoid the usual problems of sediment accumulation. However, it was soon found that these breakwaters diminish the local wave energy and the ability of waves to transport sand along the coast. As a consequence, the coastline in the shadow zone behind the breakwater advances seaward until it forms a tombolo, and cuts off the passage behind the breakwater (Fig. 2.37).

Seawalls constructed of a variety of natural or man-made materials are built roughly parallel to the shoreline to prevent direct wave erosion, or to inhibit slumping of sea cliffs. Examples include the seawall along Ross Bay near Victoria and the Stanley Park seawall of Vancouver. Experience has shown that vertical seawalls undergo unacceptably high rates of erosion at their bases, so most present day structures of this type are built with a shoreward slope. The height of the seawall is usually enough to prevent possible damage from overtopping by storm-wave run-up at high tide. Although a seawall provides protection for the land immediately shoreward, it affords no protection to the beach in front of the wall or to adjacent portions of the coastline. In fact, when a seawall is built on an initially receding shoreline, the erosion processes simply shift further along the coast. The only sure defense of a coast it appears, is a wide, high beach.

A groyne (or groin) is a barrier that projects directly seaward from the shore. Its primary purpose is to trap a portion of the littoral drift to add to an existing beach that would otherwise be eroded. As with jetties, sediment accumulates on the updrift side of the groyne whereas on the downdrift side the flow is starved of beach material and erosion may take place (Fig. 2.38). Boat launching ramps built by owners of waterfront property produce similar effects to groynes and can inadvertently cause erosion of a neighbor's beach further along the shoreline.

Aside from the usual attempts by British Columbia residents to stem erosion of a beach front, there have been organized and elaborate schemes. One involved the west-to-northwest sector of the Point Grey cliffs, whose continuing erosion into the Strait of Georgia concerns numerous agencies. During the summer of 1974, a 1000-m long sand-fill, plus cobble-core berm and groyne system was constructed along Towers Beach, University of British Columbia, to prevent marine erosion along the base of the cliffs. The berm partially protected the cliff base from weathering during the following winter but by February had failed along a 450-m section, due to storm-generated west-to-northwest waves (McLean 1975). Moreover, the groynes proved ineffective in retaining the sand fill, and over 5400 m^3 of sand were moved eastward by the littoral drift. With a useful lifespan of less than 2 yr the man-made berm had done little to limit the average recession rate for the cliffs of 15.2 cm/yr.

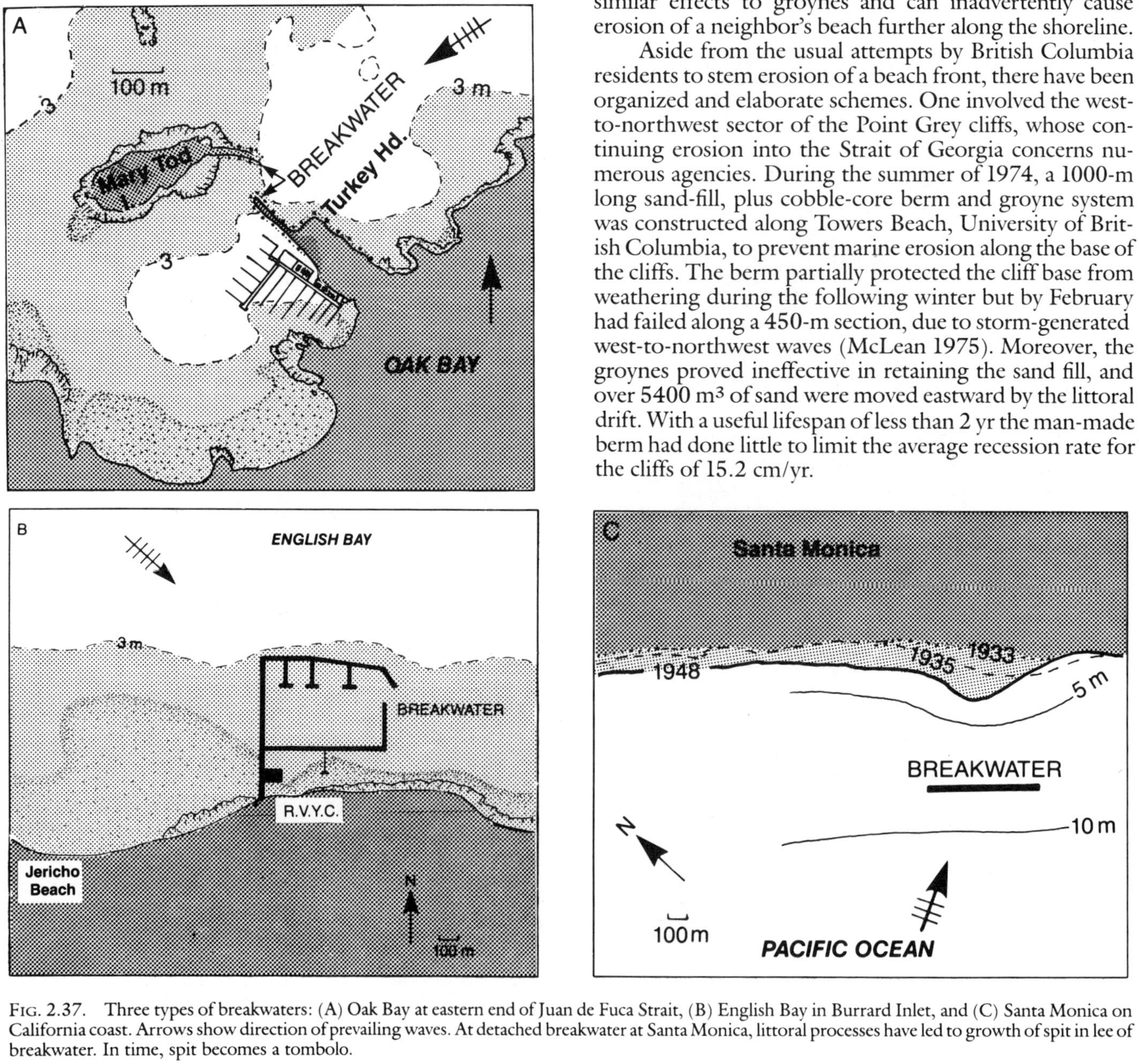

FIG. 2.37. Three types of breakwaters: (A) Oak Bay at eastern end of Juan de Fuca Strait, (B) English Bay in Burrard Inlet, and (C) Santa Monica on California coast. Arrows show direction of prevailing waves. At detached breakwater at Santa Monica, littoral processes have led to growth of spit in lee of breakwater. In time, spit becomes a tombolo.

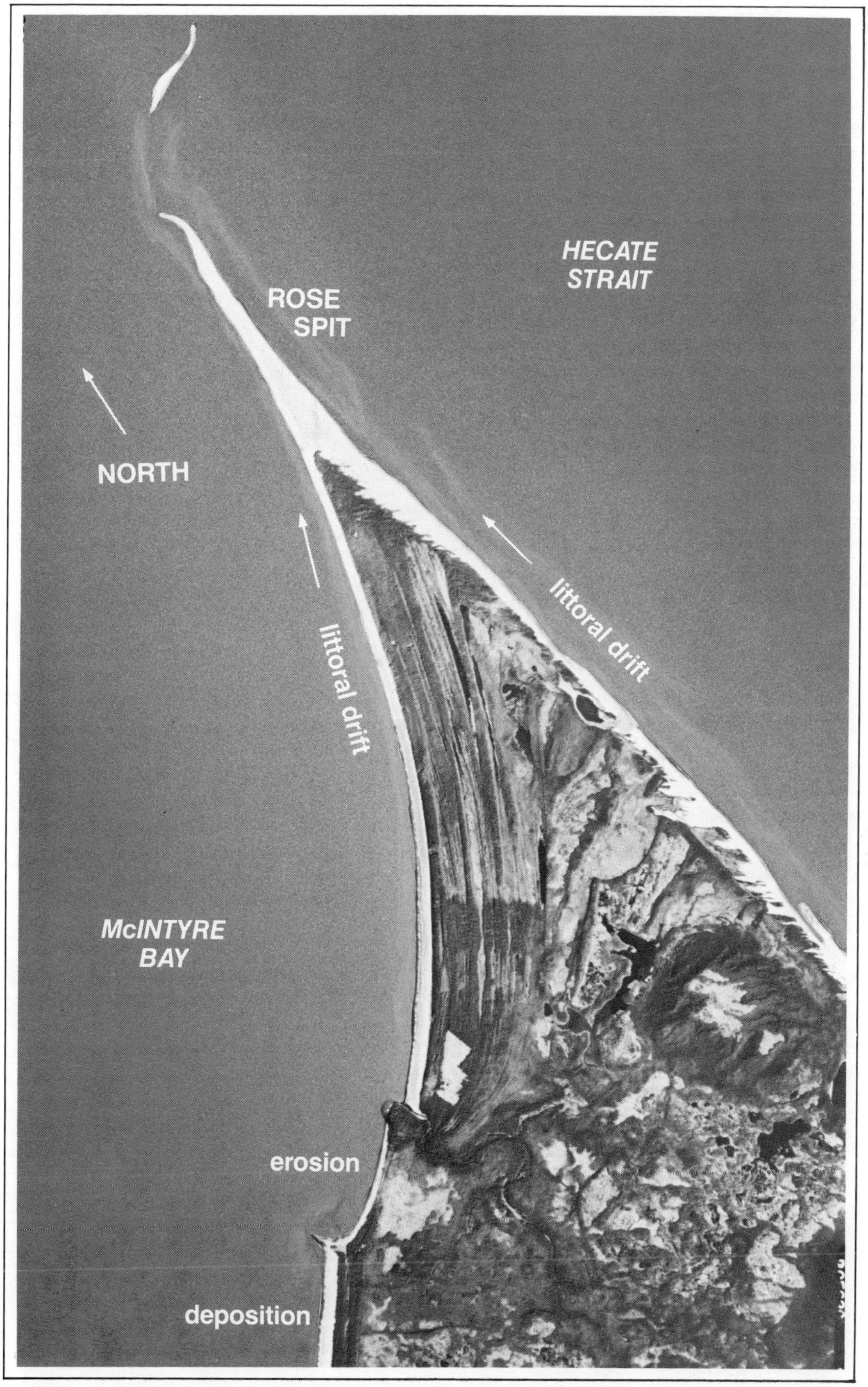

FIG. 2.38. Natural groyne system of headlands along northeast coast of Graham Island, Queen Charlotte Islands. Rose Spit at top of photograph. Northward littoral drift causes sediments to accumulate on upstream side of each promontory and erode from downstream side. Series of relict shorelines inland of present beach formed at times of higher sea level stands following last ice age. (B.C. Government Air Photo 1973)

PART II
GENERAL PHYSICAL OCEANOGRAPHY

Chapter 3. Tides and Tidal Streams

Early Knowledge

The rhythmic rise and fall of sea level, called the tide, has long been part of the seafarer's life. The early Greeks and Romans lived on the almost tideless Mediterranean Sea and learned of tides and their association with the sun and moon from voyages to the Atlantic Ocean. Pytheas of Marseille, who is reported to have circumnavigated Britain around 320 B.C., was one of the first to actually record the existence of tides and to note the close relationship between the time of high water and the transit of the moon. Julius Caesar also noted this relationship during a campaign to Britain, although his initial ignorance of tides caused his fleet to be wrecked on the beach during an invasion attempt just prior to a high tide. A somewhat similar fate befell Alexander the Great at the mouth of the Ganges River a few centuries earlier. Arab and Persian pilots who began to sail the coasts of India and China around the 9th century A.D knew of the twice daily rise and fall of sea level in certain river estuaries and the associated reversals in currents. Although the close association of the tides with the lunar phases is mentioned in Arab writings of the 9th century, one prominent scholar, Ibn al-Fakih (902 A.D.), linked tides in Canton Harbour to an angel that dips his finger in the China Sea so the sea rises and then ebbs on its removal, or to a whale that "inhales water causing ebb and exhales it causing flow."

In 1325 the Arab scholar, Al Dimiski of Damascus, published remarkably accurate tidal predictions for farmers irrigating their fields near the mouth of Shatt al Arab, the river formed by the confluence of the Tigris and Euphrates rivers, that flows into the northern Persian Gulf. He mentions for example that there were two high and two low tides each day, that for each there was slightly less than 1 h lag of both the ebb and the flood from the day before, and that the period of ebb in the river estuary persisted longer than the period of flood. Similar records from the 16th century indicate that certain individuals in England made their livelihood predicting tides in areas of the coast.

It was not until Sir Isaac Newton published his *Equilibrium Theory of the Tides* in 1687 that a scientific understanding was begun. Even today, the study of tides is far from complete despite a good understanding of their cause. Scientists are only beginning to sort out the effects of tides in the deep ocean and complicated coastal regions like those of British Columbia and Washington. There are even times when tidal theorists seem to revert to medieval thinking. A recent proposal for instance suggested that the tides exist because "the oceans initially came from the moon and are trying to return."

Datum

It has been universally agreed that the reference level from which to measure the height of the tides is the chart datum, the level to which low water can be expected to fall during normal tides (Fig. 3.1). Datum is used as the reference level for depths plotted on nautical charts. It has been chosen in such a way that it is low enough that few tides will fall below it, but not so low that a chart will show less depth than the mariner will usually find. Negative values in the tide tables, indicating unusually low tides, serve as a warning that the water level will fall below datum. One of the main purposes of tide gage installations is to obtain measurements long enough to accurately establish the datum level.

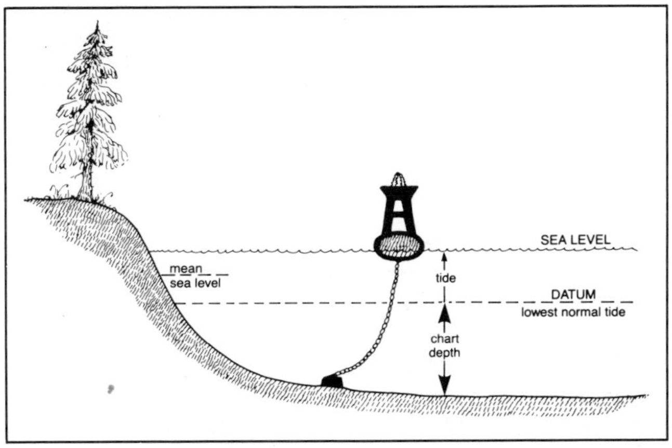

FIG. 3.1. Low water datum. Tides are measured upwards from this level.

A word of caution to those who use either American or Canadian charts — the two countries differ in their interpretation of "lower low water." Where Canadians define chart datum as the level of water at the "lowest normal tides," west coast Americans define it as "mean lower low water." Because averaged values are used, U.S. datums are slightly higher than Canadian datum. Soundings on American charts would, therefore, indicate a greater depth of water than a Canadian chart of the same area. In other words, Canadian hydrographers are more conservative when informing the mariner on the minimum depth of water he can normally expect at a particular location. No difficulty should arise if the appropriate set of tide tables is used; U.S. tables for U.S. charts and Canadian tables for Canadian charts. The main difference between the two is that U.S. tables tend to show negative tides more often than Canadian. The actual water depth in each case is found by adding the tide height for the particular area for the specified time to the chart datum; that is, actual depth = chart datum + tide height, as per the tide tables. It should be noted that datum is not the same as mean sea level. Datum is determined solely by the lowest tides whereas mean sea level is determined from the average of all tides, including high waters. Consequently, the datum level changes from place to place along the coast because of the changing nature of the tides. Datum elevations at Victoria and Vancouver, for example, are different because their low waters do not normally fall to the same levels.

Measurement of Tides

To determine the datum level and daily fluctuation in water depth at a particular location, measurements must be made. This is commonly done by tide gages — flotation devices and recorders placed within a boxlike container called a stilling well. The unit is affixed at a known height above the bottom to a permanent structure such as a wharf (Fig. 3.2a). A narrow opening near the submerged base of the well allows outside communication with the ocean and at the same time effectively damps-out rapid, unwanted fluctuations in sea level produced by surface waves. Gradual changes in water elevation due to the tide cause the float to move a pen across scaled chart paper that is slowly advanced by a clock mechanism. The resulting ink trace or "tidal" curve then gives the height of the water surface as a function of time (Fig. 3.2b). On the basis of such measurements the datum level is established and tide elevations subsequently determined.

Tidal records obtained from coastal locations deemed to be of major importance to navigation (reference ports) are analyzed by the Canadian Hydrographic Service and used to make the tidal predictions published in *Canadian Tide and Current Tables*. Predictions for these ports are generally based on a year or more of continuous observation, to ensure accurate times and heights of high and low water. Secondary ports are coastal locations of lesser navigational importance. Predicted tides for these ports are found by applying the appropriate time and height corrections given in the tide tables to the reference port predictions. These corrections are simply the measured differences in time and range for the tide between the two locations. In contrast to reference ports, tidal records at secondary ports typically only span periods of about 1 mo.

Though tide gages have been the traditional method of measurement around the world, they suffer from a variety of faults. They are susceptible to erroneous readings caused by swell waves and to partial plugging of the intake by debris and living organisms like starfish. Supporting structures are also expensive and often in danger of wave damage, despite the fact that gages are comparatively cheap. Lastly, installation of these gages is limited to shallow water adjacent to the coast. For these reasons, much more sophisticated instrumentation known as pressure gages has recently been developed. These are generally self-contained units that can be set on the seafloor at appreciable depths to record on tape the total bottom pressure of the combined weight of the air and water above. The principal component in many gages is a vibrating wire or crystal whose oscillation frequency varies with changes in the ambient pressure. Using an assumed density of sea water of 1.025 g/cm^3, this variation in frequency is then converted to a change in pressure and, hence, to a change in depth after allowance is made for any accompanying temperature alteration. The advantages of pressure gages are numerous: they can be placed at offshore locations, their records receive negligible contamination from wind waves, they are easy to handle, and they have comparatively good accuracy. Measurement resolution of these gages is typically around 1 part in 100,000, or about 10 times better than their accuracy, so a

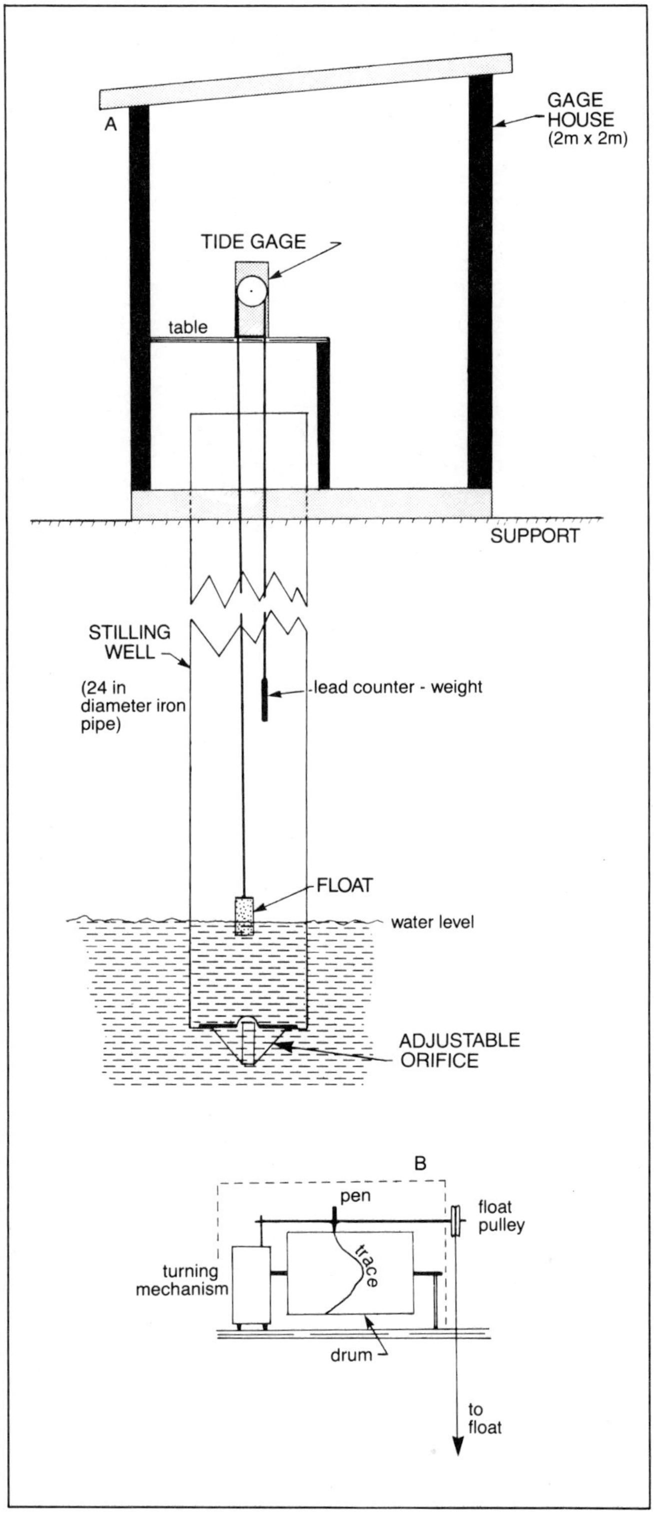

FIG. 3.2. (A) Stilling-well type tide gage. The oil-filled float adjusts with changing water level, mechanically moves pen (B) to trace a tidal curve on graph paper affixed to a drum rotating at a known rate. (Courtesy R. Brown)

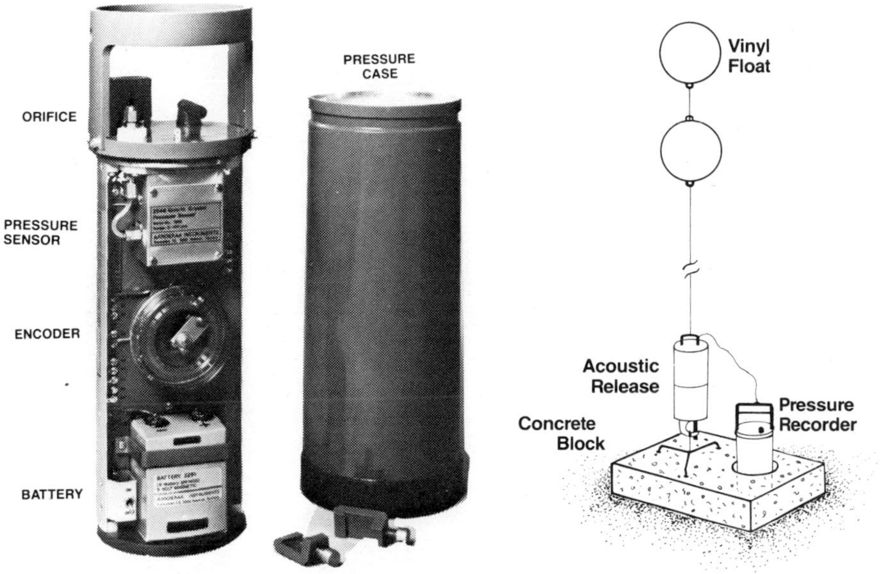

FIG. 3.3. Underwater pressure recorder and mooring arrangement. Water pressure at orifice is recorded internally and stored on magnetic tape. Recovery is effected by activating acoustic release via a surface transmitter; floats then lift recorder from concrete block to surface. (Courtesy Aanderaa Instruments Ltd., Victoria)

gage situated in 1000 m of water can detect changes in depth as small as 1 cm. The main disadvantages of these gages are their susceptibility to undesired dynamic pressure effects due to currents that flow past the orifice, pressure-to-depth conversion errors caused by departures of seawater density from the assumed value and the fact that data are not normally available in real time. As a consequence of the latter restriction, pressure gages are not employed in tsunami warning networks. Despite these limitations, pressure gages are essential to investigations of deep-sea tides and the offshore propagation of tsunamic waves. A modern pressure gage installation is shown in Fig. 3.3.

Nature of Tides

The tides nearly repeat themselves once every 24 h and 50 min. This is the lunar day and the time it takes a point on the earth to rotate back to the same position relative to the moon during each revolution. Therefore, the daily rhythm of the tidal cycle is governed by the lunar day and not by the solar day of 24 h, which paces the daily cycle of life on earth. Consequently, times of high and low water are roughly 50 min later from one solar day to the next. If the sun were the major cause of the tides they would repeat themselves nearly every 24 h, and high and low waters would occur at the same time each standard day. Henceforth, when tides are discussed "day" will usually refer to a lunar day.

The difference in depth between high and low water is called the range of the tide. Oceanic tidal ranges vary from typical low values of 10 cm or so in the Mediterranean Sea and Arctic Ocean to more than 15 m in the Bay of Fundy. Along the British Columbia–Washington coast, the range is commonly between 3 and 5 m, with greater ranges during June and December and smaller ranges during March and September.

To categorize the daily tide, the number of cycles (number of highs and lows) in a lunar day are counted. When the tide has 1 cycle (1 high and 1 low) per day it is said to be diurnal (Fig. 3.4a); when it has 2 cycles per day of nearly equal heights it is said to be semidiurnal (Fig. 3.4b). As the inserts in the two figures illustrate, tides on the west coast are purely diurnal or semidiurnal for only a few days each month. Most of the time they are a mixture of diurnal and semidiurnal tides and are said to be mixed (Fig. 3.4c). The tides at Victoria Harbour and Sooke, for example, are classified as mixed, predominantly diurnal, whereas tides at Seattle, Vancouver, Tofino, and Prince Rupert are mixed, predominantly semidiurnal. (For some examples of different types of mixed tides see Fig. 3.5.)

A property of all mixed tides on the west coast is the

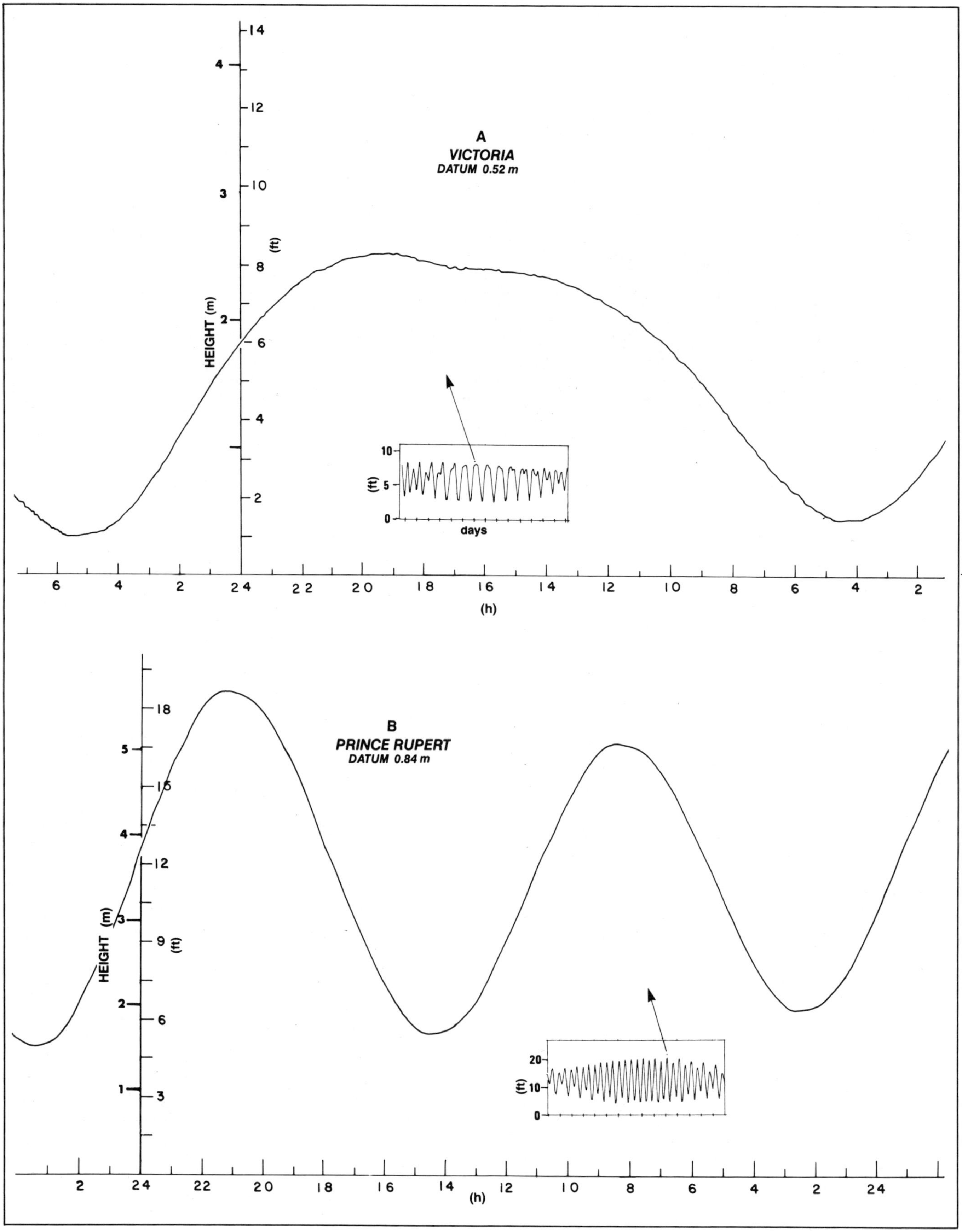

FIG. 3.4. Variations in water level over a 1-day period at three locations on British Columbia Coast. (A) diurnal tide, Victoria; (B) semidiurnal tide, Prince Rupert.

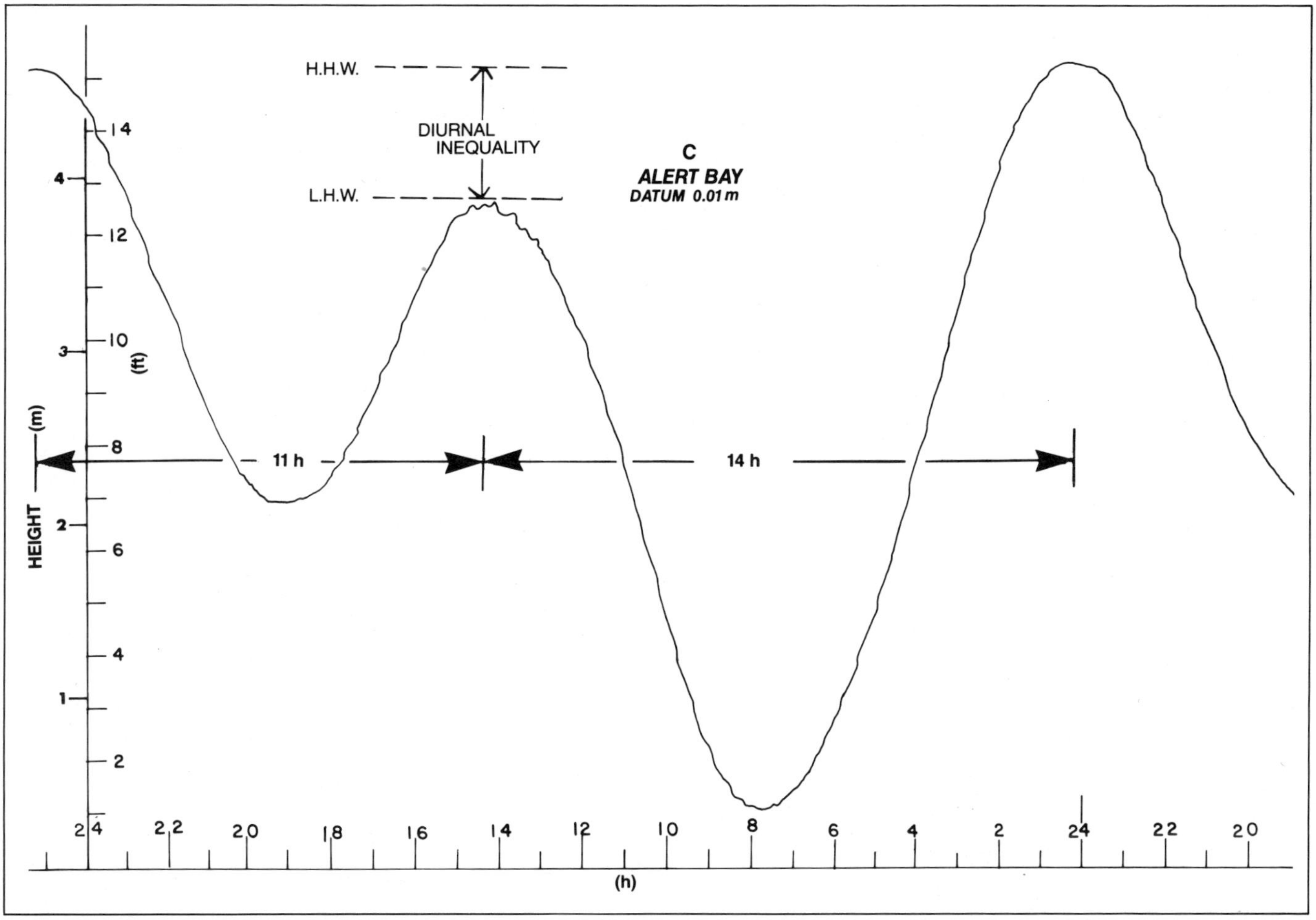

FIG. 3.4. (C) Mixed tide, Alert Bay. Time progresses from right to left in each diagram. Insets show evolution of tides at each location over 2 wk.

diurnal inequality, expressed in Fig. 3.4c as the difference in height between the higher high-water level and the lower high-water level. There is also a diurnal inequality between lower low water and higher low water. In fact, a characteristic feature of the inequality in tidal elevations within the basins that separate Vancouver Island from the mainland is that the difference in height between successive low waters is usually greater than the height difference between successive high waters. Also, whenever there is an inequality in the height between successive high tides there is a daily inequality in the time intervals between pairs of high tides, as shown by the intervals of 11 and 14 h (Fig. 3.4c). The same is true of successive low waters.

In addition to the differences in tidal range during any particular day, the range of the tide changes progressively from one day to the next with a cyclic period of around 2 wk (Fig. 3.5). Generally speaking, high tide becomes continually higher and low tide continually lower for about 7 days when the roles reverse, high tides become lower and low tides become higher for about the next 7 days. It is during such cyclic variations in range that comparatively high spring tides and comparatively low neap tides occur. These tides vary with the phases of the moon; spring tides occur near the times of a full or new moon and neap tides near the time of the moon's quarters.

The 2-wk cycle in the tidal range on the outer coast is closely linked to the moon's phases (Fig. 3.5). However, this obviously does not apply to tides within the protected waters of the southwest coast. (These differences will be discussed later.) Figure 3.5 further reveals an approximate 2-wk cycle in the magnitude of the diurnal inequality for both high and low tides. This inequality is always most pronounced when the moon is furthest north or south of the earth's equator (tropic tides), but is almost nonexistent when the moon is directly above the equator (equatorial tides). Within the Strait of Georgia, Puget Sound, and the eastern end of Juan de Fuca Strait, the diurnal inequality becomes so great that for a few days each month the tropic tides are essentially diurnal. At Sooke and Victoria, diurnal tides produced this way may persist for as long as 5 days. Along the outer coast, the diurnal inequality is smaller and the tides are never diurnal. Equatorial tides, on the other hand, are invariably semidiurnal throughout the entire world ocean. The latter is one of the few definitive statements that can be made regarding the tide without any exception or qualification.

The cyclic 14-day variation in the diurnal inequality is accompanied by a corresponding variation in the time intervals between successive high waters and between successive low waters. These time intervals alternate from equal, when the diurnal inequality is least (equatorial tides), to extreme values of 8½ and 16¼ h, when the

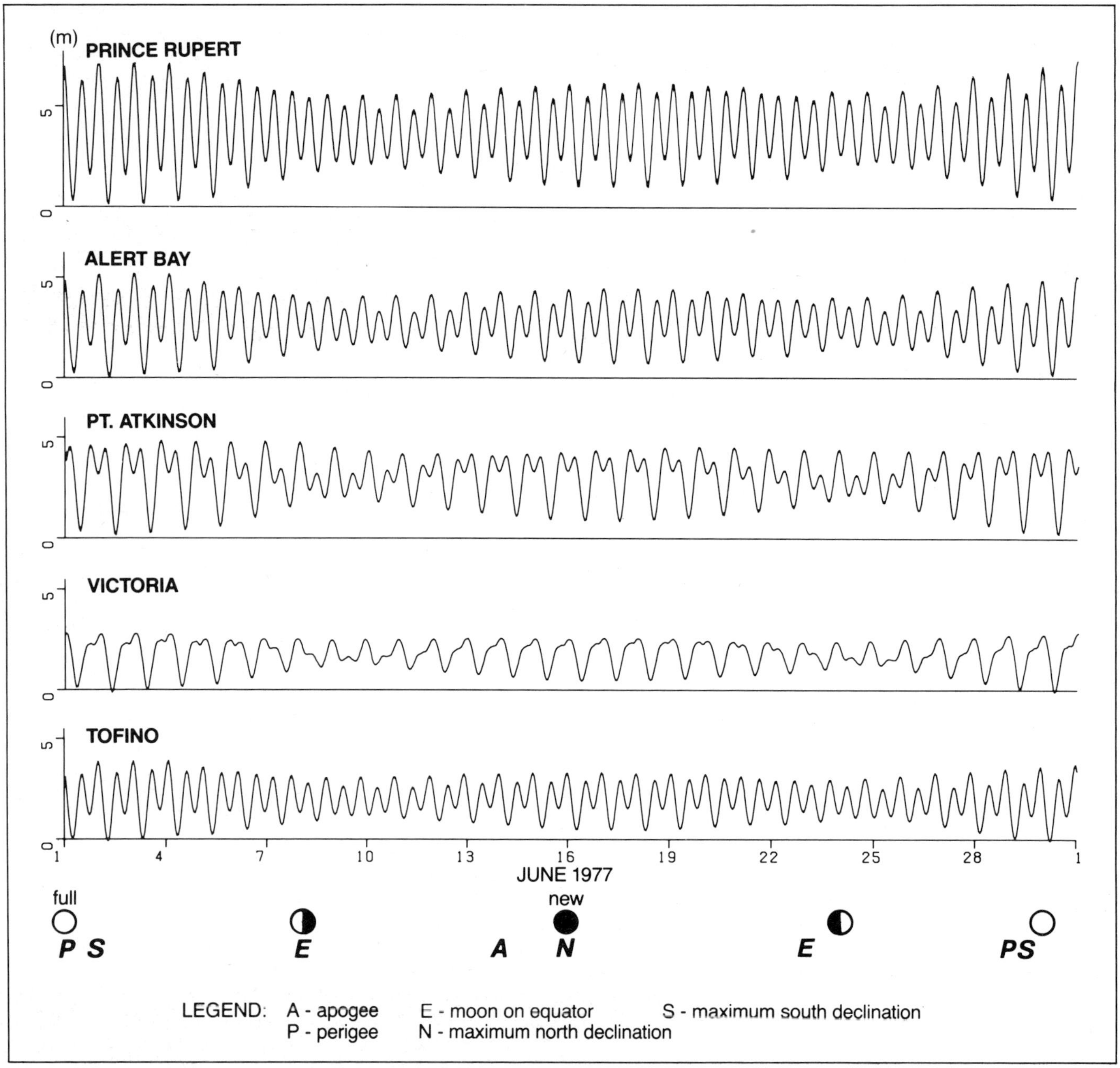

FIG. 3.5. Sea level variations over 1 mo at five locations on British Columbia coast (June 1977). Heights in metres. (Courtesy Canadian Hydrographic Service and A. Douglas)

diurnal inequality is greatest (tropic tides). In every configuration, of course, the combined time interval between the successive tides (the first interval plus the second) must always add up to the lunar day of just over 24¾ h. Thus, if the reader were to clock a 16½-h delay between two successive high tides, he would only need to wait another 8½ h until the next high tide for a total of 24¾ h.

When tidal curves like those in Fig. 3.5 are examined more closely, it can be seen that the tides are not repeated exactly every 14 days. Stages of the daily tide with comparatively large tidal ranges (or comparatively low tidal ranges) spaced 2 wk apart are not of identical magnitudes. This suggests that factors affect the tides whose cyclic variability is of longer duration than those that produce the approximate 14-day variation. If Fig. 3.5 were extended to cover an entire season, the tidal pattern would be more closely repeated every 29½ days than every 14 days. Therefore, some factor, or factors, influencing the rhythmic rise and fall of the sea must have a period of 29½ days. Even over a month, however, the tide is not repeated exactly, so the tidal record must be extended. If a full year's record were available and tides measured in January of one year were compared with those of the following January, would the tides match up exactly? Nearly, but not well enough for some purposes. Tide heights must be measured continuously for almost 19 yr before the tide would begin to repeat itself to an accuracy of a few centimetres. Although there are even longer period variations in the tidal range (see section Long-Period Tides), the whole exercise would, if extended, become somewhat academic as nontidal and nonperiodic fluctuations in sea level begin to mask the increasingly smaller differences in the measured tides. Nevertheless, the values listed in the tide tables take into consideration as many possible contributions as practical, in order to give accurate height predictions.

Equilibrium Theory of Tides

So far only the form that daily tidal variations can take has been examined; as was the situation before Newton's time. There is a set of observations but no theory to explain them. How are tides formed? Why are they almost always semidiurnal or mixed? And why does the range of the tide vary on such a regular basis about every 2 wk?

To obtain rudimentary answers to these basic questions it should be appreciated that the tide is really a combined or integrated response to a variety of natural phenomena. The single most important factor is the combined gravitational attraction of the moon and sun on the earth. Because of its greater proximity to the earth, the moon's gravitational attraction is twice as important as the sun's (Appendix D). Except for one important case, therefore, the influence of the sun can be disregarded when the main features of the tide are examined. As with Newton's Equilibrium Theory of the tides, moreover, matters are simplified by assuming that the earth is uniformly covered with water, with no continents or submarine mountains to interfere with the oceanic motions.

Because many simplifying assumptions have already been made, it is acceptable to go one step further and also ignore the presence of the moon. For the moment, imagine that the water-covered planet is drifting alone through a moonless, sunless universe! If the earth were not rotating on its own axis, its shape would then be a perfect sphere held together by its own gravity. Distances measured from its center to anywhere on the sea surface would be equal. In reality, of course, the earth spins on its axis once every 24 h and has subsequently been deformed into an ellipsoid with a diameter 42 km greater through the equator than the poles. There is a corresponding distortion of the ocean, so mean sea level decreases slightly from the equator toward either pole. An observer on earth would not notice any daily change in sea level because this distortion is uniform around a parallel of latitude. The forces produced by the earth's axial rotation, therefore, cannot generate tidelike variations in water level and, for the sake of simplicity, a nonrotating earth with a fixed orientation relative to the stars will be considered. Now the moon is returned to the picture.

The earth and moon constitute a celestial unit held in a binding partnership by their mutual gravitational attraction. It is this gravitational force that disrupts the natural tendency for each of these bodies to move in a straight line through the heavens, and constrains their individual centers to travel in nearly circular orbits around a common center of mass once every 27⅓ days (Fig. 3.6). As the earth's mass is 82 times greater than the moon's, the center of mass, like the balance point of a teeter-totter, is shifted toward the earth and actually lies beneath the surface at a

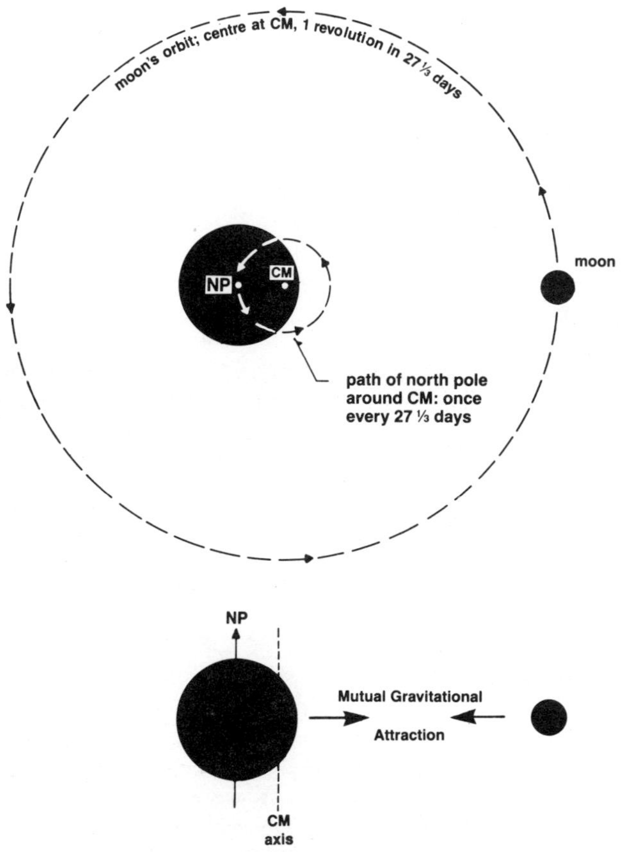

FIG. 3.6. Earth–moon system. *Top*: paths of moon and north pole (NP) around center of mass (CM) for combined system. *Bottom*: side view of center-of-mass axis and earth's axis of rotation when moon is at zero declination.

constant depth of roughly 1700 km. (The position of this imaginary point is not fixed at one particular spot for all times, but slowly circles inside the planet over a month, always keeping the same depth.) The rotation of the earth–moon system about its common center of mass causes each particle of the earth to trace out a circular path over a period of a month. Moreover, Fig. 3.7 shows the orbital radius for every particle is the same, regardless of distance from the center of mass. This has important consequences, for it means that the force associated with these circular orbits is of equal magnitude and direction for each bit of the earth. What is this force? A physicist would call it the centripetal force, the total force necessary to make each particle of the earth travel in its own particular circle in the presence of the moon's gravitational field. This centripetal force is uniform over the entire earth and is directed everywhere toward the moon at right angles to the axis of rotation drawn through the centre of mass. (To account for the tidal forces, the rotation of the earth about its own axis need not be considered). Unlike the centripetal force, however, the ever-present lunar gravitational attraction on a particle of the earth depends on that particle's distance and orientation measured relative to the center of the moon (Fig. 3.8). Only at the exact center of the earth are the moon's gravitational pull and the centripetal force identical in both strength and direction. At

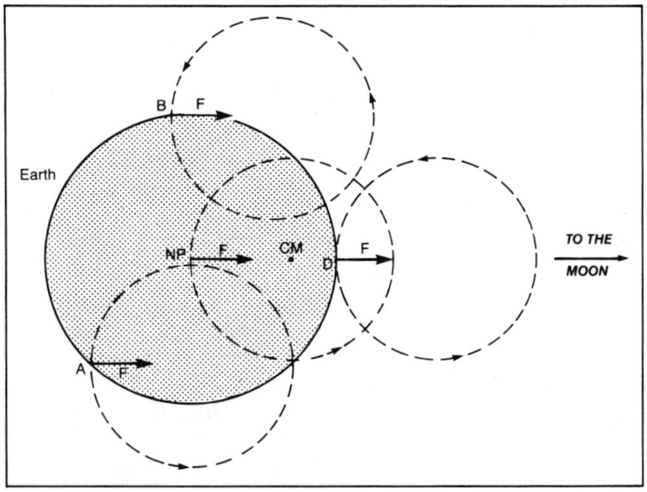

Fig. 3.7. Ignoring spin of earth about its axis of rotation, each segment of earth follows a circular path as it moves about center of mass (CM) for combined earth–moon system. Vectors (arrows) represent local centripetal force (F), which is the same over entire globe and equal to force of mutual attraction between centers of earth and moon. (see Fig. 3.6)

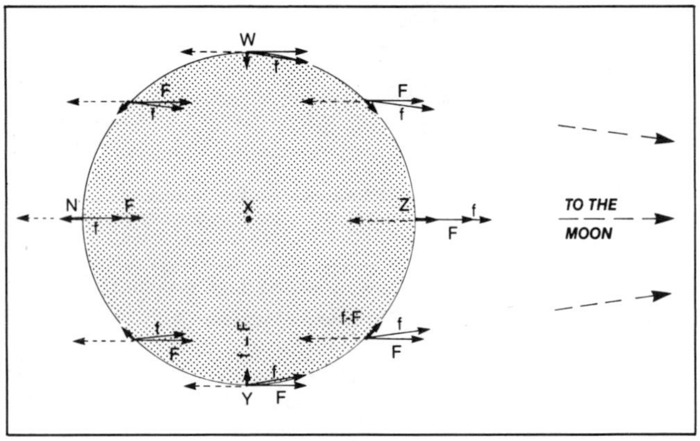

Fig. 3.8. Vector subtraction of centripetal force (F) from local attractive force (f) of moon on an element of water. Resultant vector (short stubby arrow) is tide-generating force, (f-F). Same result is obtained by vectorially adding centrifugal force (broken arrow) to f. Tide-generating force draws waters into bulges and hollows until balanced by pressure gradient associated with increasing sea surface slope. Low tides at W and Y; high tides at N (nadir) and Z (zenith). Compare to Fig. 3.9.

all other locations, there is a small difference between the two forces. It is this slight difference that causes the tide-generating force on the earth.

Before proceeding any further, a scientifically incorrect, though conceptually useful, approach will be taken to the ideas introduced so far. Instead of arguing on the basis of the inward centripetal force, each orbiting particle on the earth can be considered as experiencing an outward centrifugal force as the earth–moon system rotates around its center of mass. (In reality, there is no such force because to keep an object in a circular orbit requires an inward force not an outward one. What is commonly called the centrifugal force is nothing more than the "reluctance" of moving objects to travel in curved trajectories.) Turning a deaf ear to the critics, it is found that, because the centrifugal force originates from the same motions that produce the centripetal force, it will also be uniform in both magnitude and direction over the entire earth but will be directed away from the moon. The centrifugal force is then simply the reverse of the centripetal force (Fig. 3.8). Again the distance-dependent gravitational pull of the moon will only equal the centrifugal force at the exact center of the earth; away from the center the two opposing forces will be unbalanced and result in a net tide-generating force. The distribution of this force around a meridian of longitude is in Fig. 3.9a. It should be clear from this diagram that subtracting the moon's gravitational force from the centripetal force yields the same net force at each point as adding the moon's gravitational force to the outward centrifugal force. Similar, but weaker, tidal forces are also generated by the earth–sun pair.

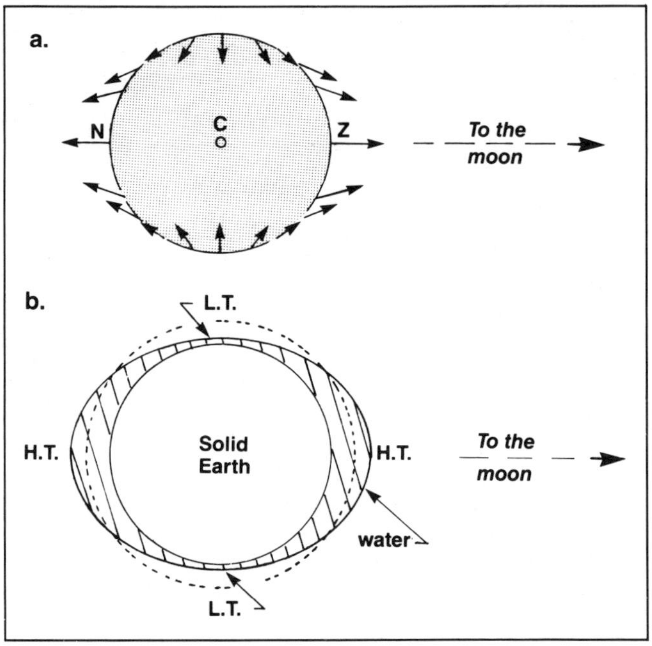

Fig. 3.9. *Top*: Simplified version of Fig. 3.8 shows distribution of tide-generating force along a meridian of longitude. *Bottom*: Tidal bulges and hollows created by tide-generating force (highly exaggerated in vertical). Broken line corresponds to mean sea level or shape of ocean in absence of tidal forces. (H.T. = high tide; L.T. = low tide)

The lunar tide-generating force, the imbalance between the centrifugal (or centripetal) force and the moon's gravitational attraction, affects the formation of ocean tides in two basic ways: on the side of the earth that faces the moon, the moon's gravitational pull is slightly stronger than the opposing centrifugal push, so the water bulges toward the moon; on the side that faces away from the moon, the centrifugal force exceeds the moon's attraction so the water bulges away from the moon (Fig. 3.9b). To put it more concisely, the water is drawn away from the earth on one side and the earth away from the water on the other. The two bulges are associated with high tides. Midway between them, sea level is lowered below normal to create a low-tide hollow that girdles the earth. Maximum high tides, therefore, occur along a line joining the center of the earth to the center of the moon, whereas low tides are distributed along an earth-encircling swath at right angles to that line. To indicate the strength of the tide-generating force, note that at points N and Z in Fig. 3.9a, where the force is upward, a person weighing 90 kg (200 lbs) loses only 10 mg (0.00035 oz), or roughly the weight of a single tear.

Relative to the sun, the earth spins on its axis once every 24 h. Because the bulges and hollows maintain a fixed orientation relative to the moon, each will "almost" travel around the earth once each solar day. "Almost," because in 24 h the position of the moon will have shifted slightly in the sky relative to its position 24 h earlier. However, in 24 h 50 min the moon will be back to its original position as seen by an observer situated at a particular location on the earth (disregarding for the moment the daily change in the moon's elevation above the horizon). This explains why the tides occur 50 min later every solar day. The fact that there are two bulges and two hollows circling the earth explains the occurrence of semidiurnal tides with two high and two low tides per lunar day.

Other features of the tides can be readily accounted for by the equilibrium theory. But, before proceeding, it is important to appreciate the limitations of this concept.

The moving bulges and hollows can be considered a traveling or progressive wave, with a large crest-to-crest separation of approximately 22,000 km at the equator and a small amplitude of a metre or so. This is a true tidal wave and should not be confused with a tsunami, a Japanese word for the destructive waves commonly generated by earthquakes and often misnamed "tidal waves" (see Chapter 9). In reality, the tidal wave doesn't really keep pace with the moon as implied, but lags behind, sometimes by hours. Some reasons are: the water's inertia makes it impossible for the ocean to respond rapidly enough to the moon's movement; friction slows the wave as it rubs along the bottom; continents obstruct the wave's passage and force it to take many complicated pathways; and the ocean isn't deep enough. Unrestricted propagation of the tidal wave, in fact, would require the wave to move at speeds up to 1650 km/h (the rotational speed of a point on the earth's equator) and the ocean to have a depth of 22 km, whereas the actual depth rarely exceeds 5 km. Moreover, according to the equilibrium theory, maximum tidal heights should only be 0.8 m at equatorial latitudes and decrease toward the poles. Because actual tides are known to greatly exceed such values over most of the world's oceans, the theory is clearly much too simplistic to explain the detailed behavior of a complex phenomenon like the tide. A more sophisticated account of the tides was presented a century after Newton by the French mathematician Pierre Simon Marquis de Laplace. Called the *Dynamical Theory of the Tides,* it takes into consideration the effects of the limited depth and extent of ocean basins, as well as the influence of the earth's rotation and friction. Other scientists since then have further advanced understanding of the tides, though it is still far from complete.

Types of Tides

The notion of an equilibrium tide is a useful concept to account for the fundamental nature of tidal fluctuations. For example, the diurnal inequality and one reason for the biweekly tidal cycle are explained if the effect of the moon's declination is considered (its angle north or south of the earth's equatorial plane that originates from the average 23.5° tilt of the earth's axis with respect to the plane of the moon's orbit). Suppose the moon is at its maximum north or south declination (Fig. 3.10a, c). As usual, there will be two high tides a day, but because the bulges are not symmetric to the earth's axis of rotation, one bulge will create a higher tide than the other along a given parallel of latitude. (Only at the equator will the high tides be equal.) This difference is the diurnal inequality. As mentioned earlier, it is greatest at the time of tropic tides, when the moon has its largest declination. In Fig. 3.10b on the other hand, there will be no inequality in the two daily tides because the tidal bulges and hollows are now symmetric with respect to the earth's axis. These are the semidiurnal equatorial tides. Notice also that the range of the tide will be above normal during the tropic tides and below normal during equatorial tides.

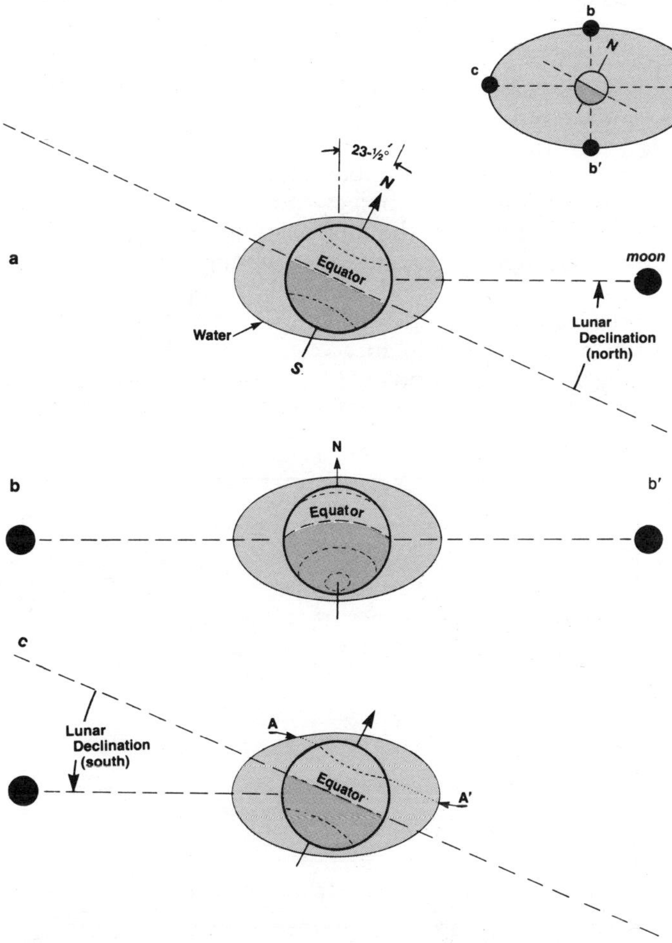

FIG. 3.10. Declinational-type tides. Except for a few days each month, tidal bulges are asymmetrical with respect to earth's NS-axis of rotation due to average 23½° tilt to moon's orbital plane (see inset, upper right). Because earth is revolving under fixed bulges, tides observed at latitudinal point, A′, will be higher than at the same point, A, ½ (lunar) day earlier; difference in tidal heights is diurnal inequality. Only when moon is over equator (b and b′) are tidal distortions symmetrical to earth's axis. Note: configurations a and c are viewed from point b′ in inset; configurations b and b′ are viewed from point c. There are roughly 15 days between a and c.

It takes the moon about 7 days to reach position b from position a in Fig. 3.10 (decreasing diurnal inequality) and another 7 days to reach position c (increasing diurnal inequality), causing a 2-wk cycle in the tidal range. Variations produced in this manner are called declinational-type tides and are an important aspect of coastal tidal patterns, particularly in Puget Sound and the eastern end of Juan de Fuca Strait. In the Strait of Georgia, declinational effects are strongest in the south and decrease northward.

There is also an important quasi-biweekly cycle in the tidal range produced by the sun's tidal pull which has been ignored until now. Its presence explains a feature of the tides that has been known for thousands of years: tides tend to have a greater range near a full or new moon than near the moon's quarters. Those with the large range are called spring tides, although they have nothing to do with the spring of the year; the term has a Saxon origin meaning "greater activity." The smaller range are neap tides, from a Saxon word meaning "inactive." Spring tides are about 20% greater than average tidal ranges, whereas neap tides are about 20% lower.

Tides of this nature, whose range varies in accordance with the lunar phases, are known as synodic-type tides. Their origin is easily explained.

During a full or new moon the sun is in line or in conjunction with the moon and earth (Fig. 3.11). The solar-generated tidal bulges and hollows then reinforce those produced by the moon, leading to a greater than normal range in the daily tides. During the moon's quarters, on the other hand, the sun is in opposition, and the tidal bulges it generates partially fill the hollows created by the moon. There is an accompanying decrease in the height of the lunar bulges by the solar-generated hollows and a subsequent reduction in the tidal range below normal. As there are approximately 15 days between a new moon and a full moon, there is a periodic 15-day variation in tidal range. Note that the period for synodic-type tides is roughly 1 day longer than that of the declinational type tides. Put another way, the synodic month is about 2 days longer than the usual sidereal month of 27⅓ days. This difference, though only one reason for the complex behavior of the tides, has an analogous origin to the difference of 50 min between the solar and lunar day. Suppose the moon and sun are in conjunction (Fig. 3.12). During the 27⅓ days it takes the moon to complete one orbit of the earth, the planet will have traveled 1/12 of its orbit around the sun. As a result, an extra 2 days are required for the sun and moon to again come into conjunction.

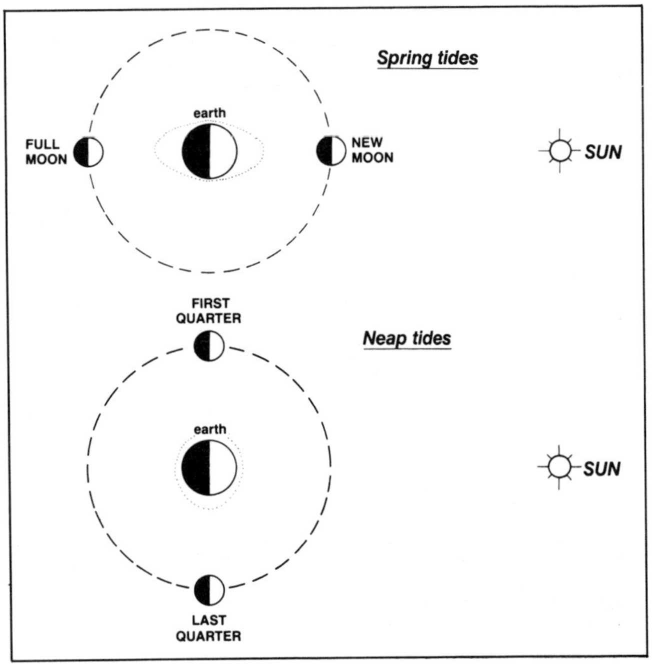

Fig. 3.11. Synodic-type tides. Plan view of relative alignment of sun, moon, and earth during spring and neap tides.

Comparatively short cyclic variations in the ranges of coastal tides also arise because of the eccentricity of the moon's orbit around the earth (Fig. 3.13). These, too, may be explained by equilibrium theory. In this case, the lunar tide-generating force causes tidal bulges and hollows to be more pronounced during the moon's closest approach to the earth at perigee and less pronounced when the moon is furthest away at apogee. Known as anomalistic-type tides, these variations in tidal range have a period of 27½ days. In protected inshore waters their importance almost equals that of the synodic tides, whereas on the outer coast they apparently only cause a 7-mo alteration in the range of the spring and neap tides.

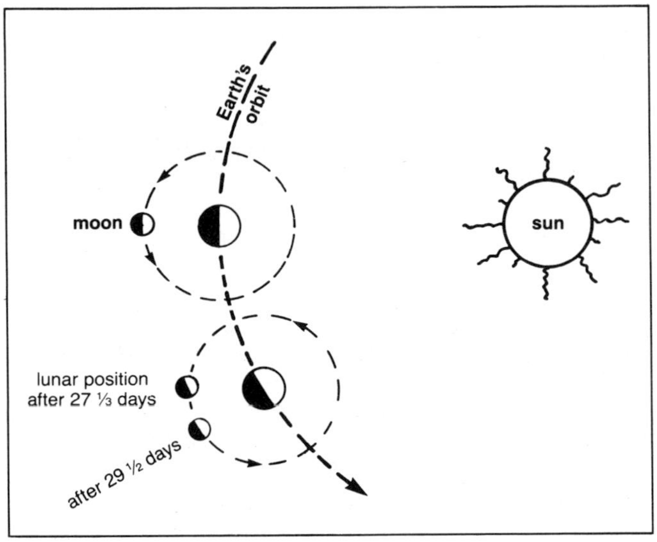

Fig. 3.12. Sidereal month (27⅓ days) versus synodic month (29½ days).

Figure 3.13 also shows that the eccentricity in the moon's orbit can lead to an inequality in the times between full to new moon and new to full moon, though the total time must equal one synodic month of 29½ days. This adds a further complication to an already complicated 2-wk tidal variation. Consider the situation when the sun is opposite point P so that the moon is new at that point. It will then be nearly full at point A; and vice versa. In either situation the length of time from new to full moon will equal that from full back to new moon, or exactly 14 days, 18 h, 22 min. But now suppose the sun is opposite point C. The moon will then be new at C and full at D, with quarters at P and A. Because the moon's orbital

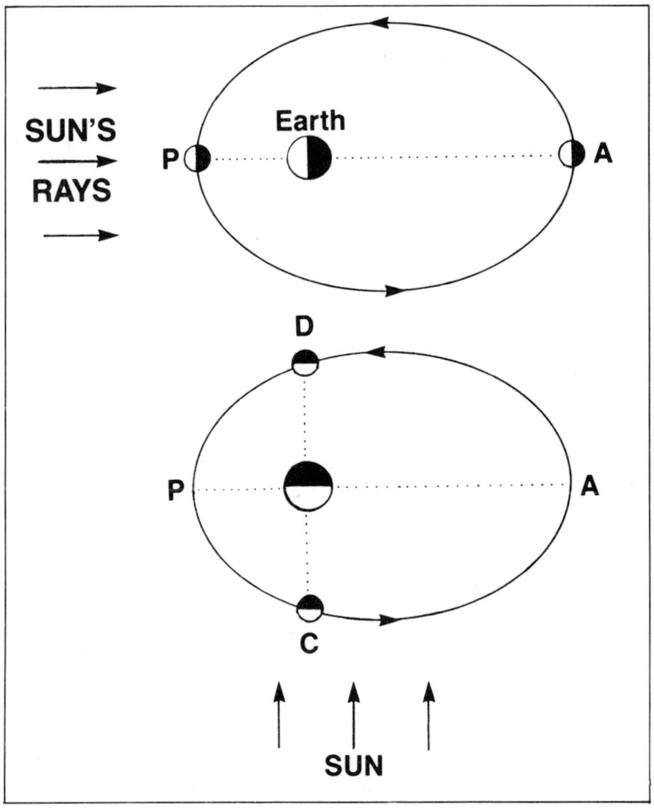

FIG. 3.13. Anomalistic-type tides. Moon's slightly elliptical orbit around earth, exaggerated for illustrative purposes, carries it closest at perigee (P) and farthest away at apogee (A).

speed is slowest near apogee (A) and greatest near perigee (P), and because the distance to A from C is greater than that from P to C, the time interval from new moon to full moon (C to D) will be longer than that from full moon back to new moon (D to C). To be precise, the intervals of time are 15 days, 14 h, 12 min; and 13 days, 22 h, 32 min respectively, the kind of extreme variability expected in the repeatability of the synodic-type tides associated with the phases of the moon.

Some concepts discussed can be applied to account for the tides in Fig. 3.5. For instance, maximum diurnal inequalities occur every time the moon has its maximum declination north (N) or south (S) of the equator while smallest inequalities occur when the moon is at the equator (E). These conditions are shown schematically in Fig. 3.10a–c, respectively. Tides along the outer coast are strongly synodic, attain greatest spring ranges immediately following the full and new moons, and smallest neap ranges follow the moon's quarters. The slight delay is due to the ever-present lag in the ocean's response to the tide-generating forces. In Fig. 3.5, one spring tide takes place when the moon is at perigee (P) and is subsequently enhanced; the other occurs near the moon's apogee and is diminished.

Within the protected coastal waters, the synodic tide is decidely less pronounced than along the outer coast, as indicated by the comparatively small range at the time of the new moon. The declinational tide, however, is more pronounced because small ranges occur just after the moon has crossed the equator (E) (Fig. 3.5). These are further augmented by the anomalistic tide. As a consequence, the greater range takes place near the moon's perigee and the lesser range near apogee. Examples from other months or locations would of course have somewhat different tidal modifications.

Long-Period Tides

In addition to the daily, biweekly, and monthly cycles in the tidal range, there are small cyclic modulations that take place over longer periods of time. Twice a year, for example, at the summer and winter solstices (June 21 and Dec. 22) the sun attains maximum declination from the equator (Fig. 3.14). When the sun and moon are in conjunction near this time, the spring tides along the British Columbia coast are at their greatest. During the fall and spring equinoxes, on the other hand, when the sun is on the equator (Sept. 22 and Mar. 21) and the moon at quarters, neap tides are at their lowest. This semiannual variation in the declination of the sun produces noticeable effects in waters adjacent to the Strait of Georgia where, as typified by Vancouver, lowest tides of the year occur near midnight just before Christmas and around noon near the end of June.

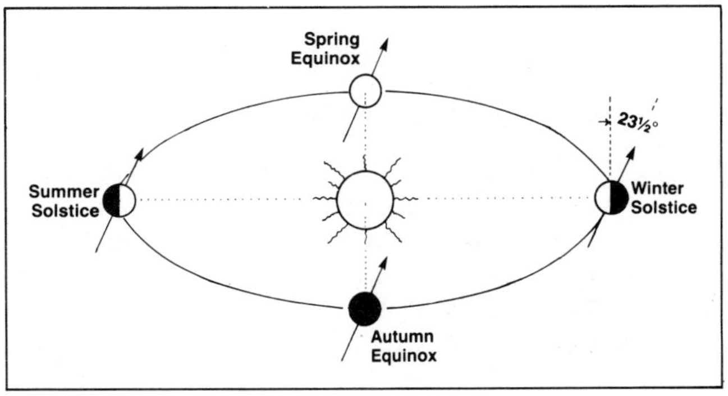

FIG. 3.14. Solar declination due to 23½°-tilt of earth's axis to orbital plane.

There is a small annual tidal range variation associated with the earth's slightly elliptical orbit around the sun and an average 14-mo cycle in the tidal range that amounts to about ½ cm (called the pole tide), caused by a wobble of the earth's axis of rotation known as the Chandler wobble. Winter–summer differences in mean atmospheric pressure and mean seawater density at the coast cause annual fluctuations in sea level of about 10–20 cm. (Effects of this type are called inverse barometer and steric effects, respectively.) A major long-period variation is caused by the 5° tilt of the moon's orbital plane with respect to the earth's (Fig. 3.15). The line of intersection formed where the plane of the moon's orbit crosses the plane of the earth's orbit slowly rotates, and requires 18.6 yr to complete one revolution. (Theoretically then, it should be possible to use 19-yr-old tide tables to accurately predict the tides today, though in practise things turn out to be more complex.) Once during each "19-yr cycle" the moon's orbital tilt adds to the 23.5° tilt of the earth's axis to permit maximum lunar declinations of 28.5° (= 23.5

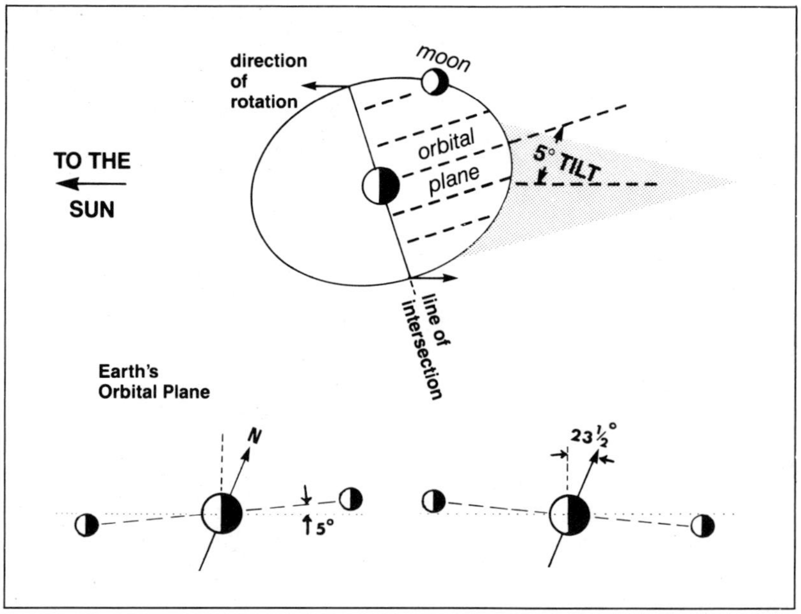

FIG. 3.15. The 5°-tilt of moon's orbital plane to plane of earth's orbit about sun. Line of intersection slowly rotates with center of earth as pivot point. Each complete wobble of lunar orbital plane takes 18.6 yr. Lower diagrams show how 5°-tilt increases maximum possible lunar declination to 28½° (right side) and 9.3 yr later decreases it to 18½° (left side).

+ 5°). At such times (the most recent 1969), the centerline of the moon's tide-generating force varies 57° north–south over the surface of the earth each month, so that diurnal inequalities attain their greatest values. After 9.3 yr, the lunar declination has decreased to only 18.5° (= 23.5–5°), monthly north–south variations are reduced to 37°, and diurnal inequalities are minimal.

Other long-period changes in the tidal range include an 8.8-yr cycle associated with alterations in the eccentricity of the moon's orbit about the earth and a 20,940-yr cycle due to a wobble in the earth's orbit about the sun. Over geological time scales, large, but irregular, variations in sea level are produced by the waxing and waning of the ice ages. In addition, prior to the fragmentation of Pangea into separate continents (Chapter 1), the supercontinent stood low with respect to sea level, and thereby allowed wide-spread inundation by the sea. Following breakup, the individual continents rose in relation to sea level and the degree of inundation decreased.

Tidal Constituents

The precise analysis of tides requires that observed records of sea level be systematically decomposed into their various constituents, each constituent or component with a specified amplitude and a specific cyclic period or frequency. Analysis of a tidal record is analogous to the decomposition of a sound wave or musical note into various harmonic frequencies. Along the Canadian west coast there are four main constituents which, when added together, account for almost all the variation in tidal range during a month. Foremost is the principal lunar semidiurnal constituent (or M_2 tide, where 2 is the number of cycles per day) associated with the moon's gravitational attraction. Except near Victoria and a few other special locations, this contribution generally accounts for around 50% of the tidal range. Next is the principal mixed diurnal constituent (or K_1 tide) with 1 cycle per lunar day which, like all diurnal effects, originates through the declination of the moon and, or, the sun. It accounts for roughly 25% of the tidal range and is largely responsible for the diurnal inequality (Fig. 3.16). Near Victoria, where it creates diurnal tides, this constituent becomes more important

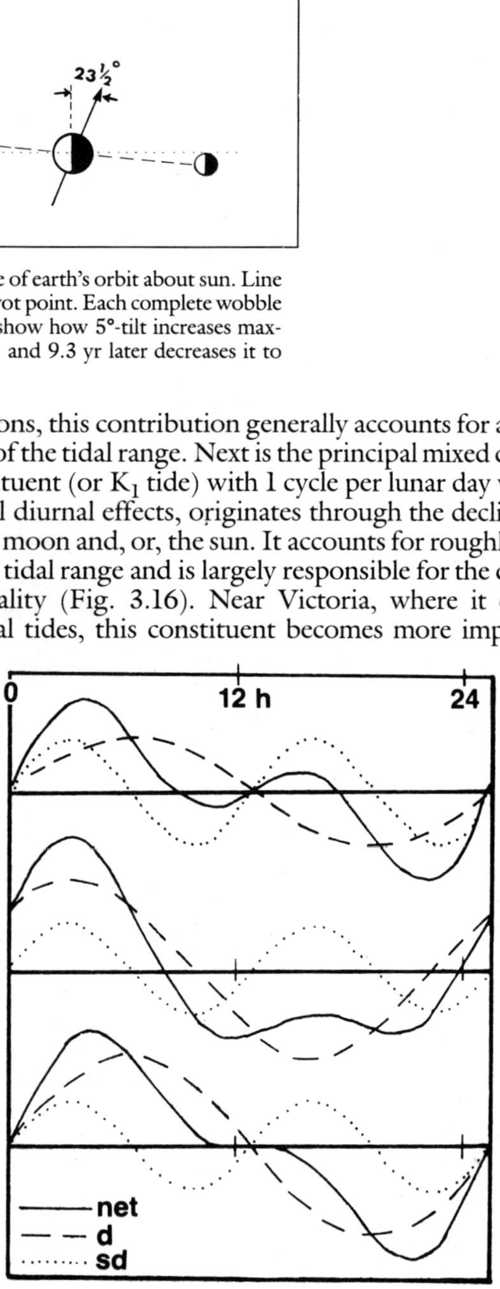

FIG. 3.16. Mixed-type tides (solid lines) result from addition of a diurnal tide (d) and a semidiurnal tide (sd). Height and diurnal inequality of net tide is different in each example due to differences in diurnal contribution relative to fixed semidiurnal contribution.

TABLE 3.1. Magnitudes (m) of the four main tide height constituents at various locations along the British Columbia–Washington coast. Each value is based upon 1 yr of tide height measurements except at Union Seamount where record was 110 days. Ratio determines type of tide. (MSD = mixed, predominantly semidiurnal; MD = mixed, predominantly diurnal). (Source: Canadian Hydrographic Survey)

Location	Constituent				Ratio	Type of Tide
	O_1	K_1	M_2	S_2	$(O_1 + K_1) \div (M_2 + S_2)$	
Union Seamount 49°35′ 132°47′	0.079	0.131	0.277	0.087	0.720	MSD
Tofino 48°09′ 125°55′	0.246	0.389	0.991	0.280	0.500	MSD
Port Renfrew 48°33′ 124°25′	0.287	0.458	0.712	0.210	0.809	MSD
Victoria 48°25′ 123°22′	0.370	0.627	0.373	0.102	2.100	MD
Port Townsend 48°08′ 122°46′	0.437	0.616	0.680	0.190	1.210	MSD
Seattle 47°36′ 122°20′	0.459	0.837	1.066	0.263	0.975	MSD
Sidney 48°39′ 123°24′	0.445	0.766	0.555	0.132	1.763	MD
Point Atkinson 49°20′ 123°15′	0.477	0.858	0.917	0.233	1.161	MSD
Comox 49°40′ 124°56′	0.489	0.885	1.002	0.253	1.095	MSD
Campbell River 50°01′ 125°14′	0.485	0.846	0.826	0.203	1.293	MSD
Alert Bay 50°35′ 126°56′	0.306	0.516	1.272	0.406	0.490	MSD
Queen Charlotte City 53°15′ 132°04′	0.315	0.511	1.975	0.651	0.315	MSD
Prince Rupert 54°19′ 130°20′	0.314	0.513	1.957	0.644	0.318	MSD

than the M_2 contribution. The next two major contributors to the tidal range are the principal solar semidiurnal constituent (S_2 tide), associated with the sun's gravitational attraction, and the diurnal (O_1) tide, linked to the moon's declination alone. These produce 15–20% of the tidal range, with about 5–10% caused by other more minor effects such as the P_1 tidal constituent due to the sun's declination alone, and the N_2 tidal constituent produced by the moon's varying distance from the earth during a month.

The ratio of magnitudes given by $(K_1 + O_1) \div (M_2 + S_2)$ determines the type of tide. When this ratio is less than 0.25 the tide is classified as semidiurnal; if it lies between 0.25 and 1.50 the tide is mixed, predominantly semidiurnal; between 1.50 and 3.0 the tide is mixed, predominantly diurnal; and finally, if the ratio exceeds 3.0 the tide is diurnal. Some values are presented in Table 3.1.

Coastal Tides

The previous explanations provide at best a bare framework for understanding the complexities of tides. Not only do continents block the free passage of the tidal wave, but variations in ocean depth alter direction and speed of propagation, and bottom friction hinders movement. In addition, the Coriolis force associated with the earth's rotation (see below) deflects tidal motions and forces them to move with the continents to the right of their direction of travel in the northern hemisphere. Consequently, the tidal wave always moves northward along the outer coasts of North America but southward along the coast of Asia. At the coasts, and within the protected coastal waters, the shape of the basin into which the tidal wave propagates plays a primary role in determining the response of sea level to the tidal wave. Each basin behaves differently. The classic example of how the shape of a basin affects the incoming ocean tide is the Bay of Fundy where a 3-m tidal range at the mouth of the bay can be amplified to a 18-m tidal range at the head of the bay.

Despite the complications, it is possible to present a general picture of tidal wave propagation into the coastal area of British Columbia and Washington. Before doing so, effects of the earth's rotation on the tide will be examined.

The Coriolis Force

Fluids such as air or water when moving freely over the solid crust of the earth's surface experience a slight deflection to the right of their direction of travel in the northern hemisphere and to the left in the southern hemisphere. Over short distances, these deflections are often too small to be noticeable, but over large enough distances they can accumulate to produce an appreciable "curving away" from the original path.

This effect is linked to the fact that motion is measured relative to a rotating frame of reference, the earth, which is rotating in space; if the earth were not spinning the effect would not exist.

If the rotating earth is compared to a merry-go-round and one person tries to throw a ball to another person on the opposite side of the carousel, Coriolis force is partially explained. When the ball is thrown, it naturally travels in a straight line (no knuckle balls allowed), yet it never reaches the receiver on the opposite side. Why not? Obviously, because the receiver has moved relative to where he was when the ball was thrown. But suppose neither player could see beyond the confines of the carousel and both were unaware they were traveling in circles, just as we are unaware that the earth is rotating. It would appear to them that the ball had taken a curved path, as if acted upon by some mysterious force! Of course, there isn't really a force, there only seems to be. Nevertheless, the effect is real enough to the players, as it is to observers of large-scale motions on the earth (Fig. 3.17). This apparent force that deflects freely moving objects away from what was thought to have been a straight line is the Coriolis force (named after the French mathematician, Gaspard Coriolis, who first described it in 1835). It has a strong influence in determining the prevailing wind patterns of the atmosphere and on the circulation of world oceans. Long-distance flying craft, such as jets and rockets, must continually adjust for it to reach their destinations. An automobile, however, does not respond to the Coriolis force because it is held at all times to the earth's surface by the friction between the tires and the road.

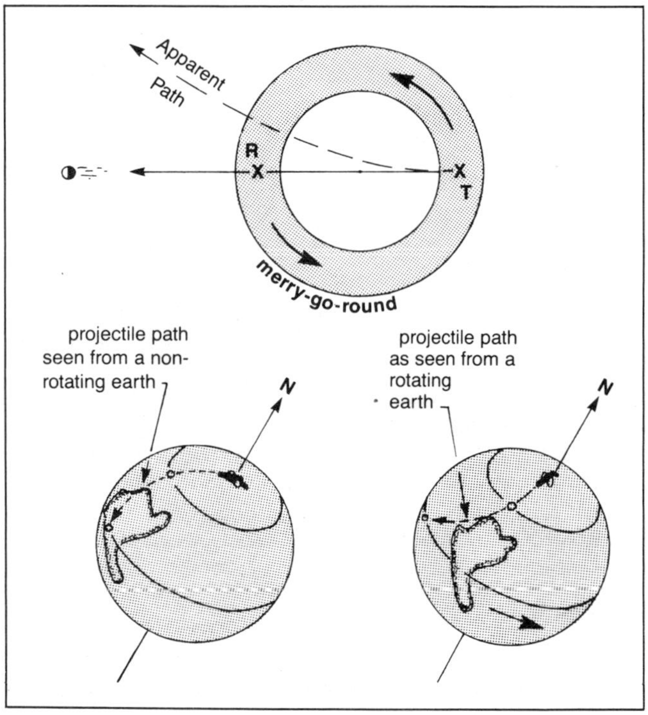

FIG. 3.17. The Coriolis force. *Top*: ball thrown in a straight line from T toward R on a merry-go-round appears to be deflected to right; i.e. to experience a force. *Bottom*: effect of earth's rotation on cannon ball fired from north pole toward equator (as seen from the ground).

Due to the deflection caused by the Coriolis force, the progressive tidal wave propagates northward along the west coast of North America "leaning" up against the coast (Fig. 3.18). This results in a seaward diminishing slope in the accompanying sea level and a corresponding decrease in the tidal range toward mid-ocean. Generally speaking, this slope amounts to an average decrease in the high- and low-tide levels by about 50 cm in 500 km, so that roughly 2500 km southwest of Vancouver Island the tidal range diminishes to a minimum. It then begins to increase again in the direction of Asia.

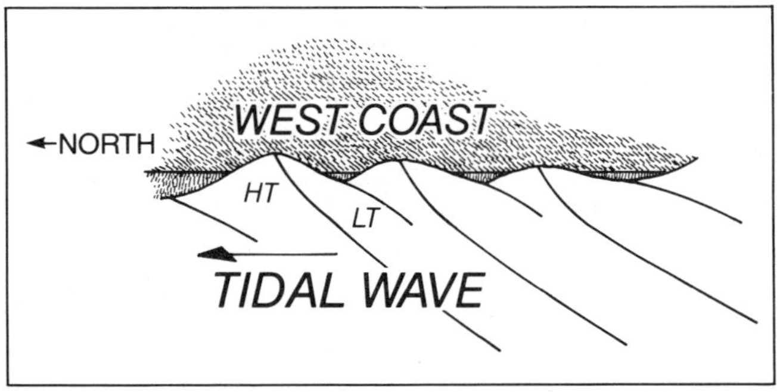

FIG. 3.18. Effect of Coriolis force on tide in Northeast Pacific Ocean. Tidal wave necessarily propagates northward, leaning against coast where the maximum tidal range takes place.

In the northeast corner of the Pacific Ocean, the speed of the tidal wave along the outer coast is typically around 740 km/h (400 kn), except over the continental shelf where it is somewhat slower due to the shallow depths. As a consequence of its great speed, there is little delay in a particular stage of the tide from one region to another and it is characterized by the following coastal features (Fig. 3.19) (modified after Dohler 1964).

From Barkley Sound to Cape Scott on the west coast of Vancouver Island, the tide occurs almost simultaneously and has an average range of about 3 m. As it moves into the numerous inlets in this region, there is a slight increase in range but no slowing down, except in the constricted passage at Quatsino Narrows where the tide is delayed by 45 min compared to the coast.

Along the west coast of the Queen Charlotte Islands and the mainland shores of Queen Charlotte Sound, the tide occurs simultaneously, but 30 min later than at Vancouver Island. The range of the tide on the mainland side of the Sound is increased even more as the tide moves into the various deep inlets, reaching 5 m at the heads. As the tide propagates northward from Queen Charlotte Sound into Hecate Strait, its range increases with the decreasing depth of the passage. This manifests itself as a south-to-north increase in the range and a time difference of the tide along the mainland side of Hecate Strait. A similar situation occurs as the tide enters Skidegate Channel that separates Graham and Moresby Island, and where a tidal range of 4.3 m at large tides at the coastal entrance becomes a range of 7.8 m at Queen Charlotte City two-thirds down the channel.

Around the northeastern end of Graham Island, the tide propagates more slowly and reaches Masset Inlet about 1 h later than Hecate Strait. At Prince Rupert, the tide arrives 1 h later than off Vancouver Island and has a mean range of 4.9 m.

It takes between 2 and 4 h for the tidal wave to sweep up Juan de Fuca Strait to the Haro Strait entrance. Along the way, the range decreases from the Cape Flattery entrance to Victoria, where the average range is around 2 m. As the tide rounds the corner between Victoria and the San Juan Islands it is partially deflected to the east, and produces slightly higher tides on the U.S. side than the Victoria side. There is then a delay of about 1 h, as the tide squeezes through the narrow channels of Haro and Rosario straits and attempts to negotiate the constricted

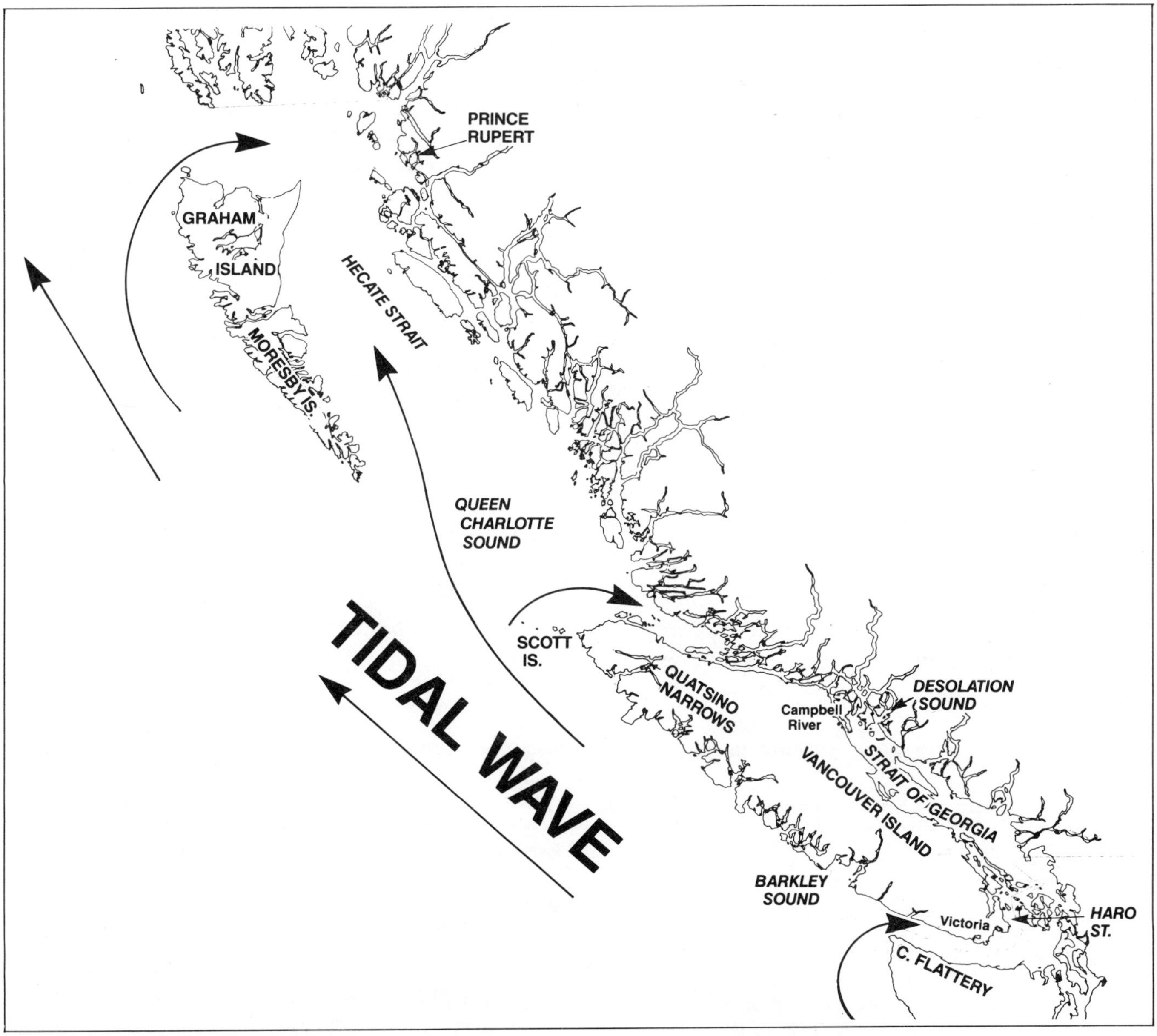

FIG. 3.19. Tide propagation along British Columbia–Washington coast.

passages between the Gulf Islands. The rapid tidal streams throughout this region are an obvious consequence of this delay. Inside the Strait of Georgia, however, water-level changes occur almost simultaneously with no more than a 30-min wait for two points to experience the same stage of the tide.

Between Cape Scott and Campbell River, the passage separating Vancouver Island from the mainland is narrow and strewn with islands. This, and the fact it takes the tide from the south almost 2 h longer to reach the narrowest parts of the channel than the tide from the north, explains the extremely rapid tidal streams associated with Seymour, Yuculta, Surge, Hole-in-the-Wall, Okosillo, and Arran rapids. This difference results in a 2-h lag in the times of the tide at Campbell River compared to Seymour Narrows.

Because propagation of the tidal wave from the south is essentially blocked by the tidal wave from the north at the northern end of the Strait of Georgia, mixing of water is minimized to the point where appreciable warming of the upper few metres can take place during the summer. In quiescent areas like Desolation Sound, surface temperatures may reach 25°C in late July.

Nonastronomical Tides

The kinds of tides described to now are known as astronomical tides as they are generated by gravitational forces. Although these are by far the most predominant type in the ocean, diurnal and semidiurnal changes in sea level can be produced by a variety of other mechanisms. Most are related to meteorological processes.

Due to the earth's daily rotation, the expansion and contraction of the ocean surface through daytime warming and nighttime cooling cause tidelike variations in sea level. (Similar fluctuations in the air are the main cause of large atmospheric tides. Though this heating–cooling cycle is primarily diurnal, the tides themselves are semidiurnal. This apparent paradox was resolved at the end of the 19th century when it was shown that the whole world atmosphere prefers to resonate with the weak semidiurnal constituent of the heating effect, but can never get into step with the much stronger diurnal component of this cyclic forcing.) Similar effects are produced by fluctuations in barometric pressure and by the daily rhythm of land and sea breezes, that move water away from and toward the coast. These so-called radiational tides are extremely difficult to separate from the astronomical tides and are usually lumped with the predictions published in the tide tables.

Storm surges are tidelike fluctuations in elevation initiated by strong onshore or offshore coastal winds during a severe storm (Fig. 3.20). Their effect is small at

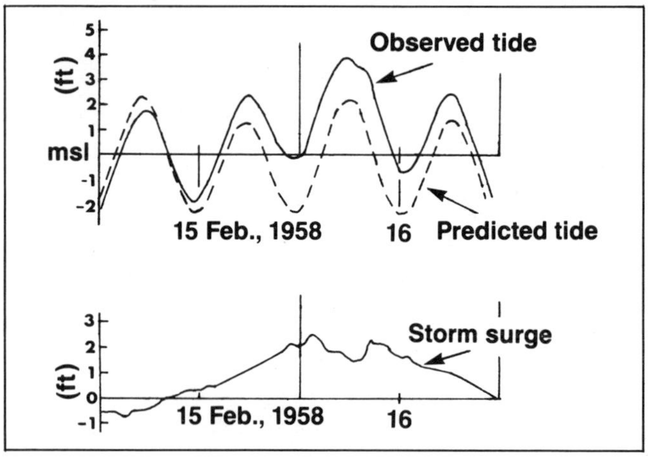

FIG. 3.20. Storm surge for Atlantic City, February 1958. Storm surge measures departure of actual tide from predicted astronomical tide (msl, mean sea level). (From Pore 1964)

steep coasts but in low-lying countries, such as Holland, or Bangladesh on the delta of the Ganges River, they can cause extensive flooding, particularly if they occur during periods of high astronomical tides. A 1970 storm surge in Bangladesh is estimated to have killed half a million people. The highest recorded tide in British Columbia of 8.1 m at Queen Charlotte City on Jan. 12, 1967, was during a strong onshore blow.

Unusually low tides can also occur because of meteorological effects. In the shallow, constricted Torres Strait that separates the northeastern tip of Australia from Papua New Guinea, for example, storm surges have been known to leave large tankers stuck firmly in the mud.

In semienclosed basins such as bays and inlets, rapid but small sea-level changes called seiches are sometimes observed in the tide-gage records. Generated by tides or local winds, or by the propagation of certain types of oceanic waves across the mouth of a basin, the magnitude and periods of these oscillations are closely tied to the geometry of the particular embayment in which they arise. Within British Columbia coastal waters, seiches can have periods of minutes to hours and heights of a few centimetres; a typical period appears to be about 30 min. They are frequently observed in Whaler Bay, Campbell River, Pedder Bay, and Port San Juan as wiggles on the smoother tidal curves (Fig. 3.21; see also Fig. 3.4 for Alert Bay). As these fluctuations can cause the water level of a basin to move up and down a few centimetres over a few minutes, it is sometimes possible to see their presence directly. While sitting at the water's edge in Esquimalt Harbour on a flat-calm day, the author once watched such a gentle rise and fall along a wide stretch of the shoreline.

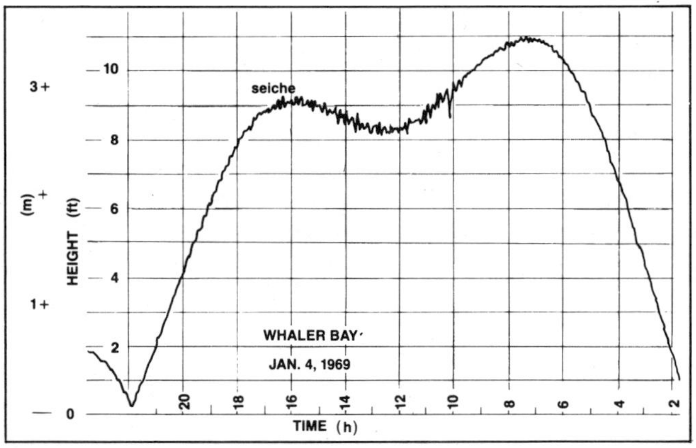

FIG. 3.21. Rapid, small-amplitude seiches superimposed on tide, Whaler Bay, southeast end of Galiano Island, Strait of Georgia. (From LeBlond 1972)

Tidal Friction

As part of a cosmological process, the dissipation of tidal energy at the open coasts of the world (and to a lesser extent the friction between tides and the ocean floor) is decelerating the earth's rate of rotation and causes a lengthening of the day by 0.001 s about every 100 yr. Studies of daily growth rings of ancient coral have shown that 370 million yr ago there were nearly 400 days a year rather than the present 365, so a day was 21.9 h long.

According to the natural laws governing the conservation of angular momentum, the reduction in the earth's spin rate has been accompanied necessarily by an increase in the orbital speed of the moon, causing it to swing out into an ever-widening orbit. This in turn has led to a decrease in the tidal range over the billions of years since the formation of the oceans. The present lunar recession of 3.3 cm/yr will continue until the earth ceases to spin on its polar axis some time in the remote future, just as earth-induced tides in the solid body of the moon have stopped the original spin and forced it to present the same side to the earth at all times. At that future date, the earth and moon will present constant faces to one another as they rotate about a common axis over a 47-day period. The solar tide will then take over, cause a steady reduction in the rotation rate of the earth–moon pair about their common center of mass, and reduce their distance of separation until, eventually, the moon will approach within the Roche Limit of 18,340 km, when its surface will be torn apart by the gravitational force of the earth. Shattered chunks of the moon will rapidly spiral to earth, and possibly halt further catastrophe by increasing the rotation rate of the earth–moon pair, allowing the moon to again recede from the earth.

Tidal Streams

Tidal streams are the horizontal currents associated with the vertical rise and fall of the tide. Unlike horizontal water motions induced by most other processes, these currents have horizontal speeds of almost uniform strength throughout the entire depth of water. In the deep ocean, speeds are only a fraction of a kilometre per hour and a parcel of water carried by the tidal flow has a typical excursion of about 1 km over each half of a tidal cycle. Near the coast, speeds exceed 1 km/h and the water moves over correspondingly greater distances. Where currents are observed to change with depth, a nontidal flow or residual current also will be present. Such flows may be associated with wind-driven ocean currents, surface-induced wind drift, internal oceanic motions generated by the interaction of the tidal streams with the bottom topography, or density currents.

Tidal streams are studied in connection with erosion, the movement of oil spills, the disposal of wastes, and navigation. The latter includes routing larger vessels that must consider time in addition to safety. The often violent tidal currents through Seymour Narrows between Vancouver Island and Quadra Island are an example of flow that must be respected by all ships. This was particularly true before the blasting of Ripple Rock on Apr. 5, 1958, by the biggest nonnuclear peace-time detonation on record at that time (Fig. 3.22). Prior to that date, the swift tidal streams and the twin peaks of the rock that protruded dangerously close to 3 m of the surface formed a deadly combination, wrecked 20 big ships, and took 114 lives.

The tidal stream associated with a rising tide is called the flood, an expression that probably derives from the flooding effect of the tide in low lying areas, whereas the current associated with a falling tide is called the ebb. Commonly used expressions like "flood tide" and "ebb tide" should be avoided, as they confuse the horizontal motions of tidal currents with the vertical displacements of the tide. (Records show that early seafarers also used the same words for tides and tidal currents.) Slack water occurs during the short time interval between the end of the flood and the beginning of the ebb, or vice versa, when the water has no horizontal motion. A further distinction is made between high-water slack at the end of the flood and low-water slack at the end of the ebb.

At any place and time, a tidal stream is specified by both direction and speed, which makes it a much more difficult and expensive quantity to measure than the tide. Measured speeds range from less than 0.5 m/s (1 kn) in the open ocean to a maximum of about 8 m/s (16 kn) in Nakwakto rapids that separate Seymour Inlet north of Vancouver Island from Queen Charlotte Sound. Speeds shown on nautical charts usually indicate the maximum values expected during spring and neap tides in the main channels only, because specification of the exact tidal current for all localities and times is an impossibility. Moreover, these values are often estimates and should be treated accordingly.

Although the connection between the tide and tidal streams is familiar (semidiurnal tides with two highs and two lows per day, for example, are associated with two floods, two ebbs, and four slack water periods a day) it

Fig. 3.22. Ripple Rock blast, Seymour Narrows, April 5, 1958. View toward north across Menzies Bay. (Courtesy R. W. Sandilands and Canadian Hydrographic Service)

must not be supposed that the strength of the current, nor the times of slack water, at any locality necessarily coincide with the corresponding vertical changes of the tide. Under certain conditions, in fact, slack water occurs midway between high and low tide, contrary to the commonly accepted notion that it occurs at high and low tide. A case in point is Nitinat Lake on the west coast of Vancouver Island (Fig. 3.23). Here, the channel that connects the lake to the ocean so constricts the passage of water that little enters or leaves the lake, the tidal range in the lake is about 0.3 m when the tidal range on the coast is 3.3 m. Therefore, because the elevation of the lake is nearly constant, slack water in the channel occurs when the coast tide has the same elevation as the lake, about midway between high and low water. Maximum currents, on the other hand, tend to develop when the difference in elevation between the lake and ocean is greatest, near low or high tide. Similar situations are common in B.C. waters, the fast currents through Nakwakto rapids and Skookumchuck rapids (Sechelt Inlet) are two of the more dramatic examples.

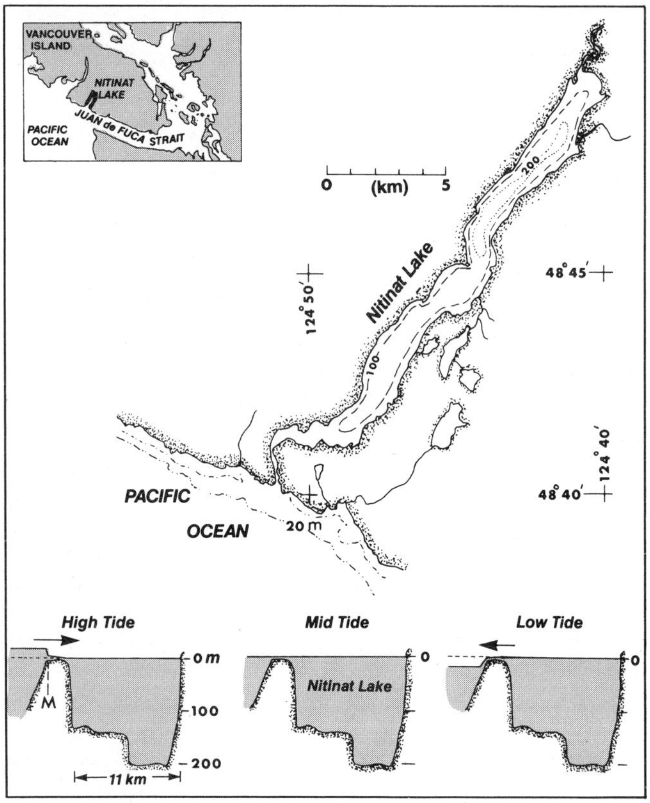

FIG. 3.23. Relationship of oceanic tide to flood and ebb streams into and out of Nitinat Lake. M = mouth of inlet.

Other indications of a complicated relationship between the stages of the tide and stages of the currents include the often nonexistence of "slack" water, and currents that continue to ebb when they should have turned to flood.

To understand the reasons for such complications requires knowledge of the processes that govern the currents. Tides and tidal streams are related in three basic ways.

A) As explained earlier, the oceanic tide propagates as a very long progressive wave, high tides correspond to the wave crests, and low tides to the wave troughs. Tidal streams in the ocean are the horizontal motion associated with the passage of this wave, floods correspond to the forward motions under the crests, and ebbs the backward motions under the troughs (Fig. 3.24). Given enough room to maneuver, this is how tides and tidal streams would be related. Once the motions become restricted by coastlines, however, the situation is usually altered, although tides and tidal currents in the main channels of larger regions like Queen Charlotte Sound, Hecate Strait, the Juan de Fuca Strait retain much of the behavior of a progressive tidal wave.

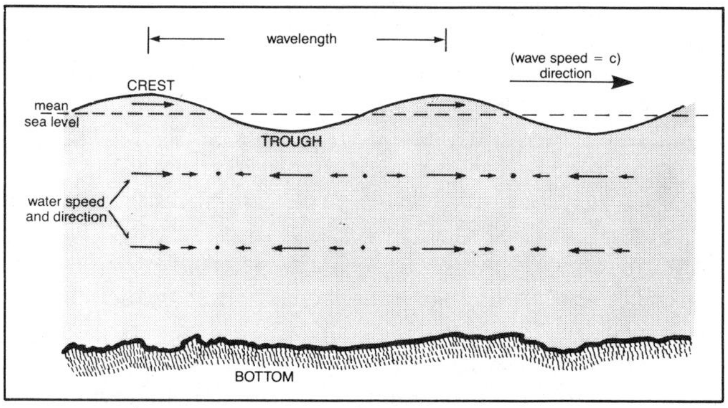

FIG. 3.24. Cross-section through a progressive wave with crests and troughs moving from left to right at speed, c. Dots indicate zero current speed at points midway between crests and troughs. Wave speed greatly exceeds current speeds associated with wave-induced motions.

B) In certain large and partly enclosed regions, the tide no longer behaves as a progressive wave. Within the Strait of Georgia proper, this is because the tidal wave that propagates northward from Juan de Fuca Strait is reflected when it reaches the northern end, where the many islands and narrow passes act as a barrier. This reflection produces a standing wave (formed when two identical waves traveling in opposite directions become superimposed on one another) so that sea level sloshes up and down as if in some giant bathtub (Fig. 3.25). The relationship between the tides and tidal streams for such a wave is completely opposite to that for a progressive wave in that maximum tidal streams are midway between high and low tides. More will be said on this subject when the circulation of the Strait of Georgia is discussed in Chapter 10. In deep inlets, the tide behaves somewhat like a standing wave, although all inlets are much too short to allow anything but a small segment of the wave to occupy them at any one time. Therefore, a variation in tidal elevation at the inlet mouth is reproduced throughout the inlet almost simultaneously, high tide at the mouth typically occurs only a few minutes before high tide at the head (Table 3.2). Current speeds are generally less than a few kilometres per hour, with maximum speeds midway between high and low tide, and slack waters at high and low tide (Fig. 3.26).

C) In passes and narrows, the difference in water level between the two ends of the channel induces currents as the water seeks a common level. This difference, called the hydraulic head, is caused by a difference in arrival time of the tide at either end or by a difference in the range of the tide at the two ends. The numerous passes within the San Juan and Gulf Islands have tidal streams produced in this way. In Active Pass, for example, the tidal streams are determined strictly by the hydraulic head, and have no resemblance to the ebbs and floods in the Strait of Georgia. Seymour Narrows is another region where tidal streams are produced by a delay in the tide between two ends of a channel. In this case, the time difference between the tide at the northern and southern ends of the Narrows amounts to as much as 2 h, with accompanying differences in elevation that can exceed a metre or more, and currents that can churn through the constricted passage at over 7 m/s (13 kn).

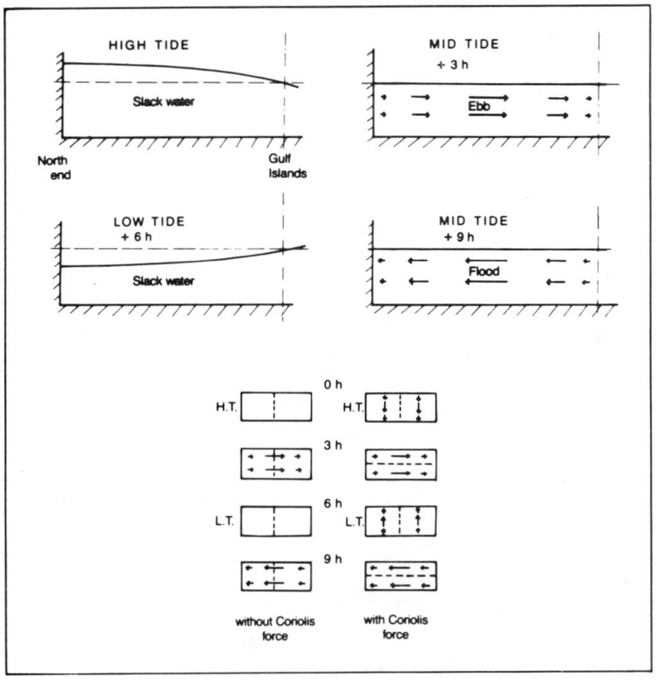

FIG. 3.25. Highly simplified view of standing wave pattern within Strait of Georgia. *Top*: sea level and tidal stream relationship for four stages of a semidiurnal tide; arrow length is proportional to speed of current. *Bottom*: corresponding surface currents with (right side) and without (left side) effect of Coriolis force. With Coriolis force, times of "slack water" are coincident with weak cross-channel currents.

TABLE 3.2. In deep inlets, times of high water (HW) and low water (LW) occur only a few minutes later at the head than at the mouth. Tidal heights were taken from simultaneous tide-gage records operated day and night; time kept accurately at the mouth and head by chronometers. (Adapted from Dawson 1920)

Long inlets (avg. tide range 4 m)	HW	LW
From Whaletown on Cortes Island to head of Bute Inlet, 84 km; from comparison of observations in two different seasons with same reference station	3 min later	9 min later
From Namu to Bella Coola by Burke Channel and Bentinck Arm, 111 km; from 144 simultaneous observations	2 min later	7 min later
From Hartley Bay in Wright Sound to Kitimat, by Douglas Channel, 79 km; from 222 simultaneous observations	4 min later	4 min later

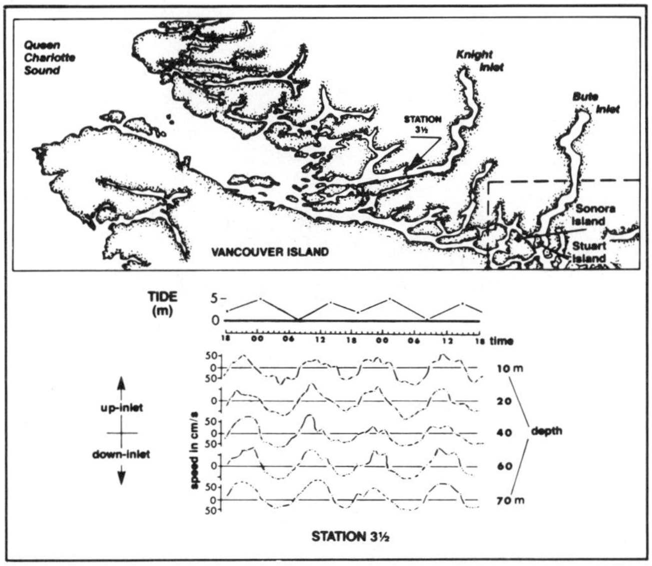

FIG. 3.26. Tides and observed along-channel currents at Station 3½ in Knight Inlet July 6–8, 1956. Speeds were measured almost simultaneously at the five different depths and plotted versus time in hours. Slack waters occur at low and high tides, maximum speeds midway between tides. (From Pickard and Rodgers 1959)

Factors that Influence Tidal Streams

Once the water begins to move, it is influenced by a number of mechanisms that tend to alter direction and speed.

Bathymetry

Tidal streams tend to flow in accordance with the general configuration of the sides and bottom of a basin. In the Strait of Georgia, for example, floods move toward the northwest while ebbs move toward the southeast. Even the tidal flow along the outer west coast is forced to conform somewhat to the orientation of the coastline. The similarity between swift, narrow channels and rivers accounts for backeddies in the flow, when nearshore water becomes separated from the main tidal stream, and for water piling up on the outside of curved passages due to the centrifugal force (the comparatively higher tidal heights on the Washington State side of the Strait of Georgia arise in this way (see Fig. 10.12).

Eddies and whirlpools are usually generated in narrow channels by swift currents. In Seymour Narrows, two lines of whirlpools originate during the flood where the rapid, southward flowing current comes in contact with the nearly motionless waters to the east and west (Fig. 3.27). In the Cordero Channel and Yuculta rapids area, between Sonora Island and Stuart Island at the entrance to Bute Inlet (Fig. 3.28), dangerous whirlpools are formed west of Little Dent Island and over certain shoals. The most dangerous whirlpools, however, occur during the flood where the rapid, 3 m/s (6 kn) eastward flow through Gillard Passage meets the equally strong southward flow

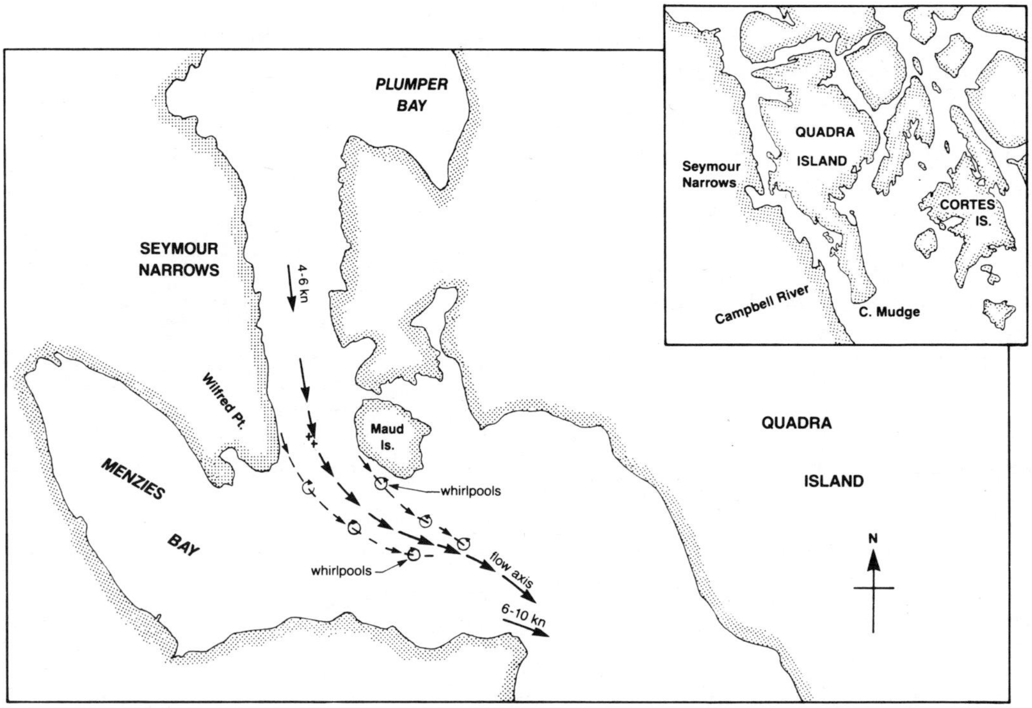

FIG. 3.27. Flood currents, Seymour Narrows. Whirlpools are formed where jetlike flow contacts slowly moving waters to right and left. Backeddies also formed in lee of shoreline promontories within the Narrows. Arrows denote typical maximum flood speeds. Plus (+) signs mark location of Ripple Rock peaks.

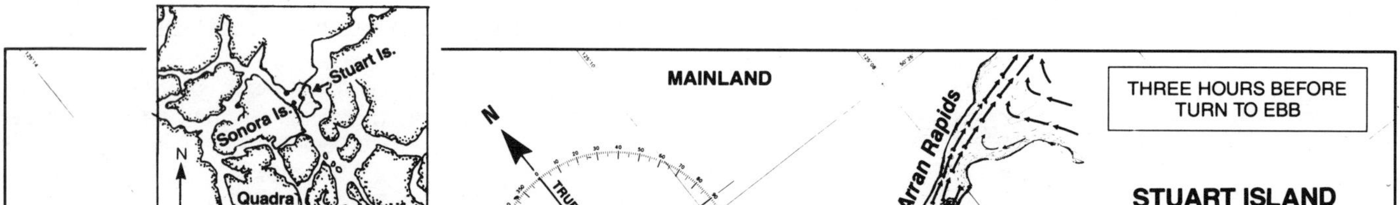

FIG. 3.28. Tidal streams 3 h before the turn to ebb in Cordero Channel region near entrance to Bute Inlet (see Inset: also Fig. 3.26). Dangerous whirlpool over 6.4-m (3½-fa) shoal (S) northwest of Dent Rapids on large tides. Violent eddies and whirlpools form in Devil's Hole and southeast of Gillard Passage. (Modified from Tidal Current Publication 23, Canadian Hydrographic Service 1970.)

toward Yuculta rapids. Hydrographer Stan Huggett recalls the unnerving feeling of being able to look down 4 m into the hole formed by one of these whirlpools as his survey launch struggled away from its edge!

Opposite to whirlpools are the smooth, dome-shaped surfaces called upwellings, produced when water is forced upward by features on the bottom of the tidal channel (see Fig. 5.2).

Finally, the bathymetric configuration of a channel may result in the flood moving in on one side while it is still ebbing on the other side. Although the ebb will eventually turn to flood, slack water within the channel will be virtually nonexistent.

Friction

A layer of water that is moving relative to another layer of water or over a solid surface is acted on by frictional drag, which opposes the motion. Consequently, the speed of a current decreases near the bottom and sides of a body of water, and eventually becomes zero at solid surfaces, a condition of the "no-slip" requirement for fluids in immediate contact with solids. Therefore, greatest speeds tend to be at the upper surface in midchannel where the effect of friction is least. A wide, deep channel will have more rapid tidal streams than a shallow, narrow one of the same hydraulic head, as exemplified by the adjacent, but different, Dodd Narrows and False Narrows between Gabriola and Vancouver islands.

If water is moving over or along a "smooth" surface (a sandy bottom, straight shoreline, or newly polished hull) the flow will also be smooth and friction minimal. On the other hand, if the surface is "rough" (a rocky bottom, an irregular shoreline, or a fouled hull) the flow will be irregular and friction large. Thus, rocky, island-strewn passages will slow tidal streams more than smooth obstacle-free ones. Irregular or turbulent motions such as whirlpools and upwellings also behave as effective frictional drags, and further reduce the water's speed. Without them, the average speed of the currents in places like Dodd Narrows or Active Pass would be greater. Obstacles such as log booms, bridge supports, and kelp beds also increase friction; the net drag on the Fraser River is significantly increased by the numerous man-placed obstacles within it.

Inertia and Momentum

Because it has mass, water possesses a certain inertia or indisposition to a change in motion; if stopped it tends to remain so, if moved it tends to continue moving, friction notwithstanding. Therefore, the speed of tidal currents associated with hydraulic heads will not adjust immediately to changes in water elevation along the channel, which partly accounts for the lag between tide and current in such areas (Fig. 3.29). Moreover, the speed of the water determines momentum in the same way that the speed of a boat determines momentum. The greater the momentum the greater the effort needed to bring the water to rest. Thus, water that runs out of a harbor during a falling tide will not immediately come to rest simply because the tide has begun to rise again; momentum will carry it against or around the advancing tide. Because the current speed is typically greatest in the deeper sections of

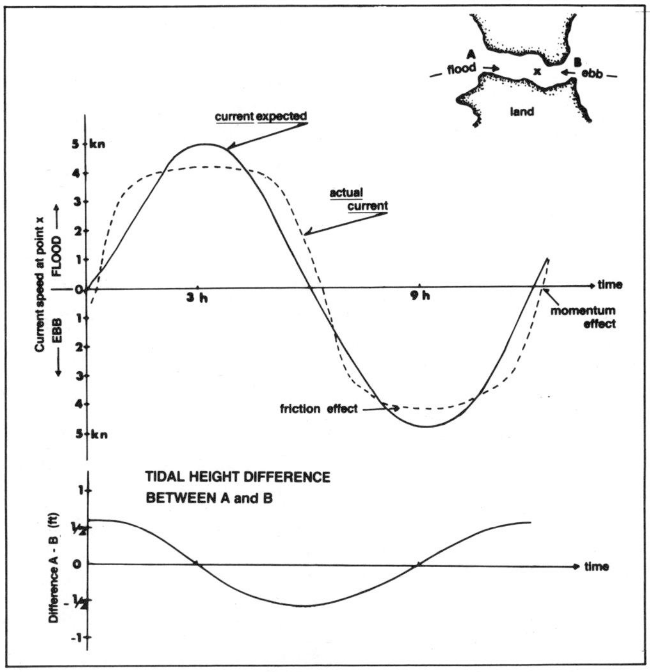

Fig. 3.29. A hypothetical example of expected and actual currents measured at a location X in a channel. Lower curve shows difference in sea level elevation (hydraulic head) between ends of channel, A to B, as it varies over a semidiurnal tidal cycle; upper curves give associated currents. Frictional effects reduce maximum possible speeds generated by hydraulic head, and momentum delays time of slack water from expected. Also, actual currents (broken line) reverse direction much more quickly than in absence of friction and momentum effects.

the channel, the water there could still be ebbing after it had begun to flood elsewhere in the harbor. On the other hand, the momentum of any current is quickly lost once it moves into more open water from the confines of the channel, as Porlier Pass, where even strongest floods rapidly weaken a few kilometres into the Strait of Georgia.

Coriolis Force

Provided there are not too many topographical restrictions on the flow, this force produces a continual deflection on the current to the right of its direction of motion, unlike friction drag and the hydraulic head that act along the direction of motion. In the open North Pacific, Coriolis force causes tidal streams to constantly change direction, whereby they turn in a complete revolution of 360° over each tidal cycle. More will be said on this topic shortly. In more confined regions, the Coriolis effect is often suppressed, and becomes negligible within the smaller passes of British Columbia. Nevertheless, its influence is marginally present in larger areas like the Strait of Georgia, Hecate Strait, and Juan de Fuca Strait where floods tend to be slightly stronger on the mainland sides (to the right of the direction of flood) and ebbs slightly stronger on the island sides (to the right of the direction of ebb).

River Runoff

Water brought into a region on the flood can be more dense (heavier) than water taken out during the ebb. If so, currents associated with the flood will be strongest closer to the bottom while currents associated with the ebb will be strongest nearer the surface. Vertical tidal mixing of Puget Sound and Strait of Georgia water with the more dense water of Juan de Fuca Strait at the turbulent passes connecting them produces such a situation. Thus, ebb currents are strongest in the upper half of Juan de Fuca Strait while flood currents are strongest in the lower half (see Fig. 11.12). This bias is, of course, nothing more than the estuarine circulation described in Chapter 2, which has been superimposed upon the "true" streams directly associated with the tidal forces of the moon and sun.

Winds

Wind-generated currents may upset the natural rhythm of tidal streams. Opposing winds can delay the turn of the tide near the surface, particularly in regions with a distinct brackish upper layer, and reduce the maximum speed of the tidal current. Winds that blow along the direction of flow on the other hand can augment the speed of the tidal current. This wind-effect may be appreciable; in some cases it accounts for the discrepancy between predicted and observed tidal currents.

Current Ellipses

Where the direction of tidal streams is constrained by the sides of a channel, the currents tend to be rectilinear. That is, ebbs and floods at all stages of the tide are directed parallel to the trend of the channel and there is negligible cross-channel set. Under such conditions, the stream simply decelerates without changing orientation, reaches slack water, and then accelerates in the opposite direction (Fig. 3.30a). Tidal streams of this kind are common to the narrow, regular basins within the inshore waters such as Johnstone Strait, Juan de Fuca Strait, and Puget Sound.

Where the tidal streams have room to maneuver, however, the influences of the earth's rotation, centrifugal forces, friction, and inertial effects cause the current's direction to turn with time as it changes between flood and ebb (Figs. 3.30b, c). For such rotary currents, the shape of the curve traced out by the tip of the current vector over a tidal cycle depends on the type of tide producing the flow. This effect can be shown by following the path of an imaginary ship riding at a deep-sea anchor,

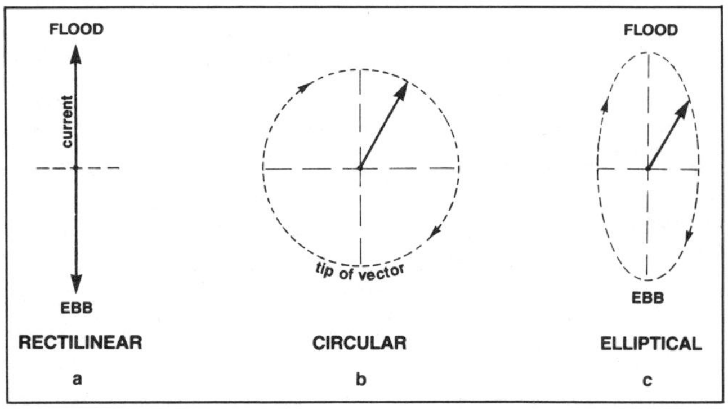

FIG. 3.30. Rectilinear versus rotary tidal streams, as measured at a fixed location. Vectors (arrows) give direction of flow at center point, and broken line gives location of tip over a complete tidal cycle. Length of arrow is proportional to speed of current.

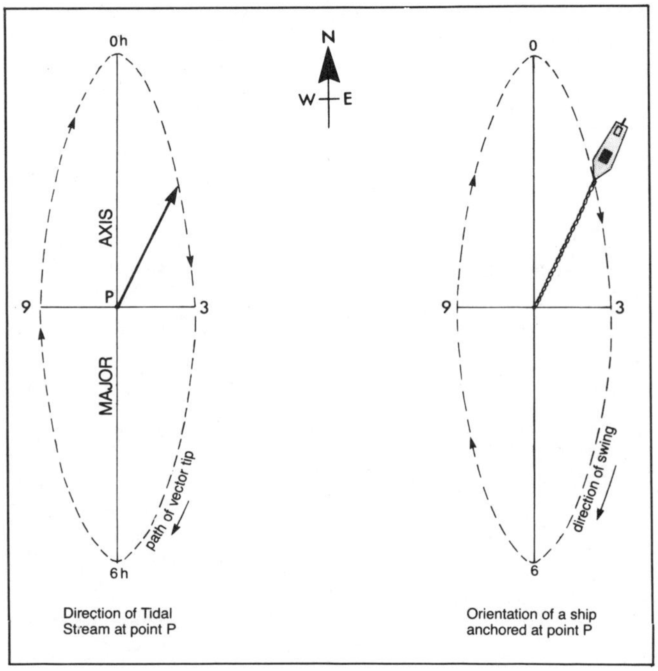

FIG. 3.31. Rotary tidal currents. Changes in speed and direction of tidal stream over a tidal cycle as measured at Point P are compared to length and orientation of an anchor line pivoted at Point P. Length and orientation of anchor line vary in accordance with strength and direction of flow.

so that its distance from the pivot point is directly proportional to the speed of the tidal stream (Fig. 3.31); the stronger the current the farther the ship will be from the center, and vice versa. The direction of the current, of course, will be parallel to the anchor line. Suppose the tidal streams are purely semidiurnal and there is no change in their speed with time, only a change in direction. Then, over one complete cycle (12 h, 25 min), the ship would swing in a complete circle as it rotated with the top of the current vector. The swing would be counterclockwise if the tidal streams "backed" with time or clockwise if "veered." If tides are diurnal, the ship would complete the circle in just under 25 h.

The sense of rotation of the tidal stream in a given area can only be determined by direct observation. Moreover, the current usually has a preferred direction of flood and ebb, with appreciably lesser speeds at right angles to these directions. Instead of tracing out a circle, therefore, the tip of the current vector generally traces an ellipse, its major axis along the direction of maximum ebb and flood and its minor axis along the direction of weakest flow (Fig. 3.30). (A rectilinear tidal stream is one where the ellipse is so flattened that it has no minor axis.) As an example, consider a tidal stream with maximum flood to the north (0°True) and semidiurnal clockwise rotation, as in Fig. 3.31. As the flood swings to the east its speed gradually diminishes until, about 3.1 h later, it flows weakly to the east (90°T). The speed then begins to increase until the stream attains maximum ebb to the south (180°T), 3.1 h later. This in turn is followed by a reduction in speed to a minimum westerly flow (270°T) after an additional 3.1 h, and a return to maximum flood after a total time of about 12.4 h.

Because tides in the ocean are typically of the mixed variety, the picture presented here is an oversimplification. It is, nevertheless, valid under certain circumstances (a) on those days of the month when B.C. tides are almost purely diurnal or purely semidiurnal, (b) in inshore waters where the ellipses are constrained by channel topography, and (c) if attention is concentrated on each separate tidal constituent that makes up the mixed tide. In isolation then, the K_1 and M_2 constituents of the tidal streams behave as described. Therefore, tide analysts prefer to work with the "tidal current ellipses" for each constituent as a way to visualize the variation of currents with time in a particular location. However, the mariner should realize that the path traced out by the actual current vector associated with mixed tides is more complex than the simple ellipse traced out by each constituent, though from one day to the next the overall pattern is nearly repeated.

Tidelines

The assortment of driftwood, uprooted seaweed, foam, and other debris floating on the water is constantly shifted around by the different current regimes. Wherever these currents meet or "converge," accumulated material can form into discernible lines or rows called tidelines, although their formation may have nothing whatever to do with the tides.

Essentially four distinct mechanisms produce tidelines through a convergence of surface waters.

A) They may be formed where water in a particular current regime is sinking underneath or riding over top of the surface layer of another regime. In this case, the debris is skimmed off the advancing flow as it sinks below the surface of the opposing water mass, or alternatively the advancing current skims off debris as it moves over the adjoining water mass (Fig. 3.32). The former process is somewhat analogous to the subsidence of the oceanic crust beneath the continental margin discussed in Chapter 1. Tidelines of this type are commonly found in inlets with rivers at their heads, and at the boundary between clear Strait of Georgia water and the silty runoff from the Fraser River. In Pl. 2, tidelines of accumulated foam can be seen along the edges of the Squamish River runoff as it moves down Howe Sound. Tidelines at the entrances to tidal channels, or downstream of an island where different current regimes make contact, fall into this category, as do the tidelines south of Rosario and Haro straits and off Kelsey Bay in Johnstone Strait.

Mariners should note that these tidelines often mark an abrupt change in the speed and direction of the near-surface currents, as illustrated by Fig. 3.33, based on measurements taken about 5 km west of Sturgeon Bank off the Fraser River delta. As the tidelines slowly passed

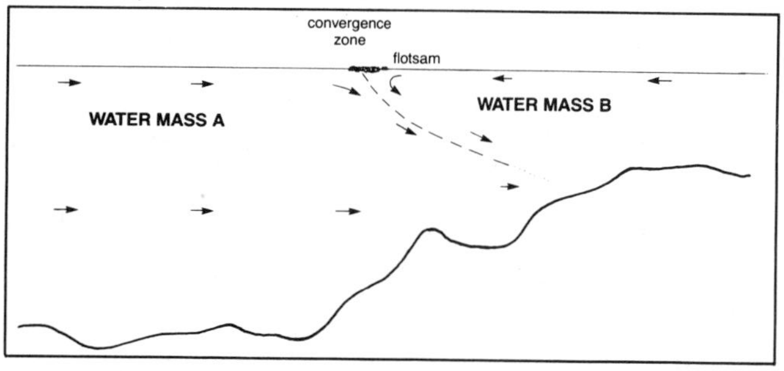

FIG. 3.32. Formation of tidelines at boundary between two opposing bodies of water. Surface debris gathers at convergence zone to form distinct lines that shift with changes in winds and tides.

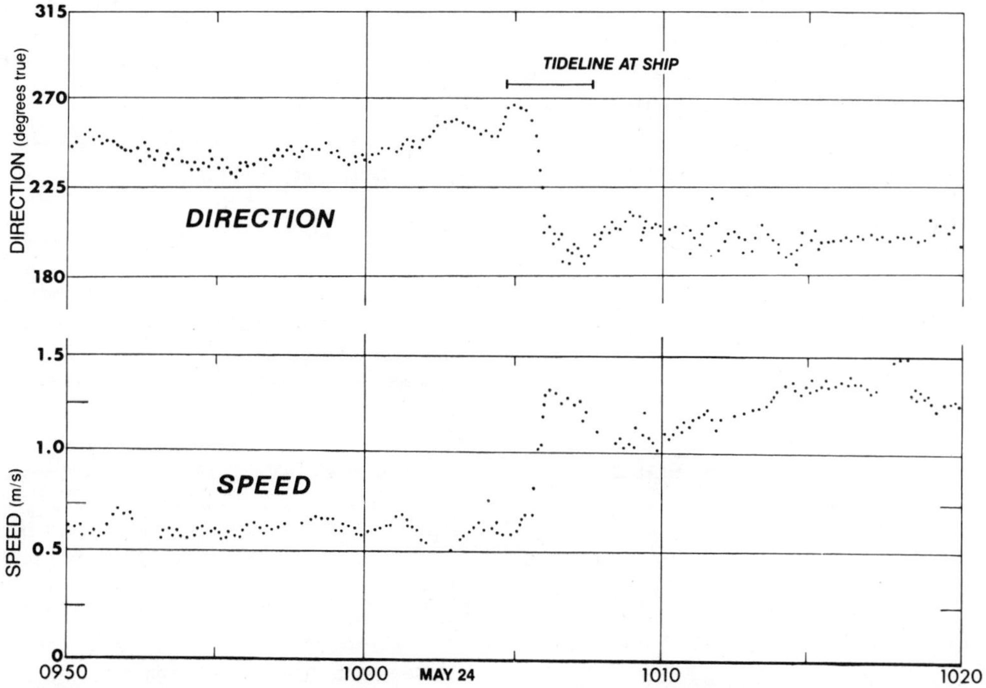

FIG. 3.33. Abrupt change in surface current speed and direction associated with passage of a tideline. Measurements taken from a research ship 5 km west of Sturgeon Bank, Fraser River, May 24, 1967. (Courtesy S. Tabata)

under the research ship, the current speed increased rapidly from 50 cm/s (1 kn) in the clear salty water to nearly 150 cm/s (3 kn) in the silty brackish water; the direction of the current swung almost 90° as it shifted from a westward flow to a southward flow.

B) Backeddies may also produce tidelines. In this case, the debris trapped within the eddy tends to become strung-out at the region of intersection between the slowly moving currents in the eddy and the more swiftly moving water in the main channel. Once established, this situation will persist until there is a reversal in flow or a disruption by the winds.

C) Surface convergence zones associated with internal gravity waves (Chapter 6) are another common source of tidelines. These subsurface waves are responsible for the well-defined series of bands or "slicks" that often characterize the surface of estuarine-type regions like the Strait of Georgia and mainland inlets during the summer. Even when there is little debris to accumulate, these bands may still be visible as relatively smooth stretches of water horizontally separated by many tens of metres of rippled water (and vice versa) that extend for many kilometres in twisting, disjointed ribbons. From the air they also appear as rows of light silty water separated by areas of dark blue oceanic water (Pl. 7). Sun glare associated with the relatively flat areas of the pattern may further help to delineate certain bands (Pl. 8). Logs caught in such convergence zones align themselves along the slicks while small ripples steepen rapidly and disappear as they approach the edge of the slick.

It should be noted that cat's paws, produced during periods of light winds, also will create rippled and nonrippled areas of water. These are usually patchy and without the lateral extent characteristic of internal wave slicks. Quite often the smooth regions of the slicks are due to a thin oily film of marine organisms, which damps the ripples and gives the sea a glossy look. Oil from motorboats has the same physical effect.

D) Lastly, winds can produce convergence zones by inducing cell-like circulation patterns in the surface water. Tidelines of this type are called windrows, and occur where the wind-blown foam and other material gathers at the junction of two adjacent cells set up at right angles to the wind direction (Chapter 4).

Red Tide

A visible feature of coastal areas in summer is the appearance of patches of reddish-brown water known as red tide (Pl. 9). Although it is essentially a biological phenomenon, the occurrence of red tide does bear some relation to the type of physical oceanic processes discussed. As in the formation of tidelines, red tide is usually associated with converging currents, except that the material brought together is not surface debris but a particular species of freely floating microscopic marine plants called phytoplankton (phyto meaning plant, and plankton meaning drifter). Found everywhere throughout the world's oceans, phytoplankton photosynthesize carbon dioxide and water into oxygen and carbohydrates in the presence of sunlight just as land plants do. In the spring, the increase in the amount of sunlight and accompanying increase in surface temperatures in coastal areas encourages multiplication of these plants, a process further hastened in regions of converging currents where nutrients are more abundant. Eventually, conditions become so favorable that their numbers exceed millions per litre of water and produce a display of color called a "bloom." In B.C. waters there are two such phytoplankton blooms, one associated with the rapid multiplication of diatoms (a form of phytoplankton with over 20,000 world species) and the second with the rapid multiplication of dinoflagellates (dinos, a whirling; and flagella, whips). The main diatom bloom takes place in spring and is responsible for the greenish tinge to the water in the Strait of Georgia at that time.

The dinoflagellate bloom takes place in early to late summer in coastal British Columbia and Washington, where it produces red tides. For the most part, this phenomena is beneficial to aquatic life because the microscopic plants provide both oxygen and food for other organisms. But there are times when its presence can lead to toxic effects. In the Gulf of Mexico, for example, a particular species of dinoflagellate releases a poison into the water which, during a red tide, becomes harmful to man and leads to massive fish kills. British Columbia species retain the toxin and there is no such problem. On the other hand, dinoflagellates are food for clams, oysters, mussels, and other shellfish that subsequently tend to concentrate the toxin within their tissue. Although immune to the poison, a healthy looking shellfish may in fact be a highly lethal morsel if red tide has been a persistent feature of the animal's habitat, and, if eaten, may lead to paralytic shellfish poisoning. The B.C. coast is potentially one of the worst areas in the world for such poisoning. Butterclams are one of the most serious offenders because they can retain the toxin for up to 2 yr. Mussels, on the contrary, quickly lose any toxin they take up and are usually safe to eat in late winter and early spring. Despite the possibilities, shellfish poisoning is rare. However, the prudent mariner wishing to live off the sea in the summer and autumn would be wise to heed all official shellfish poison warnings and to check with the Fisheries Inspection Branch, Fisheries and Marine Service, Fisheries and Oceans, 1090 West Pender St., Vancouver, B.C., or a local fisheries office before feasting on steamed clams and raw oysters. Symptoms of paralytic shellfish poisoning are generally noticed within 10 min of ingestion and include a tingling of the lips, tongue, and inside of the mouth. The tingling sensation gradually spreads to the fingers and toes, and leads to numbness in the legs and arms. Anyone experiencing these symptoms after eating shellfish should contact a doctor immediately. (On May 17, 1980, a resident of Health Bay, a small village on Gilford Island, died in Alert Bay hospital of paralytic shellfish poisoning after eating butterclams. Two others became seriously ill, but recovered within a few days.)

If the concentration of phytoplankton in a red tide becomes so great that the food supply becomes exhausted, the entire population will rapidly die out, sometimes in the short period of 1 day. As the decaying process uses up available oxygen, the water may become incapable of supporting life, which in turn can lead to mortality of marine life unfortunate enough to reside in the region, or to have had this poorly aerated water moved to their area by a change in the wind and current.

Bioluminescence

Anyone who has been boating or swimming in coastal waters at night will have noticed the brilliant "phosphorescence" that occurs whenever the water is disturbed. On moonless summer nights, a wake may become a faintly glowing path that extends far behind the ship, and each wave crest that laps against the hull produces a glittering display of brilliance against the darkness of the ocean.

Early mariners and scientists attributed these flashes of light to a wide variety of causes and it wasn't until the end of the 18th century that luminescence in the sea was found to be a biological process. It is now known that it originates with the same dinoflagellates that form red tides; each microscopic organism responds with a bright rapid flash lasting for 1/10 s whenever there is a disturbance in its aquatic environment. Contrary to popular belief, the processes are unlike phosphorescence (which depends on the prior absorption of light from an external source) as the organism generates the light internally through a chemical reaction. Moreover, it is "cold" light unaccompanied by the emission of heat that typifies most chemical reactions. Why these animals possess the capability for luminescence is unclear, but would appear to be some sort of alarm mechanism yet to be understood.

Chapter 4. Secondary Currents

Tidal currents dominate the flow patterns in most coastal areas. Nevertheless, there are secondary currents set into motion through a variety of forces that lead to variations in circulation quite distinct from the cyclic tidal oscillations. The estuarine-type circulation discussed in Chapter 2, for example, is a secondary flow pattern maintained by horizontal differences in water temperature and salinity, for which discernible fluctuations in speed occur over periods of weeks to seasons. Wind-generated currents are secondary flow that may alter speed and direction over time spans as short as a few hours.

Although at any particular instant the presence of secondary currents may be completely hidden by the stronger tidal flow, their influence can be quite pronounced over long periods of time and can be of considerable importance in determining the oceanography of a particular coastal area.

Wind Drift

The direct effect of the wind's drag on a nearly smooth water surface is felt only in the top few centimetres. This thin layer is then made to move down-wind at about 3% of the wind speed. (Smooth is used here in the aerodynamic sense in that the wind conforms to any bumps on the sea surface and does not break up into turbulent patches.) In a 5 m/s wind a thin "skin" of water would move at approximately 15 cm/s (3% of 5 m/s), but would have no effect on a boat with a discernible draught. The shallow penetration of the wind drag can be readily observed in a pond or sheltered bay where small submerged particles, such as pollen, can be seen drifting just beneath the water surface with the wind, while suspended particles a metre or more below the surface remain almost motionless.

The ability of the wind to produce currents to greater depths is significantly enhanced if it is putting energy into surface gravity waves (wind waves). Waves effectively increase the wind drag by increasing the roughness of the surface, and make it more difficult for air to flow smoothly over the water. Under these conditions, approximately 40% of the wind energy goes into the waves, of which 5% is lost to breaking crests in the form of whitecaps. The increased wind drag, together with the momentum transferred to the water by whitecapping, leads to substantially deeper wind drift. As the amount of energy in the waves themselves depends on the wind duration and speed as well as its fetch (the unrestricted length of water surface over which the wind blows), the speed and extent of the wind current will depend on these factors also. Therefore, the state of the sea, and not the wind directly, determines the speed and depth of penetration of these currents.

Rain, especially heavy rain, associated with storm winds may further augment the wind's ability to drive a surface current. This can happen in a number of ways. First, the falling rain drops gather horizontal momentum because they are carried along at some fraction of the wind speed. When they strike the surface of the sea this momentum is then imparted to the water which, like the wind drag itself, causes the water to be driven forward in the direction of the wind. In addition, rain drops splashing onto the sea produce an increase in the roughness of the surface, which effectively enhances the wind drag. Lastly, the natural tendency of the wind speed to diminish to zero very close to the water surface due to friction is partly offset by rain drops. Because they lose only 10–20% of their horizontal speed as they fall through the lower few metres of the atmosphere, rain drops may actually be moving faster than the air and, therefore, transfer to it some of their momentum. This in turn strengthens the wind close to the sea surface and increases its force on the water.

Wind-generated surface waves produce a weak transport of water in the direction of the waves called the Stokes drift. This is not a wind-current, but is associated with the fact that orbital motions under a wave are not completely closed, and allow the water to advance slightly forward with the passage of each wave. The speed of this drift is less than 1/10 the wind-drift and is usually unimportant.

Winds may generate more subtle circulations within the surface waters. For instance, the streaks of foam and surface debris that align as windrows along the direction of the wind are associated with cell-like, circling patterns in the water at right angles to the wind direction (Fig. 4.1). Looking downwind, the water to the right of the windrow is circulating anticlockwise while that to the left

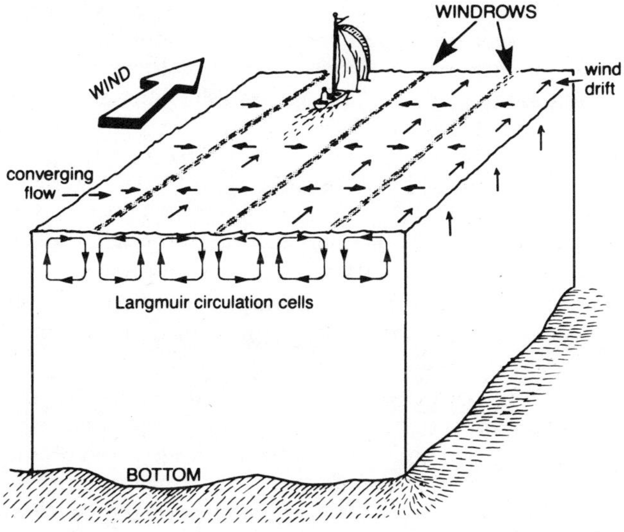

FIG. 4.1. Cellular circulation patterns associated with windrows. Foam and surface debris gather in streaks where currents of two Langmuir cells converge. Combined effect of separate wind drift (→) and Langmuir cells produces corkscrewlike flow pattern aligned in direction of wind.

is circulating clockwise; surface debris gathers or converges where these two circulations meet to produce sinking water. These so-called Langmuir circulations combine with the stronger currents generated in the wind direction to produce an overall downwind water motion that somewhat resembles a corkscrew. The spacing between adjacent windrows is roughly equal to twice the length of the dominant surface gravity waves.

Although the exact reason for the formation of Langmuir circulations is not completely understood, evidence now suggests they result from a complicated interaction between the Stokes drift assocated with the waves, and the wind-driven current directly formed by the wind drag, or alternatively, through the action of breaking waves and the wind-driven current. Whatever the cause, a number of observed features of these circulations can be of practical use. For example, the tendency of oil slicks to collect along the convergence bands has been successfully exploited in cleaning up oil spills at sea. And, as lines of greatest wind-directed surface drift are found along windrows, whereas lines of weakest wind-drift lie midway between adjacent windrows, yachtsmen have a natural indicator to set their course. Clearly, best advantage on a downwind run is riding along a windrow. When beating to windward it is best to stay midway between the dominant windrows as much as possible.

Perhaps some of the most important types of wind-induced secondary flows are those associated with near vertically propagating disturbances called inertial (or gyroscopic) waves. Generated within the upper ocean by abrupt changes in wind direction, these inertial currents are rotary flows whose direction constantly changes over a specified period of time, somewhat akin to the rotary tidal streams discussed in the previous Chapter. Unlike tidal streams inertial currents are invariably circularly polarized in that the current vector always rotates clockwise (northern hemisphere) and maintains a uniform speed over a single rotation. Put another way, the tip of the current vector, in the absence of other types of current, traces out a circle (see Fig. 3.30b). (In the southern hemisphere, the sense of rotation is counterclockwise.) Set in motion by a pulse of momentum from the winds, the currents are maintained by a balance between the rightward turning effect of the Coriolis force (northern hemisphere) and the centripetal (or centrifugal) force due to the curvature of the water's path. Therefore, the time for the current to swing once round the circle, the inertial period, is determined by the local value of the earth's vertical component of rotation. At about 50° lat., this corresponds to 15½ h, part way between 12½ and 25 h periods of the tides. (At 30° latitude the inertial period equals the diurnal period of the tides, about 24 h.)

For reasons that are only partly understood, inertial motions appear to be mainly confined to the upper 100 m or less of the ocean's surface and undergo rapid attenuation after only a few periods of oscillation amounting to several days. They tend to occur in open regions of the oceans away from the interference of coastal boundaries and not in confined basins such as the Strait of Georgia or Juan de Fuca Strait. Not surprisingly, inertial currents are highly intermittent, often appearing suddenly to disrupt the normal tidal streams for a few days following the passage of a storm front. Current speeds commonly reach 25 cm/s off the west coast of Vancouver Island and at the entrance to Queen Charlotte Sound, with a high degree of coherence between the current motions over tens to hundreds of kilometres.

Relaxation Currents

As well as driving surface drift currents, winds can indirectly affect other types of flow. Where the wind drift is restricted by a lee shore, persistent onshore winds will cause the water to pile up against the coast and tilt the sea surface, an effect that can be simulated by blowing on a bowl of water. By the same token, offshore winds will produce a sea-level tilt by moving water away from the coast. On a large enough scale this can lead to the formation of storm surges on low-lying coasts mentioned earlier in connection with tides. When such winds weaken or reverse direction, the raised (or depressed) water surface at the shore will seek its equilibrium level. The relaxation currents associated with this readjustment of water level may persist for hours or days, depending on the area affected, and result in perceivable deviations of the surface currents from those expected on the basis of tides and local winds alone. Wickett (1973) suggested that the unusually strong southerly currents of nearly 1.5 m/s (3 kn) observed at a drilling rig at the southern end of Hecate Strait on Sept. 25, 1968, were due to such currents after a period of strong onshore winds. In this particular case, the outflow was apparently augmented by 7 cm of rain on the eastern shore of the Strait and by runoff from adjacent inlets.

In partially enclosed basins like harbors and certain inlets, disturbance of the surface level by passing storms can set up oscillations that cause currents to slosh back and forth several times before the system returns to equilibrium. Sea level and current oscillations of this kind, called seiches, have periods of minutes to hours and amplitudes of 5–10 cm, depending on the depth and geometry of the basin and the nature of the disturbing mechanism. Generally speaking, the seiche-currents attain maximum speeds midway between the two opposite directions of tilt (see Fig. 3.25). Seiches can also be initiated by tides, tsunamis, and other types of oceanic waves at the mouth of the basin. In many instances the exact cause of observed seiche activity is not clearly understood.

Slippery Water

As noted in the discussion of inlets and other bodies of water that receive considerable amounts of river runoff, a relatively thin brackish layer often overrides the deeper oceanic water. For yachtsmen who want to use surface currents to advantage, the presence of this layer can be of fundamental importance. In the absence of such vertical stratification, only the top few centimetres are dragged along by wind directly. With vertical stratification, the whole brackish layer, many metres thick, can be made to slide downwind like a lubricated "slab" with only a minimum of resistance from the deeper water beneath (Fig. 4.2). In the open ocean also, wind-mixing of relatively warm, low-salinity surface water can, over a period of many days, lead to the formation of a thin layer of comparatively light water over large areas of the sea surface. In the open ocean, the slab may also be set into inertial motion, whereby a large region of water moves in a circular path without change of orientation.

The phenomenon of "slippery water" was used successfully by British sailors in the 1968 Summer Olympics in Acapulco, Mexico. To quote David Houghton, the meteorological advisor to the British crews, "For the first two weeks the sea surface . . . behaved just like the slippery layer which had been postulated. In fact it was too good to be true. The surface water moved almost directly downwind at speeds of up to about 2 knots, depending on how long the wind had been blowing from that particular direction. When the wind dropped to a calm the water continued in the same direction with little change in speed and it took a wind from the opposite direction from 24 to 36 hours to stop the water and get it moving the other way." It is possible to estimate the speed attainable by a brackish layer by rough approximations for a given wind

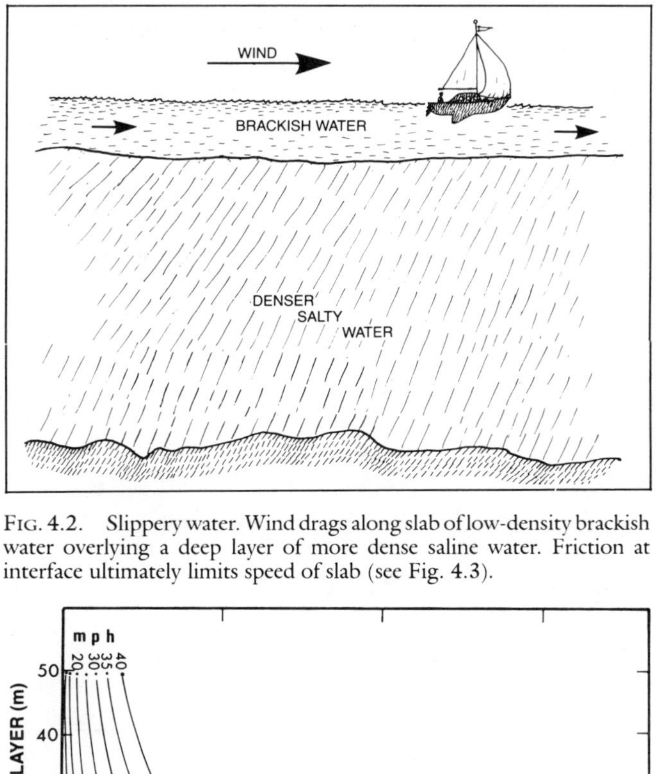

FIG. 4.2. Slippery water. Wind drags along slab of low-density brackish water overlying a deep layer of more dense saline water. Friction at interface ultimately limits speed of slab (see Fig. 4.3).

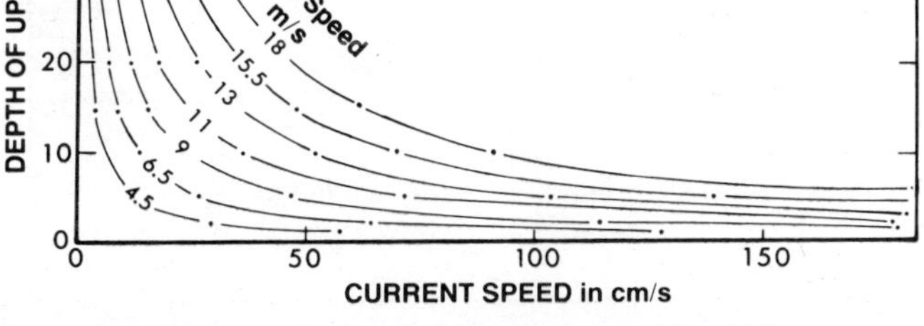

FIG. 4.3. Maximum down-wind speed a layer of brackish water can theoretically attain for a specified layer thickness and wind speed. (From Waldichuk 1957)

speed. These are shown graphically in Fig. 4.3, which indicates for example that a 15-m/s (30 kn) wind can move a brackish layer 10 m thick downwind at about 0.6 m/s (1.2 kn). The thinner the layer the greater the speed for a particular wind. Anyone using this relationship, however, should bear in mind that wave action can destroy the layered structure of the water, particularly if the winds are strong or the upper layer is thin (less than 1 m or so).

Ekman Spiral

In open areas of the ocean, frictional effects associated with turbulent motions assist in transmitting the influence of the wind downward through the sea surface. The relationship between wind and current that is often used, though with some reservation, is based on the theory of the Swedish oceanographer, Ekman (1905), who attempted to explain the behavior of the ice pack in the Arctic Ocean. As Fridtjof Nansen observed in his famous 1893 expedition when his specially designed ship the *Fram* was locked into the ice for 3 yr, the arctic ice drifted at about 45° to the right of the wind at a few percent of the wind speed. According to Ekman's explanation, a combination of wind drag, internal friction, and the Coriolis force were responsible for this behavior. He further showed that, with increasing depth, the wind-induced current should turn more and more away from the direction of the wind and decrease in speed, to produce the so-called Ekman spiral (Fig. 4.4). Present information indicates that the angle for surface currents is closer to 20° to the right of the wind at about 2–3% of the wind speed, provided that (1) the water is actively "roughed up" by wind-wave action; (2) the water depth exceeds the maximum depth of direct wind influence, around 100 m; and (3) the flow is not confined horizontally by the presence of land. If the water is shallower than the depth of wind influence or close to shore, the entire water column tends to move downwind with a speed that decreases with

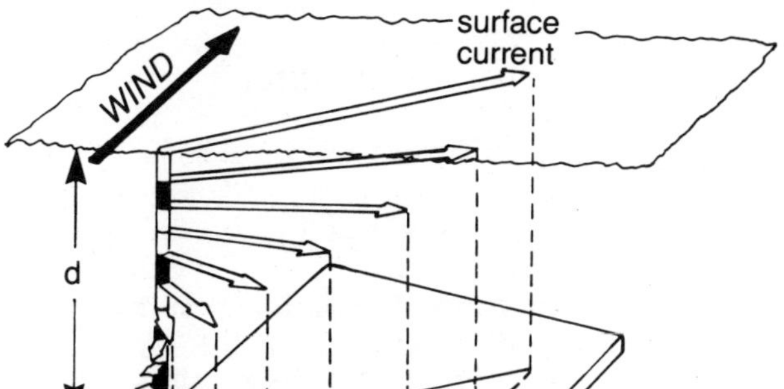

FIG. 4.4. Classic Ekman spiral. Arrows and their projection onto lower surface (broken line) indicate direction and relative strength of current at various depths in relation to surface wind in northern hemisphere. (In southern hemisphere currents are to left of wind.) Depth, d, varies but is typically between 30 and 100 m in open ocean. Deeper currents are weak and in the opposite direction to wind.

depth. In light airs with little or no wave action the downwind flow becomes mostly confined to the top few centimetres.

As is so often the case with simple explanations for the generation of currents, the full Ekman spiral is rarely observed in the ocean, either because conditions aren't quite right or because other effects overshadow it. Nevertheless, the direction and speed of the surface current within the limits of the theory's applicability appear to be good enough as a first approximation to wind influence in the open, deep-sea regions of the ocean.

The current associated with the Ekman spiral, the Ekman drift, is thought to play a vital role in establishing the large wind-driven circulation patterns of the world oceans. Briefly, the mechanism works as follows. First the wind sets up the Ekman drift pattern with currents to the right of the wind in the upper 100 m or so of the ocean surface (northern hemisphere). (In the southern hemisphere, drift currents are to the left of the wind.) Then, because of the north–south changes in the strength of the Coriolis force and horizontal variations in the wind speed and direction, some regions of the world oceans experience a long-term convergence of the surface water while other regions experience a divergence of these waters. In large areas where the top layers are diverging there must be a slow upward flow from beneath to compensate for the horizontal spreading of the water. The upward motion in turn induces horizontal currents of its own within the top 1000 m or more of the water column. The currents form into the cyclonic gyres so characteristic of the major oceans. Similarly, convergence of the surface waters by long-term influence of the winds necessitates downward flow, and the formation of the equally characteristic anticyclonic gyres. The cyclonic subarctic gyre in the Northeast Pacific Ocean, for example, is associated with the counterclockwise winds of the Aleutian low-pressure system that dominates this region much of the year. The anticyclonic North Pacific gyre west of California is associated with the equally persistent clockwise winds of the North Pacific high-pressure system.

Density Currents

These currents are formed when a volume of water with a different density than its surroundings seeks its appropriate level. That is, it seeks a depth where the water above is lighter (fresher or warmer) and the water underneath is heavier (saltier or colder). Although such currents are generally slow and, therefore, have little direct effect on boaters, they are extremely important to the long-term dispersal of pollutants and to maintaining the great variety of flora and fauna in deep portions of certain basins. Without the continual replenishment of the deep and intermediate depth waters in the coastal regime, the dissolved oxygen in these waters would soon be used up by aquatic plants and animals and decaying materials. If this were to happen, most life in the deeper waters would eventually disappear.

Despite their sluggish nature, density currents also make an important contribution to circulation patterns in the world oceans. It is now believed that the small density differences in mid-ocean associated with the poleward decrease in the solar heating of the earth's surface are largely responsible for the slow currents below the depths of large-scale wind influence (500–1000 m). Known as the thermohaline (heat-salt) circulation these density-induced flows appear to play an equal role to the wind in maintaining the motion of the sea. Moreover, density currents that sink and then flow northward from the shallow continental seas of Antarctica are responsible for the formation of the slowly drifting bottom waters below depths of 3000 m in the Atlantic and Pacific oceans. As a final example, the tongue of warm salty Mediterranean water that flows through the Straits of Gibraltar contributes significantly to the immediate-depth circulation in the North Atlantic between 20–40° N.

Under certain circumstances, density currents can be pronounced and create atypical circulation patterns. One of the most dramatic examples of this on the B.C. coast occurs in the Rupert–Holberg Inlet system at the northern end of Vancouver Island (Fig 4.5). Oceanographic observations in this area have shown that intense vertical mixing of the flood over the sill within Quatsino Narrows frequently results in water that is more dense than that of Rupert Inlet, into which it eventually flows. Driven by this horizontal difference in density, the flood current is then able to dive down the side of the basin as a rapidly moving density current with speeds in excess of 1 m/s (2 kn). Sweeping along the bottom of the inlet, the current

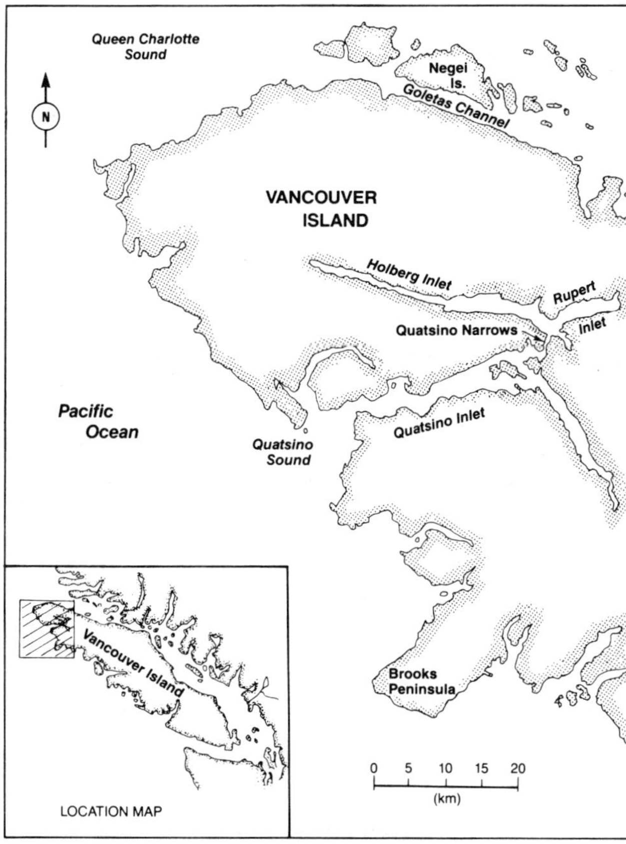

FIG. 4.5. Map of the Rupert–Holberg Inlet system northwest coast, Vancouver Island. (From Goyette and Nelson 1977)

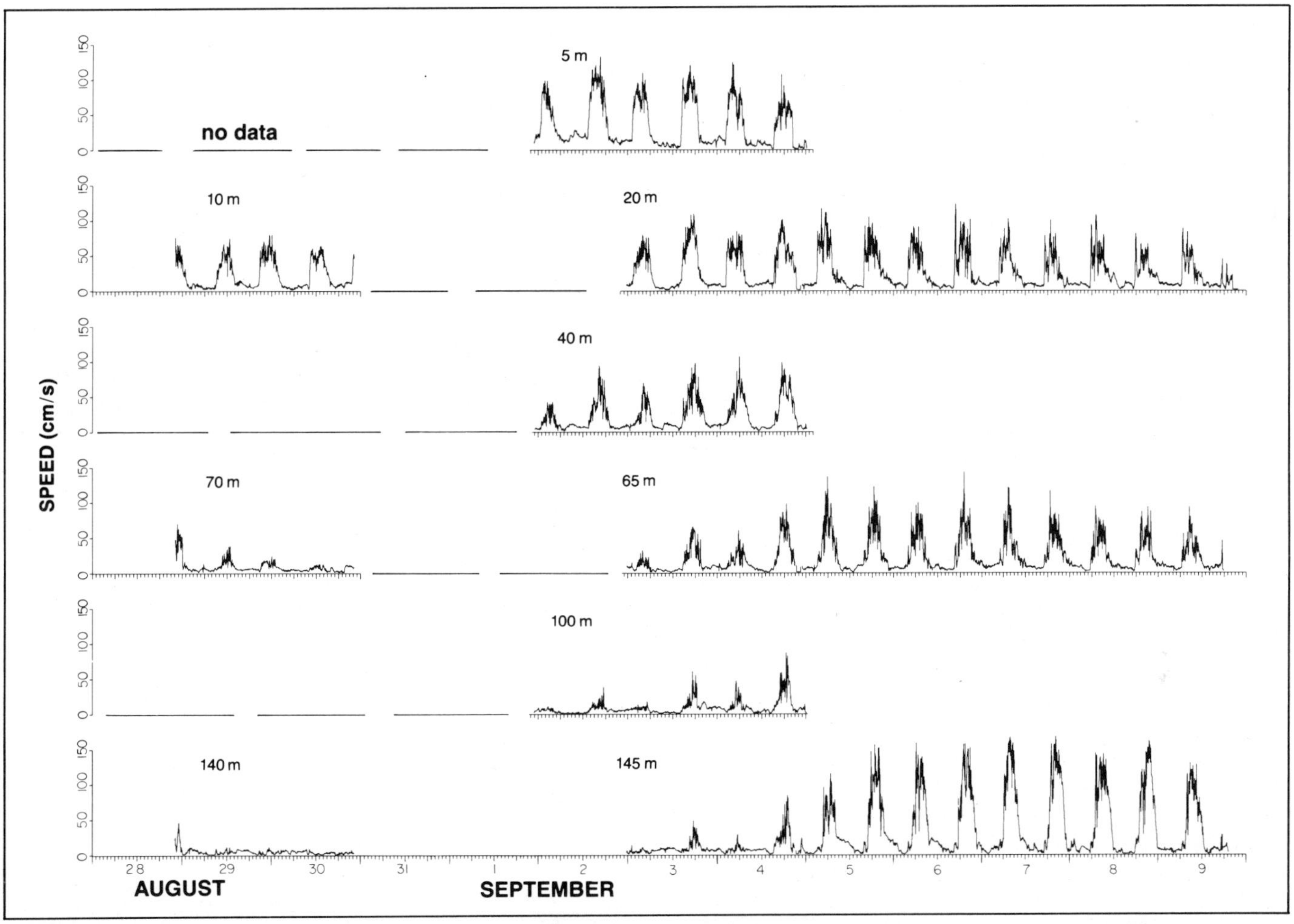

FIG. 4.6. Measured currents in Rupert Inlet immediately north of Quatsino Narrows (see Fig. 4.5). Plots show variation in current speed at various depths, August–September 1976. In late summer, positively buoyant jetlike flood current is confined to near-surface depths; in autumn, as surface waters become more dense, entrant jet becomes negatively buoyant and can penetrate to bottom. (Courtesy D. Stucchi)

displaces the less dense water upward, and causes an increase in the salinity and dissolved oxygen content of the deep portions of the basin. This process appears to be particularly well established during spring and summer, when precipitation and runoff are low and dense water of oceanic origin is present in Quatsino Sound. During the fall and winter, the larger runoff and precipitation lead to decreased density contrasts between the tidally mixed water of Quatsino Narrows and Rupert Inlet, so there is more intermittent formation of the bottom density current. (Ebb currents in Rupert Inlet are always weak.)

There are other aspects of density current structure worth noting. At the beginning of the flood, for instance, the water mixed in Quatsino Narrows is water that went out with the preceding ebb. Because its density is essentially the same as that of the upper layer of Rupert Inlet, it must at first begin to flood into Rupert Inlet as a surface flow. (Maximum ebbs and floods of the surface current at the northern end of Quatsino Narrows can reach 3 m/s). As the tide continues to rise, however, more and more comparatively dense water is brought in from the oceanic side of the Narrows, with a subsequent increase in the density of the mixture produced near the sill. As a result, the core of the high-velocity flood stream entering the inlet begins to appear at deeper and deeper depths until finally it reaches the bottom, having adjusted its level of intrusion into Rupert Inlet in accordance with changing density (Fig. 4.6). It takes roughly 2 h for the core of the density current to reach the bottom, corresponding to the time required for the flood to transport cold saline ocean water in Quatsino Inlet to the vicinity of the Narrows.

In many respects, the strong density current in Rupert Inlet during the flood is beneficial. Unlike some inlets on the coast (e.g. Saanich Inlet), there is a constant renewal of bottom water and no stagnation or depletion of oxygen. Unfortunately, Rupert Inlet has been used since 1971 as a submarine tailings dump for the copper-molybdenum wastes of Utah Mines, presently over 30,000 t/day, which it was argued, would gather placidly on the bottom and not affect the aquatic environment. The decision to use Rupert Inlet in this manner was clearly based on ignorance. Subsequent studies have shown that the density current that sweeps along the bottom on the flood picks up large quantities of fine tailings and carries them to the surface along the northern side of the basin, where they become the light colored water seen in aerial photographs (Pl. 10). From there, the currents distribute the suspended wastes over large areas of the inlet before

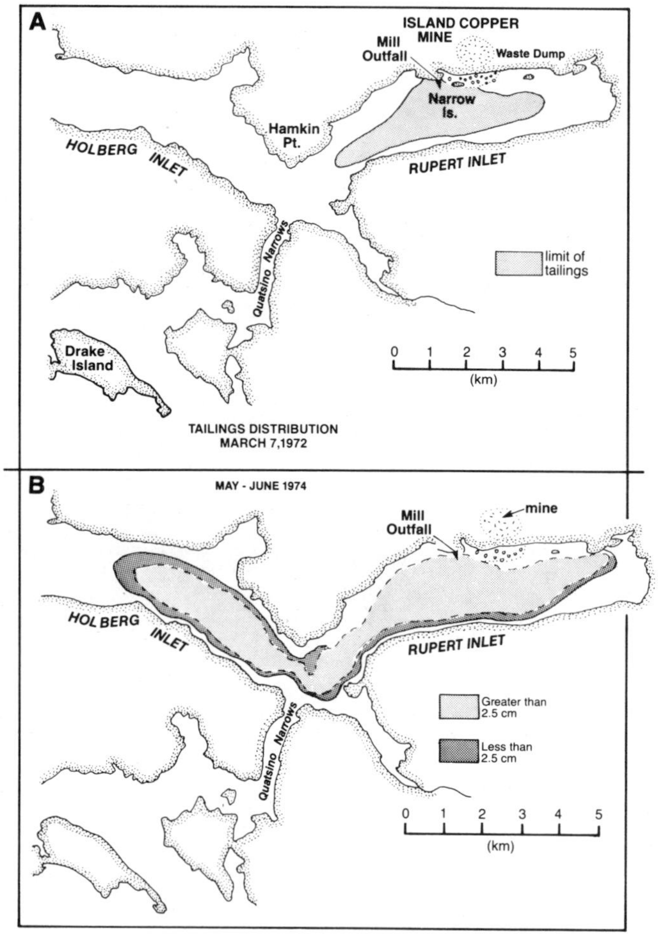

FIG. 4.7. Distribution of mine tailings on bottom of Rupert and Holberg Inlet at two different times, (A) March 1972 and (B) May–June 1974. The effluent (large concentrations of copper, manganese, chromium, zinc, molybdenum, lead, etc.) is normally diluted with seawater and then discharged at a depth of 50 m via a marine outfall located immediately west of Narrow Island. Since dumping began in October 1971 tailings have been redistributed over a large area of bottom by tidal currents. (From Goyette and Nelson 1977)

they again settle to the bottom (Fig. 4.7). Although there is no unequivocal evidence as yet to indicate a detrimental effect on marine life, the fine suspended particulate matter of the wastes destroys the clarity of the water, reduces the amount of sunlight penetrating the surface waters, accelerates the corrosion of marine equipment, and may even induce secondary flows of its own by altering the density of seawater. If nothing else, the situation of Utah Mines emphasizes the need for comprehensive oceanographic studies prior to granting pollution permits.

Sea-Slope Currents

The extent to which the difference in atmospheric pressure between widely spaced regions in the ocean influences the currents is not well known. Presumably such motions are of negligible importance to navigation, although they may have long-term effects yet to be discovered.

Simply stated, the water surface over which there is a high atmospheric pressure (that is, higher than normal) will tend to be depressed, whereas that over which there is a low atmospheric pressure will tend to be elevated. Thus, any difference in the pressure between two regions will produce a tilt in sea level toward the region with the lowest atmospheric pressure. For this to occur, however, there must be a redistribution of water via a current from the region of high pressure toward the region of low pressure (Fig. 4.8). The speed of such a current is usually small and distributed over the entire depth of the water column. Suppose a barometer at Campbell River registered a high pressure of 1010 mb when a barometer on Saturna Island in the southern Strait of Georgia registered a low of 990 mb, an uncommon but not unlikely situation. The tilt of the water level along the Strait of Georgia would then become about 20 cm or about 1 cm for each 1 mb difference in pressure. But even if this difference were to build up over a short period of only 1 day and then hold steady, the current generated would only be around a few centimetres per second. The same is true of currents generated when the pressure systems break down and the water surface once again becomes level. Nevertheless, in the deep ocean the amount of water transport associated with small sea surface slopes can be extremely large when the total depth is taken into consideration, and they are, therefore, important to the redistribution of water in the world oceans. (Due to the Coriolis effect, the direction of large-scale mean ocean currents is perpendicular to the sea surface slope.)

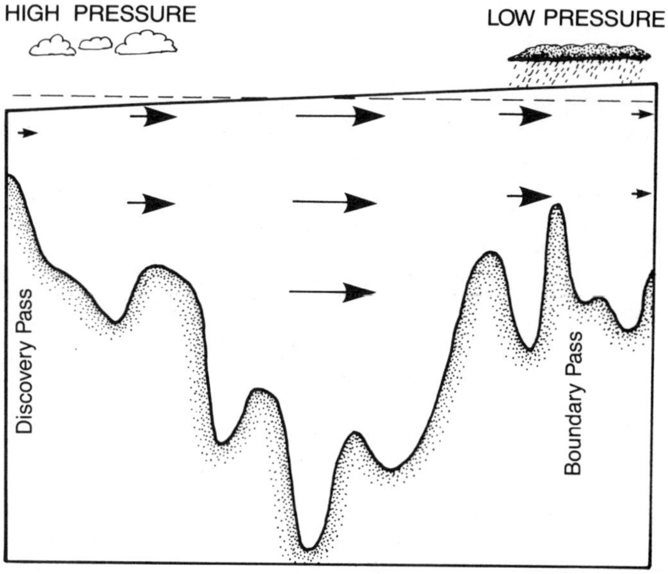

FIG. 4.8. Longitudinal cross-section along Strait of Georgia from Discovery Pass to Boundary Pass. Change in sea-surface slope produced by along-strait difference in atmospheric pressure, exaggerated for illustrative purposes. Arrows show direction of (weak) currents that accompany alteration in slope; broken line is equilibrium sea level.

Jets and Eddies

Although they are primarily associated with tidal currents, jets and eddies are secondary flows in that they tend to be localized phenomena, which do not persist throughout a complete tidal cycle.

Tidal streams from a narrows or passage may maintain their inertia for many kilometres in the wider basin into which they flow. Under such conditions, they appear as a jet of water with large unidirectional speeds relative to the surrounding water. Currents attain maximum speeds along the axis of the jet and decrease rapidly toward either side, where the flow becomes more irregular (Fig. 4.9). Tide rips accompany these jets whenever they oppose the propagation of surface waves. Tidal currents that flood into the Strait of Georgia through Porlier, Active, and Boundary passes take the form of spreading jets and can penetrate to about 2 km. There also is a tendency for these entrant flows to follow curved trajectories as they encounter the tidal flow of the Strait. The jetlike flood of up to 100 cm/s (2 kn) that enters the northern end of Johnstone Strait from Weynton Passage penetrates the entire 400-m depth and moves diagonally to the shoreline. The jet flooding out of Porlier Pass often turns northward to hug the coast of Valdes Island.

Rivers enter coastal waters in the form of surface jets whose direction is affected by bathymetry, local tidal currents, and the Coriolis force. In the main arm of the Fraser River, entrant speeds can be as great as 2.5 m/s (5 kn) during periods of large discharge and large ebbs. As with tidal jets, the momentum of the entrant water is quickly diffused by the processes of lateral spreading, vertical mixing, and friction. This is particularly true of small rivers whose seaward influence is rapidly lost a short distance from the delta front.

Intense jets associated with tidal passes are generally accompanied by eddies and small whirlpools, which form where the rapidly flowing water of the jet "rubs" against the slower moving waters on either side. When viewed in the downstream direction the eddies turn in a clockwise sense on the right hand side of the jet and in a counterclockwise direction on the left hand side of the jet (see Fig. 3.27). These eddies often detach themselves from the edge of the mainstream and wander into the surrounding regions where they decay.

Backeddies form downstream of promontories where the water has been cut off from the main current, or along the sides of strong tidal channels due to the retarding effect of the shoreline. In certain cases, such flow reversals may intensify into narrow jets along the banks of the channel, accompanied by overfalls of sharp drops in the water level in the direction of flow. In Seymour Narrows during a flood, strong northward counter flows are found along both sides of the channel. There is a similar

FIG. 4.9. Strong ebb current flow southward through Dodd Narrows from Northumberland Channel near Nanaimo, Aug. 13, 1962. Flow is relatively laminar (smooth) through neck of narrows but becomes turbulent in broader basin to south. (Courtesy R. H. Herlinveaux)

flow reversal to the north of Cape Mudge on a flood that has become a favorite fishing haunt of the Campbell River sport fishermen. There is a highly visible nearshore counterflow in Active Pass just east of Mathew Point on the ebb that is often emphasized by a change in the nature of the surface waves.

One of the more pronounced backeddies in B.C. waters is between Victoria Harbour and Race Rocks during the flood. The main portion of the eastward tidal stream continues toward Haro Strait rather than turning northward into the harbor. A particularly interesting example of a backeddy is south of James Island near Sidney during an ebb. Aerial photographs show that large eddies peel off both sides of the downstream end of the island and curve into the shadow zone behind (Pl. 11). James Spit, a shoal that covers an area greater than the island, may be partly maintained by these eddies, which would allow transported sand from the upstream cliffs to be dumped into the more quiescent waters behind the island on the ebb. (However, the presence of three prominent spits along Cordova Channel to the west of James Island attests to predominantly northward littoral drift in this region.) As a final example, there is a strong counterclockwise eddy during the flood at the southern end of Haro Strait (see Fig. 11.13). Surface currents associated with this eddy, measured in 1979 by hydrographic launches and aircraft-tracked floats, may be in excess of 75 cm/s (1.5 kn), and are partly responsible for the confused and variable tidal currents that characterize the region.

Chapter 5. Upwelling: Bringing Cold Water to the Surface

Causes of Upwelling

It is well known that the ocean surface becomes increasingly warmer toward the equator. If it didn't, tourist havens like Hawaii and Fiji would have little appeal. However, there are important exceptions to this general rule. For example, in summer the surface water temperatures along the coast between California and Oregon are often lower than those off the west coast of British Columbia. It is sometimes not until the latitude of Baja California that coastal water temperatures become as high as their more offshore counterparts (Fig. 5.1).

The reason for these regions of anomalously cold sea water is upwelling, the process by which subsurface water moves upward to the surface. It is an extremely important phenomenon that affects the fisheries, weather, and current patterns in many parts of the world. It also can be a very localized process confined to bays or tidal channels. To understand this, the two fundamental reasons for upwelling will be explained.

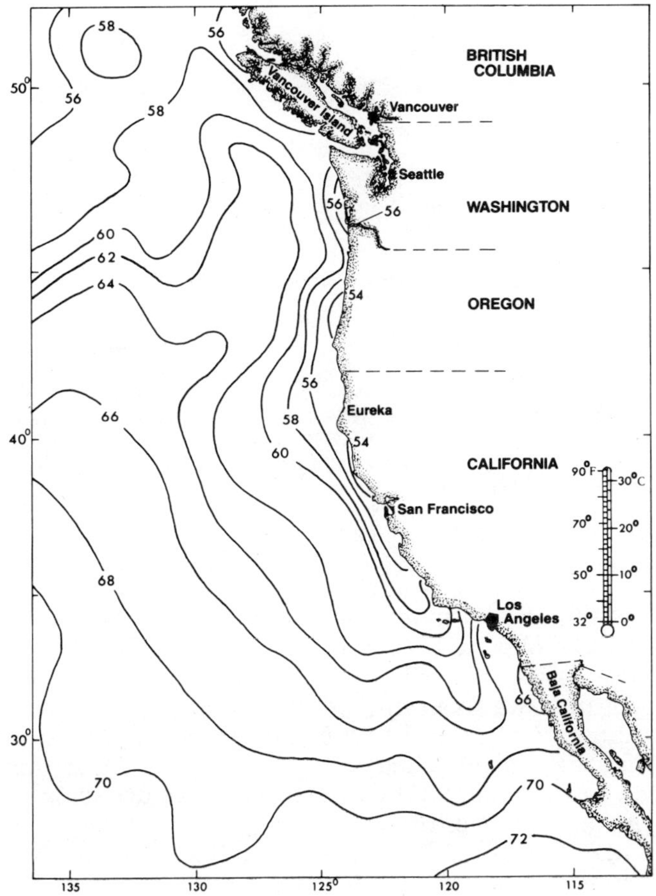

FIG. 5.1. Average sea surface temperatures off west coast North America, Aug. 1–15, 1977. Temperatures in °F. (From Natl. Mar. Fish. Serv. Bull. 1977)

First, it will occur whenever colder water from below forces its way upward and pushes away the warmer surface water (Fig. 5.2). Extremely vigorous examples of this process take place in many B.C. passes where tidal currents are deflected upward by underwater ridges, shoals, and other bumps on the channel bottom. The smooth, dome-shaped upwellings in strong tidal channels like Active Pass and Seymour Narrows are formed in this manner. (Any boater who has been at the helm as his boat

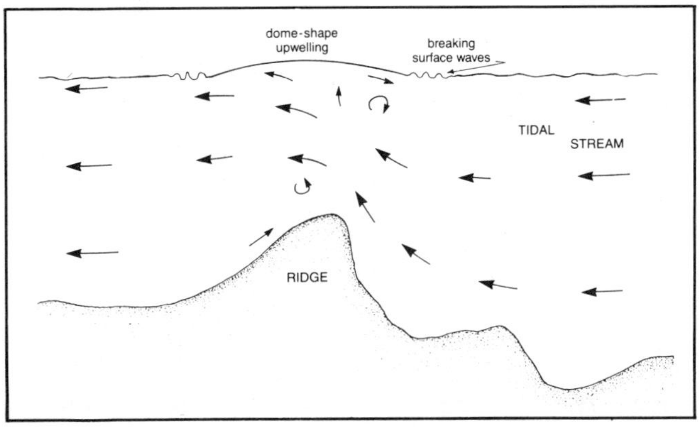

FIG. 5.2. Upwelling of cooler water produced by deflection of tidal currents over underwater ridge. Extent and location of upwellings may fluctuate over periods of a few minutes due to variations in current velocity.

passed over an upwelling can testify to the strength of the water motions.) This type of upwelling helps keep the water in Juan de Fuca Strait and the southern Strait of Georgia cold throughout the year. Although much less vigorous, a similar process takes place in mid-ocean, where slowly moving currents are deflected upward by submarine mountain peaks or ridges. Cobb Seamount, for instance, about 500 km southwest of Vancouver Island, rises abruptly from 2400 m to a depth of only 30 m in less than 35 km, and has a profound effect on the currents in its vicinity.

The second cause of upwellings is slightly more subtle and has more widespread importance. The deeper water doesn't force its way upward. Rather, the classical explanation is that winds push the surface water away from a particular area so the deeper water must rise to replace it. The subsurface water is responding, therefore, to what takes place near the surface, whereas, in the first case, the surface water is responding to what takes place below. It is this second type of upwelling that regularly produces the very cold summer water along the west coast of the United States. (A change in the coastal circulation by other means can produce the same effect.) Other major west coast regions similarly affected are Peru, Southwest Africa, Portugal, and Morocco (Fig. 5.3). Upwelling also occurs off the west coast of Vancouver Island during summer but tends to be more intermittent and less well developed. The only major upwelling region on an eastern

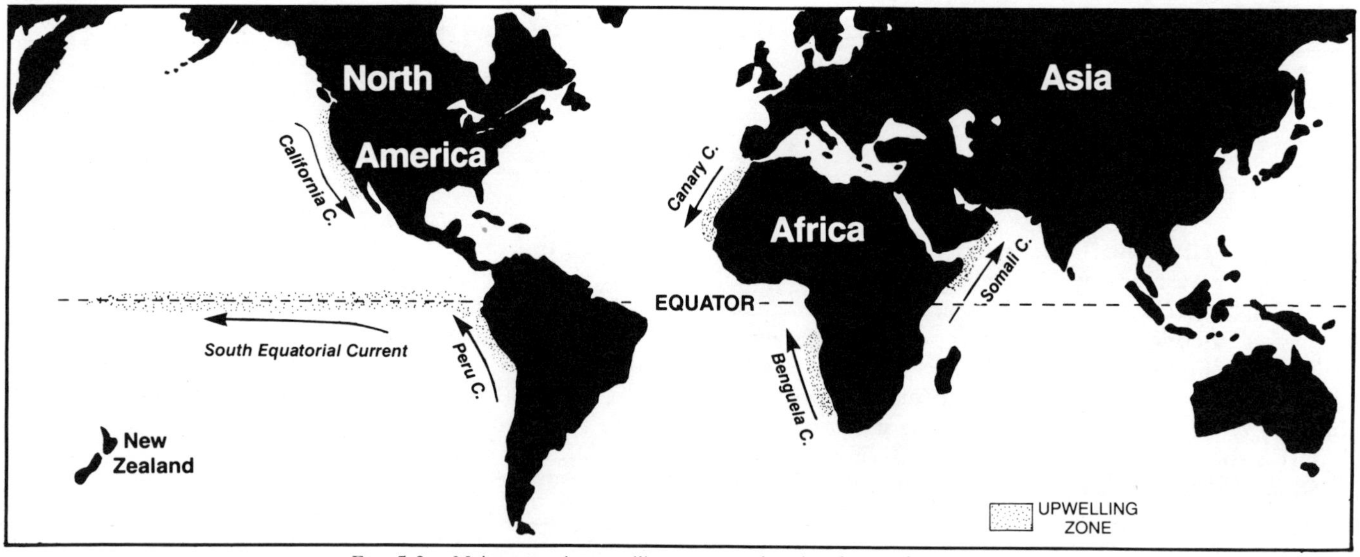

FIG. 5.3. Major oceanic upwelling zones and regional coastal currents.

shore is along the Arabian coast during the southwest monsoon. Finally, a very important belt of upwelled water is stretched out along the equator between the International Dateline and the Galapagos Islands in the Pacific Ocean.

For the most part, extensive wind systems, in conjunction with the Coriolis effect associated with the earth's rotation, create these upwelling regions. Off the west coast of North America, for instance, the summer northwesterlies initiate the whole process (Fig. 5.4). When these winds blow down the coast, the stress they exert on the water produces slow drift currents to depths of about 100 m. As discussed in Chapter 4, the earth's rotation then causes the drift currents to deflect to the right of the wind. As the water is pushed offshore, cold, nutrient-rich water from depths of 100–300 m wells up to replace it. Unlike the situation in B.C. tidal channels, such wind-induced upwelling is extremely slow, with upward speeds between 1 and 10 m/day. But, because thousands of square kilometres may be influenced at one time, the overall effect is of great consequence.

As fishermen in offshore coastal regions discovered long ago, the cold, upwelled water appears in the form of patches, tongues, and plumes that continually change their shapes. The extent of any particular cold spot may be less than 1 km or as much as 30 km across, and 20 m or so thick. Moreover, the lateral fronts or boundaries between the colder upwelled water and the warmer water it is replacing may be very distinct. Even without a thermometer, an observant fisherman can distinguish between the deeper blue color of the newly upwelled water and the murky green of the older warmer water. (Figure 5.5 shows a more modern way to observe upwelling.)

In the equatorial Pacific, upwelling occurs almost the entire year, but is most intense from June to October when the southeast trade winds blow across the equator. Equatorial uprising arises because the earth's rotation deflects the trade wind-produced currents northward on the north side of the equator and southward on the south side of the equator. This diverging of currents induces the belt of upwelling along the equator.

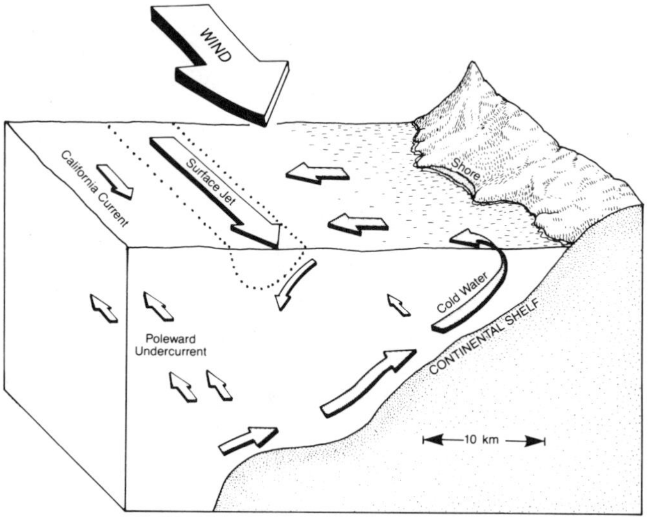

FIG. 5.4. Main features of classical wind-induced upwelling in northern hemisphere. Equatorward winds combined with Coriolis force move nearshore surface water offshore; colder subsurface water rises to replace it. Temperature fronts form at confluence of upwelling region (seaward of edge of dashed area). Secondary currents associated with upwelling are strong, narrow equatorward surface jet and weak poleward undercurrent. Dotted line delineates core of surface jet.

At a more local level, winds that push water directly away from a lee shore will induce upwelling (Fig. 5.6). The Coriolis effect is unimportant because of the limited area of water affected. In the Strait of Georgia, westerlies will produce weak upward motion along the eastern shores of Vancouver Island and some of the larger islands. Similarly, the prevailing northwest winds in summer over the northern Strait of Georgia will induce upwelling on the leeward sides of Texada, Lasqueti, and Denman islands. The same winds push the sun-warmed surface waters out of partially enclosed regions like Departure Bay and make swimming much less pleasant. Lastly, limited areas of upwelling can form on the down-current side of prominences that project into a current. As a current flows past a point of land, for example, it will draw some surface

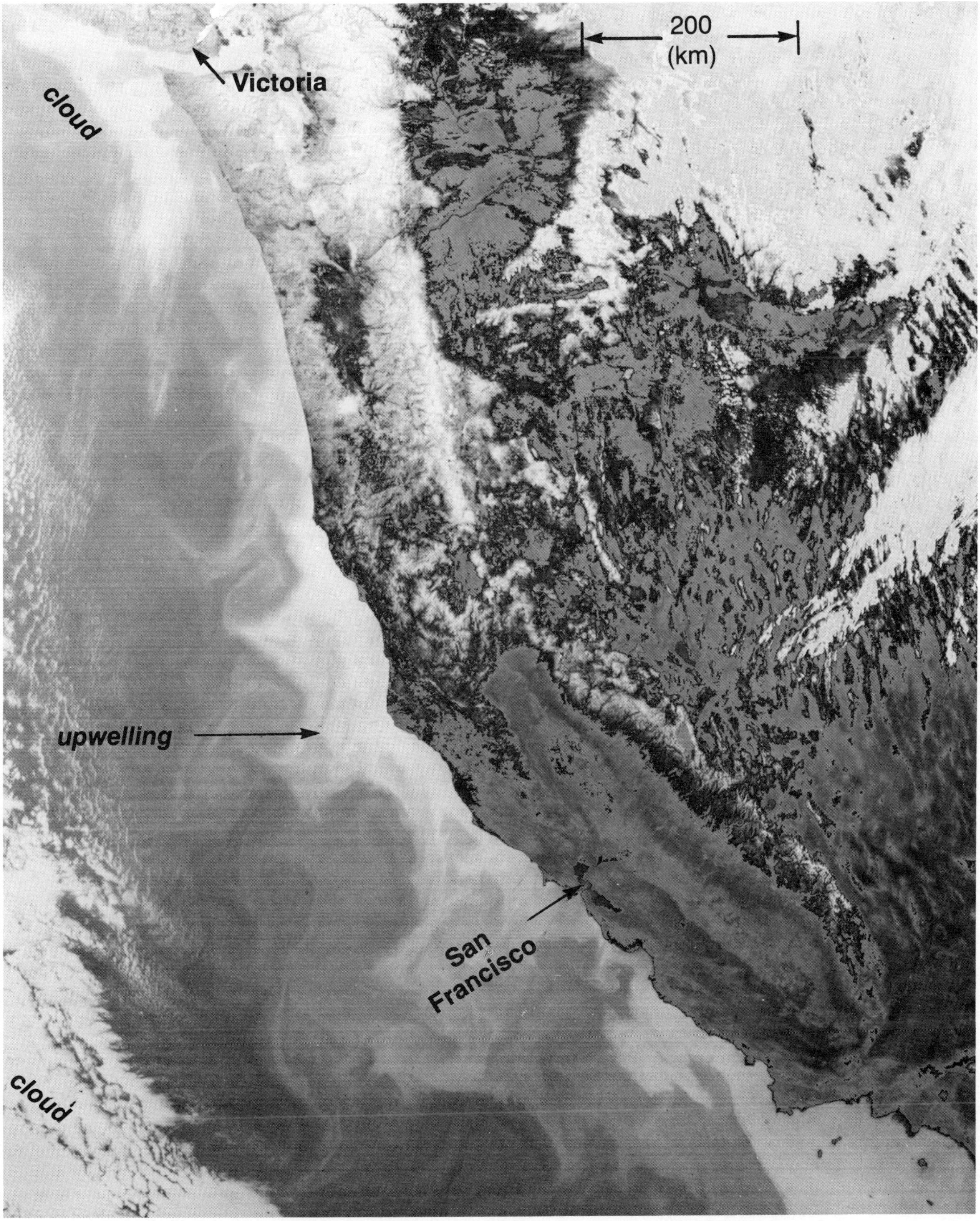

FIG. 5.5. NOAA satellite infrared scan of Pacific coast from Vancouver Island to southern California (Sept. 11, 1974, from 1500 km). Warmer areas darker. Relatively cold upwelled water adjacent to Oregon–California coast whitish compared to warmer water of midocean (See Fig. 5.1). Clouds and snow appear white, forests and land dark. (Courtesy S. Tabata)

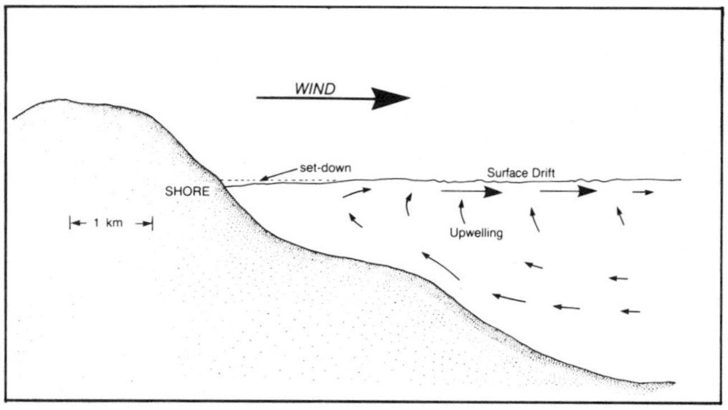

FIG. 5.6. Localized upwelling generated along a leeward shore. Set-down is small drop in mean sea level, result of net offshore transport of water.

water behind the point along with it, which in turn causes subsurface water to upwell.

Localized Effects

The most obvious feature of upwelling is the presence of colder than normal temperatures at the sea surface. Upwelled water along an open coast may have originated from as deep as 300 m so these temperatures may, in fact, be lower in summer, when upwelling is most intense, than in winter, when it is weaker or nonexistent (Fig. 5.7). One effect of this cold water is to produce sea fog where it comes into contact with moist warm air. Dense fogs along the Pacific coast of North America during summer are generally created this way and may persist for many days, regardless of wind strength. The comparatively cold surface water of Juan de Fuca Strait is also a cause of frequent sea fog, particularly during atmospheric inversions when stable air conditions exist near sea level. Daytime heating of the surrounding land, however, will usually dissipate the fog by late morning following the initiation of vertical air currents. In a sense, these fogs, too, are associated with upwelling processes — the tidal mixing in the passes of the San Juan and Gulf islands brings cold water to the surface in the first place.

Summertime recreational activity in the sheltered waters of British Columbia and Washington may be adversely affected by upwelling. One day the water may be pleasantly warm, the next day uncomfortably cool, even though daytime air temperatures are unchanged. Water temperatures along the beaches on the western side of the Strait of Georgia are often subject to rapid alterations in swimming conditions; conditions that can change abruptly within a matter of hours. Good examples of this are at Departure Bay and Piper's Beach near Nanaimo (Fig. 5.8) where the author noticed a deterioration of swimming conditions following a shift of local winds.

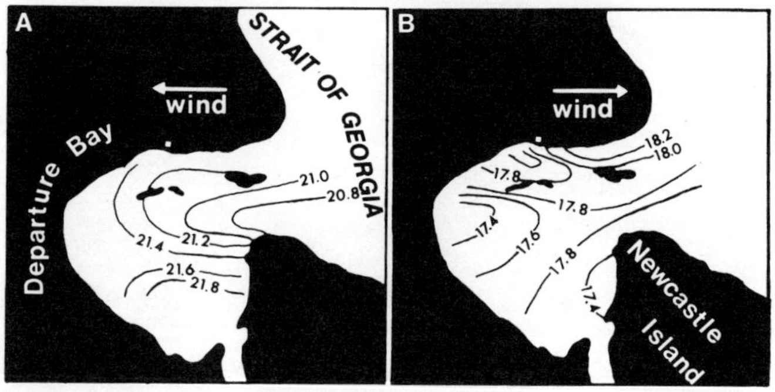

FIG. 5.8. Surface temperatures (°C) Departure Bay, Nanaimo. (A) July 4, 1968 during easterly winds; and (B) July 5, 1968 during westerly winds. Pacific Biological Station marked on north side of bay. (From Henry and Murty 1972)

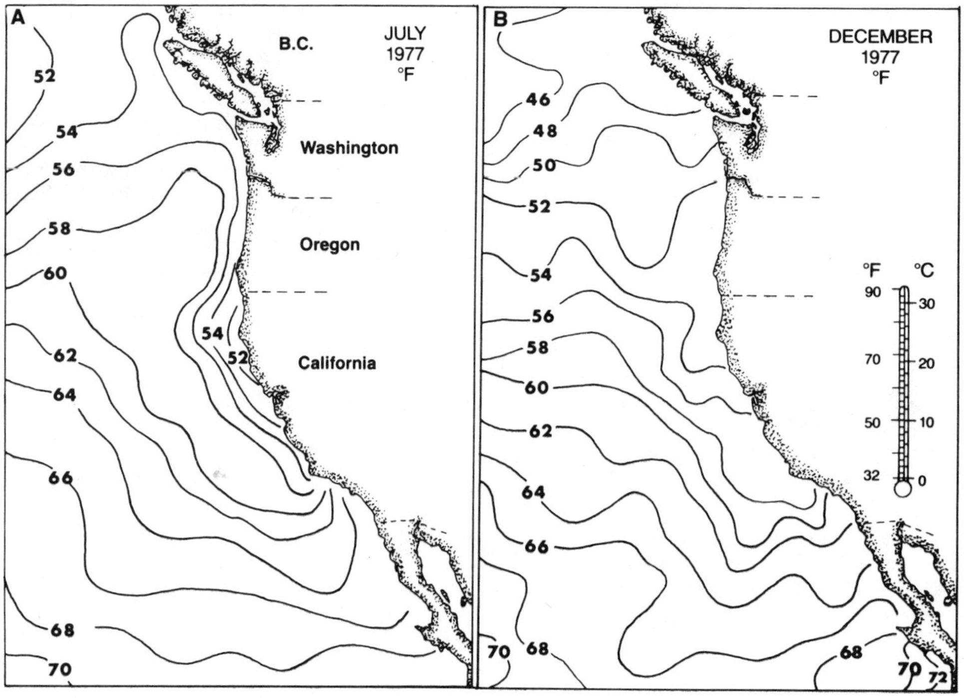

FIG. 5.7. Average sea surface temperatures (°F) off west coast North America, (A) July–15, and (B) Dec. 16–31, 1977. (From Natl. Mar. Fish. Serv. Bull. 1977)

More scientifically, Fig. 5.8a shows surface water temperatures for Departure Bay on July 4, 1968, during easterly (onshore) winds; Fig. 5.8b shows water temperatures a day later after the winds shifted to westerlies (offshore winds). The drop of almost 4C° in water temperatures in less than 24 h was caused by the upwelling of colder water from a depth of about 5 m to replace the warmer surface layer that was moved out of the bay by the westerly winds. The onset of afternoon westerly winds via the sea-breeze effect is a common occurrence over the Strait of Georgia region during summer and, consequently, a bather should not be surprised to find a drop in the water temperature at his or her favorite beach by late afternoon.

Climate

Another aspect of upwelling is its effect on coastal climates. In regions where the prevailing winds have an onshore component, the cold surface water will cool the air, and thereby moderate the coastal temperatures and humidity. Tofino, for example, has much lower summer temperatures and higher humidities than Kamloops. The contrast is even more striking in Oregon where the rather cool, moist, seaboard conditions can change to the sweltering dry heat of the countryside only a few kilometres inland. However, the reverse is true of the coastal area of Peru, which is flanked by one of the world's major upwelling regions on one side and by one of the worlds highest mountain chains (the Andes) on the other. In this case, the plantless sand desert at the hot, humid coast gives way to the pleasant warmth of the foothill grasslands less than 30 km inland. But the influence of upwelling is not confined to local climatic regimes. Recent studies indicate that changes in the intensity of the upwelling along the equator lead to alterations of the weather pattern over the whole northern hemisphere. Even remote places like Iceland and Siberia are affected by events at the equator.

Fishing Grounds

Probably the most directly important aspect of upwelling is that it acts like a "biological bump" that brings deep, nutrient-rich water upward to the sunlit surface layer. It is here that one-celled marine plants called phytoplankton use these nutrients, together with solar energy for photosynthesis, to grow into vast oceanic "crops." These in turn provide the basic food source for all marine animals from the simplest microscopic animals called zooplankton to the largest of living creatures, the blue whale. The oceanic food chain is of course, exactly analogous to the one on land where all animals depend, either directly or indirectly, on plants for their existence. In fact, the situations are so alike that aquatic animals that feed directly on the phytoplankton crop are said to be "grazing." By bringing large amounts of nutrients to the surface, upwelling encourages the growth of lush oceanic pastures where marine life can thrive in abundance. Fishing yields in these regions are at least a thousand times greater than in other oceanic regions. Although coastal upwelling areas comprise only 1/10 of 1% of the total area of the oceans, they are estimated to yield more than 1/2 the fish harvest. Even weak upwelling regions like the B.C. coast are noted for their ability to support great numbers of fish, especially within biological "hot spots" such as the entrance to Barkley Sound on the west coast of Vancouver Island.

Although upwelling occurs regularly year after year in specific areas, it does not continue nonstop throughout the whole of the upwelling period. After all, the prevailing coastal winds often reverse direction for periods of days or weeks due to changes in the weather patterns. Thus, despite the fact that the summer winds along the Pacific shores of North America are usually from the northwest, they can often shift and blow from the south. When this happens, the winds no longer move the surface water offshore and upwelling comes to a halt. So does the unfettered growth of phytoplankton. As a consequence, fish that were feeding in the rich upwelling ecosystem disperse elsewhere in search of food or, if the stoppage of upwelling occurs suddenly, die of starvation. This is followed by a sharp drop in fish catches. Although fishing fleets along the Pacific coast from California to Alaska are often affected by such events, their losses are nowhere as drastic as those that befall the Peruvian fishing industry.

El Niño

Although seasonal off the B.C. coast, along the Peruvian coast upwelling essentially occurs throughout the year. The exception is during February and March when upwelling ceases off northern Peru and a tongue of the warm eastward-flowing equatorial and countercurrent pushes southward to about 6°S. Called El Niño, the Christ Child, because it occurs in the Christmas season, this event is not usually of dire consequence to the rest of the Peruvian coast. On the average of every 7 yr, however, there is a general break-down of the trade winds system over the equatorial Pacific Ocean, which leads to a weakening of the westward flowing South Equatorial Current and northward flowing Peru Current (Fig. 5.3). Within a period of a few months, this in turn results in a cessation of upwelling along the entire coast of Peru, and to a southward penetration of warm, low-nutrient equatorial water as far as 12°S. This extreme form of El Niño can lead to the catastrophic mortality of phytoplankton and fish or, at best, forces the fish to scatter offshore or into deeper water in search of food, with disastrous consequences for fishermen and sea birds alike. Recorded by fishermen for 180 yr, the most recent major El Niño events were in 1965, 1972–73, and 1976. It appears that an El Niño type event also occurred in 1982–83 and influenced the weather and oceanography of the Equatorial region of the Pacific Ocean. In 1970 when this effect did not take place, the Peruvian waters yielded 22% (12.3 million t) of the total world fish catch, mostly anchovies. In 1972–73, the catch plummeted to only 4.7 million t, a catastrophe for Peru. To add to the problems, the scarcity of accessible fish concentrations near the coast also leads to the mass starvation of sea birds and, therefore, to depletion of the Peruvian foreign monetary reserve, which depends heavily on the export of fertilizer produced from the excreta (guano) from these birds. Moreover, the rotting phytoplankton on the beaches produce hydrogen sulphide gas, which combines with the moist sea breezes to blacken the paint on ships. Dubbed the "Callao Painter," this occurs each year but is most intense during a catastrophic El Niño. Lastly, there are often torrential rains associated with the cessation of upwelling that produce severe flooding and crop damage. Considering all these disasters, it is easy to sympathize with Peru's 200-mile limit and the government's tenacious desire to protect the fish stocks from foreign fishermen.

Coastal Currents

A recent investigation off Oregon and Washington by American oceanographers has shown that upwelling leads to an intensification of the southward coastal current. Known as the California Current, this southward flow is usually broad and slow, but during sustained upwelling the portion of the current within 20–40 km of the coast accelerates into a narrow "jet," which can attain speeds in the neighborhood of 100 cm/s (Fig. 5.4). Off the coasts of British Columbia and Washington, upwelling is much less intense than off Oregon so the coastal "jet" should be proportionally weaker. From this discussion, it is obvious that sail boats journeying southward off these coasts in summer will generally have the best of conditions — winds from the northwest and southward setting currents — while northbound boats will be tacking against opposing ocean currents. Just to complete the picture, the investigation off Oregon further showed that the upwelling also induces a slow northward flow at depth, to compensate for the southward flow near the surface. Similar conditions are generated in other upwelling regions in the world although not all have been investigated in great detail.

PART III
OCEAN WAVES

Chapter 6. Nature of Ocean Waves

The study of ocean waves is one of the most challenging and sophisticated branches of oceanography. Studies originated in the early 19th century with pioneering work on surface waves, and now encompass research on the generation, propagation, and decay of a diversity of wave types. Yet, despite the considerable advances in understanding, many fundamental questions related to ocean waves remain unanswered. It is obvious, for example, that winds generate waves on the surface of the ocean, but precisely how is still open to considerable debate.

Terminology

All wavelike variations in the ocean arise in the same basic way. To begin with, some mechanism such as wind or a boat disturbs the water from its equilibrium or stable, undisturbed state (Fig. 6.1). Then, a restoring mechanism such as the gravitational attraction of the earth takes over, and forces the water back toward its equilibrium position, which it overshoots because of its inertia, to form a distortion in the opposite sense. Again it is forced back toward its equilibrium level, but again inertia causes it to overshoot, so that the whole process repeats itself at regular intervals of time. (A swinging pendulum undergoes a similar kind of action.) The net result is a propagating disturbance or signal called a wave.

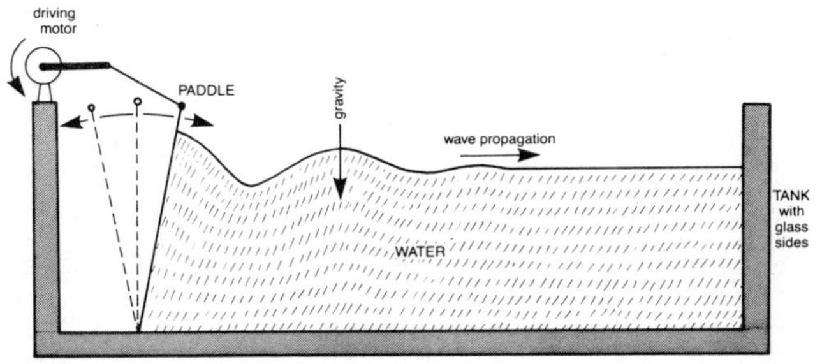

FIG. 6.1. Generation of surface gravity waves in a wave tank. Motorized paddle is "disturbing" mechanism and gravity "restoring" mechanism.

Characteristics of a wave depend on three factors: (a) the type of disturbance initially applied to the water and whether it is continually applied to produce a forced wave or whether it is quickly removed to allow the wave to propagate away as a free wave; (b) the type of restoring mechanism that forces the water back to equilibrium; and (c) the properties of the water itself. Free surface waves where gravity is the restoring mechanism will be principally discussed.

Figure 6.2 defines parts of a wave. A wave period is the time that it takes two successive crests to pass a fixed point. Wave speed is the speed of a wave relative to a fixed point; also called the wave celerity or phase speed.

In addition, it is useful to distinguish between deep, intermediate, and shallow waters, compared to the hori-

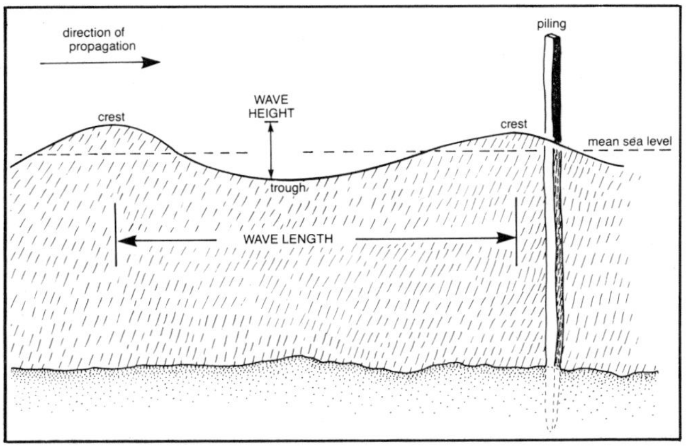

FIG. 6.2. General wave terminology. Wave period is time it takes one wavelength to pass a fixed point. Wave speed (phase speed or celerity) measures rate of travel over bottom.

zontal separation between adjacent crests. If the total water depth exceeds ¼ of the wavelength, as in Fig. 6.3a, the water is said to be deep and the waves are deepwater waves or short waves (shortening the waves has the same effect on their properties as deepening the water); waves actively generated by the wind are usually this kind. On the other hand, if the water depth is between ¼ and 1/20 the wavelength, the waves are intermediate-water waves; swell at the continental shelf from the open ocean typically falls into this category. Finally, if the water depth is less than 1/20 of the wavelength (Fig. 6.3b) the water is shallow and the waves are shallow-water waves or long waves; tides and tsunamis are always this kind as their wavelengths exceed hundreds of kilometres whereas the ocean depth never exceeds 13 km.

Deepwater waves are unaffected by the bottom, so wave motions are identical to those in infinitely deep water. Intermediate-water waves, on the other hand, are partly affected by the bottom topography and shallow-water waves are strongly influenced by variations in depth. The following table provides a summary of the depth range valid for each type of wave (D is mean depth of water, L is wavelength in deep water). It should be pointed out that the choice of ¼, as the depth-to-wavelength ratio at which waves change from deep to intermediate type, is somewhat arbitrary. It is based on the fact that the actual wave speed at this depth differs from that in deep water by less than 5%, the maximum acceptable error permitted by most oceanographers when categorizing the wave types. At still shallower depths, this difference increases and it becomes unacceptable to use wave speeds based on deepwater formulae. Some oceanographers prefer to place the transition at a ratio of depth to wavelength = ½, for which the error in actual wave speed is only 0.2% of the deepwater value. However, this is much too stringent a requirement for most practical purposes. The choice of 1/20, for the depth-to-wavelength ratio that marks the transition from intermediate to shallow-water waves, gives a

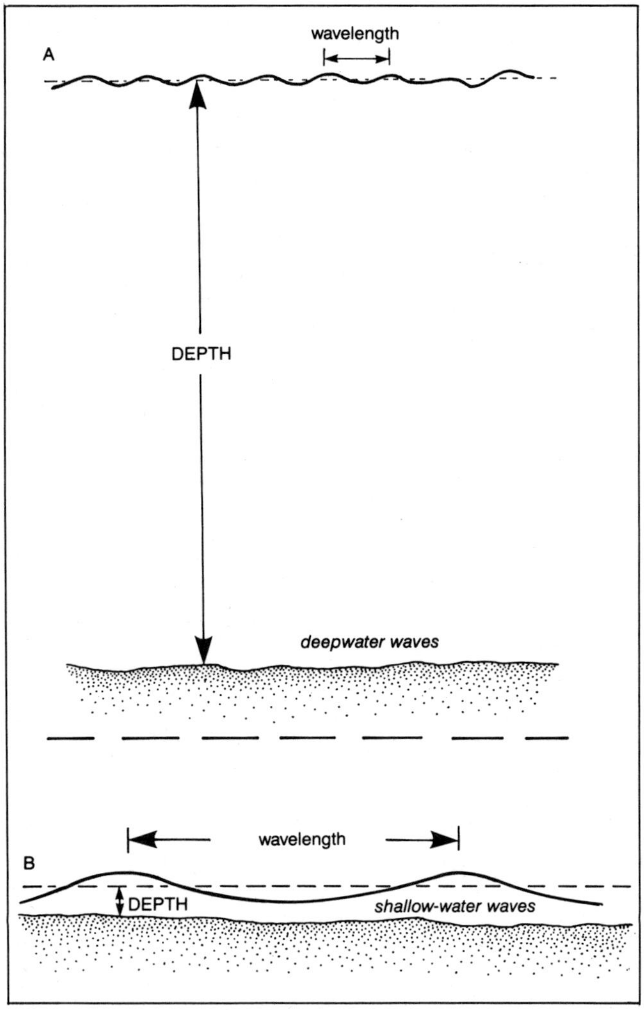

FIG. 6.3. (A) Deepwater or short waves versus (B) shallow-water or long waves.

phase speed error of 2%. For still shallower depths, the relationships for shallow-water waves provide a more than acceptably accurate description of the wave properties.

Feature	Deep water	Intermediate water	Shallow water
Applicability range	D > ¼ L	¼ L > D > 1/20 L	D < 1/20 L
Examples	sea and swell in open ocean	long swell over shelf, sea outside surf zone	tsunamis, tides, swell nearshore

Water Motion

To anyone familiar with the sea, it is obvious that water doesn't move at the speed of waves. If it did, marine travel would be impossible. For example, the Alaska earthquake of Mar. 27, 1964, generated very long waves (tsunamis) that traveled over the Pacific Ocean at speeds approaching 850 m/s (440 kn). It is not too difficult to imagine what would happen to a ship if the water too were moving at that speed.

To see how the water really moves, watch a floating object like a wood chip. As the wave crest approaches, the chip moves forward, pauses as the trough approaches, then moves backward to nearly its original position as the trough moves by. At the same time the wood chip is moving back and forth, it is also moving up and down with the water level. For deepwater waves, this combined motion makes the chip, and the water, move in a nearly circular path (Fig. 6.4a). (Actually the circles are not completely closed as the water drifts ever so slowly in the direction of wave travel to produce the Stokes' drift mentioned in Chapter 4.) Observations below the surface would show that the diameters of these circles decreased with depth. At a depth equal to ½ the wavelength, the diameters would only be 1/23 of their surface values, and at

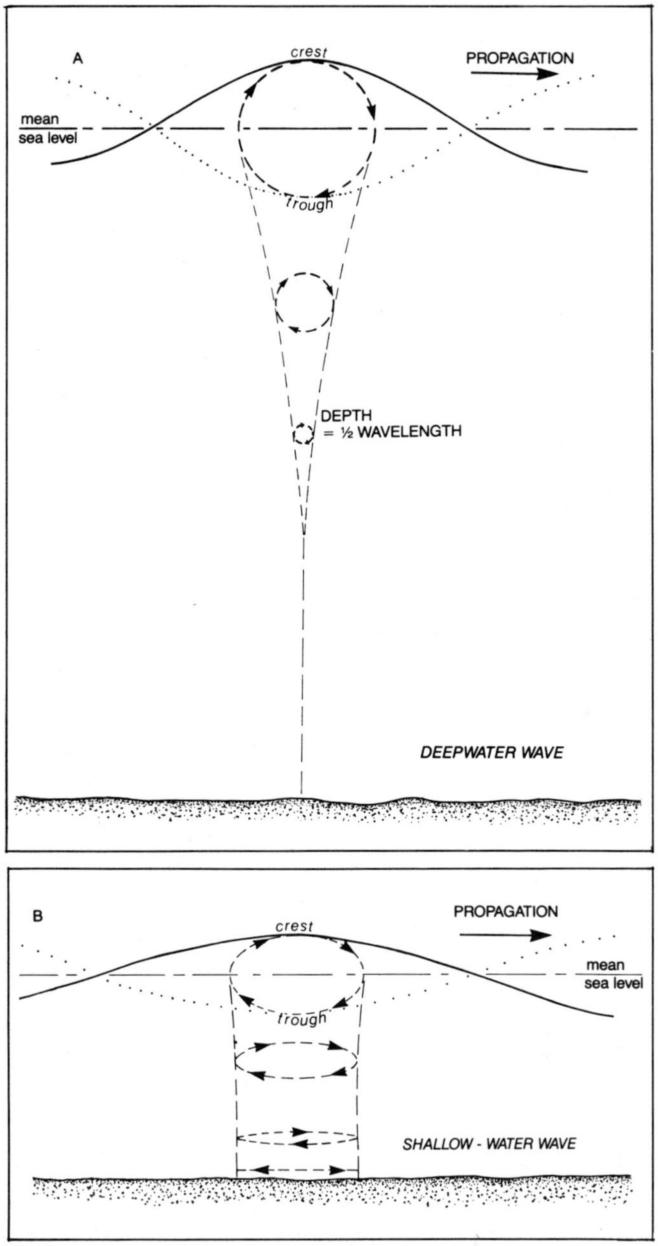

FIG. 6.4. (A) Circular paths followed by water particles at different depths as one complete deepwater wave moves left to right. Orbital diameters decrease rapidly with depth. Motions are forward under crest and backward under trough. (B) Elliptical paths traced out by water particles at different depths during passage of one complete shallow-water wave. Strictly back and forth motions at bottom. In both diagrams only crest and trough of wave form are shown.

a depth of one wavelength, only 1/535 of their surface values. It's not surprising that a ride below the waves on a submarine is so smooth.

Not only does the water not travel with the waves, but its circling speed is much less than the wave speed. An approximate formula shows

water speed ÷ wave speed = 3 × (wave height ÷ wave length).

Therefore, the water directly under a wave 1 m high and 90 m long will have circling water speeds of 1/30 the wave speed at most. Because a wave this long would travel 12 m/s in deep water, the water itself would only circle around at 0.4 m/s (about 0.8 kn).

The situation is somewhat different for intermediate-water waves as the bottom restricts the up and down motion of the water and the circles become flattened into ellipses. Vertical motions associated with these waves decrease uniformly with depth until they become zero at the bottom. For shallow-water waves (Fig. 6.4b), the ellipses essentially flatten to straight lines, so that motions are almost entirely in the horizontal plane. Furthermore, the horizontal motions decrease only slightly from top to bottom, contrary to their counterparts in deepwater waves. This is why currents associated with the tide, a decidedly shallow-water wave, are almost uniform strength from top to bottom in the open ocean.

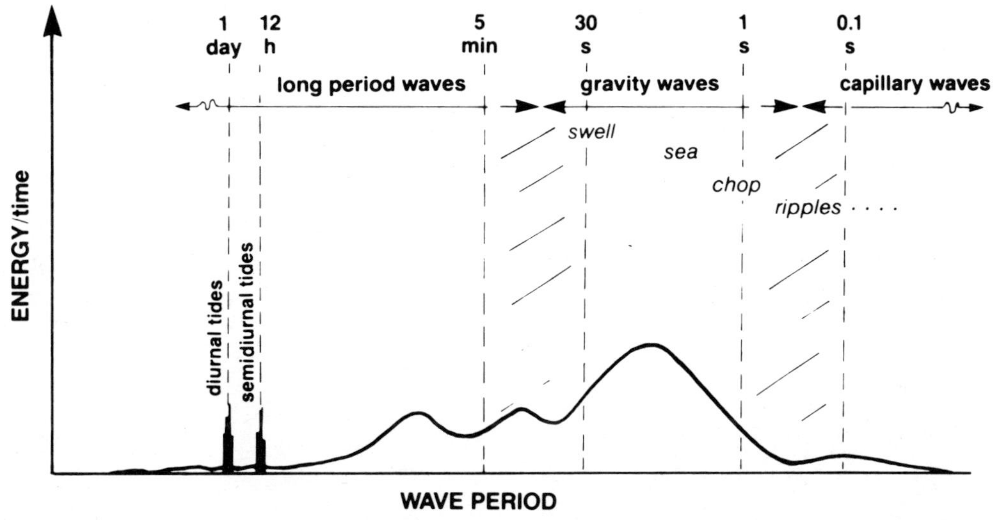

FIG. 6.5. Power spectrum. Plot of relative amount of energy contained by waves over a range of wave periods for particular time and location. Broken lines show overlap between wave types. Tidal energy is concentrated within a narrow band of periods near diurnal and semidiurnal periods. (After Kinsman 1965)

Classification

Besides the distinction between deep, intermediate, and shallow there are other ways to classify ocean waves. A useful classification is by period, which can range from fractions of a second for ripples to 12 h and longer for tides. The waves are then placed into groups with similar characteristics. The custom at present is to plot the period of the wave versus the relative amount of energy contained by waves at each specific wave period (Fig. 6.5). Such a plot is called a power spectrum because it indicates the amount of energy contained within various segments of a spectrum of wave periods, during the time that the particular observations were made. Alternatively, Table 6.1 shows the various wave types and the primary generating mechanism and primary restoring mechanism.

TABLE 6.1. Basic types of surface waves with approximate range of periods and wavelengths. Also primary generating and restoring mechanisms for each type of wave.

Name	Periods	Wavelengths	Generating Mechanism	Restoring Mechanism
Capillary waves (ripples, wavelets)	less than 0.1 s	less than 2 cm	wind, pressure fluctuations	surface tension
Gravity waves (chop, sea, swell)	0.5–30 s	10 cm–1000 m	wind	gravity
Infragravity waves	minutes	hundreds of metres to hundreds of kilometres	storm systems (winds and atmospheric pressure gradients)	gravity
Tsunamis	tens of minutes to 1 h	hundreds of kilometres	submarine earthquakes, shoreline slumping	gravity
Tides	mainly 12½ and 25 h	thousands of kilometres	gravitational attraction of sun and moon	gravity and Coriolis force

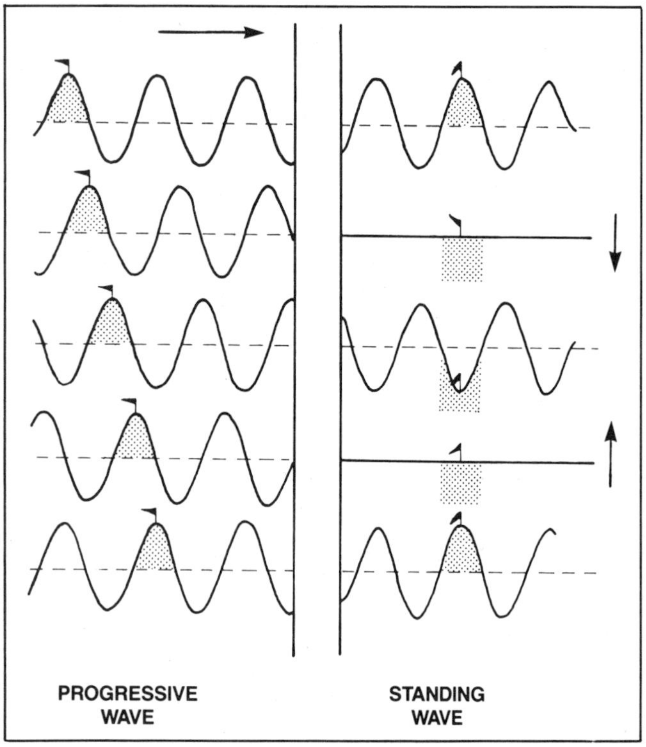

FIG. 6.6. Progressive waves versus standing waves. In standing waves, water oscillates up and down at fixed locations and there is no advance of the wave form as in progressive wave.

FIG. 6.7. Westerly swell at Shell exploratory drill rig 35 km west of Estevan Point, Pacific coast Vancouver Island, December 1968. Vessel about 61 m (200 ft) long and distance between vertical supports approximately 85 m (280 ft). Swells have wavelengths about 60–100 m and heights 2–3 m. (Courtesy R. H. Herlinveaux)

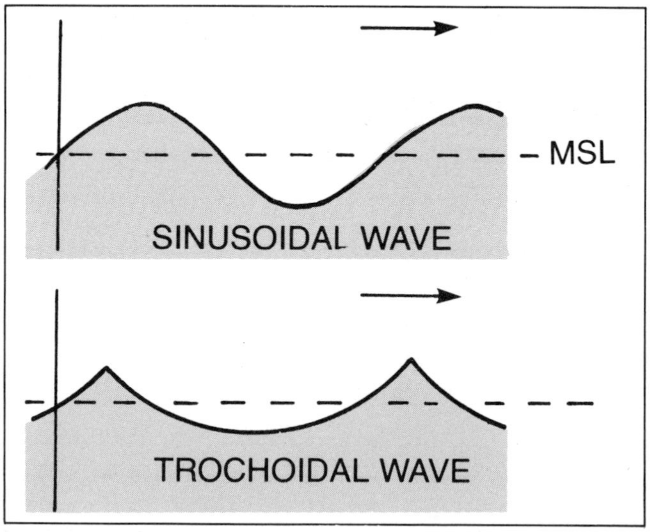

FIG. 6.8. Two possible forms for surface gravity waves. These are idealized; most wave shapes lie somewhere in between. (MSL, mean sea level)

FIG. 6.9. A hydraulic jump or bore on gradually shoaling, sandy beach. Leading edge of incoming wave steepens as it propagates against backrush of previous waves and eventually collapses. Laminar flow in front, and highly turbulent waters immediately behind leading wall of water. (Photo by author)

Other classifications include progressive waves (waves that propagate); and standing waves (nonpropagating waves formed when two identical waves traveling in opposite directions move through one another). Figure 6.6 illustrates how they differ. A further distinction is made between surface waves (which attain their maximum amplitude at the air–sea boundary), and internal waves (which attain their maximum amplitude within the interior of a body of water and produce but negligible distortion of the surface). Surface wave shapes can vary from sinusoidal forms typical of swell (Fig. 6.7) to trochoidal forms typical of chop or seas (Fig. 6.8), to steep-faced bores or hydraulic jumps characterized by waves on an opposing current and nearshore breakers (Fig. 6.9). Similar shapes are attained by internal waves as verified by laboratory experiments and numerous oceanographic measurements.

Oceanographers also like to distinguish between waves of the first class, which do not depend on the earth's rotation for their existence, and waves of the second class, which would cease to exist if the earth were to stop rotating about its axis. Gravity is the main restoring force for first-class waves; the main restoring mechanisms for second-class waves are the Coriolis force and the slope of the sea floor. Among first-class waves are wind-waves, swell, internal gravity waves, tsunamis, and the tides. In contrast, second-class waves are not directly visible occurrences because of their long periods (days to years), low heights (centimetres), and very extensive wavelengths (tens to thousands of kilometres). Moreover, the water motions associated with these waves are predominantly time-varying current systems, rather than vertical variations in sea level. The presence of these wavelike back and forth motions is largely responsible for the slow cyclic reversals in currents often observed in the oceans (Fig. 6.10).

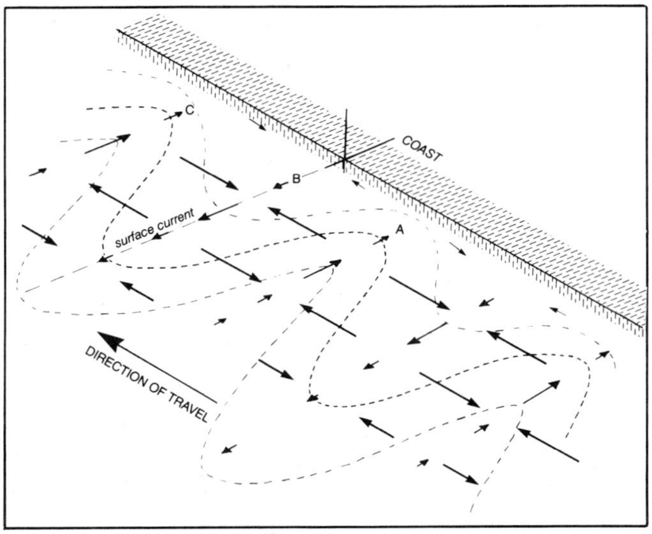

FIG. 6.10. A propagating system of surface currents. Alternating pattern of onshore–offshore currents is wavelike. Pattern as whole moves parallel to coast. If the cyclic period of the current oscillation were 3 days, then after 1½ days, onshore flow at A would have moved to B, where flow would be onshore; offshore flow at B would have moved to C. After 3 days, onshore flow regime from A would have advanced to C, completing cycle.

Wave Growth

The hopes of the becalmed yachtsman waiting impatiently for the wind usually are heightened by the sight of "cat's paws" rippling the otherwise smooth surface of the water. Apparently produced by wind speeds of ¼–1 m/s (½–2 kn), these ripples have short wavelengths of less than 10 cm in which both surface tension and the earth's gravity act as restoring forces for the motions; for lengths less than about 1.5 cm, the ripples are dominated by surface tension alone and are called capillary waves. If the wind dies, friction dampens the wavelets and they quickly disappear. If the wind continues to blow, however, the ripples grow until, for wavelengths over 10 cm they have evolved into surface gravity waves. The earth's gravitational attraction provides the primary restoring mechanism for such waves, and surface tension is negligible. As the wind increases in strength, the waves grow higher and longer, and begin to move at greater speeds. Eventually, the wind will so saturate the sea with energy that the waves will begin to break and the roughness of the surface will reach a limiting state called "a fully developed sea." This maximum limit depends on three factors: the wind's force, or speed; the wind's duration, or time it has been blowing at a particular speed; and the fetch, or unobstructed distance over which the wind has been blowing in the same direction at the same speed. For example, it is believed that a minimum duration of 42 h is required for a 20 m/s (40 kn) wind blowing over a 1300–km (700 nm) fetch to produce a fully developed sea with average wave heights of 8.5 m (28 ft). Table 6.2 shows other examples.

TABLE 6.2. Minimum fetch and duration to produce fully developed seas at various wind speeds. Significant height is average height of highest ⅓ of observed waves. Maximum probable waveheight is about 1.8 × significant height.

Wind speed		Fetch		Duration	Avg. height		Significant height		Avg. highest 10% waves		Period greatest energy concentration
(*kn*)	(*m/s*)	(*nm*)	(*km*)	(*h*)	(*ft*)	(*m*)	(*ft*)	(*m*)	(*ft*)	(*m*)	(*s*)
10	5	10	19	2.4	0.9	0.3	1.4	0.4	1.8	0.6	4
15	8	34	63	6	2.5	0.8	3.5	1.1	5	1.5	6
20	11	75	139	10	5	1.5	8	2.4	10	3.1	8
25	13	160	296	16	9	2.7	14	4.3	18	5.5	10
30	16	280	518	23	14	4.3	22	6.7	28	8.4	12
40	22	710	1315	42	28	8.5	44	13.4	57	17.4	16
50	27	1420	2630	69	48	14.6	78	23.8	99	30.2	20

Waves within restricted regions like the Strait of Georgia are limited by the wind strength and by the fetch; maximum wave heights in this case are around 2.7 m even during the winter, with average heights of only 0.6 m. These maximum waves occur during winds from the northwest and southeast, as both the force and fetch of the winds are greatest in these directions.

Although theory shows that waves begin to break when the ratio of their wave height to wavelength reaches 1/7 (Fig. 6.11), waves in the open ocean seldom get as high as ⅒ their lengths, and ratios 1/15 to 1/50 occur most

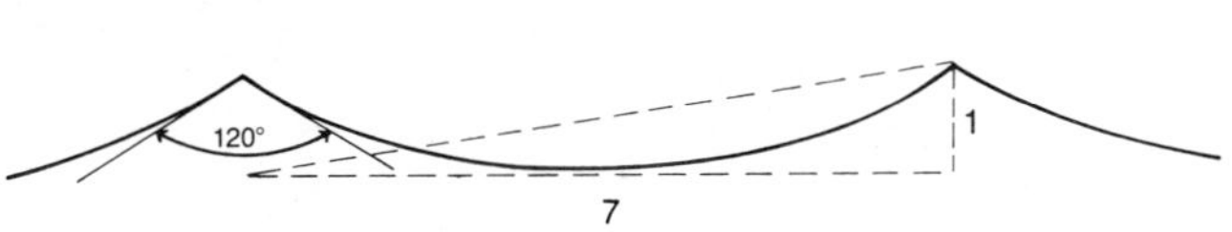

FIG. 6.11. Maximum ratio 1:7 height to length that a surface gravity wave can sustain before it begins to collapse at crest. Ratio corresponds to interior angle of 120° at wave crest.

often, even in rough weather. Generally, waves break because their tops get blown off or because they are propagating against a current. This current may be an ocean current, circling motions associated with the presence of much longer waves, or a tidal current.

Sea and Swell

The gravity waves worked on by wind in the generating area are called sea, whereas those that have escaped the generating area are called swell (Fig. 6.12). Sea is shorter,

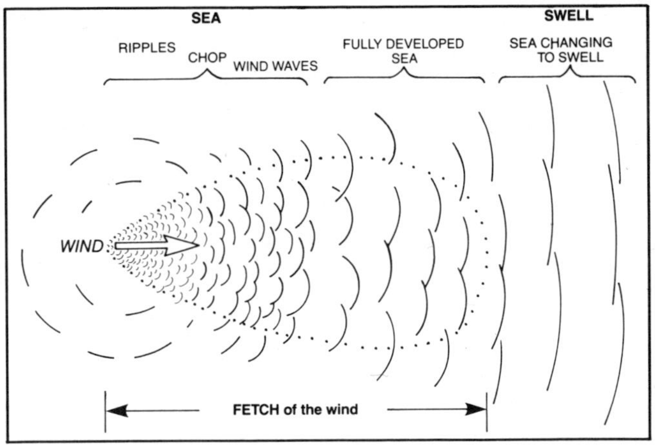

FIG. 6.12. Development of wave regimes during storm. Winds are confined to fetch delineated by dotted line. Sea changes to swell outside storm area. (Modified after Bascom 1964)

steeper, more rugged, and of more concern to the average mariner, particularly when the lengths of the waves are nearly the same as his boat. Sea is also much more confused; waves run off in all directions, some even propagate against the wind! How, then, do the smooth, regular waves associated with swell come about?

This question brings to light an important facet of waves in deep water: the longer the wave, the faster it travels. Waves from the wake of a passing ship, for example, arrive in order of their length, longer ones first, followed by the next longest and so on. This is called dispersion, because waves of different lengths tend to move away from one another, or "disperse." Therefore, the longer waves will lead the way out of a storm area. Because these waves are also less easily damped by friction than shorter waves, and shorter waves tend to transfer their energy to longer and longer waves, it's no surprise that the longer swell is the final product of sea. In the Pacific Ocean, swell has typical maximum wavelengths of around 200 m and is usually never shorter than 30 m unless the winds have recently blown strongly for a brief period in a nearby area. The corresponding speeds for such typical swell lengths range from 7.5 m/s (15 kn) to 30 m/s (60 kn), while the time it takes an individual wave to pass a given point (the wave period) ranges from 5 to 20 s. These results follow immediately from the fact that once the wave speed, or the wavelength, or the wave period is known the other two values can be calculated, as in Table

TABLE 6.3. Approximate wavelengths (L) and wave speeds (C) of low-amplitude sinusoidal waves in deep water. For a specified wave period (T) use simple relationship $L = (g/2\pi)T^2$ and $C = L/T = \sqrt{gL/2\pi}$, where g is acceleration of earth's gravity. Using values of g found in Appendix A, formula for wavelength becomes $L = 5.16\ T^2$ ft or $L = 1.56\ T^2$ m. At depth (D) each wave becomes a shallow-water wave.

Period	Wavelength		Wave speed			$D = L/20$ Depth
(s)	(m)	(ft)	(m/s)	(ft/s)	(kn)	(m)
5	39	128	7.8	25.5	15	2
10	156	512	15.6	51.1	30	8
15	351	1161	23.4	77.1	46	18
20	624	2048	31.2	102.1	61	31
25	975	3200	40.0	128.0	76	49

6.3. This interrelationship holds equally for sea waves generated by winds in the deep ocean. However, there is no basic formula for the height of a wave, and only after thousands of observations over many years has it been possible to obtain generalized information about the relationship of height to other characteristic properties of a wave. As might be expected, the length and period of an average wave in deep water tend to increase as height increases, which in turn is directly related to the strength, duration, and fetch of the wind. Unfortunately, such data are necessarily restricted to fully developed seas simply because these alone represent a natural limiting state that is always the same throughout the oceans of the world. This is not the case for swells. Their properties at a particular time and place depend heavily on the history of their birth and factors that have influenced them during their advance across the ocean.

Because of the greater fetch, Pacific Ocean swell is typically longer than Atlantic Ocean swell, although the maximum recorded wavelength of 820 m was measured in the equatorial Atlantic. Because of its great length, swell can cross great stretches of the ocean with little energy loss; swell generated in the Antarctic Ocean has been detected on the Alaska coast. Within shallowing regions, on the other hand, swell quickly loses energy. Once it enters Juan de Fuca Strait, for example, its amplitude begins to diminish rapidly and by the time it is opposite Victoria it has nearly disappeared. Swell, like all gravity waves grows in amplitude and decreases in wavelength over rapidly shoaling regions before it breaks as surf on the shore. (Surfers, by the way, aren't "carried along" by the wave, they slide down it like a skier on a hill. The surfer doesn't lose altitude because the wave front is continually moving up as he slides down!)

Group Speed

Another important aspect of wave propagation is group speed. Contrary to common belief, each wave is not an individual that propagates alone. It is always part of a group or "train" of similar waves, whose extent is limited simply because what produced them in the first place had a limited extent. A storm, for example, only blows over a confined area of the ocean. Often the grouplike nature of the wave train goes unnoticed because of the confusion of surface motions in open waters, or because the extent of the group is so large that its beginning or end cannot be determined from the mariner's vantage point. However, there are instances when wave groups are quite noticeable. Ripples made by a pebble cast into a still pond move outward as a distinct train of concentric wavelets. Boaters within Juan de Fuca Strait or Queen Charlotte Strait on relatively calm days will often note the passage of groups of 10 or so long, low undulations associated with swell wave trains entering from the Pacific Ocean. Or, consider the wake made by a passing motor boat with a displacement hull and suppose the wake approaches a drifting dinghy. First, the advancing wake consists of a group of individual surface gravity waves, usually few enough in number to be counted. But concentration on one particular wave in the group shows it eventually propagates up to the leading edge of the group and disappears (Fig. 6.13).

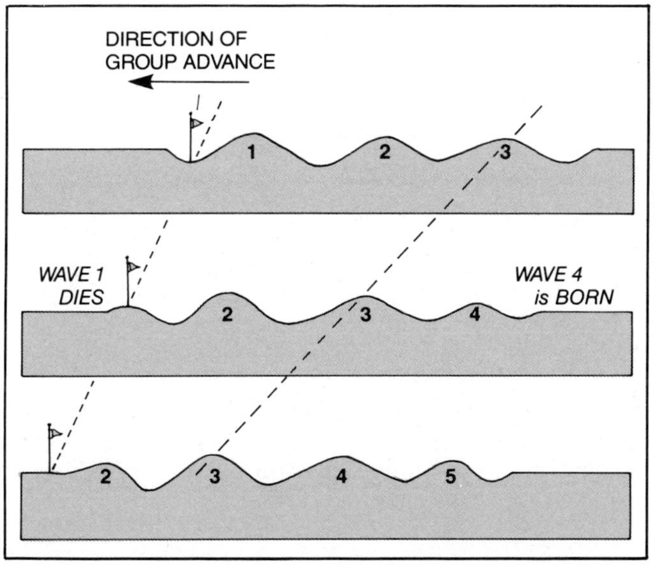

FIG. 6.13. Progression of train (or packet) of deepwater gravity waves over sea surface. Advancing flag is measure of group velocity, shown also by slight tilt of line drawn through flag's position. The wave (or phase) velocity is measured by advance of wave 3 and sharply tilted line drawn through it. In deep water, group moves at half rate of individual wave.

In fact, each successive wave does the same thing: it is "born" at the rear of the advancing group, propagates through it, then "dies" at the front. An individual wave only exists for the time it takes it to move from the back to the front of the group. This means that an individual wave travels more quickly than the group as a whole. That is, the group speed is less than the wave speed. For deepwater waves, group speed is exactly half as fast. Thus, the energy of the waves is carried toward the dinghy at half the speed of the individual waves. (As the water shallows, however, the group speed begins to keep pace with the waves, and the two become equal for shallow-water waves.)

Capillary waves, it should be pointed out, behave somewhat differently from surface gravity waves. The influence of surface tension causes the shorter waves to travel faster than the longer ones and the waves are said to experience anomalous dispersion. As an example, the wake generated by a fishing line has a different structure from that generated by a boat in that the shorter waves, rather than the longer waves, form the leading edge of the wake. Moreover, the group speed exceeds the wave speed for capillary waves, although this is almost impossible to see in every-day experience as the wavelets only propagate a few centimetres before being destroyed by friction.

Freak Waves

Freak or giant waves occur every so often when the various randomly moving groups of waves add up, or "get into step" to produce a mountain of water. Their life is short, only a minute or two, as the waves soon get out of step again. As a consequence, they are totally unpredictable and can come out of nowhere. The compilation of 40,164 extracts from ship's logs (below) by Schumacher (1939) shows how rare these giant waves really are; only 10% of all waves were over 6 m high.

Wave height (ft)	0.3	3–4	4–7	7–12	12–20	over 20
(m)	0–0.9	0.9–1.2	1.2–2.1	2.1–3.7	3.7–6.1	over 6
Frequency of occurrence (%)	20	25	20	15	10	10

In fact, statistical theory indicates that only 1 wave in 23 is twice the height of the average, 1 in 1175 is over 3 times the average, and only 1 in over 300,000 exceeds 4 times the average height. Nevertheless, freak waves 20–30 m high have been reliably recorded. One wave, recorded by the officers of the U.S.S. *Ramapo* in the North Pacific in 1933, was estimated to be 34 m (112 ft) high. Observations taken for a number of years at Ocean Weather Station P, 1500 km off the west coast of British Columbia, indicate that 20-m (65-ft) waves are not uncommon in the North Pacific, especially during November through February.

During the stormy night of the 1976 Swiftsure Yacht Classic the Columbia 30 *Native Dancer* was struck by three devastating waves as she sailed westward off Bonilla Point on the coast of Vancouver Island. Her skipper and one crewman were thrown overboard as the waves crashed over the deck. In the nightmare that followed, skipper William Willard was drowned. The crewman, through a combination of luck and resolve, managed to ride the breakers to shore, while the remainder of the crew stayed with the disabled boat until her eventual beaching near Carmanah Creek. William Willard's body was found 2 km northwest of Carmanah Point 1 mo later. The tragedy marked the first loss of life in the history of the Swiftsure.

Although these cresting waves may have been large wind waves generated by the gale-force winds, there is the possibility they were freak waves that originated far out to sea during a distant storm. Many vessels have been swamped by large waves off the southwest coast of Vancouver Island, and not all during storm conditions. On July 29, 1975, for example, a hydrographic launch, the *Barracuda*, was hit broadside by a huge breaking wave off Barkley Sound. According to the official report, it was one of the calmest days of the season with only a 2-m swell, yet the boat was suddenly broached in about 20 m of water by a large cresting wave. It is possible, of course, that the launch had actually ventured unknowingly over a nearby reef, which amplified the normal swell to breaker height. On the other hand, the survey crew may have been dumped into the ocean by a very long swell that originated well out in the open ocean. Waves of this nature are ocean-wide phenomena; large winter storm waves generated in the South Pacific, for instance, are observed at the California coast after propagating for many days across thousands of kilometres of ocean. If *Native Dancer* was indeed the victim of the chance arrival of a group of long, low swells generated far out in the Pacific days before, she may have actually been in water as deep as 30 m when she was hit, and more than 2 km offshore.

Giant waves are also generated through the interaction of large swell with intense ocean currents. Such conditions frequently arise along the western margins of most of the world oceans where strong, narrow western boundary currents with speeds of 1–3 m/s flow parallel to the coast over the continental slope. Examples of these currents include the northeastward flowing Gulf Stream off the east coast of the United States, the northeastward flowing Kuroshio off the coast of Japan, and the southward flowing Agulhas Current off the southeast coast of Africa. On the eastern sides of the world oceans, conditions are less favorable to the formation of these giant waves. Eastern boundary currents such as those off British Columbia and Washington are broad, of variable direction, and weak, with typical speeds of less than 0.5 m/s (1 kn).

As with rips (next section), the height and steepness of large storm-generated waves that propagate against an opposing flow increase by as much as 25%. Once the waves enter the current regime they may become trapped and thereafter tend to propagate within a comparatively narrow band centered about the region of maximum flow. The cross-stream variations in the current speed behave as a "lens" that focuses the wave energy. At the edges of this confined band the front face of the waves steepens and, depending on the strength of the current and the height of the incoming swell, the wave may undergo a further amplification by as much as a factor of 4. The results can be disastrous to even the largest ships. Of 11 recent marine disasters off the southeast coast of Africa, for example, all but one could be linked to giant waves formed by the interaction of storm-generated swell from the southern Atlantic Ocean and the opposing Agulhas Current; in most instances matters were aggravated by locally generated seas. Freak waves in this area have been reported with heights in excess of 18 m and, as is common in the ocean, were associated with a long, deep trough, known as "a hole in the sea," in advance of the mountainous crest (Fig. 6.14). As a consequence, a ship steaming at full speed into the seas first encounters a hole, and is unable to rise in time to ride over the steep ensuing crest. Green water breaks over the vessel's superstructure with enormous forces capable of inflicting severe or fatal damage.

On average, one large ship a day meets an untimely end somewhere on the world oceans. Part of the blame lies with man's ignorance of the sea and his faith in the ability of modern vessels to endure any kind of wave condition. Masters of large deep-sea commercial vessels, moreover, are encouraged to shorten passage times to increase profits. They regularly take advantage of the strong western boundary currents and drive their ships at top speeds regardless of the weather. Westbound tankers from the Middle East for example ride the Agulhas Current on their way around Africa during conditions that would normally send oceanographic research vessels scuttling to safer offshore or onshore areas. This attitude is even more surprising considering that, unlike more conventional V-shaped hulls, the long flat-bottomed bargelike tankers respond rather poorly to the rapid changes in wave height associated with giant waves, and consequently suffer more damage.

In future, it may be the high cost of marine insurance that will eventually force all such ships to heed storm warnings and have more respect for the power of the sea.

Rips

When waves encounter an opposing current, they become shorter and steeper. If the speed of the waves is much greater than that of the current, these changes will be small. But if the two have nearly the same speed the waves bunch up into rough and often dangerous patterns called rips. Their danger is twofold: first, the lengths and

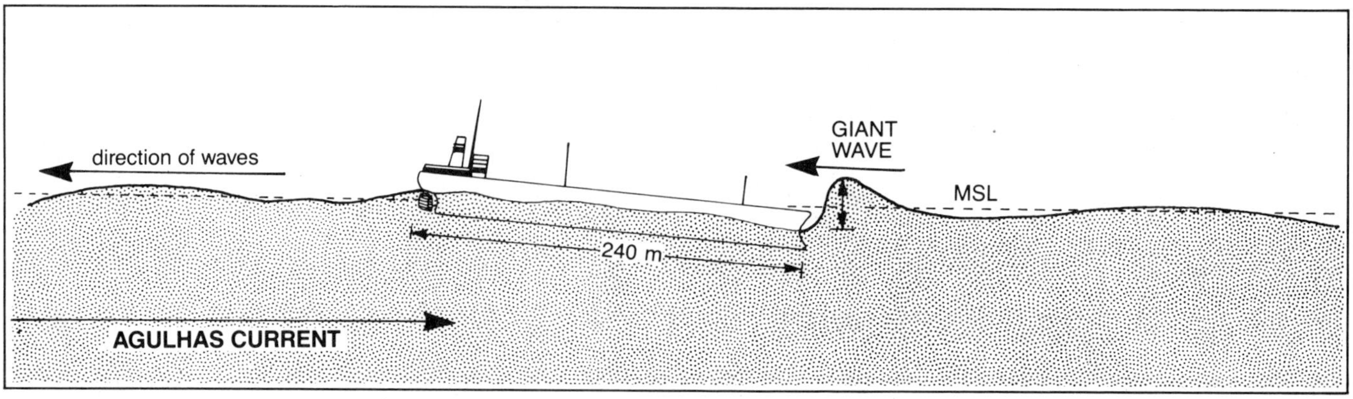

FIG. 6.14. Giant wave formed by combined effect of three wave trains with wavelengths of 260 m, 150 m, and 50 m. Waves become in-phase for short period and create abnormal wave about 21 m high. Long, deep trough or "hole in the sea" precedes wave. (msl, mean sea level) (Adapted from Mallory 1974)

heights of waves in the rip may make it impossible for a boat to fit comfortably between them, and cause it to pitch and roll awkwardly; secondly, rips are often associated with strong currents in confined passes and narrows, thereby increasing the possibility of groundings or collisions. Most rips are formed where waves generated by a wind in more open water begin to propagate into the entrance to a pass or narrows from which a strong tidal current is flowing. Rips in the Strait of Georgia entrance to Active Pass and Porlier Pass occur in this way, and strong southeast winds along the Strait produce violent rips south of Cape Mudge during the flood. In addition to the wind, wakes from larger vessels moving against a tidal current will produce rips, so the small boater is advised to give them as much berth as possible.

One of the most striking displays of rip formation seen by the author was from the Seattle–Victoria tourist vessel *Princess Marguerite*. Steaming toward Victoria on a calm summer day the course crossed a distinct tideline formed where the ebb stream from Haro Strait was converging with the initial stages of the flood stream in Juan de Fuca Strait. When the waves of the ship's wake encountered the tideline they slowed, steepened, and began to break, indicating the presence of a sharp change in the speed of the surface currents. These breaking waves remained trapped near the tideline, unable to stem the flow. The ship waves generated on the northern side of the tideline were again smooth and undulate and followed the vessel in the normal fashion.

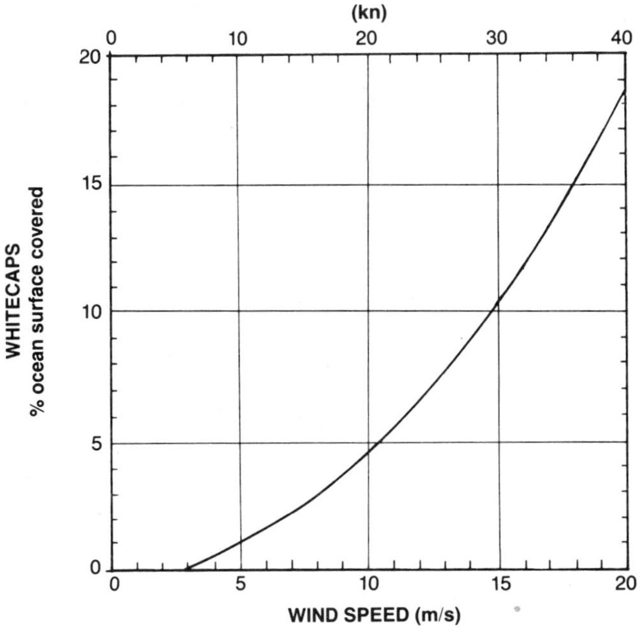

FIG. 6.15. Percentage of sea surface covered by whitecaps for different wind speeds. Whitecapping begins at around Beaufort wind force 3 or 3 m/s (roughly 6 kn) and increases rapidly for winds greater than 6 m/s (12 kn). At winds of 15 m/s approximately 10% of sea is covered with whitecaps. (From Blanchard 1963)

Whitecaps

The shape of a wave is determined by such factors as wind, surface currents, and the degree of interference between the waves themselves. If these combine to make the wave unstable, it will break in an attempt to regain its stability. Air bubbles trapped within the turbulent water of the breaking crest form the foaming crown called a whitecap. The foam and bubbles then become part of the physical surface of the water instead of moving with the wave. The percentage of the sea (or lake) surface covered by white water from this process is markedly dependent on fetch, wind duration, and wind speed, as well as certain atmospheric conditions. The rate at which the trapped bubbles effervesce depends on the surface tension of the water. As surface tension can be significantly altered by substances in the water like salt, oil, or marine organisms, the longevity of a patch of white water can vary from one region to another. Thus, for identical conditions of sea-state and wind, the whitecap coverage of a freshwater lake will be less than that for the salty ocean.

Observations at sea show that there are virtually no whitecaps for wind speeds less than 3 m/s (6 kn) but there is an abrupt increase in whitecap activity when winds exceed 6 m/s (12 kn). A study by Blanchard (1963) suggested that the percentage of the sea surface covered by whitecaps increased in direct relation to the square of the wind speed for winds exceeding 5 m/s (10 kn) (Fig. 6.15). For 4 m/s (8 kn) winds, about 0.5% of the sea surface is covered by whitecaps, whereas at 10 m/s (20 kn) this increases to about 5%. Extrapolation of Blanchard's graph suggests that about 100% of the sea surface is covered with foam for 50 m/s (100 kn) winds. Studies of lakes, on the other hand, show a somewhat different situation. For one thing, the abrupt increase in whitecapping doesn't occur until winds reach 8 m/s (16 kn), and for winds below 7 m/s (14 kn) less than 0.1% of the water surface has whitecaps. Furthermore, the percentage of lake surface covered doesn't increase with wind speed as rapidly as it does in the ocean.

A point worth emphasizing is that the observations apply to exposed waters and not to more sheltered areas; the wind speed may be great enough to cause a large percentage of whitecapping in midocean yet generate only a limited degree of whitecapping in more confined coastal areas. The amount of whitecapping will also vary somewhat with the strength and direction of the surface flow. For a given wind speed, whitecap activity will be greater than average if the waves oppose the flow and less than average if they move with it.

Wave Dissipation

Because the total energy of waves in the ocean is not increasing with time, energy that is fed into waves via external mechanisms like the wind must eventually be lost to other forms of motion. The breaking of waves, either at sea or on the shore, is one of the major processes by which waves lose energy to their surroundings. Wave energy is also diffused throughout the ocean by frictional effects and by geometrical spreading, when a group of waves gradually spreads out to cover an ever-increasing area of the sea surface. Although the total energy associated with the group may remain unchanged, the increased area means there is a local reduction in energy and wave height. As the opposite of the generation process, winds will quickly knock down a countering sea by frictional drag on the waves. And rain effectively attenuates the shorter wind waves by continual battering against the rising crests, perhaps explaining why seas appear to be a little less severe afer the rain has begun.

Internal Gravity Waves

Discussion to this point has been mainly devoted to surface gravity waves generated by the wind, that is, to relatively short waves that form at the boundary or "interface" between the air and the water, and whose restoring force is the earth's gravitational pull. Just as these waves exist where there is a change from the higher density water to the much lower density air, gravity waves also can occur within the interior of a body of water, provided its density increases with depth. The origin of the restoring force for such waves is illustrated in Fig. 6.16. In the simplest case of a homogeneous layer of fresh water overlying a homogeneous layer of slightly heavier salt water (Fig. 6.17), internal gravity waves can be made to propagate along the fluid boundary that separates the two layers. (In a frequently used laboratory demonstration, internal gravity waves are generated in a large fish tank by disturbing the equilibrium level between a layer of light lubricating oil floating over a layer of salty water.) When the density increases more gradually with depth, the waves are no longer confined to a specific boundary but can propagate upward or downward at a variety of angles to the horizontal; the angle of propagation in this case depends on the density structure of the water and the period of the waves. Moreover, for internal waves there is a "cut-off" period such that waves of certain periods cease to be possible wherever the rate of vertical increase in water density is less than a specified value; in the limit of homogeneous water, of course, no internal gravity waves can exist.

Internal gravity waves are ubiquitous features of the world's oceans. Oceanographic measurements, and more recently satellite imagery of surface slick bands, show them to be particularly common over continental shelf regions, where they are thought to be generated by tidal currents that move back and forth over the continental margins. They are commonly measured in the deep ocean where they presumably originate from the flow of oceanic currents over abyssal hills and submarine mountains. (As an analogy, internal gravity waves in the atmosphere often take the form of orographically induced lee waves that may be made visible by a series of lenticular clouds. The break-up of vertically propagating internal gravity waves in the atmosphere is believed to be a major cause of clear-air turbulence (CAT), which frequently rattles jet aircraft cruising high within the stratosphere.) On the B.C. coast, internal gravity waves become especially well pronounced in certain silled inlets and protected basins when they are covered by a thin layer of brackish water during times of large river discharge. Such a condition occurs in the southern Strait of Georgia and the middle portion of Knight Inlet.

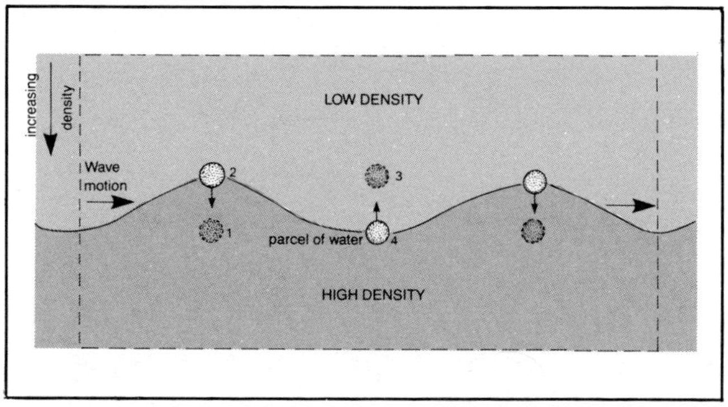

FIG. 6.16. Origin of restoring mechanism for internal gravity waves in a vertically stratified fluid. A water parcel displaced from positions 1 to 2 has a slightly greater density than surroundings and has a negative (downward) buoyancy force. Similarly, parcel moved from positions 3 to 4 is less dense than surroundings and has a positive (upward) buoyancy force. In homogeneous water there is no such restoring force. The density at one level is the same as another.

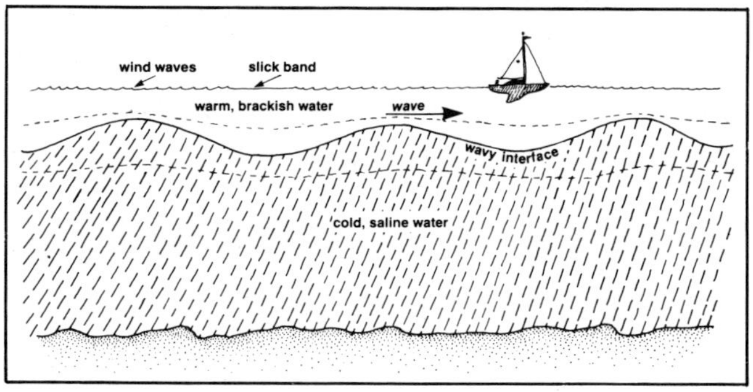

FIG. 6.17. Internal gravity waves propagate along the boundary, or interface, between layer of low-density (warm, brackish) water and layer of higher-density (cold, saline) water. Slick bands have developed over wave troughs.

Properties

Contrary to surface gravity waves, which possess maximum vertical water displacements at the surface and diminishing vertical displacements with depth, internal gravity waves produce maximum vertical displacements within the water column and negligible distortion of the water surface. Therefore, the up and down motions associated with the waves are essentially confined to the interior of the fluid. Consequently, the wave crests flatten near the surface whereas the troughs deepen (Fig. 6.18); the reverse is true for internal gravity waves near the seafloor where motions are restricted by the presence of the solid bottom.

The vertical excursion of an interior parcel of water during the passage of an internal gravity wave can range from a fraction of a metre for waves with periods of seconds and wavelengths of metres, up to 100 m for waves

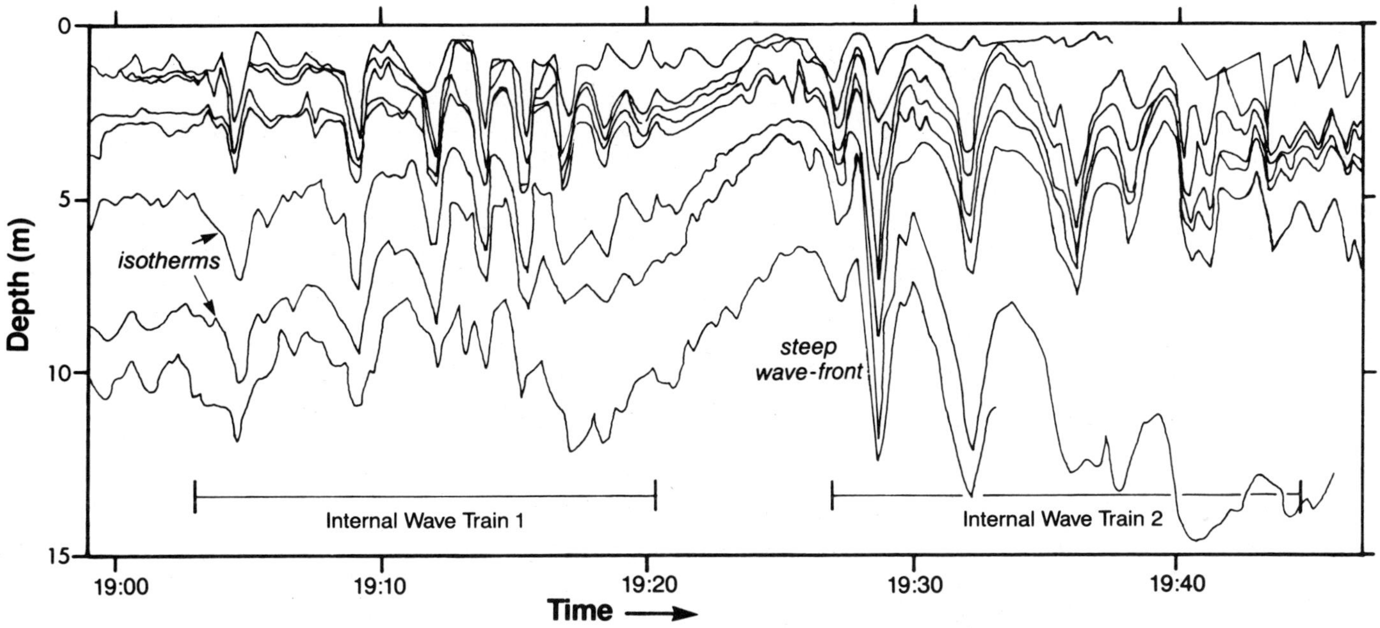

FIG. 6.18. Displacement of isotherms during passage of two packets of internal gravity waves. Waves originated at Boundary Passage in southern Strait of Georgia and propagated northward within shallow thermocline (June 1966). Observations made by shipborne temperature profiling device. (Modified after Hughes 1969)

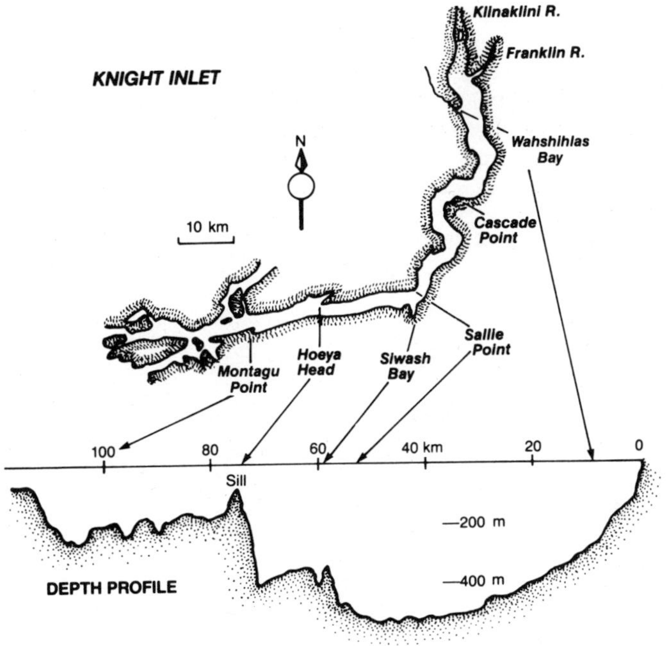

FIG. 6.19. Map and mid-channel depth profile of Knight Inlet (see Fig. 3.26). Cross-channel sill off Hoeya Head is 75 km seaward of head of inlet and 70 m deep. Most fresh water enters fiord from Klinaklini and Franklin rivers.

in the deep ocean with periods of hours and wavelengths of tens of kilometres.

Due to the rather small changes in water density within the ocean, the magnitude of buoyancy-restoring forces that act to create internal gravity waves (Fig. 6.16) are about 1/1000 those of surface gravity waves. As this means an effective reduction in the "bounce rate" or oscillation period of an interior water parcel, gravity waves tend to be longer and to move much more slowly than their surface counterparts. Typical internal wave lengths in the Strait of Georgia, for instance, are about 25–100 m and typical propagation speeds around 0.5 m/s (1 kn). The few surface waves in the Strait that ever attain a wavelength of 100 m travel at speeds of close to 13 m/s (25 kn). A further comparison shows that "high" surface waves in the southern Strait have heights of 2 m and periods of approximately 8 s; internal waves in this region generally attain maximum heights of 5–10 m and periods range from 5 s to 10 min.

One feature common to both types of waves is their tendency to propagate as wave trains or packets led by the faster-traveling, longer, and higher waves of each group. This dispersive effect is illustrated in Fig. 6.18, which shows the passage of two internal wave trains past a fixed location north of Boundary Pass in the southern Strait of Georgia. The waves are revealed through their distortion of lines of equal temperature (isotherms) with time. Slick bands similar to those in Pl. 7, 8 may also accompany the passage of such wave trains. (More will be said on surface manifestations of internal gravity waves later in this section.) Some striking examples of internal wave trains have recently been obtained from an oceanographic study of Knight Inlet (Fig. 6.19). During the flood, a well-defined wave packet frequently propagates up-inlet from the vicinity of Hoeya Head midway along the channel (Fig. 6.20). The leading edge of the packet is commonly in the form of a steep-faced internal bore whose associated surface slick (Fig. 6.21) is accompanied by a low hissing noise created by the breaking of surface ripples at the slick boundary. Wave crests are about 70–100 m apart and the up-inlet speed of the group is roughly 50 cm/s (1 kn).

As with surface gravity waves, the passage of a group of internal gravity waves is accompanied by a sequence of horizontal current reversals, although the surface wave

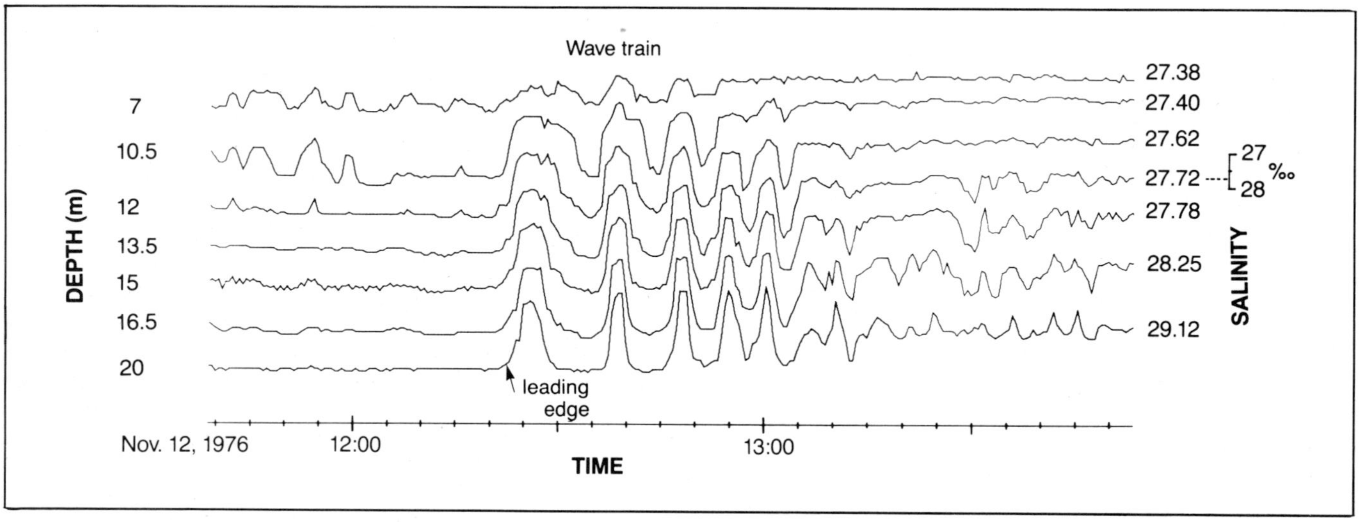

FIG. 6.20. Example of train of seven internal gravity waves propagating past a point 7.6 km upstream of Hoeya Head, Nov. 12, 1976. Waves are oscillations in depth of observed isohalines. Leading wave was largest of group and, unlike successive waves, produced a marked ship drift as it passed under hull. Wave heights exceeded 10 m, wavelengths were 70–100 m, and there was a decrease in periods between successive waves. (After Farmer and Smith 1978)

FIG. 6.21. Series of surface bands created by internal gravity waves in Knight Inlet; aerial view toward west from a location eastward of Hoeya Head (see Fig. 6.19). Waves were generated during ebb in lee of sill across channel from Hoeya Head (H) and propagated toward head of inlet (toward the reader) on flood. (See Fig. 6.23 for explanation of wave generation.) (Courtesy D. Farmer)

sequence of forward motions over the crests and backwards motions over the troughs does not always apply to internal waves. These currents can be appreciable. In the Strait of Georgia, short internal waves with periods of 2.5 min have produced horizontal surface current speeds as great as 25 cm/s (0.5 kn). A boat riding at anchor would experience a flow in one direction, then in the opposite direction, as each internal wave advanced slowly and silently past its location.

Generation

The mechanism most responsible for the generation of internal gravity waves is the interaction of the "everyday" surface tide with rapid changes in bathymetry. For example, it is now generally accepted that oceanic internal tides (long internal gravity waves with nearly the same daily period of rise and fall as the surface tides) are generated at the edge of the continental shelf through the action of tidal currents that move up and down the continental slope. These tidal currents disturb the equilibrium levels of the density surfaces and set them into wavelike oscillations. A portion of the waves then travel onto the shelf while the rest are radiated seaward within a downward propagating beam of energy (Fig. 6.22). Observations off the coasts of British Columbia and Washington verify the existence of these large internal oscillations and indicate how the waves carry seaward a fraction of the energy

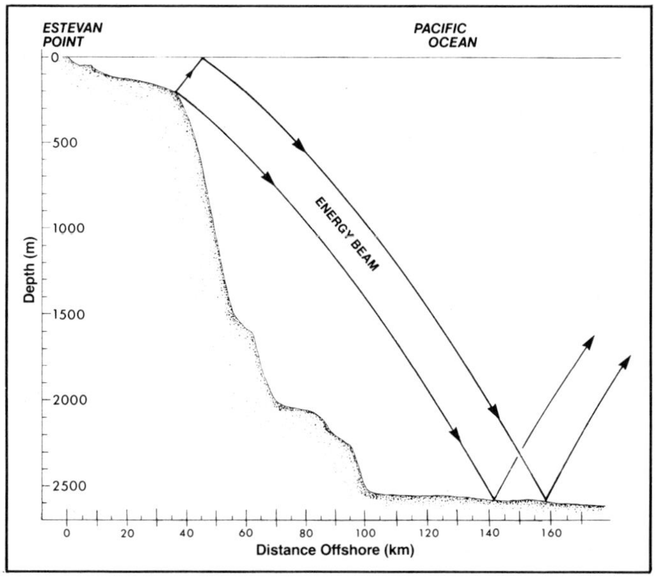

FIG. 6.22. Possible orientation of beam of energy associated with internal tides emanating from edge of continental shelf seaward of Estevan Point, west coast Vancouver Island. Beam characterized by relatively large semidiurnal oscillations in depth of isopycnal surfaces and by pronounced variations in strength and direction of tidal currents with depth. Distortion of beam occurs as a result of water density variations, ocean currents, and seafloor roughness. Internal tidal energy is eventually dissipated through frictional damping within a few hundred kilometres of coast.

contained within the surface tide. This conversion of surface tidal energy into internal tidal energy at continental margins has been considered by some to be an important mechanism for the dissipation of the lunar and solar tides within world ocean basins, although there is now growing evidence to refute this concept. Recent measurements of bottom pressures and temperature variations off the B.C. coast further suggest that internal tides originate with tidal flow over guyots and seamounts. Abyssal hills and rough seafloor terrain are also expected to be sources of a variety of internal wave motions.

Groups of internal gravity waves covering a wide range of wave periods and amplitudes are generated by the ebb and flood of tidal currents over shallow sills within certain B.C. coastal waters. The sills at the entrances to Boundary Passage and Active Pass are particularly vigorous sources for these waves, which move slowly into the Strait of Georgia on the flood. The ferry between Swartz Bay and Tsawwassen affords one of the better vantage points from which to view developing internal wave bands. On a fairly calm summer day, it is quite common to see as many as 10 near-concentric bands spreading into the Strait from the eastern end of Active Pass on the turn to flood. The waves and associated slicks then spread over much of the southern and central portions of the Strait, and some eventually reach Burrard Inlet before they collapse along the shorelines.

Internal gravity waves are also generated by tidal flow over shallow sills within certain coastal fiords such as Knight Inlet. During the summer freshet, when the milky colored river outflow forms a shallow brackish layer over the salty oceanic water below, the maximum vertical displacements by these waves occur close to the surface, and produce distinct slick bands. The currents associated with the waves are often quite conspicuous. One vivid example of this was seen by the author one calm summer day on a cruise to Knight Inlet on the CSS *Vector* (39.6 m, 505 t). As it rode at anchor in about 70 m of water over the sill near Hoeya Head (Fig. 6.19), the ship swung rapidly about 90° on her anchor chain as the leading edge of an approaching group of ruffled surface bands began to pass beneath the hull, and then swung back again as the trailing edge of the band passed by. It was an impressive display of the powerful horizontal currents produced by large internal gravity waves, which usually go unnoticed by vessels proceeding under power.

In Johnstone Strait, strong internal tidal motions appear to be generated by the swift tidal currents that surge over the partial sill near Kelsey Bay. Recent research in this area suggests that, because the seaward propagating branch of these waves appreciably distorts the "normal" tidal flow for approximately 10 km west of the sill, near-surface currents are only half as intense as those at depth. There are also internal tides in Knight Inlet, although they are considerably weaker than those in Johnstone Strait.

The way packets of short internal gravity waves are formed in the southern Strait of Georgia and in Knight Inlet is noteworthy. The mechanism is similar in the two regions so Knight Inlet, where oceanographic studies of the problem have been made by scientists of the Coastal Zone Section of the Institute of Ocean Sciences, will be the example. At a certain stage of the larger ebb currents, water flowing over the sill near Hoeya Head forms a pattern of long lee waves (Fig. 6.23a). This in turn leads to the formation of an internal hydraulic jump in the lee of the sill and to the generation of a series of short internal waves, with maximum amplitudes at depths comparable to the top of the sill. Because the waves are unable to propagate against the ebb, their position remains fixed relative to the sill, a situation not unlike surface gravity waves that attempt to advance against an opposing current in a rip. Approximately 1.7 h before the turn to flood, a packet of 5–10 waves has been generated and is ready to propagate up-channel (Fig. 6.23b). With the turn to flood, the wave packet (with maximum amplitudes along the boundary between the shallow brackish layer and the deeper saltwater layer) crosses the sill and advances forward as an internal bore manifested at the surface by a sequence of slicks (Fig. 6.23c, d). Similarly, a weaker group of internal gravity waves may sometimes form on the up-channel side of the sill on the flood. Such waves will, of course, travel seaward on the ensuing ebb.

Aside from the interaction between tidal currents and bottom topography, internal gravity waves are also thought to be generated by winds, by atmospheric pressure fluctuations, through the mutual interaction of waves

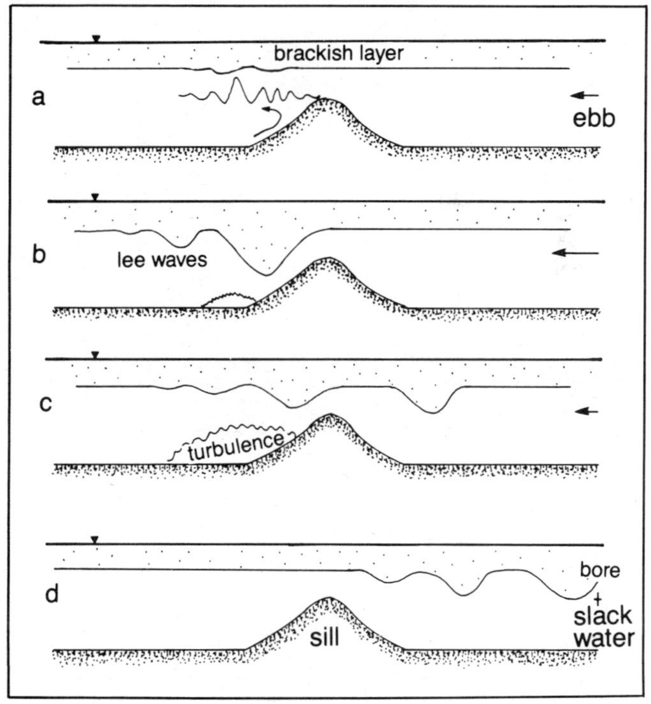

FIG. 6.23. Schematic diagram shows generation of internal gravity waves over sill off Hoeya Head, Knight Inlet. During first stages of ebb (a), flow separates from crest of sill as series of growing waves. Near maximum ebb (b), flow is deformed into a series of internal lee waves analogous to atmospheric lee waves downwind of a mountain range. Developing wave system cannot propagate against current and remains tied to sill as standing wave pattern; flow separation is suppressed. Near low-water slack (c), leading wave escapes upstream, followed at slack water (d) by additional lee waves that form a train of steep internal waves (internal bore). Processes may be repeated for both ebb and flood currents. Surface slicks in Fig. 6.21 are associated with stage (d). (Courtesy D. Farmer)

in a sea, and by swell. Within coastal regions inundated with freshwater runoff, deep-draught ships with hulls below the upper brackish layer can generate internal waves in the form of an internal wake. Unless the vessel outruns these waves, its progress will be markedly impeded as part of its engine power goes into creating the wake. Once the ship speed appreciably exceeds that of the waves they are no longer generated. This phenomenon of "dead water" was first studied in Norwegian fiords at the turn of this century, though it is quite conceivable that the Vikings knew of its effect 1000 yr before. Scientists at the Defense Research Establishment Pacific in Esquimalt, B.C. have used the dead water effect to study the interaction between internal waves and surface waves under controlled conditions in Bute Inlet. First, a 4-m draught research vessel was put in reverse to generate a pattern of internal gravity waves in the upper brackish layer (Fig. 6.24); the experiment was then conducted as the ship was driven forward through the wave field. The advantage of Bute Inlet in this case is that it has no shallow sill, and internal waves are absent even during summer when density structure is favorable for their formation. Finally, the collapse of a submarine's wake will form an outward radiating pattern of internal gravity waves.

FIG. 6.24. Surface slicks in Orford Bay, Bute Inlet, associated with a group of ship-generated internal gravity waves (an internal wake), July 1972. RV *Endeavour* (draught, 3.7 m; length, 65.0 m), generated internal waves by steaming astern along a fixed track, adjusting engine power to obtain maximum "dead water" effect. The resulting wave group, formed at interface of 4-m thick surface brackish layer and deep saline water beneath, was studied by reversing ship direction and transversing waves. (From Hughes and Grant 1978)

Dissipation

Internal gravity waves lose energy at a proportionally more rapid rate than surface gravity waves of comparable periods, and are incapable of traveling long distances within the ocean. In Knight Inlet internal wave packets generated near Hoeya Head appear to be completely dissipated within 30 km up-channel, their destruction perhaps aided by an inability to negotiate the sinuous portion of the inlet north of Sallie Point. Observations in laboratory experiments and in lakes, and by scuba divers in the warm Mediterranean Sea, using dye to trace the water motion, show that within the interior of the fluid waves roll into billows and then collapse in confusion when they break. This type of wave breakdown, known as Kelvin-Helmholtz (shear) instability, occurs when there is a large enough change in the speed of the current with depth to cause the waves to become unstable. Internal waves also collapse in interior regions of the ocean where their speed of propagation matches the local speed of the current (critical layer absorption); in this case, the waves give up their energy to the mean current, which is then accelerated.

As with surface waves, internal waves break when they run up a sloping bottom or when their heights become too large to support their weight. Moreover, internal waves that propagate downward to the seafloor are thought to be scattered and, subsequently, dissipated by irregular bottom topography. The turbulent agitation generated in this way is now considered to be a mechanism for maintaining well-mixed bottom waters. In Knight Inlet, the energy contained within the induced internal wave fields is more than adequate to account for the observed degree of turbulent mixing. The initially strong, seaward propagating internal tides generated over the sill near Kelsey Bay in Johnstone Strait (see Chapter 12) appear to be dissipated through turbulent frictional effects within 20 km of their origin. The dissipated energy of these waves would then in turn augment the turbulent mixing processes in the channel.

Internal Wave Slicks

Internal gravity waves that propagate at shallow depths frequently are revealed through the effect of their currents on the surface waters. Such surface manifestations of the waves may take a variety of forms, from distortion of the surface wave field to differences in water color. To explain the basics of what is taking place, it is simply assumed that the surface waters are converging over the internal wave troughs and diverging over the crests. Winds are light and the form of the waves resembles Fig. 6.18. Slick bands can then arise in the following ways: (a) floating surface debris such as wood chips, seaweed, and litter collects in regions of converging surface flow to create a series of "tidelines" (Fig. 6.25a). Logs trapped in these regions are aligned parallel to the trend of the slick and advance with the waves. Wavelets are usually absent from the slicks because of the dampening effect of the debris; (b) during summer in particular, a thin oily film of organic material may gather on the sea surface. When this film becomes concentrated over the wave troughs and remains cohesive, it effectively dampens any surface ripples and leads to a series of glassy slicks (Fig. 6.25b). Ripples disappear on reaching the edge of the slick; (c) in river-fed basins like the Strait of Georgia, the deeper oceanic waters are covered by a thin layer of grayish silty water. When viewed from the air, the thicker concentrations of silty water over the convergence zones retain their

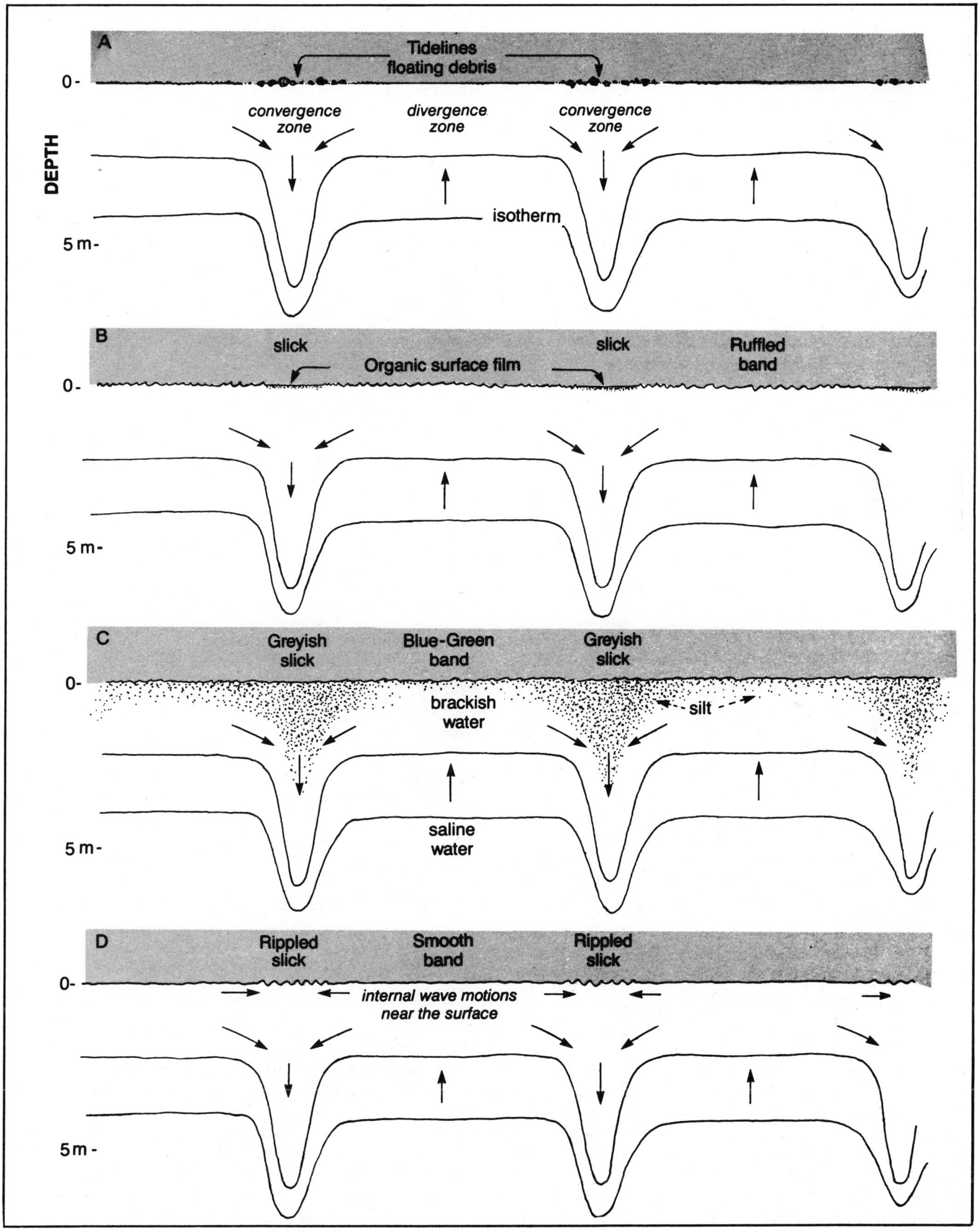

FIG. 6.25. Types of slick bands generated by shallow depth internal gravity waves in coastal waters. For illustrative purposes, currents associated with waves are assumed to converge over the troughs and to diverge over crests, although this is not always the case in nature. Lines represent depths of isotherms or lines of equal temperature (compare with Fig. 6.18).

dull gray appearance, whereas the relatively thin layer over the crests allows the blue-green of the deeper waters to show through (Fig. 6.25c); (d) currents of the internal gravity waves alter the wavelengths and, hence, the slopes of the shorter surface gravity waves. Over the converging zones, wavelengths are decreased and the roughness of the sea surface enhanced; over the diverging zones wavelengths are increased and the sea made smoother (Fig. 6.25d). Viewed from afar, however, it is not the roughness per se that shows up the slick bands, but the relative amount of light reflected from the different areas. In Fig. 6.26a, for example, sunlight is reflected directly to the eyes

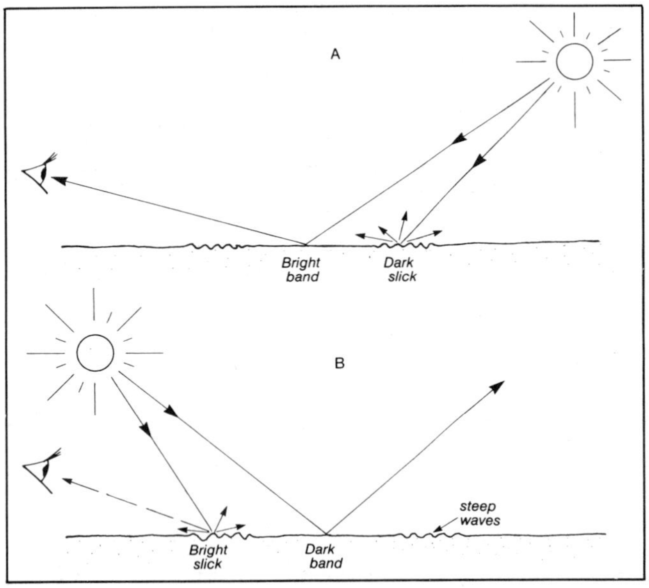

FIG. 6.26. Dependence of relative brightness of slicks and intervening bands on orientation of sun to observer. In (A) light is reflected directly to viewer from smooth sea-surface regions between slicks; same light scattered from irregular wavelets over slicks makes them appear dark. In (B) situation is reversed; slicks appear relatively bright because only scattered light reaches observer.

from the comparatively smooth areas of water but only a portion of the scattered light from the rougher slick bands reaches the eyes. Therefore, the bands appear dark and the intervening regions bright. When the sun is behind (Fig. 6.26b), the reverse is true. It is now the scattered light from the rippled slick bands that makes them appear comparatively bright. Little light is reflected from the zones of smoother water and they appear dark. (Of course, everything is reversed if an organic film has dampened out the surface waves in the slick band!)

Ship Waves

Many factors oppose the advance of a ship through water. Among these are the dynamic pressures of winds against the superstructure, the dynamic force of waves against the hull, the frictional drag of water on the wetted surfaces, and the turbulent eddies shed by the hull. In addition, there are the vessel-generated waves which, under certain conditions, can account for as much as half the total resistance force on a ship. Ensuring that this wave-making drag is kept to a reasonable level is a fundamental consideration in ship design.

Aside from their practical aspects, ship waves in their own right are an interesting phenomenon. One of the more conspicuous features is that the region occupied by a deepwater ship-wave pattern has the same general V-shape whether it is generated by a duck or an oceanliner (Fig.

FIG. 6.27. Ship-wake pattern generated by submarine in Strait of Georgia. Wake dominated by long waves with crests nearly perpendicular to path of vessel. Two diverging wave systems visible with angles in wedge-shape wake roughly equal to 140°. (Photo by author)

6.27). The spread of the V-pattern depends ultimately on the speed and length of the vessel, but for a wide range of vessels, traveling at constant speed on a steady course, the interior angle formed between the two leading edges of the wake (i.e. between the two arms of the V) is always around 39°. This angle is close to the 38°56′ predicted by the theory for a Kelvin ship-wave, first put forward by Lord Kelvin in 1904 for a pointed object or small displacement vessel moving over the surface of still water (Fig. 6.28). When the ship's course is curved, the wake pattern too is curved and the angle less well defined. An exception to the 39°-angle is when the vessel's length is large compared to the maximum wavelengths its hull can generate at its particular speed as the submarine in Fig. 6.27 (expressed mathematically as $L >> S^2/g$ where L is the length and S the speed of the vessel, and g the acceleration of gravity; ratio S^2/gL, the Froude Number, determines the nature of the ship-wave pattern). In this case, the pattern is dominated by the longest possible hull-generated waves whose direction of travel is aligned more closely with the vessel's path. Other exceptions are when the length of the vessel is small compared to the maximum wavelengths it can generate, as in the case of speedboats. Then the pattern is dominated by short waves whose direction of travel is nearly at right angles to the vessel's path, so the interior angle of the V-pattern is considerably less than 39°.

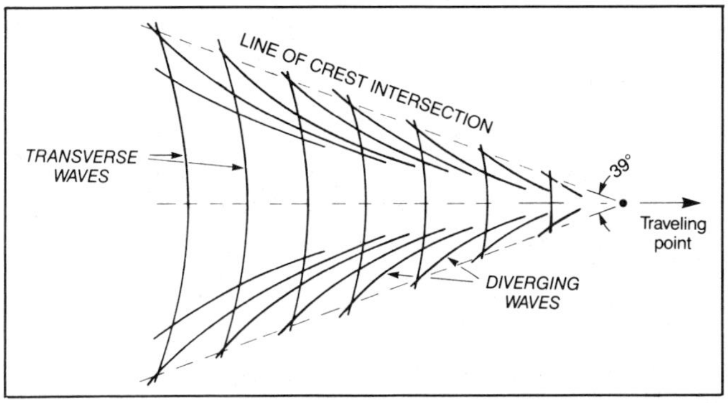

FIG. 6.28. Crests within a Kelvin ship-wave pattern generated by traveling pressure point (or small boat) in deep water. Point moves at constant speed in straight line. Lines of crestal intersection form interior angle of 39°. In shallow water, angle broadens to 180° as diverging waves coincide with transverse waves.

The V in the simple Kelvin wave-pattern (Fig. 6.28) corresponds to the line of intersection between crests of two separate wave systems: (1) the diverging wave system of a train of curved crests and troughs concave outward from the line of travel, and (2) the transverse wave system of a train of curved waves, convex forward and perpendicular to the line of travel. The interference between the crests and troughs of the two patterns accounts for the fact that the heights of hull-generated waves are always maximal at the leading edge of the V-pattern and minimal within the confines of the pattern. Anyone who has plowed his craft across the wake of another vessel will have noticed the near absence of waves after the initial line of crests had passed. Moreover, the leading waves will have curved crests that only extend a short distance laterally before they disappear. For powerful displacement vessels such as tug boats, these short wave-crest segments are often steepened to the point of breaking, indicating that a considerable percentage of engine power is lost in the generation of the wave pattern.

Because of their size, ships generate a considerably more complex wake pattern than the moving pointlike disturbance described above. Basically, there are now two nearly identical wave patterns, one from the bow and the other from the stern, each pattern with diverging and transverse wave systems (Fig. 6.29). Maximum wave heights again occur at the edges of each of the combined patterns where reinforcement takes place; minimum heights occur where the wave sets cancel one another. Transverse waves that originate at the bow of the ship begin with a crest, whereas those from the stern begin with a trough. The wavelength of these transverse waves increases as the square of the ship speed, so there may be several waves from the bow system along the waterline at low speeds but only a portion of a wave at high speeds.

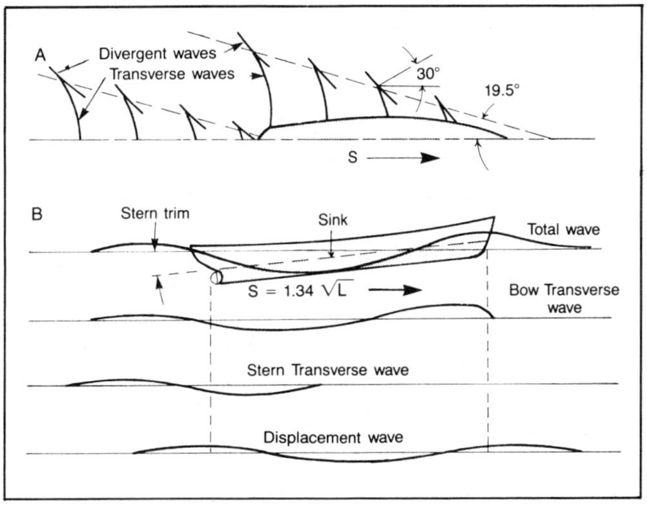

FIG. 6.29. (A) Plan view of ship-wave pattern shows sets of transverse and diverging wave systems originating at bow and stern; (B) total wave pattern along a hull (top) consists of transverse bow and stern waves, and permanent displacement wave. Speed is for $S/\sqrt{L} = 1.34$, where troughs of bow and stern waves coincide, producing maximum sink and stern trim. Ship speed, S, in knots; waterline length, L, in feet. (From Van Dorn 1974)

As well as wave sets generated by the bow and stern, secondary wave patterns also originate when there is a shoulder or bend in the ship's hull, each shoulder wave beginning as a trough. Depending on the shape of the hull and speed through the water, these various wave-sets cancel or reinforce one another to create a modified system of diverging and transverse waves within the wake.

The ship waves described so far are free waves, in the sense they become detached from the vessel following their generation by the pistonlike action of the hull. In addition, however, there is a displacement wave, which remains fixed with the vessel and does not contribute to the trailing wake. This wave begins as a crest under the bow after rising from a distance ahead of the forefoot, forms a trough amidship, and then rises to a second crest at the stern (Fig. 6.29). The wave is a direct result of dynamic pressure differences created by the diversion of the flow around the hull in accordance with the venturi effect. The trough, for example, results from a rapid drop in the internal water pressure due to the acceleration of the flow around the beam of the ship, whereas the crests originate with decelerated flow near the bow and stern of the vessel.

For displacement hulls, the heights of all ship waves increase as the square of the vessel speed and become steepest near its limiting hull speed. Limiting hull speed is roughly given by $S = 1.34\sqrt{L}$ kn for waterline length, L, in feet, or $S = 1.25\sqrt{L}$ m/s for L in metres. At this speed, the trough of the first bow wave has extended far enough aft thac it coincides with the initial trough of the stern wave. The two troughs reinforce, and produce maximum stern trim as the stern sits deeper relative to the raised bow. (In actual practise the range of the ratio $S/\sqrt{L}$ may lie between 1.0 and 1.5 (English units) depending on a variety of factors including vessel displacement, hull form, and the amount of fuel the skipper wishes to expend to get that fractional increase in speed). Bow waves are generally higher than stern waves so, as the vessel moves faster, her bow has a tendency to ride higher on the growing bow crest. At the same time, her stern sits lower in the accompanying bow trough that has begun to extend aft with the enhanced ship speed. This, combined with a further sinking of the hull into the trough of the displacement wave, increases the total wave resistance and saps engine power needed to provide the forward thrust. (The total wave resistance increases about in proportion to the square of the ship wave heights.) In effect, then, the ship works its way deeper and deeper into a wave-resistance trap as it attempts to increase its speed. Planing hulls are designed to avoid such a predicament by high power and light displacement to raise the hull out of the water. At planing speeds most drag on the hull comes from friction on the wetted surface and air resistance.

The ship wave pattern alters if the water beneath the hull shoals to less than about 25% of the vessel's waterline length. The accompanying reduction in speed may be quite dramatic, as though someone applied brakes, if the depth change is sufficiently abrupt. This alteration to shallow-water conditions can be explained. First, the proximity of the bottom helps accelerate the flow under the hull, increases the height of the displacement wave and, therefore, wave resistance. Second, all hull-generated waves undergo a transition from deepwater to shallow-water waves so their speeds become directly determined by the water depth rather than the speed of the vessel. The V, formed by the divergent bow and stern wave patterns, spreads from a deepwater angle of 39° to an angle of 180° in the limit of shallow-water conditions. The divergent waves then augment the transverse waves to form a single transverse wave crest beneath the bow and a single trough at the stern, both of which fade rapidly away from the hull. As for the vessel, it becomes caught in an ever-deepening wave resistance trap as her bow continues to rise and her stern to sink. There is also danger of grounding by the stern in a rapidly shoaling region if these effects become excessive. Only if a great amount of power is expended can the vessel exceed the critical wave speed, in this case the shallow-water wave speed, $\sqrt{gD}$, which will lead to a rapid reduction in the excess wave resistance as the speed is further increased. (D is the water depth.) This latter situation is akin to a jet breaking the "sound barrier" by traveling faster than MACH 1, the maximum speed of sound for the particular altitude at which the jet is flying. No divergent soundwave pattern is capable of keeping up with the jet in this case and it is only accompanied by a bow shock wave that causes the familiar sonic boom at the ground. Like the jet, moreover, a boat must expend a large amount of energy to reach "aquatic MACH 1" speed but once it has done so, proportionally less effort is required to maintain or increase the craft's speed. As an indication of how high these speeds must be, consider a tugboat in 2 m of water. Because g = 9.8 m/s/s, the formula $S = \sqrt{gD}$ shows that the boat must attain a speed, S, if 4.5 m/s (9 kn) before it can begin to overtake the wave-resistance barrier. The tugboat at this stage would be generating a steep breaking wake, much to the consternation of anyone who might happen to be nearby in a canoe or dinghy with little freeboard.

Deep draught ships that move relative to the water within shallow, confined channels such as the lower Fraser River (mean depth = 10 m) may experience a further modification to their wake pattern as they enter from deeper water. When the base of the hull comes within a few metres of the channel bottom, the volume of water squeezed between them must speed-up in order to pass beneath the ship. As a result of the venturi effect, there is a drop in water pressure proportional to the square of the change in current speed and a subsequent reduction in the buoyant force that keeps the vessel afloat. The ship is then said to squat as it sinks a little deeper in the water. Due to the shallow depths, the fore and aft bulges of the wake are disproportionally heightened by the squat and can lead to large surges of water at the shore that exceed those associated with the ship waves. Picnics on the Fraser River banks have been ruined by unexpectedly large run-ups following the passage of loaded cargo ships bound for New Westminster. Ships progressing downriver are less likely to induce such effects because of their lower speeds relative to the river currents, assuming, of course, that the ship has about the same speed relative to the shore it had when proceeding up-river.

Chapter 7. Generation of Wind Waves

Winds generate waves. This has been apparent to man for many thousands of years, and a factor he has had to contend with since he first ventured onto the seas. Waves are also generated by other mechanisms but, without question, it is the wind that contributes most to the ruffled veneer of the waters on the face of this planet. However, exactly how the energy of the wind finds its way into the waves is not yet completely understood. The problem of wind-wave generation continues to be an important aspect of oceanographic research.

A Difficult Problem

Besides trying to decide the quantities to be measured in the first place, some main roadblocks to investigating wave generation are the formidable engineering problems. It is extremely difficult to design instrumentation that measures changes in the wind and pressure at the water surface and, at the same time, measures the up and down motion of waves. To begin with, sensors must be exposed to the air that flows over the waves, yet they must remain dry, because the smallest drop of water on the delicate sensors is like a sledge hammer blow compared to the more gentle puffs of winds. Moreover, the instrument must float on the surface in such a way that it tracks the wave's vertical motion but doesn't slop around like a boat in a chop. Another factor is that the presence of the measuring device disrupts the airflow over the wave. So, unless the oceanographer knows exactly how his floating instrument is disturbing the natural process of wave generation, the information gathered may lead to erroneous conclusions. In addition, a problem haunts marine investigators everywhere — how to make sensitive electronics operate effectively in salt water. For some reason, salt water has the uncanny ability to seep into the most well-sealed device. This, and the fact it is both a conductor of electricity and a corroder of metals, is a common cause of nightmares among experimentalists. Finally, there is the role of the theoretician. Simply stated, his task is to construct a mathematical model, based on the laws of physics, which will explain in detail the processes of wind-wave generation. To be acceptable, the mathematical model must stand the test of observations.

What must be explained? First of all, any theory needs to explain how waves can be formed on an initially flat surface. To do this, the theory must take into account the role of surface tension in the early stages of wave development. Observations show that this force at the water surface controls the first ripples (capillary waves) that appear when the wind begins to blow. Next, any theory must account for the growth of these ripples to gravity waves. As mentioned in the previous chapter, there are three principal factors governing this growth: average strength of the wind, duration of the wind at a given speed, and fetch of the wind. The longer and harder the wind blows the greater the amount of energy transferred from the wind to the waves. (Wave energy is proportional to the square of the wave height.) The wind fetch is important because it limits the length of time a group of traveling waves will directly receive energy from a particular wind system. High, long-period waves only can develop under conditions of strong winds for a long duration over an extensive area of the ocean. Therefore, the generation of such waves is confined to mid-ocean regions, where the fetch is limited solely by the finite dimensions of the storm. They do not occur in restricted waters like the Strait of Georgia where the fetch is limited by the presence of land.

It is fairly obvious that the height of the waves will increase as the strength, duration, and fetch of the wind increase. But, for a particular set of conditions, the waves don't continue to grow indefinitely. There is a limit. Called a fully developed sea, it marks the point at which the waves become saturated with energy and where any further addition of wind energy goes into increasing the chaotic and turbulent nature of the sea surface (Table 6.2). Although the way this process comes about is of interest to oceanographers, it is the growth of the waves between the extremes of no waves and a fully developed sea that is of prime concern to experimentalists and theoreticians alike.

Wave Generation Mechanisms

The first attempt to put the concept of wind-wave generation on a firm mathematical footing was in 1874 by Sir William Thomson (Lord Kelvin). Drawing on the ideas of H. Helmholtz, he was able to describe wave growth through a mechanism called the Kelvin-Helmholtz Instability. Although it has long ago been abandoned as an explanation for surface waves, this mechanism is accepted today as one of the main causes of clear-air turbulence (mentioned in Chapter 6 in connection with internal gravity waves). A somewhat similar argument for wave growth was presented in 1932 by the English physicist, Lamb. His theory, unfortunately, required a wind speed of at least 6.5 m/s (13 kn) before waves would begin to grow and, therefore, described neither an effective nor realistic process. A better idea was put forward by Jeffreys (1925). Known as the sheltering theory, it only requires wind speeds of 1 m/s (2 kn). Simply put, the wind was thought to pass over a wave in such a way that it separated from a wave crest on the leeward side, and left a sheltered region or backeddy (Fig.

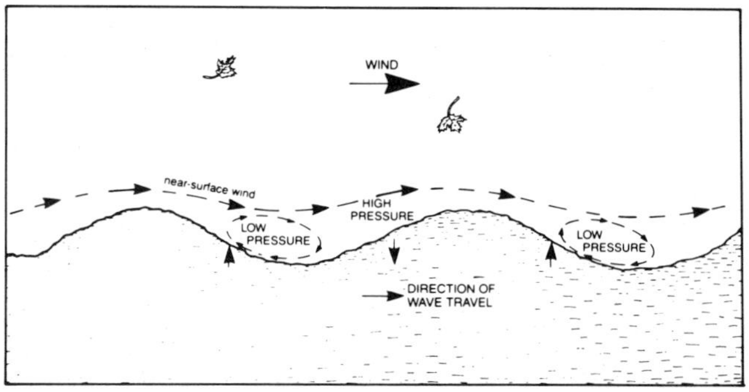

FIG. 7.1. Effective near-surface pressure distribution over growing wind waves according to Jeffreys sheltering theory. Airflow separates from wave crest and creates a backeddy (or sheltered region) of low wind speed and pressure on leeward slope. This, combined with high-pressure region created at the windward slope, adds energy and height to wave.

7.1). Air pressure in the backeddy would be less than normal air pressure and would enhance the upward motion of the leeward slope of the wave crest. Similarly, the greater than normal air pressure on the windward side of the crest would press more strongly on the windward slope of the crest to enhance the downward motion. A continuation of this effect would cause the waves to grow. What the theory didn't and couldn't explain was how the waves got there in the first place. With a flat surface, there's nothing to distort the airflow. Besides, when the whole concept is looked at more closely, it has an even more important oversight. The wind "at" the water surface, in fact, moves much more slowly than the wave and, therefore, produces a somewhat different pressure distribution than the sheltering theory suggests. Because this fact forms the basis for modern wave generation theory, it's worth elaborating. To understand it, it is necessary to realize that air that touches the water surface must be moving at exactly the same speed as the water, friction between the two dictates this (Fig. 7.2). If there are no currents, the water, and, therefore, the air on the water surface has no net horizontal motion as it rises up and down with the passage of the waveform. With increasing elevation above the water surface, however, the situation begins to change. The air moves faster and faster the farther it is from the water until, at a height of 10 m, it reaches the speed meteorologists call the "wind speed." But this means that at some small height above the water, usually less than a few centimetres, the air and the wave are moving at the same speed. This is known as the critical height. It is below that important level that the air is moving more slowly than the wave.

Ursell (1956) reviewed the various ideas about wave generation and concluded that ". . . the present state of our knowledge is profoundly unsatisfactory." His critique led to a new flurry of activity. Phillips (1957) put forth a theory on the generation of ripples on a flat sea. He showed how air pressure fluctuations within a gust of wind could ruffle the surface as they moved along with the wind by getting into step with the very ripples they were making. Unfortunately, the mechanism is incapable of forcing the ripples to grow into larger waves. Even more disconcerting is the fact that the pressure fluctuations needed by this concept are 100 times greater than ob-

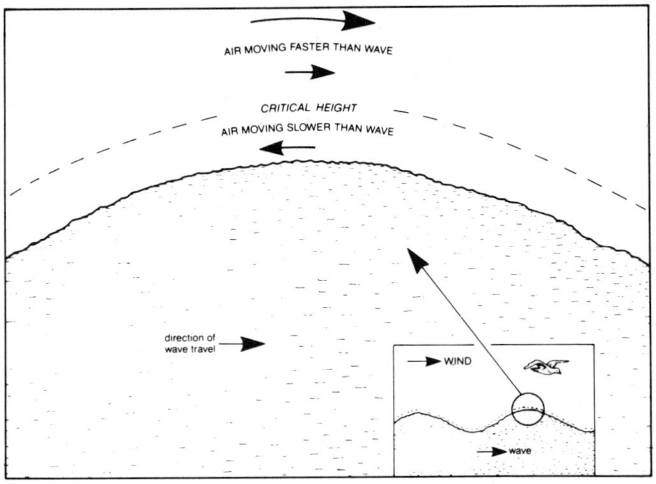

FIG. 7.2. Critical height. Frictional forces cause air in contact with sea surface to move at horizontal speed of water so near-surface wind speeds are less than wave speeds. Critical height marks transition from wind speed less than wave speed to wind speed greater than wave speed and distorts the over-the-wave pressure pattern so it is capable of pumping energy into wave from wind. (Upper arrows show direction of air motion relative to fixed point moving with wave crest.)

served. Another American, Miles (1960) developed a theory that took over where Phillips' left off. The idea is complicated and based on the existence of the critical height discussed in the previous paragraph. According to Miles, the structure of the air flow at the critical height determines the force exerted by the wind on the sea surface. Deformation of this flow over existing waves produces a low pressure on the leeward face of the wave and a high pressure on the windward face. This is just what is needed to add energy to the waves, so they grow with wind duration and fetch. Lumped together then, these two recent concepts form the so-called Miles–Phillips Theory of wave generation. The curious fact that it is known as the "classical theory" shows the relative newness of the subject.

Since 1960, attempts have been made to refine the Miles–Phillips Theory to make it agree more closely with experimental findings. Generally, these efforts have met with limited success and it seems that the generation mechanism the theory describes cannot transfer enough energy from the wind to the waves to make them grow as quickly as observations indicate. Moreover, the theory cannot explain a phenomenon long observed by mariners, that a sudden reversal of the wind at sea literally knocks the waves flat. This leaves scientists somewhat in a quandary, but it does open the door for fresh ideas. One recent explanation for example, suggests that downdrafts associated with the wind, and not the horizontal winds, pump energy into the waves. "Cat's paws," which ripple the smooth surface of the water on nearly windless days, are created in this way. A downdraft carried forward with the wind could slide down the forward face of a wavecrest and then push on the trailing face of the adjacent wavecrest (Fig. 7.3). In this way, energy would be transferred from the wind to the waves by downward bursts of air. As local downdrafts are inherent to the general turbulence in the wind, their numbers and strength would increase as the average wind speed increased. Consequently, the height of

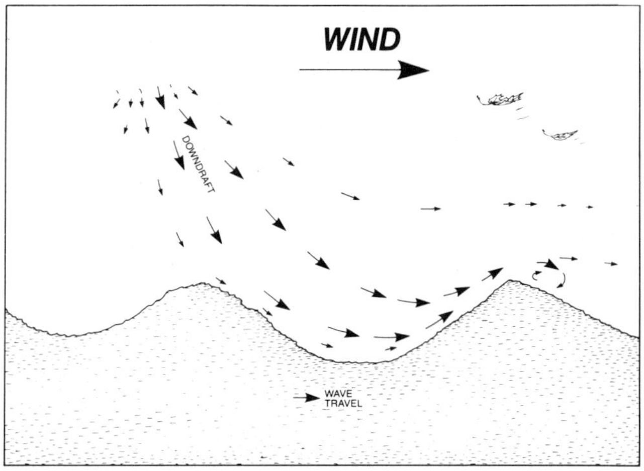

FIG. 7.3. How downdrafts that originate within main wind field could slide down leeward face of wave to push on windward face of adjoining crest. Wave grows in accordance with strength, duration, and repetition of downdrafts.

the waves would be determined by the strength, duration, and fetch of the wind. Wave knockdown would occur after a wind reversal because the downdrafts would naturally oppose the waves generated by the previous wind. Although this is all speculation, one thing is certain; once the winds begin to input some of their energy to the sea, it is the transfer of energy between the waves themselves via wave–wave interactions that is fundamental to the growth to higher and longer wind waves.

Growth of Wind Waves

Ripples first develop on a calm water surface when the wind reaches a certain threshold velocity. How large this velocity must be depends on the degree of surface contamination and, to a lesser extent, on the vertical stability of the air. Seawater is always heavily contaminated with various organic and inorganic materials that tend to stiffen the surface tension so that, compared to a glacial-melt lake for example, greater wind speeds are needed to deform the surface (1.0–1.5 m/s or 2–3 kn, as measured 10 m above the water). Furthermore, it is thought that the pressure fluctuations carried along by the mean wind, rather than the wind drag, initially cause the ripples.

Capillary waves controlled by the combined forces of gravity and surface tension are always the first ripples to appear on calm water once the wind begins to blow. These wavelets propagate at the minimum possible speed for such waves (23 cm/s) and form a criss-cross pattern of two sets of wave crests, each set moving at an angle 70–80° to the wind direction (Fig. 7.4) with regular spacings of about 1.8 cm between individual crests and with periods of 0.073 s. Because of their low heights, these wavelets are best seen where sunlight reflects from the water surface or as "dancing" patterns of light on the bottom of a shallow. The large angle between the wind direction and the direction of wave-crest travel is due to the fact that the slow propagation speed of the capillary waves makes it impossi-

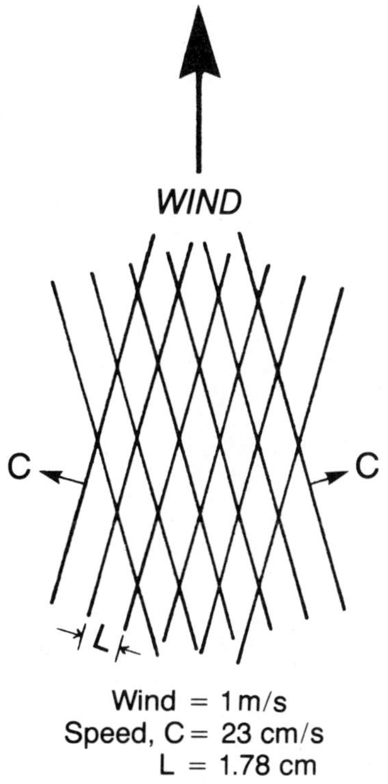

FIG. 7.4. Initial ripple pattern formed at onset of wave growth at low wind speeds. Two intersecting sets of long-crested wavelets move at minimum wave velocity (23 cm/s) at nearly right angles to the wind. (Modified after Van Dorn 1974)

ble for them to travel at the downwind speed of the pressure fluctuations. Therefore, instead of trying to keep direct pace with the wind, the wavelets simply head off in just the right direction where their speed matches the speed of one of the wind's velocity components. Actually, there are two such preferred directions, one to the left and one to the right of the main flow of air along which the waves continually receive energy from the pressure fluctuations.

If the wind dies, frictional effects dampen the ripples and they rapidly disappear, but if it strengthens, the lengths, heights, and periods of the ripples increase and there is a subsequent increase in their speed of propagation. This is accompanied by a decrease in the angle of wave travel with respect to the wind, until for winds of 2–3 m/s (4–6 kn) the wavelets move at 30° to the wind. The reason for this is that the component of the wind velocity which now matches the propagation speed of the waves lies ever closer to the true wind direction. At this stage, the initial criss-cross pattern of intersecting sets of long-crested wavelets has deformed into a chicken-wire type pattern (Fig. 7.5), with smooth wrinkles at the intersecting region traveling in the direction of the wind. The air flow over the wave still remains unaffected by the waves and the surface is said to be "hydrodynamically smooth."

For wind speeds above 3 m/s (6 kn), the growing waves are independent of surface tension and their heights become large enough to affect the air flow. The water

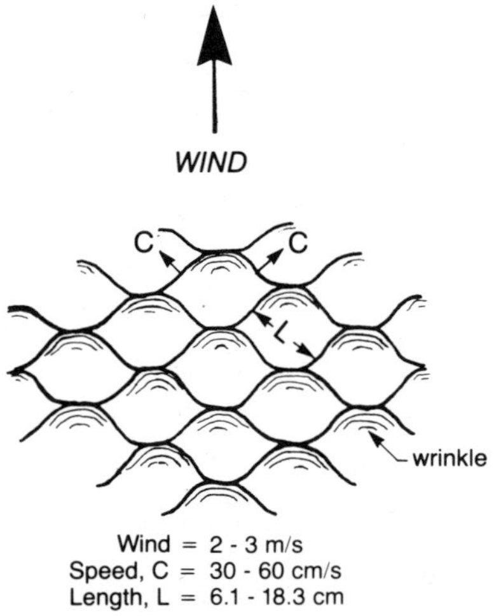

FIG. 7.5. Ripple pattern formed by long-crested wavelets at wind speeds 2–3 m/s (4–6 kn). Speed and length of waves increased from Fig. 7.4 and waves move more in direction of wind. (Modified after Van Dorn 1974)

surface has now become "hydrodynamically rough." This induces turbulent pressure fluctuations in the wind, which in turn increase the amount of energy fed into the waves, and accelerate their length and height. A sort of feed-back mechanism is set up by which disturbances in the wind caused by the wavy sea surface enhance the amount of energy transmitted from the wind to the water, and so on. In addition, the mechanism favors the growth of waves whose crests are propagating nearly in the wind direction, thereby narrowing the off-wind directional spread of wave energy. With increased wind speeds above 3 m/s, the wave field becomes progressively more irregular as differences in the heights, lengths, speeds, and periods of the waves are further exaggerated by the wind. In the generation region, moreover, wave–wave interactions effect the transfer of energy from shorter to longer waves and the dominant wavelength increases.

Once the growth of wind waves has progressed beyond the capillary stage it is no longer possible to distinguish between individual wave groups. Instead, there is an ensemble of waves that moves within approximately 50° to the left and right of the wind direction, whose description is only meaningful in a statistical sense. Rather than a collection of identifiable groups of waves, there is a continuous spectrum of waves, whose heights, periods, etc., range from the smallest capillary wavelets undergoing dissipation by friction, to the largest waves whose properties are limited by the eventual balance between the rate energy is supplied by the wind and the rate it is removed by dissipative processes such as breaking (the fully developed sea). Only on rare occasions is it possible to find well-defined wave patterns associated with intense winds. From a becalmed sailboat in Juan de Fuca Strait for example, the author once witnessed the approach of steep, breaking gravity waves roughly 30 cm high aligned in remarkably parallel rows for a considerable distance along the leading edge of a squall line moving from the south. The first crest in the advancing wave pattern struck only seconds after the arrival of a wicked downdraft, accompanied by horizontal winds of over 10 m/s. Following the passage of the front, the regular series of waves soon gave way to the confused pattern of seas typical of strong winds.

Wave Statistics

To describe and forecast the sea state, remarkably successful techniques have been developed, which relate certain statistical properties of the wave field to the wind strength, fetch, and duration. Although theory forms an integral part of such formulations, their backbone is based on actual observations obtained from a variety of sources. These include ship logs, waverider buoys (instruments tethered to a ship or to the bottom whose accelerometers record wave heights as they ride the surface), ship-borne accelerometers mounted in the hull, bottom-mounted pressure sensors, airborne laser altimeters, capacitance wave staffs that make use of electrical conductivity of seawater, and, more recently, satellite laser altimeters and over-the-horizon radar. With the continued improvements in collection methods and the amount of wind-versus-wave data, improved accuracy of the forecast techniques is assured.

One way to forecast information about wave heights within the wind-generating region is to compare known values of wind speed, duration, and fetch with a number called E, computed from a wave record obtained at a fixed location in deep water. Once the value of E is established for every type of wind condition it is simply a matter of drawing up graphs or tables, which permit the user to relate the wind to E, and then to the corresponding wave height, period, or length. A typical wave record (Fig. 7.6) from a waverider accelerometer buoy or from a ship-borne accelerometer consists of a series of 20-min recordings obtained every 3 h throughout the duration of the experiment. Computation of E for each 20-min segment is straightforward. First the record is divided into smaller equally spaced segments, say 1 s apart. Then, after the average height of all these incremental values has been determined (approximately corresponding to sea level in the absence of waves) the differences in height between the actual water elevation and the average level at each point is determined. Each difference value or amplitude is then squared and added together. The resulting sum is divided by the number of values used and multiplied by 2; in the present example the number of values used is 1200 s ÷ 1 s = 1200. Mathematically

$$E = \frac{2}{N}(a_1^2 + a_2^2 + \ldots + a_N^2)$$

where the amplitudes a_1, a_2, etc., are deviations of the water surface from its mean level, including both troughs and crests, and N is the number of increments ($N = 1200$ for 1-s increments in a 20-min record).

E has the dimensions of height squared (ft^2 or m^2) and is a measure of the average wave energy during the time the record was taken. Provided the wave heights were

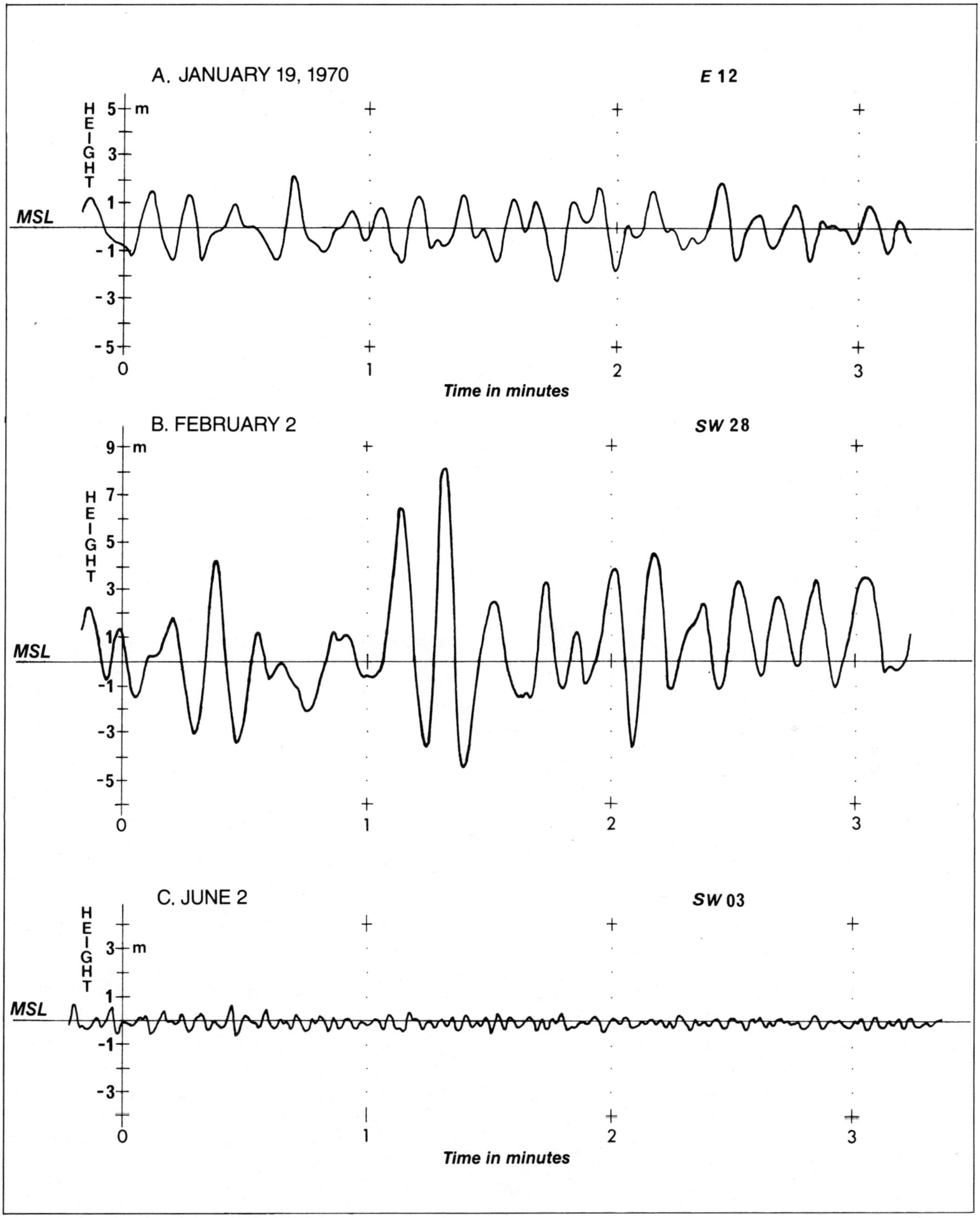

FIG. 7.6. Three examples of wave records obtained at Station P from ship-borne accelerometer wave meters onboard Canadian weatherships, 1970. Each segment is 3 min long, heights in metres relative to approximate mean sea level (msl). Wind direction and speed (m/s) 3 h prior to each record shown at right (in B, winds were southwest at 28 m/s or 54 kn). Large wave Feb. 2 had a crest-to-trough height 12.5 m (41 ft).

randomly distributed in space and time during the measurement period, $\sqrt{E}$ can be used to make certain predictions regarding the probability of encountering waves of a particular height. ($\sqrt{E}$ can have the dimensions of feet or metres. To convert its value in metres to feet multiply by 3.29.) For example, 10% of the waves will have heights less than 0.64 $\sqrt{E}$ and 10% will have heights greater than 3.04 $\sqrt{E}$. Other ranges are shown in Table 7.1.

TABLE 7.1. Approximate wave height distributions for a given value of E, in 10% steps. Heights are measured from trough to crest. (After Pierson *et al.* 1971)

10% ranges	Cumulative ascending values	Cumulative descending values
0.00–0.64 $\sqrt{E}$ m(ft)	10% less than 0.64 $\sqrt{E}$ m(ft)	10% greater than 3.04 $\sqrt{E}$ m(ft)
0.64–0.94 $\sqrt{E}$ m(ft)	20% less than 0.94 $\sqrt{E}$ m(ft)	20% greater than 2.54 $\sqrt{E}$ m(ft)
0.94–1.20 $\sqrt{E}$ m(ft)	30% less than 1.20 $\sqrt{E}$ m(ft)	30% greater than 2.20 $\sqrt{E}$ m(ft)
1.20–1.42 $\sqrt{E}$ m(ft)	40% less than 1.42 $\sqrt{E}$ m(ft)	40% greater than 1.92 $\sqrt{E}$ m(ft)
1.42–1.66 $\sqrt{E}$ m(ft)	50% less than 1.66 $\sqrt{E}$ m(ft)	50% greater than 1.66 $\sqrt{E}$ m(ft)
1.66–1.92 $\sqrt{E}$ m(ft)	60% less than 1.92 $\sqrt{E}$ m(ft)	60% greater than 1.42 $\sqrt{E}$ m(ft)
1.92–2.20 $\sqrt{E}$ m(ft)	70% less than 2.20 $\sqrt{E}$ m(ft)	70% greater than 1.20 $\sqrt{E}$ m(ft)
2.20–2.54 $\sqrt{E}$ m(ft)	80% less than 2.54 $\sqrt{E}$ m(ft)	80% greater than 0.94 $\sqrt{E}$ m(ft)
2.54–3.04 $\sqrt{E}$ m(ft)	90% less than 3.04 $\sqrt{E}$ m(ft)	90% greater than 0.64 $\sqrt{E}$ m(ft)
greater than 3.04 $\sqrt{E}$ m(ft)		

Figure 7.7 indicates the kind of range for $\sqrt{E}$ expected for various wind speeds in a fully developed sea. At wind speeds of 15.5 m/s (30 kn), for example. $\sqrt{E}$ = 1.82 m (6 ft) so that 10% of the waves will be below 0.64 × 1.82 m = 1.16 m (3.8 ft), whereas 10% will exceed 3.04 × 1.82 m = 5.53 m (18.2 ft). The value of E for a fully developed sea is given in terms of the wind speed, U, by

$$E(\text{m}^2) = 0.616\ (U/10)^5 \text{ for } U \text{ in m/s}$$
$$E(\text{ft}^2) = 0.240\ (U/10)^5 \text{ for } U \text{ in kn}$$

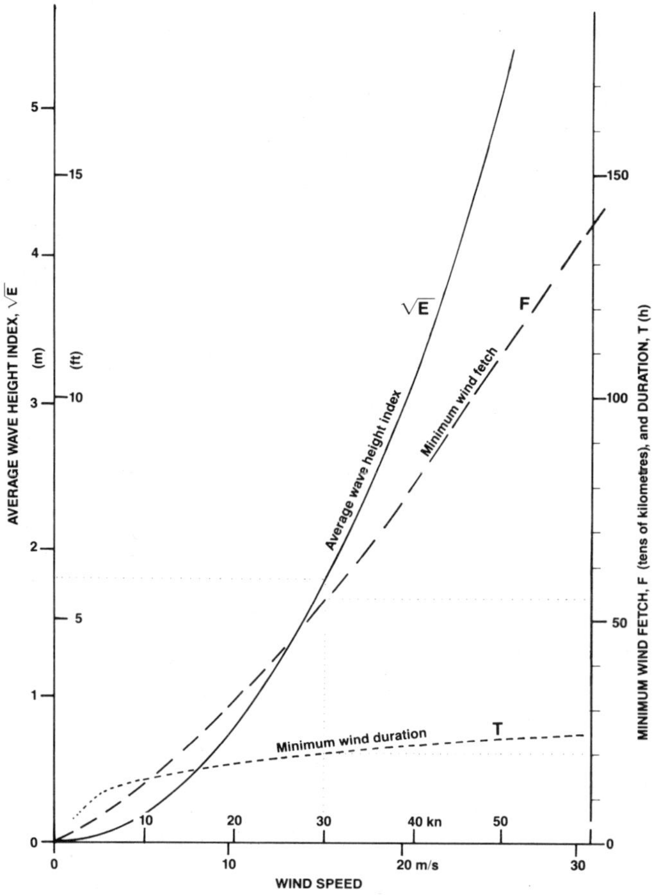

FIG. 7.7. Average wave height index $\sqrt{E}$ versus wind speed for fully developed sea. Minimum duration, T, and minimum fetch, F, of wind needed to generate fully developed sea at various wind speeds. For a 15.5 m/s (30 kn) wind, $\sqrt{E}$ = 1.82 m (6 ft), T = 20 h, F = 547 km (dotted lines).

Four of the most commonly used indices of average wave height are:

most frequent probable wave height	H_f	$= 1.41\sqrt{E}$
average height of all waves	$\overline{H}$	$= 1.77\sqrt{E}$
average height of highest ⅓ of all waves	$H_{1/3}$	$= 2.83\sqrt{E}$
average height of highest ⅒ of all waves	$H_{1/10}$	$= 3.60\sqrt{E}$

The average value, $H_{1/3}$, called the significant wave height, is perhaps the most useful measure of the larger waves for the mariner. Its value from the above formula is equivalent to adding the heights of the highest wave, the second highest wave, the third highest wave, and so on, until one-third of all the waves in a record have been taken into account, and then finding the average of the sum. (Similarly $H_{1/10}$ is found by truncating the sum after ⅒ the waves have been used.) As there is roughly a 15% chance of encountering a significant wave in a sea, it is somewhat akin to the saying that "every seventh wave is the highest." It should always be borne in mind, however, that probability theory only yields the "possibilty" of encountering a particular wave height in a large sample of waves; thus, although on average one in seven waves will be a significant wave, two significant waves might occur one after the other or be separated by hundreds of waves. As an analogy, a gambler will on average throw a pair of sixes once for every 36 attempts with the dice (odds: 1 in 36) but may throw a pair of sixes on the first roll of the dice, or on the second roll, or on the hundredth roll.

Use of the significant wave as a sea state forecast tool was initially developed during World War II and led to the so-called S-M-B Method of prediction, after Sverdrup, Munk, and Bretschneider who helped perfect it. Graphs like

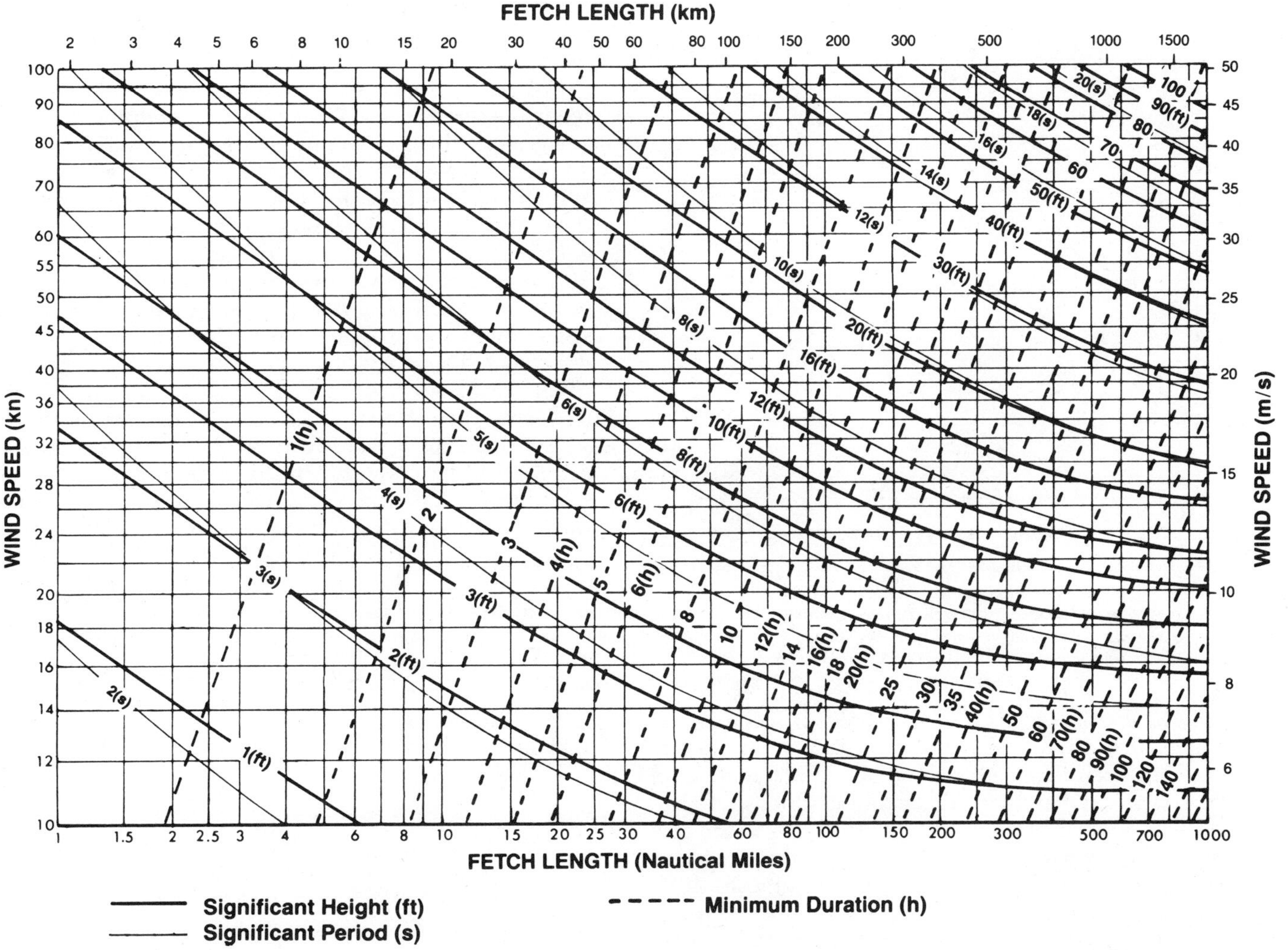

FIG. 7.8. Deepwater wave-forecasting curves give significant height, $H_{1/3}$, period, $T_{1/3}$, for different wind speeds, fetches, and durations. From known wind speed on left-hand axis, move across to where line intersects appropriate fetch line (if wave generation is fetch limited) or appropriate duration line if duration limited; read off values of $H_{1/3}$, $T_{1/3}$. A 15.5 m/s (30 kn) wind over fetch of 140 km for more than 10 h is fetch limited, so $H_{1/3}$ = 3 m, $T_{1/3}$ = 7 s. (Modified after U.S. Army Coastal Engineering Research Center 1977)

Fig. 7.8 were devised to estimate the significant wave height and its associated period, $T_{1/3}$, at the end of a fetch length for a specified wind speed, duration, and fetch. Such graphs also permit determination of the significant wave characteristics under fetch-limited and duration-limited conditions, if either the fetch or duration of the wind is insufficient to generate fully developed seas. Later modifications took into consideration the effects of shallow-water refraction, bottom friction, and percolation on the growth of wind waves. Once the significant wave height is known other wave statistics such as $H_{1/10}$, $\overline{H}$, and H_f can be readily determined.

Extreme Waves

The height of the most probable maximum wave, H_{max}, the highest wave likely to be encountered during a particular sea state, is a function of both $\sqrt{E}$ (or the significant wave height) and the total number, N, of waves encountered. That is, the chances of meeting a giant, breaking "rogue" wave are increased as greater energy is fed into the ocean waves by the wind and as more waves pass by. Figure 7.9 indicates for instance that for every 200 waves (broken line) there is a 5% chance one will exceed 5.8 $\sqrt{E}$ and an equal probability one will be below 4.1 $\sqrt{E}$. The most probable extreme wave height in the 200-wave sample will be 4.9 $\sqrt{E}$. For a fully developed sea in 15-m/s winds, the most probable maximum wave height will be 4.9 × 1.82 m = 8.9 m, with a 5% chance that one wave in 200 will exceed 10 m. Moreover, because all three curves in Fig. 7.9 slope upward to the right, the probability of a ship encountering progressively higher waves increases in a predictable fashion for every passing wave, provided of course that the sea conditions remain nearly constant in time. Consequently, there are limitations to the applicability of this figure. It applies best to slowly decaying swell and provides good estimates for fully developed or fetch-limited seas, but will under estimate extreme wave heights in a rapidly growing sea. In reality, only fully developed sea conditions can be reproduced with consistent reliability if wind-speed is the only known factor.

The length of time required to establish the extreme-wave probabilities in Fig. 7.9 depends on the average wave period, $\overline{T}$, for the particular sea state, and the number of

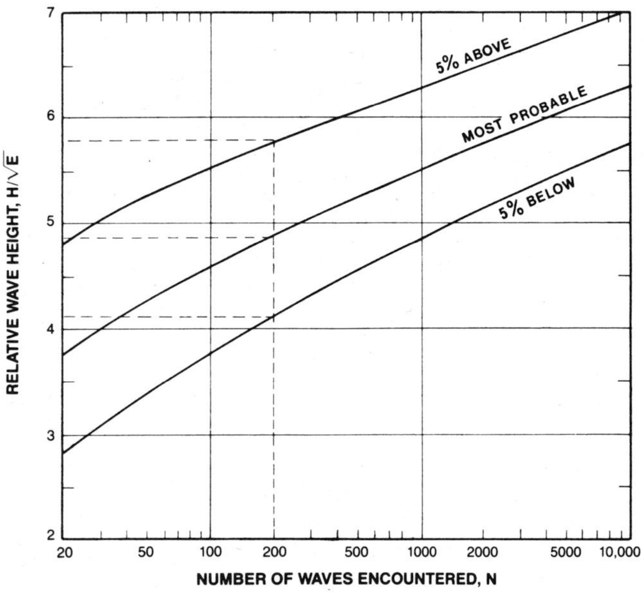

FIG. 7.9. Extreme wave probability increases with number, N, of waves. There is a 5% chance highest of 200 successive waves will exceed $H = 5.8\sqrt{E}$, the same chance it will be lower than $H = 4.1\sqrt{E}$, 90% probability it will be between. (From Van Dorn 1974)

waves, N, considered. If $\overline{T}$ is measured in seconds, for example 10 s (a value likely to occur during a moderate blow in the open ocean), then the above estimates are valid provided that the average sea conditions do not change appreciably over the time $\overline{T} \times N = 10 \times N$ s. In the case of $N = 200$ waves traveling past a fixed observation point, sea conditions would need to stay about the same for 2000 s or about ½ h. Clearly, the greater the number of waves under consideration the longer the sea state must remain unchanged before simple estimates of maximum probable wave heights can be made by graphs as in Fig. 7.9.

Lastly, the maximum probable wave height can be estimated from the significant wave height by

$$H_{max} = b \times H_{1/3}$$

where b is in the range 1.53–1.85. As with the previous discussion, the choice of b depends on the average wave period and the length of the wave record (i.e. the number of waves), with the larger value applying to many thousands of waves.

Wave Spectra

The wave spectrum concept provides an alternative, more modern approach to the description of wind-generated seas. It differs from the previous statistical approach in that the sea state is described in terms of a spectrum of wave energies and periods rather than in terms of discrete values, such as the significant wave height and period. Though its practical application is more difficult, it yields better insight into the nature of wave growth and distribution of energy among the various wave components. This is not to say that the earlier method is obsolete. On the contrary, those who have a vested interest in wave prediction still use it to determine significant, and related, wave properties.

Figure 7.10 shows a spectrum obtained from wave records collected onboard CCGS *Quadra* near Cobb Seamount, 500 km southwest of Vancouver Island. In essence, it is a plot of the observed wave energy (height^2) at each possible wave frequency, f, where f, the inverse of the wave period, T, is measured in cycles per second. (If 5 equally spaced waves move past the observation point in 20 s, then their "frequency" is ¼ of a wave, or "cycle," per second and their period is 4 s.) The total area under the spectral curve is proportional to the total energy in the wave field. The shape of the curve tells how the energy is partitioned among the various periodic waves of different frequency (or period), which have combined to form the sea state. Except at very low wind speeds, spectra like that of Fig. 7.10 have a well-defined maximum with a corresponding peak period, T_{peak}, where the concentration of wave energy is greatest. The number of waves in a particular wave record can be obtained by dividing the length of time of the record by the peak period. For instance, the 28-min record in Fig. 7.10 had a peak period of 13 s and, therefore, consisted approximately of 129 waves.

For increasing wind speeds, the main concentration of energy in a wave spectrum shifts toward the lower frequencies (longer periods), and indicates that energy within the growing sea is transferring to longer and

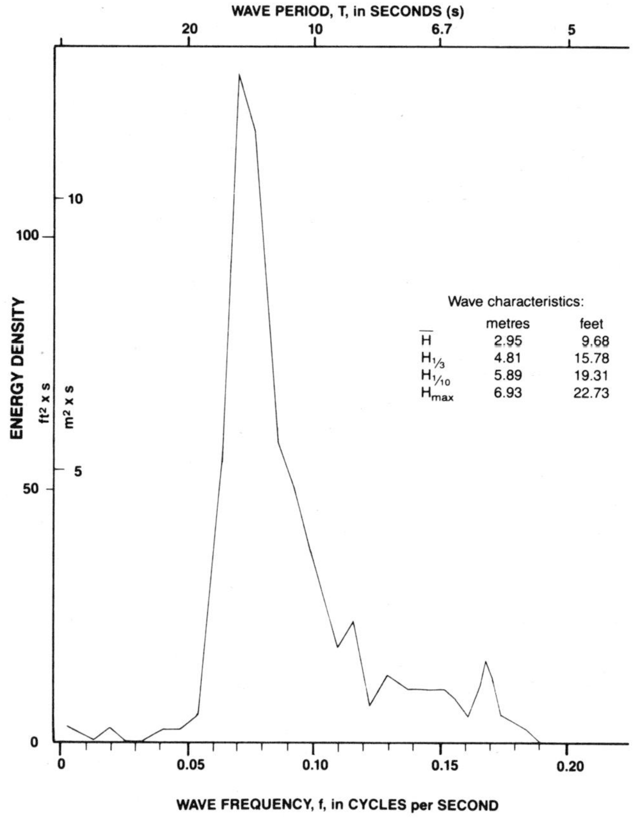

FIG. 7.10. Wave spectrum based on wave records collected at Cobb Seamount Feb. 17, 1973, from recorder in hull of CCGS *Quadra*. Peak period is roughly 13 s. Other characteristic features of waves are in Table.

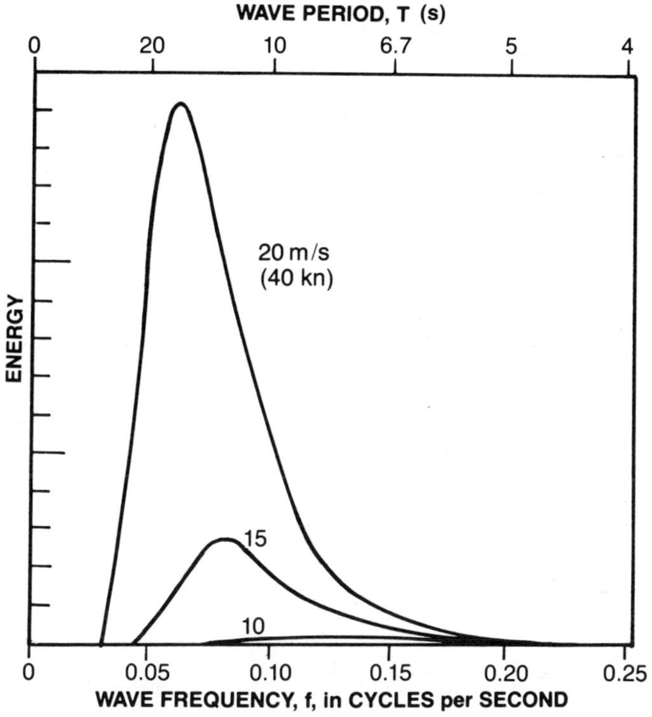

FIG. 7.11. Wave spectra for fully developed seas at wind speeds of 10, 15, and 20 m/s (roughly equivalent to winds of 20, 30, and 40 kn, respectively). Peak height increases as square of wind speed and shifts to lower frequencies (longer periods) as wind strengthens. Area beneath each curve is proportional to total energy of all waves and increases as fifth power of wind speed.

longer waves (Fig. 7.11). This feature is further reflected in the fact that the overall range of frequencies (or periods) increases with increased wind speed, and most extension in range occurs at the lower end of the frequency scale. As an approximation, the period with the maximum energy density is related to the wind speed, U, by $T_{peak} = 0.787\,U$(s) if U is in m/s, or $T_{peak} = 0.404\,U$(s) if U is in knots. Table 7.2 gives the peak period and peak frequency for various wind speeds based on these formulae. The average wave period, $\overline{T}$, in a fully developed sea (the period in practice measured by timing a succession of wave crests traveling past a fixed point) is related to the wind speed via $\overline{T} = 0.554\,U$ (s) when U is in m/s (= 0.285 U for U in knots). The average period can be used to estimate the average wave length, $\overline{L}$, in a fully developed sea via the equation

$$\overline{L} \simeq 3.41\,\overline{T}^2 \text{ (ft)}$$
$$\simeq 1.04\,\overline{T}^2 \text{ (m)}$$

for $\overline{T}$ in seconds (see Table 7.3). However, wave lengths are much more variable than periods because of the presence of short waves and generally wavelengths are considerably shorter than would be expected using $\overline{T}$.

TABLE 7.2. Peak period (T_{peak}) and peak frequency (f_{peak}) in wave spectrum as a function of wind speed (U) for fully developed sea. Each value is at midpoint of band of frequencies (or periods) where most spectral wave energy is concentrated. $T_{peak} = 0.787\,U = 1/f_{peak}$ where f_{peak} is in cycles per second, cps

Wind speed (*m/s*)	Wind speed (*kn*)	T_{peak} (*s*)	f_{peak} (cps)
5	9.7	3.9	0.254
10	19.4	7.9	0.127
15	29.2	11.8	0.085
20	38.9	15.7	0.064
25	48.6	19.7	0.051
30	58.3	23.6	0.042

TABLE 7.3. Average period ($\overline{T}$) and average wave length ($\overline{L}$) as a function of wind speed (U) for fully developed sea. Values based on equations $\overline{T} = 0.554\,U$ (s) and $\overline{L} = 1.04\,\overline{T}^2$ (m). Wavelength distributions are presently unknown, so latter represents only a crude approximation to average conditions in a sea.

Wind speed (*m/s*)	Wind speed (*kn*)	$\overline{T}$ (*s*)	$\overline{L}$ (*m*)
5	9.7	2.8	8
10	19.4	5.5	32
15	29.2	8.3	72
20	38.9	11.1	128
25	48.6	13.9	200
30	58.3	16.6	287

In addition to the shift in frequencies associated with a growing sea, there is a rapid increase in the total energy of the wave field as indicated by the greater successive areas under the spectral curves. For a fully developed sea, the total energy, E_t, is proportional to the fifth power of the wind speed; mathematically $E_t \propto U^5$. Hence, a doubling of the wind speed, say from 10 to 20 m/s, leads to a factor of 32 increase in the energy contained in the waves. As E_t is in turn proportional to the height squared (H^2), the significant wave height derived from the spectrum for a wind of unlimited fetch and duration is given by $H_{1/3}$ = constant × $U^{5/2}$, whereby a doubling of the wind speed leads to a factor of 5.7 increase in the significant wave height. The constant value is determined by actual wind-versus-wave data. A sea is fully developed for a given wind speed when all possible wave components that make up the spectrum are present with their maximum amount of energy. Storm seas at wind speeds in excess of about 25 m/s (50 kn) rarely reach the fully developed stage because, even if the fetch is sufficient, the duration of the storm usually falls short of the required time to bring the waves to maximum fury. The main exception is in the Southern Ocean which encircles Antarctica where fetches are long, and where numerous heavy storms in rapid succession have an integrated effect comparable to a single heavy storm of long duration. This effect is responsible for the well-known extreme storm conditions in winter off Cape Horn (56°S). In fact, winds in the Antarctic region can be so violent that masters of large sailing ships assigned them names according to the degree of rigging noise they produced. The zone between 40–50°S became known as the "roaring forties," between 50–60°S as the "howling fifties," and that above 60°S the "screeching sixties."

Another feature in Fig. 7.11 is that the significant part of each wave spectrum becomes concentrated into an ever-narrowing range of frequencies as the wind strengthens. For wave forecasting, therefore, spectra can be cut off at the high and low ends of the frequency scale for energy levels below a certain minimum percentage (about 5%) of the total wave energy.

The Beaufort Scale

The Beaufort Scale of Wind Force (Table 7.4) relates wind speed to the physical appearance of the sea surface by considering such factors as apparent wave height and the prominence of breakers, whitecaps, foam, and spray. Originally devised by Admiral Beaufort in the early 19th century to simplify the signaling of wind and weather conditions between sailing vessels, it has since been repeatedly modified to make it more relevant to modern navigation. As with any subjective observation method, the Beaufort Scale is far from perfect. For example, one person's estimate of the sea state may differ from another's, and steep breaking waves in an opposing current can lead to an overestimation of the actual wind speed. Studies also have shown that estimates of the Beaufort Number from coastal locations differ from those in the open ocean at comparable wind speeds. Furthermore, the method has less applicability to fetch-limited protected waters where fully developed seas cannot develop. Proposal of a new Beaufort Scale somewhat different from that in Table 7.4 is at present under consideration by the World Meteorological Organization in Geneva.

TABLE 7.4. Beaufort scale of wind force.

Beaufort	Mariner's Description	Wind Speed (*kn*)	(*m/s*)	Effect of Wind at sea	on land
0	calm	0–1	0–0.5	like a mirror	still, smoke rises vertically
1	light air	1–3	0.5–1.5	ripples form with the appearance of scales, but without foam crests	smoke drifts, vanes remain motionless
2	light breeze	4–6	2.1–3.1	small wavelets, crests appear glassy, no breaking	leaves rustle, vanes move, wind can be felt on face
3	gentle breeze	7–10	3.5–5.2	larger wavelets begin to break, glassy foam, perhaps some scattered white horses	constant movement of leaves and small twigs, flags begin to stream
4	moderate breeze	11–16	5.7–8.2	small waves predominant but fairly frequent white horses	dust and loose paper are lifted, thin branches move
5	fresh breeze	17–21	8.7–10.8	moderate waves, distinctly elongated, many white horses, chance of spray	small trees in leaf begin to sway
6	strong breeze	22–27	11.3–13.9	long waves with extensive white foam breaking crests begin to form, spray likely	large branches move, power lines whistle, stop lights sway, umbrellas difficult to control
7	moderate gale	28–33	14.4–17.0	sea heaps up, white foam breaking waves start to be blown in streaks, beginning of spindrift	entire trees sway, some resistance to walkers, car feels force of wind
8	fresh gale	34–40	17.5–20.6	moderately high waves with extensive crests, tops of crests break into spindrift; foam blown into well-marked streaks, spray blown off crests	twigs break off trees, difficult walking against wind
9	strong gale	41–47	21.1–24.2	high waves, rolling sea, dense streaks of foam, spray may affect visibility	roof tiles lifted off, windows may be blown in, trees may top
10	white gale	48–55	24.7–28.3	very high waves with long overhanging crests, foam in great patches blown in dense white streaks downwind; heavy rolling sea causes ships to slam, visibility reduced by spray, sea surface takes on whitish appearance	trees uprooted, considerable structural damage to some buildings
11	storm	56–66	28.8–34.0	exceptionally high waves, sea covered with long white patches of foam blown downwind, everywhere wave crests blown into froth, visibility impeded by spray	widespread damage, extensive flooding in low lying areas if winds onshore
12	hurricane	above 66	above 34	air filled with foam and spray, sea entirely white, visibility seriously impaired	

Wave Attenuation

For the most part, energy imparted to waves by the wind is eventually lost as breakers at the seashore. In addition, however, other processes within the ocean provide for both a localized as well as global loss and redistribution of wave energy long before it reaches the coast.

Waves in a mature or fully developed sea state have limited energy due mainly to the dissipative effects of wave breaking. And, because most fetches in the open ocean have a finite extent determined by the wind field, a portion of the energy also continually radiates out of the generation region in the form of swell. Moreover, not all waves are generated exactly in the direction of the wind, hence, there will always be some leakage of wave energy through the sides of a fetch. The amount escaping in this manner decreases rapidly with increased angle off the wind, and less than 10% of the total swell energy propagates at angles greater than about 50° from the wind direction. For an abating wind, the decay of wave height within the fetch is largely governed by the rates at which the wind speed diminishes and energy escapes the generating region.

Outside the confines of the wind fetch, the spectrum of wave energy in a particular direction is slowly eroded through the combined effects of internal viscous dissipation, geometrical spreading, dispersion, opposing winds, and wave–wave interactions. Each effect will be considered separately.

Internal dissipation, a result of the inherent friction (or viscosity) of the fluid and surface film effects, rapidly attenuates wave periods of less than 2–3 s (short wavelengths) but has negligible effect on long period waves. Viewed in terms of the wave spectrum, it tends to erode the high frequency end but leaves the low frequency end unscathed. Due to viscous damping alone, a 4-s period wave will travel for about 1000 h, a deepwater distance of some 23,000 km, before its height is reduced to half its original value. But a 1-s wave travels for only 4.3 h, a distance of 12 km, before its height is reduced by one half.

Geometrical spreading occurs because not all waves generated within the fetch propagate precisely downwind. With increased decay distance then, the narrow beam of wave energy initially centered along the wind direction gradually fans out over an ever-broadening area, with a subsequent reduction in wave height. A common example of geometric spreading is the widening of a light beam from a flashlight in which the intensity of light (or energy per unit area) diminishes with distance. As the flashlight beam spreads in all three directions its intensity falls off as the square of the distance; surface waves, on the other hand, spread in only two directions so their "intensity" diminishes in direct proportion to distance.

As discussed earlier, dispersion is a kind of filtering process whereby waves are sorted out according to their period. Longer period waves have greater group velocities in deep-water than do shorter period waves and, therefore, lead the way out of the fetch. An assemblage of different wave periods will more and more resolve itself into groups of smooth undulating swell with increased distance from the generation area. As a consequence, fewer wave components and, therefore, less total wave energy, will be present over a specified duration of time the farther an oceanic area is from the original fetch. In other words, the total wave energy is spread ever thinner in the direction of swell propogation because of the difference in travel speeds of the various wave components. Thus, long, low-energy swells with periods of around 20 s often act as forerunners of storms by preceding the arrival of the main body of waves from a storm. The familiar ground swells are groups of long waves that originate from a distant storm that gradually decrease in height and period as the storm abates or passes by.

Although a sudden shift in the wind effectively reduces waves in a developing sea, contrary winds cause only a minimal reduction in swell heights. It seems that under most circumstances the long, low, gentle swell profiles offer little resistance to the opposing wind.

Provided their periods, wavelengths, and propagation directions simultaneously conform to a set of prescribed conditions, waves can effect an exchange of energy with one another. Through such wave-wave interactions, a portion of the energy of a particular wave is transferred to another wave of almost identical period, but propagating in a slightly different direction. The latter results in a transfer of wave energy into off-wind directions, and makes the process somewhat similar to geometric spreading. But unlike geometric spreading, wave-wave interactions are most effective within and near the generating region and become increasingly unimportant away from the fetch as dispersion separates the wave components on the basis of period.

To summarize, the short-period waves within the storm generation area are continually attenuated by viscous damping and wave breaking. Longer period waves on the other hand are only slightly modified by these damping processes and escape the fetch as slowly undulating swell. As their distance from the fetch increases, the swells spread out by dispersion, the longest period waves propagate the fastest. Wave-wave interactions are important in the dissipation of swell energy in and near the generation region, and geometric spreading causes a slight decrease in wave height with distance from the storm. Once waves have been transformed into regular swell they may travel many thousands of kilometres across the world oceans with only minor energy loss. Snodgrass et al. (1966) traced the path of storm waves generated near Antarctica as far as Yakutat, Alaska, over 10,000 km. The waves traveled at the deepwater group speed appropriate to their period, and followed great circle routes, which indicates a negligible effect on their path by the earth's rotation. Wave spectra at 0, 1000, and 10,000 km from the end of the storm fetch (Fig. 7.12) indicated that most wave energy dissipation occurred in the vicinity of the generation region, and that dissipation was most pronounced for short wave periods. After 1000 km, little additional energy loss took place. Using observed wave data, Bretschneider (1952) prepared a series of empirical graphs that give the change in wave height and period as a function of both the decay distance from the end of the fetch and the length of the fetch. These demonstrate, for instance, that the higher and steeper the waves the more

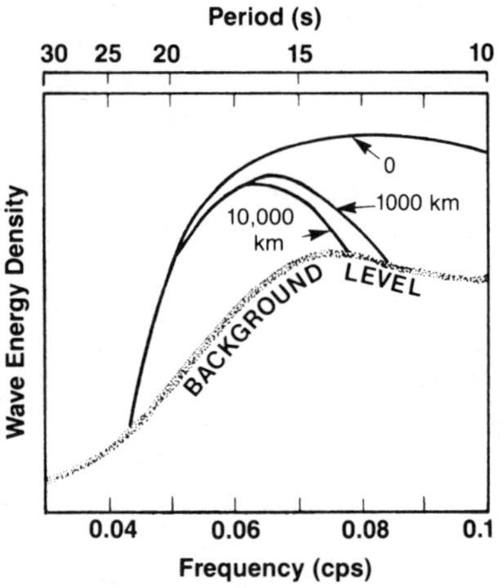

FIG. 7.12. Spectra associated with waves generated by storm in South Pacific Ocean, at 0, 1000, and 10,000 km from storm front. (From Snodgrass et al. 1966)

rapid their attenuation as they radiate out of the fetch. As an example, for an initial significant wave height of 4.6 m (15 ft) at the end of a 740 km (400 nm) fetch, the graphs show that the height would be reduced to 2.3 m after 740 km, and to about 1.4 m after a further 2800 km. The significant wave period on the other hand would be increased, an initial period of 10 s would become approximately 12.6 s at 2800 km.

Winds and Waves in a Moving Fetch

No description of oceanic waves would be complete without consideration of the wave patterns generated by a traveling atmospheric depression. For the sake of expediency, a mature storm that moves at constant velocity over the open ocean will be considered, bearing in mind that in any practical situation storms commonly alter both direction and speed, and the intensity and fetch of accompanying winds also undergo continual modification as the depression grows and decays.

A typical fully developed extratropical depression in the Northeast Pacific consists of cyclonic winds that blow around an eastward migrating low-pressure core. To a first approximation, the region of the storm can be subdivided into four sectors each with different wind directions, speeds, and fetch (Fig. 7.13). The wind in a given sector will generate its own particular wave field which (depending on the speed and duration of the storm center relative to the waves) may be contaminated with swell that have propagated in from one or more of the adjoining sectors.

As the storm approaches, an observer situated east of sector A (Fig. 7.13) first experiences light, steady winds from the south. These are accompanied by choppy seas, whose wave heights and periods grow with time as if generated by a sudden, isolated wind fetch from the south. Provided the storm's speed is less than the group speed of the longest waves radiating eastward from sector D, the observer would also be conscious of a low ground swell from the west. During the later stages of development, however, this swell is overshadowed by the northerly swell of sector A that becomes the longest and highest of the storm.

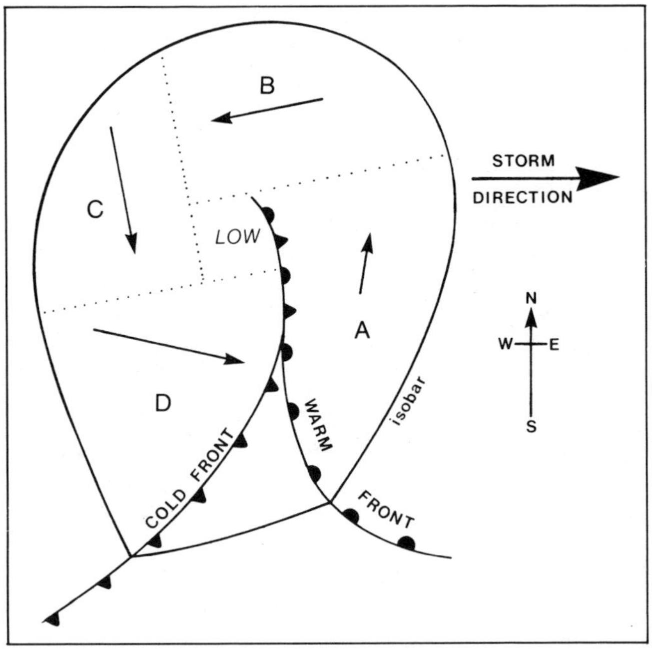

FIG. 7.13. Eastward advancing storm subdivided into four major wind sectors. Arrows denote direction and relative strength of winds characterizing each sector (compare with Fig. 2.15d).

With the arrival of the frontal edge of sector D, winds veer rapidly to the west, seas become steep and chaotic and heavy rain begins to fall. It is in this sector that winds, seas, and storm move in the same direction, so that the waves remain longest under the influence of the strongest winds of the storm to attain their maximum heights. Wave growth is considerably more rapid than for the stationary fetch of comparable duration and strength. (In the worst possible situation of a storm moving at the same speed as the highest waves in sector D, the unfortunate mariner would encounter a sudden 90° shift to gale-force winds accompanied by fully developed seas from the west.) Conditions are made even more chaotic by remnant seas from the south and by large swell radiating inward from sector C. Sloppy seas may persist long after the storm has passed. The sole consolation is that swell are lowest and shortest of any sector of the storm due to alignment of the winds along the direction of storm travel.

An observer initially located to the east of sector B fares much better than his counterpart east of sector A. For one thing, swell waves generated by the gradually increasing winds of sector B are radiating away from his position. Even more importantly, the waves and storm are moving in opposite directions. As a result, energy continually escapes the fetch, and limits the growth rate and maximum height of waves within the sea state. In addition, the relatively high swell that propagate northward from sector A are somewhat dispersed by the time they travel the width of the storm and are generally not a problem. With the passage of the storm center to the south, the second observer would note a gradual backing of the wind to the north, but without any appreciable change in speed. Seas would again be limited by the relatively short fetch. Following the complete passage of the depression, the relaxing mariner would only be aware of some swell activity through his area from sector B.

To summarize, wave conditions to the north of an advancing storm in the northern hemisphere are generally less extreme than to the south, despite the fact that both regions are exposed to similar average wind speeds and durations and total fetch. The southern location is more dangerous because winds in sector D are stronger than in other sectors, and because the fetch and waves move with the storm, to effectively trap the wave energy. Not only do seas become highest in this reigon but also their growth rate is greatest. On the other hand, when a storm moves much more slowly than the waves it generates, the swell is able to radiate considerable energy out of sector D. Seas in this sector would still be highest, but their growth rate and final energy levels would be reduced compared with a rapidly moving storm.

Chapter 8. Shallow-Water Waves

When the water depth shoals to less than ¼ the crest-to-crest distance between successive deepwater waves, the waves begin "to feel the bottom" and to adjust their speed in accordance with changes in the local bathymetry. This influence of depth on the propagation of waves increases with continued shoaling until, for shallow-water depths of less than 1⁄20 the wavelength, it becomes the dominating factor.

As a deepwater wave travels into shallow water, it undergoes two important alterations in its speed of propagation. First, the wave propagates at exactly the same speed as the group as a whole, compared to deep water where its speed is twice the speed of the group. Second, the wave speed becomes directly proportional to the square root of the water depth, whereas in deep water its speed is totally independent of depth. More specifically, in deep water, the wave speed $C_d = \sqrt{gL/2\pi}$, where L is the wavelength and g is the acceleration of gravity, whereas in shallow water $C_s = \sqrt{gD}$, where D is the total depth of water beneath the wave. The first alteration means that in shallow water it is possible to stick with an individual wave as it advances over the bottom, much as surfers do. In this case, the waves behave more like a collection of solitary waves, each independent of the other, than a closely tied group within a wave train. The second alteration leads to an important phenomenon called refraction, the bending or curving of the direction of travel of a wave as it encounters changes in the factors that affect its speed. (As an analogy, mirages and the shimmering of hot pavement are produced by the refraction of light "waves" as they pass through layers of air of different temperature. The bending occurs because the speed of light increases as the air becomes warmer. Refraction causes the apparent bending of an oar sticking out of the water, as shown in Fig 8.1). Intermediate-water waves are also refracted by the bottom but to a lesser degree than shallow-water waves.

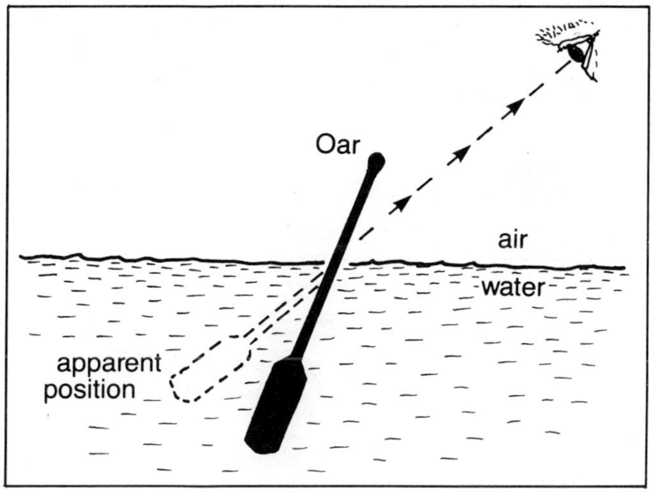

FIG. 8.1. Apparent bending of oar due to refraction of light as it passes from water to air. Speed of light increases as density of medium decreases, so light ray emerging from water is curved toward sea surface. Projection of ray paths back from observer's eye, makes oar appear bent and slightly smaller.

Refraction

The bending of wave patterns in shallow water stems from the direct dependence of wave speed on depth. If one part of a wave's crest (or trough) is in deeper water than an adjoining part of the same crest (or trough) the former will move more rapidly and the wave front will curve round. The bending always occurs in such a way that the wave becomes more aligned with the contours of the bottom (Fig. 8.2). This is the reason waves on a beach are always closely parallel to the shoreline just before they break, despite the fact that they may initially approach the beach at large angles. The old saying that "points of land draw the waves" derives from the fact that refraction tends to concentrate waves at headlands but to diffuse them in bays. Refraction also causes waves to concentrate over submarine ridges and to spread out over submarine canyons, and accounts for the existence of criss-cross wave patterns in the lee of an island where one would normally expect to be shielded from any wave action (Fig. 8.3, 8.4). Finally, longer waves such as swell are bent more than shorter waves simply because they feel the bottom sooner and have more time to be influenced by the changing bottom contours.

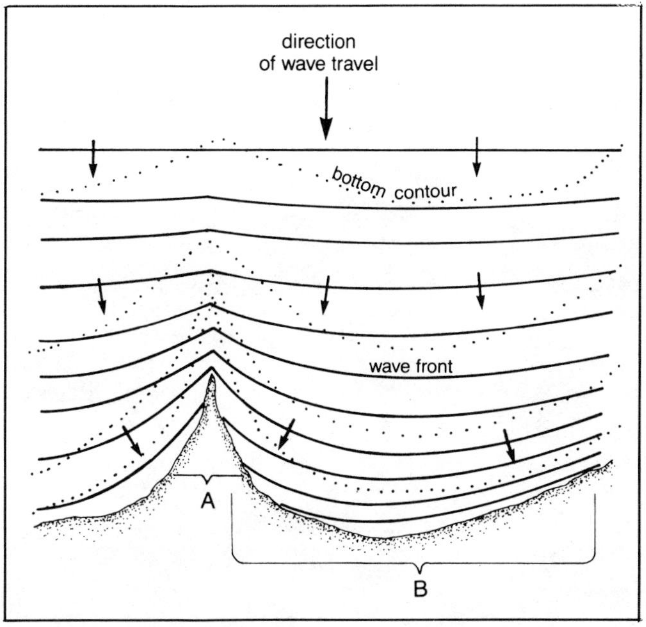

FIG. 8.2. Refraction of ocean waves approaching a coast. Bathymetric contours (dotted lines) shoal shoreward. Wave crests (solid lines) propagate in direction of arrows or "rays." Waves converge toward submarine ridge and associated point of land, A, and diverge outward within embayment, B.

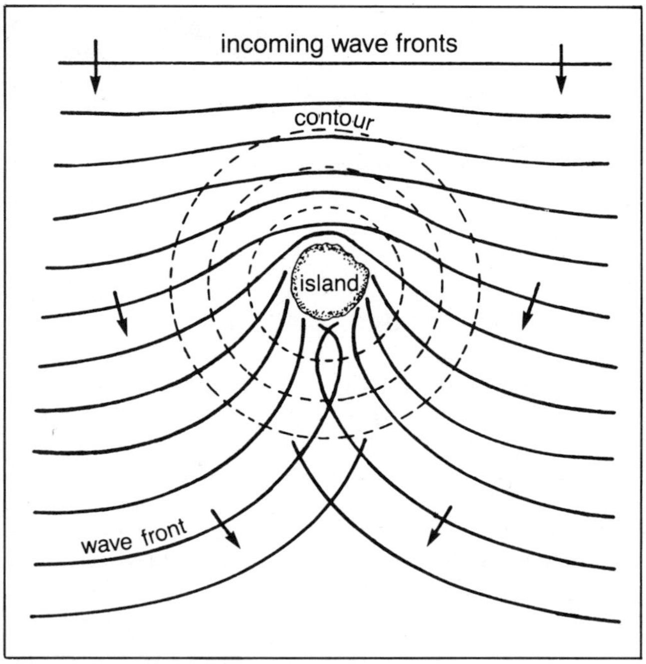

FIG. 8.3. Refraction of swell around nearly circular island forms criss-cross interference pattern in lee geometric shadow zone. Broken lines represent bottom contours with depths increasing gradually outward.

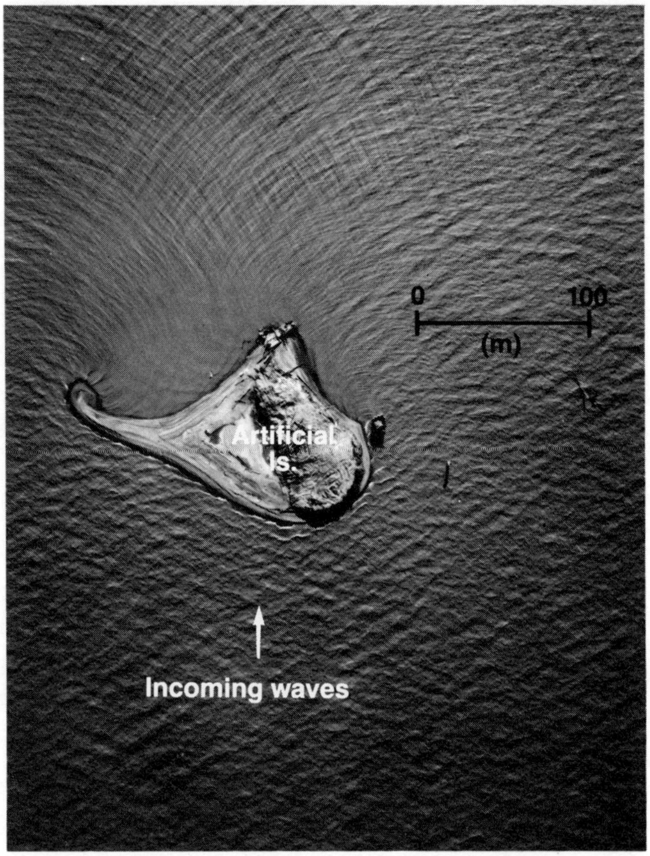

FIG. 8.4. Refraction of wind waves around artificial island (Netserk F-40) in Beaufort Sea immediately seaward of Mackenzie River delta (July 1978). Typical water depths 10 m in area. Refracted waves break parallel to shoreline and form criss-cross pattern in lee of island. Refracted waves curve sharply around to lee shore. (Courtesy F. Stephenson)

Refraction is somewhat responsible for the sloppy conditions around offshore shoals such as Swiftsure Bank. For example, swells with wavelengths exceeding 120 m and from the southwest would swing around as they passed over the Bank (minimum depth about 30 m). This in turn would lead to a concentration of wave energy on the northern side. In a sense, then, Swiftsure Bank acts like a giant imperfect "lens," which tends to focus the energy of the swell toward the Vancouver Island shoreline between Pachena and Carmanah points.

The tendency of swell heights to decrease within Juan de Fuca Strait is in part a refraction effect. Because the Strait is a submarine valley, it gradually deepens toward midchannel, and thereby causes the swell waves to bow outward as they advance. As a result, the energy of a particular swell is spread out and wave height diminishes. Eventually, of course, the waves become nearly aligned with the bottom contours and break at the shore where they give up their energy. Continuation of this process generally leads to ever-improving wave conditions inward along the axis of the Strait.

Diffraction

When waves impinge on a protruding barrier such as a jetty, breakwater, or sharply pointed promontory, a portion of the wave energy makes its way into the shadow zone behind the barrier by the process known as diffraction. As a consequence, there can be appreciable wave action in the lee of a natural or man-made breakwater that would seemingly have afforded a well-protected anchorage. The diffracted waves in such circumstances have not turned the corner in the sense of being refracted but have originated from the tip of the barrier, which acts like a source of waves as it scatters the original incoming waves in all directions (Fig. 8.5a). (Diffraction of light waves at the edge of an obstacle is the reason shadows are inherently fuzzy.)

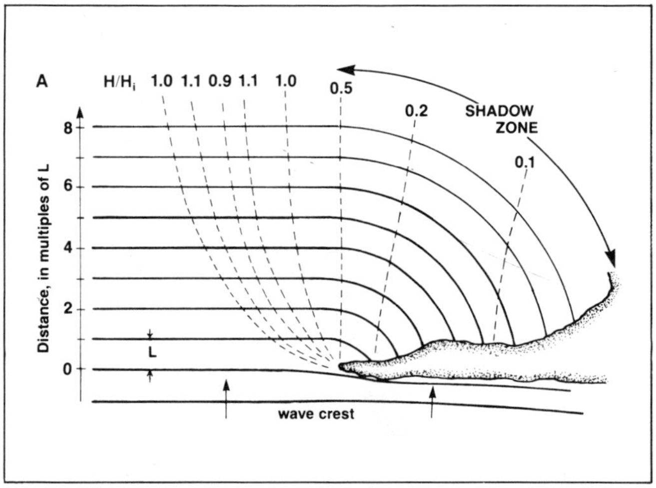

FIG. 8.5. (A) Diffraction of series of incoming waves by headland or jetty, waves penetrate into geometrical shadow zone behind barrier. Broken lines give ratio of local wave height (H) to incident height (H_i). Inward distances measured in multiples of incoming wavelength, L.

Although wave heights generally diminish rapidly in the lee of a barrier, there are certain areas close to the "line of sight" of the incoming waves where the amplitude of the waves is actually increased. The gap in an across-harbor breakwater or between two sand spits can also generate a complicated diffraction pattern (Fig. 8.5b). An example of diffraction occurs within the harbor of Oak Bay Marina near Victoria, which opens into the waters of Juan de Fuca Strait via a narrow gap in an artificial barrier (see Fig. 2.37a). Wave diffraction patterns can also be seen behind the Steveston, Iona, and North Arm jetties at the mouth of the Fraser River and in the lee of the Ogden Point breakwater at the entrance to Victoria Harbour.

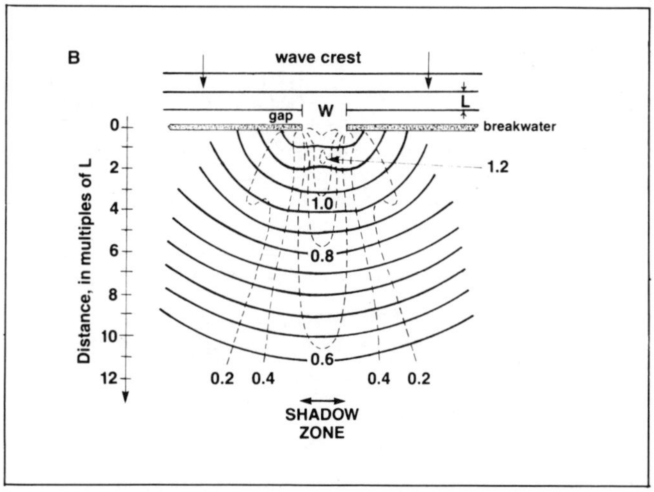

FIG. 8.5. (B) Diffraction of waves at breakwater gap width, W, equal to twice the wavelength, L. Broken line as in (A) (From U.S. Army Coastal Engineering Research Center 1977)

Reflection

In addition to refractive and diffractive effects, waves impinging on a coast may undergo total or partial reflection back to sea. The proportion of incoming wave energy reflected from a coast depends on the beach slope and the wavelength of the waves. The longer the waves for a specific beach slope the greater the reflection, because longer waves experience a greater change in depth over each wavelength than shorter waves; in a relative sense the beach appears steeper to long waves than to short waves.

For swell and wind waves that propagate over beach slopes flatter than about 10%, almost all the wave energy is lost to breakers and little returns seaward as reflected waves. Gently sloping beaches behave as almost perfect absorbers of waves. The energy given up by the breaking waves goes into the formation of currents, turbulence, noise, and the movement of beach sediments. As the beach face steepens, an ever-increasing amount of wave energy is reflected away from the shore. In the extreme case of a perfectly straight vertical wall or cliff that extends several wave heights beneath the water surface, the waves are totally reflected and initially reverse direction with only a slight decrease in height. This is common along the rugged coasts of British Columbia and Washington, and can further be seen where wind waves bounce off the hull of a large ship.

When reflection takes place, the reflected waves propagate away from the point of contact with an equal but opposite angle to that of the incident waves, like light from a mirror (Fig. 8.6). Therefore, waves that approach head-on to a smooth vertical cliff will be reflected along the same path as the incoming waves, leading to the formation of standing waves and a choppy sea within a short distance from the cliff. Confused, choppy sea conditions tend to exist close to steep shores when the waves strike at an angle. In this case, the inherent irregularities of the shoreline scatter the arriving waves in all directions, rather than along a single direction.

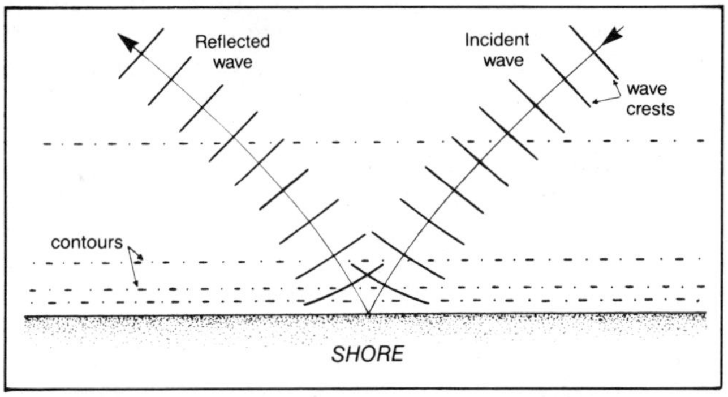

FIG. 8.6. Reflection of waves at comparatively steep coastline. Waves are refracted slightly before reflected, then refracted again as they propagate seaward.

Under certain conditions of beach slope and wavelength, refraction can combine with reflection to "trap" some of the incoming wave energy along a beach, shown schematically in Fig. 8.7. The reflected wave attempts to return to sea but is curved back toward the shore by the bottom topography and the process is repeated. In most situations, the trapping is usually far from perfect as irregularities in the bathymetry and wave patterns allow a fraction of the wave energy to leak seaward after each reflection. This, combined with other effects such as friction, causes the height of the trapped wave to diminish as it travels parallel to the beach and the wave eventually

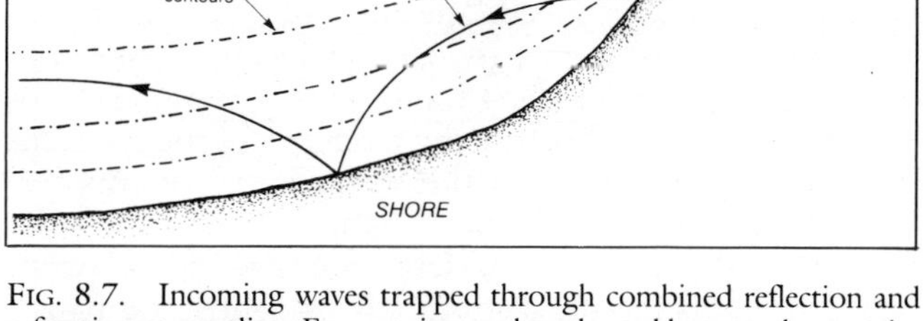

FIG. 8.7. Incoming waves trapped through combined reflection and refraction at coastline. For certain wavelengths and bottom slope, series of incident waves cannot return seaward after reflection and is confined to nearshore region. Friction and some wave energy leakage seaward gradually reduce wave height alongshore.

disappears. There is now observational evidence to show that ocean swell is commonly trapped over broad coastal beaches and that the energy of tides and tsunamis is often captured by continental shelf regions. Mid-ocean islands, seamounts, and banks also act as energy traps for tidal motions, and thereby cause complex tide and current patterns around their periphery.

Steepening

In addition to altering the speed, the transition from deep to shallow water changes the height and length of the waves. Only the period remains the same. Contrary to popular misconception, the height of a wave does not begin to increase immediately after it advances into a shoaling region. In fact, there is an initial lowering of the wave, although at most the decrease is usually less than 10% of the deepwater height (Fig. 8.8). Studies have shown that this reduction in wave heights takes place in water depths between ½ and 1/17 as large as the wavelength. A wave 170 m long would begin to decrease in amplitude when the water depth fell to 85 m and would not regain its original height until the depth shoaled to 10 m. Once the ratio of depth-to-wavelength becomes smaller than 1/17, there is a rapid increase in the height until break-point is reached.

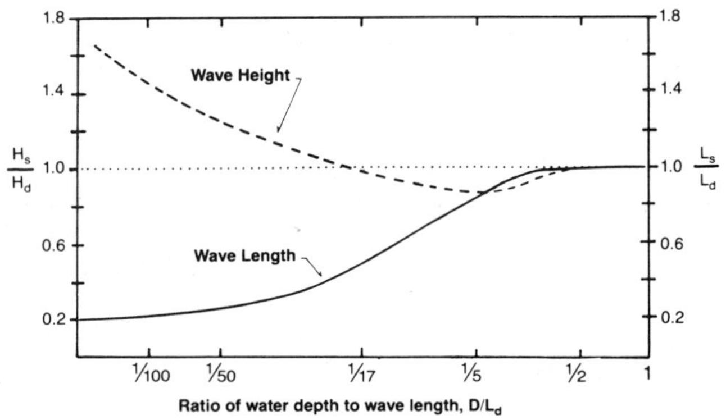

FIG. 8.8. Change in wave height (broken line) and wavelength (solid line) accompany decrease in water depth. Ratio H_s/H_d is height of wave in shoaling region divided by height in deep water; ratio L_s/L_d is wavelength in shoaling region divided by wavelength in deep water. (From King 1966)

Unlike the height, the length of a shallow-water wave continually decreases as the depth decreases (Fig. 8.8), and the decrease proceeds at a rate proportional to the lowering of the wave speed. The effect of this reduction in crest-to-crest separation is to increase the steepness of the waves as they enter shoaling water. This increase in steepness is at first small, because the tendency of the shortening wavelength to steepen the waves is partly counterbalanced by their initially lower amplitudes. When the ratio of the depth to the wavelength reaches 1/17 the steepening proceeds rapidly as both effects contribute to enhanced wave slopes.

Associated with the change in steepness is a change in wave form; the crests become narrower and more peaked, while the troughs become broader and flatter. This alteration is most noticeable for long low swells, which deform from smooth undulating crests in deep water to sharp conspicuous crests in very shallow water.

To what degree the offshore height of a wave will increase as the shore is approached depends on a number of factors, such as the shape of the wave, the slope of the beach, and the smoothness of the bottom. As a crude estimate, the height of an incoming wave will have increased by a factor of approximately 1.5 by the time it is ready to break. More exact values are given in Table 8.1, which shows growth of the incoming wave heights will be greater for greater beach slopes or for longer wavelengths.

TABLE 8.1. Ratios of breaker heights to the deepwater swell heights for four different beach slopes and for two values of the ratio of incoming wave height to wavelength. Example: value of 1.60 indicates that, for beach slopes of 5% (or 1 in 20) and for swell waves with lengths 100 times their heights, breakers are a factor of 1.6 higher than wave height in deep water. Values are applicable for low swell but not for steep sea waves whose growth would be less. (From Iversen 1952)

Slope (%)	Breaker height ÷ deep-water height	
	$\frac{\text{height}}{\text{length}} = 0.01$	$\frac{\text{height}}{\text{length}} = 0.02$
2	1.31	1.12
3.3	1.43	1.23
5	1.60	1.31
10	1.76	1.40

Long waves, therefore, amplify more than short ones and steeply sloping beaches enhance the wave heights more than gradually sloping beaches.

An increase in steepness occurs anywhere a wave begins to feel the bottom and is responsible for the amplification of long swell over shallow continental shelf regions.

Breaking

A shallow-water wave will break for two basic reasons: (1) the wave becomes so steep it can no longer support its own weight and the crest collapses. As mentioned in Chapter 6, this generally occurs when the wave height reaches 1/7 of the wavelength; or (2) the speed of the water that circles around with the crest of the wave exceeds the speed of the wave, so the water in the crest overtakes the wave form, and causes it to fall over and break. A rule of thumb used by oceanographic engineers is that a wave will break when its height exceeds ¾ of the local water depth. Thus, a 3-m swell would break in 4 m of water (water depth is always measured at a point half way between the trough and crest). As previously mentioned, the height of the wave when it breaks will be about 1.5 times larger than its height in deep water.

For the sake of definition, breakers can be placed into four categories: spilling, plunging, collapsing, and surging, although in reality these often grade into one another. Spilling breakers move forward with a foaming turbulent crest (Fig. 8.9). The wave does not lose identity

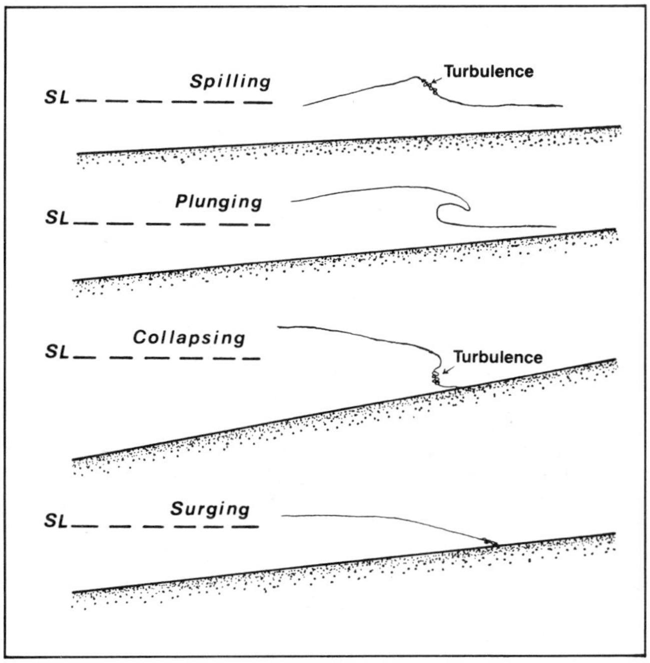

FIG. 8.9. Cross-sections of the four types of breakers (SL, approximate mean sea level). (From Galvin 1972)

but gradually diminishes in height until it becomes a swash on the shore. Breakers of this type are formed when steep, deepwater waves advance over a gently sloping, often sandy, beach. They are responsible for the rows of breakers that generally characterize many wider shoals and beaches along the exposed outer coast. Plunging breakers collapse with a splash as the crest curls over and falls into water receding from the previous wave. This type of breaker occurs when relatively low, deepwater waves move up a steep beach; shingle beaches commonly have plunging breakers. A collapsing breaker is somewhat similar to a plunging breaker. The difference is that the initial point on the vertical wave front from which the curling tongue of water originates is well below the point of maximum height of the wave. Moreover, the wave essentially collapses right onto an exposed beach face rather than into a depth of water associated with the previous wave as with plunging breakers. Lastly, the waves may not really break at all, but instead retain relatively smooth profiles that slide up and down the beach with only a minor display of foam and turbulence. These are surging waves.

Observations indicate that, for a specific beach and constant wave period, the type of breaker changes as the incoming height of the wave increases. As the wave height increases, the breakers change from surging to collapsing to plunging to spilling. The same sequence is found if the wave height and wave period are kept the same but the slope of the beach is continually decreased; it also occurs if the height and beach slope are kept fixed but the wave period is decreased.

The zone between the breaker line, delineated by the first series of breakers, and the shore is called the surf zone. The zone between the last breaker and dry land is called the swash zone (see Fig. 2.16).

Longshore Currents

One important aspect of breaking waves is their ability to generate currents that flow parallel to the shoreline within the surf zone. These longshore (or littoral) currents occur because each wave thrusts the water forward when it breaks. They form (1) when crests of the breaking waves approach the shoreline at an angle, or, (2) when there is a gradation of the breaker heights from one part of the surf zone to another.

In the first case, the current is produced directly through the accumulated effect of many breakers over a period of time. For this current to flow parallel to the shoreline the waves must break at an angle to the shoreline, so each breaker has the tendency to push the water along the beach. The onshore component of each breaker's thrust, on the other hand, does not create an onshore current because the effect is counterbalanced by an offshore pressure gradient. In this particular process, shorter waves produce stronger currents than longer waves of the same height because they suffer less refraction and are more likely to approach the beach at an angle. Currents in excess of 50 cm/s can be generated in this way during prolonged periods of high waves.

In the second case, the higher breakers pile up more water in the surf zone than lower breakers, regardless of how they approach the beach. As a consequence, a current is set up parallel to the shore to transport water away from the region of the larger breakers. This process together with the previous one, is a regular cause of longshore currents along the North American west coast. Currents of this nature are also generated over broad expansive shoals in the Strait of Georgia, such as Roberts Bank and Sturgeon Bank off the mouth of the Fraser River. The transport of sediments and pollutants and the erosion of shorelines are important engineering aspects of these currents. Ediz Hook and Dungeness Spit on the southern shore of Juan de Fuca Strait are extensive sediment deposits created in part by the longshore currents set up by predominantly eastward propagating waves that break in the Strait.

Rip Currents

Actually, the along-the-shore extent of a longshore current is always limited and the water that has accumulated in the surf zone through the action of the breakers must eventually return to sea. It often does this by forming a series of strong narrow currents called rip currents that flow directly seaward through the surf zone. The longshore current is said to "feed" the rip current. (A typical surf zone current regime is in Fig. 8.10). Rip currents can attain speeds of 1.0–1.5 m/s (2–3 kn) within the surf zone, but spread out and decelerate once outside the zone. The location of these currents is often marked by deep channels through the sand bars, which give the water a darker appearance. Because of the enhanced depths, moreover, waves rarely break in rip channels, although the effect of the opposing rip current may cause very short waves to form a chop similar to a wind chop. Along the

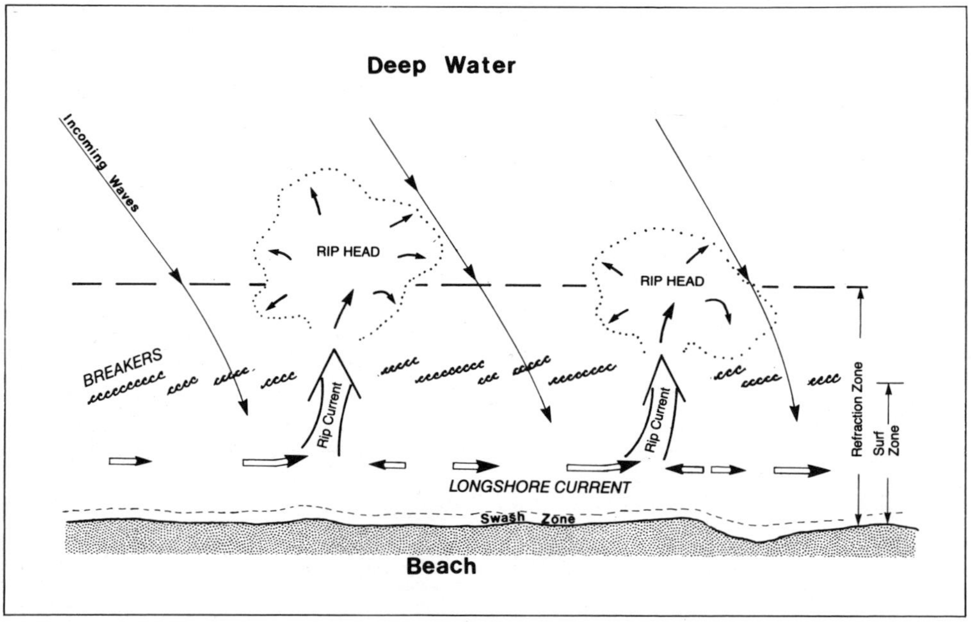

FIG. 8.10. Schematization of nearshore circulation regime generated by breaking waves that advance shoreward at an angle to beach. Refraction begins when waves reach refraction zone at ½ depth of incoming wavelength. Rip current diverges and weakens at rip head. Similar pattern occurs if waves approach beach head-on, but breakers on left higher than on right.

beaches of southern California, there is always a noticeable longshore current produced by the large ocean swell that breaks at the exposed Pacific coast. Rip currents can be seen as intense flows that lead seaward via deep, narrow, breakerless channels. While body surfing off Long Beach (Vancouver Island) and Kalaloch Beach (Olympic Peninsula), the author has been aware of fairly strong longshore currents, and distinct rip currents in the lee of offshore rocks and shoals.

It should be pointed out that why rip currents form is still not fully understood, although the location and intensity of these currents apparently depend on the submarine topography, the configuration of the coast, and the height and period of the waves. Anyone venturing into the surf zone should be prepared if they unexpectedly find themselves swept seaward. According to all authorities, the best plan is not to struggle against the current but to swim across it. Because rip currents are always narrow, a swimmer will soon find himself in a region of reduced flow, where the shoreward propagating waves will tend to carry him shoreward. From a precautionary point of view it is worth noting that rip current and longshore current velocities strengthen on a falling tide, greatest increases take place immediately prior to low tide when water draining from the beach is directed toward the deeper rip channels where it adds to the seaward flow. This effect is particularly relevant to beaches along the coasts of Washington and British Columbia, where the tidal range is large. But not everyone treats rip currents as sinister manifestations of the circulation pattern in the surf zone. Experienced surfers, boaters, and scuba divers, for example, will sometimes use these seaward currents to get a free ride through the breaker zone into deeper water. Rip currents may also be essential to the removal of pollutants dumped into the nearshore region.

The study of rip currents began in the early 1940s on the beaches of California, where it was discovered they were an integral part of nearshore circulation cells similar to Fig. 8.10. Often there were a series of cells and associated rips strung out along the length of the beach. The alongshore spacing between the rip currents was aproximately 4 times the distance from the shoreline to midpoint of the breaker zone. Early evidence further showed that the velocity and seaward extent of the rip currents was related to the height of the incoming waves and that rips were located away from the largest breakers. Where the waves were breaking at an oblique angle to extensive beaches, the location of the rip currents slowly migrated along the beach in the direction of the longshore current. It is now thought that rip currents are linked to an alongshore variation in wave set-up, the lumping up of sea level above the still water depth due to breaking waves. Longshore currents flow away from regions of greater set-up (higher breakers) and feed into seaward flowing rip currents at alongshore locations of lesser set-up (lower breakers). Thus, the problem of accounting for the locations and strengths of rip currents is reduced to an understanding of the longshore variation in wave set-up.

The most obvious mechanism capable of producing such variations is wave refraction, whereby the offshore bottom contours cause wave energy to concentrate into one area to produce high waves, and to diverge at an adjacent area to produce low waves. Submarine canyons and ridges are capable of generating rip currents in this way (Fig. 8.11). Refraction may also be responsible for the rip current pattern along Long Beach on the west coast of Vancouver Island (Fig. 8.12). The inward curvature of the bottom contours causes incoming waves to diverge away from three main areas that mark the locations of major rip currents: the center of the beach off Green Point, the large

FIG. 8.11. Modified rip cells (horizontal circulation cells) near Rose Spit, Queen Charlotte Islands, July 1979. Rhythmically spaced cells caused by waves breaking over bars that project from shoreline; circulation pattern is shoreward flow along transverse bars and seaward rip current mid-way between bars. Large dunes in background over 10 m high and driven into forest by strong southeasterly gales. (Courtesy J. Harper)

rock offshore of the Airport road, and the area in front of the former Wickannish Inn. Midway between these areas, and at the prominences at either end of the beach, the waves converge to produce comparatively large breakers that drive the longshore currents feeding into the rip currents.

In many instances, the offshore beach profiles are uniform so refraction is the same along the entire beach and there is no differentiation in wave set-up, yet rip currents still persist. This fact has recently led to a theory based on the combined effect of incoming waves with another type of surface wave called an edge wave. Unlike the shoreward propagating sea or swell waves, edge waves depend on the beach or "edge" for their existence, and lean against it for support much like an ordinary wave traveling along a wall or the side of a pool. Thus, the crests

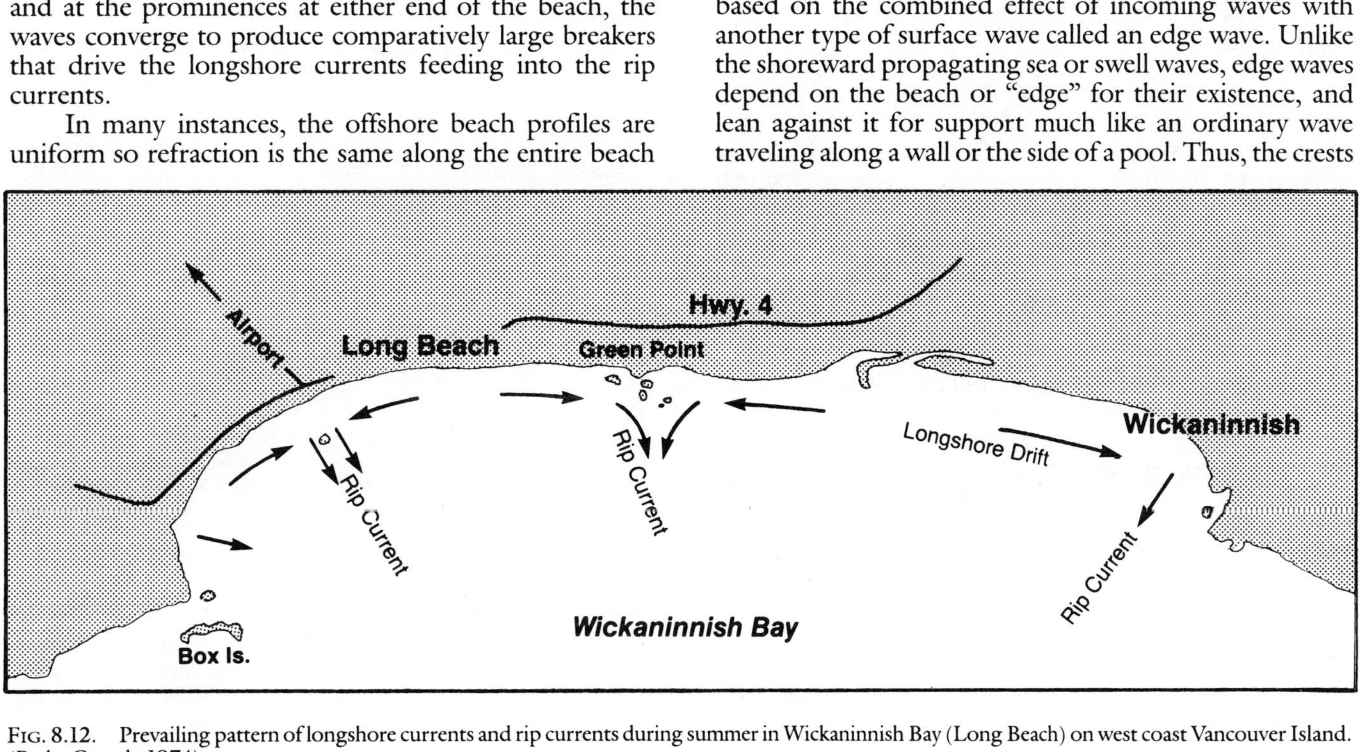

FIG. 8.12. Prevailing pattern of longshore currents and rip currents during summer in Wickaninnish Bay (Long Beach) on west coast Vancouver Island. (Parks Canada 1974)

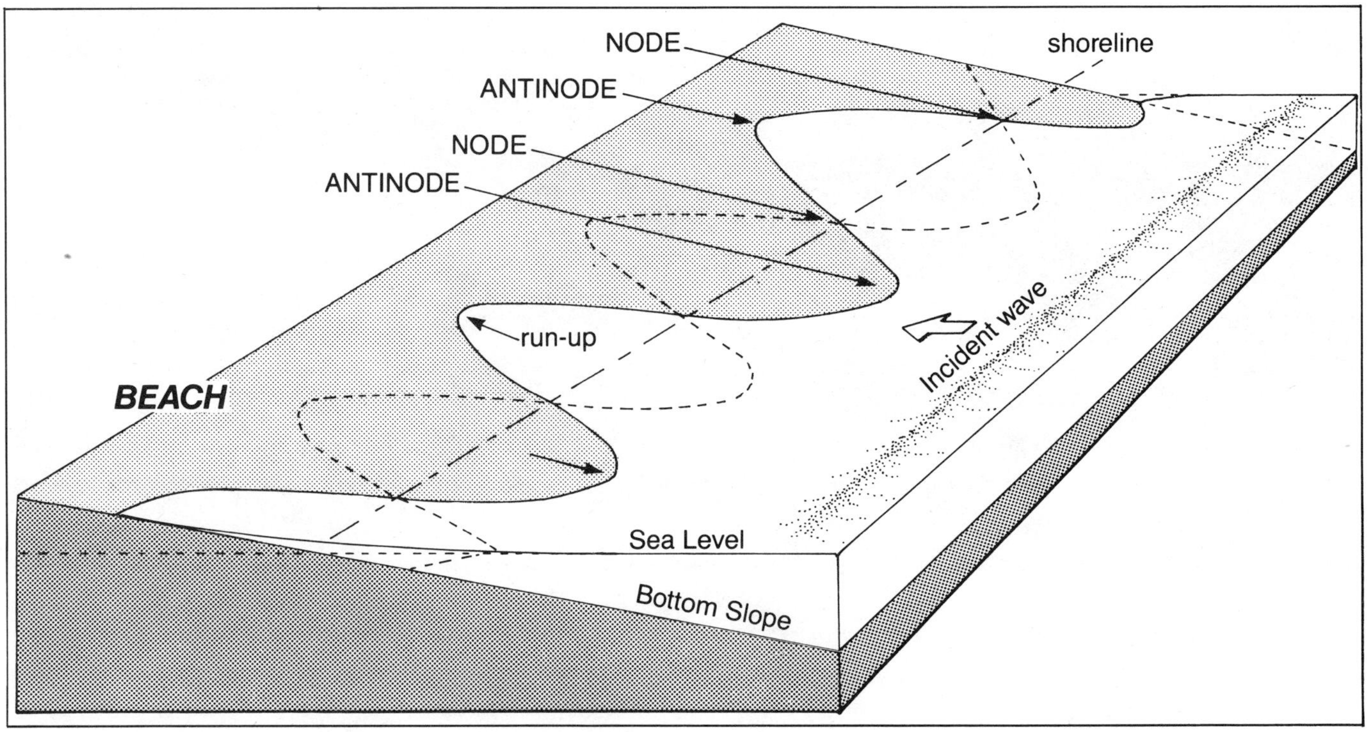

FIG. 8.13. Schematic diagram of simple standing edge wave pattern at gradually sloping shoreline. Solid line and broken line mark water's edge along beach at two different times (½ cycle apart) of an edge-wave cycle. Motions at antinodes are cyclic reversals of run-up and backwash; no edge-wave induced motions at nodes. Combined effect of breaking incident surface gravity waves and edge wave are responsible for nearshore circulation patterns as in Fig. 8.14.

and troughs of edge waves are at right angles to the shoreline rather than parallel to it. In addition, they attain maximum heights or depressions close to shore, accompanied by a rapid decrease in height seaward where they become negligible just outside the breaker zone (Fig. 8.13). Edge wave motions oscillate between run-up and backwash along the length of the sloping beach face. To produce the observed patterns of rip and longshore currents, standing edge waves of the same period as the incoming waves must be present along the beach. At the position of the nodes there will be no up and down motion of the water elevation, while midway between the nodes, at the antinodes, maximum up and down motion will occur. Because the up–down motion at the antinodes has the same rhythm as the arrival time between successive breakers (e.g. 10 s for one complete up–down cycle and 10 s between incoming waves), every second antinode along the beach will be higher than the still water level when an incoming wave breaks. The two effects combined cause higher than normal breakers at fixed locations. Adjacent antinodes, on the other hand, are always below the still water level at the instant of breaking, and result in lower than normal breakers midway between regions of higher breakers. Only at the nodal points of the edge waves will there be no additional effect on the height of the incoming breakers. This persistent generation of large and small breakers along the beach produces a consistent alongshore variation in breaker height, which in turn, sets up a regular pattern of nearshore circulation cells, with rips at the positions of the low breakers at every second antinode of the edge wave (Fig. 8.14).

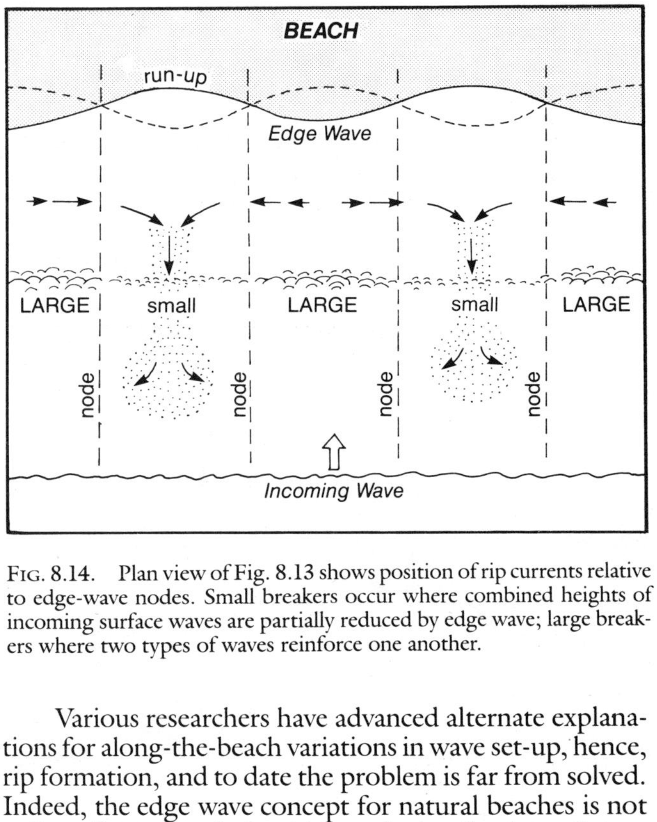

FIG. 8.14. Plan view of Fig. 8.13 shows position of rip currents relative to edge-wave nodes. Small breakers occur where combined heights of incoming surface waves are partially reduced by edge wave; large breakers where two types of waves reinforce one another.

Various researchers have advanced alternate explanations for along-the-beach variations in wave set-up, hence, rip formation, and to date the problem is far from solved. Indeed, the edge wave concept for natural beaches is not universally accepted, though its existence has been shown to be responsible for rip formation in laboratory model basins. The basic question about the origin of the standing edge waves on beaches has yet to be answered satisfactorily.

Once a nearshore circulation cell is created, it tends to stabilize the bottom configuration through scouring. This can then act to stabilize the rip current positions, and make them less susceptible to changes in the edge wave or incoming wave patterns. With the establishment of a series of submarine bars and rip channels there need no longer be an alongshore variation in breaker height to maintain the longshore-rip current circulation.

A large-scale cuspate shoreline may also mark the presence of well-defined rip currents. The rips flow seaward from the embayments between the rhythmic topography of sediment-formed prominences and underwater ridges, although certain evidence suggests that once the cuspate structure is fully established the rip currents that initially produced it weaken and disappear. Smaller scale beach cusps of ridges or mounds of coarse sediments that extend down the beach face will deflect swash motions toward the interspacing embayments. The return flow concentrates into something resembling a "rip current" though, unlike a true rip current, velocities are not steady with time and do not depend on a gradation in offshore breaker height.

Man-made structures, such as jetties and groynes, sometimes initiate riplike currents by deflecting the natural littoral drift away from the shore. Once, when swimming on a deserted beach in Fiji over which there was a perceptible alongshore current, I suddenly found myself being swept seaward as I drifted close to a rock-work structure that extended about 10 m across the littoral zone. Though the strong offshore flow presumably terminated a short distance past the end of the groyne, my instinctive reaction was to swim across the current and to speculate on its extent from the safety of the shore!

Chapter 9. Tsunamis (Tidal Waves)

On March 27, 1964, at 7:36 p.m., Pacific Standard Time, an intense earthquake originated near the western side of Unakwik Inlet approximately 102 km east of Anchorage (Fig. 9.1). It was one of the strongest earthquakes ever recorded on the North American continent, and registered a magnitude of 8.5 on the Richter Scale[1]. A massive uplift over 250,000 km^2 of the seafloor adjacent to the Alaska coast accompanied the tremor, which sent a series of "tidal waves" at high speed toward all corners of the Pacific Ocean. Less than 4 h later, the first waves reached the outer coast of Vancouver Island where they produced abnormally large and rapid changes in sea level (Fig. 9.2). Although the villages of Hot Springs Cove and Zeballos suffered major wave and flood damage, Port Alberni was the hardest hit when a 7-m wave swept up Alberni Inlet. It was the largest wave of its kind recorded in British Columbia and, though it caused no deaths, was responsible for $10 million damage to ships, residential property, and industry (Fig. 9.3, 9.4). The USA was less fortunate. In Alaska, tidal waves surged onto the land and killed 107 people, while in California and Oregon, nearly 2000 km from the source, 15 people were killed. Damage exceeded $104 million and many were left injured and homeless.

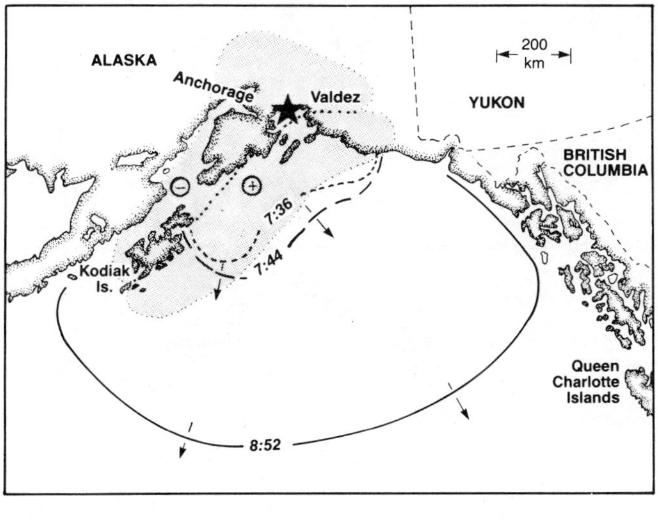

FIG. 9.1. Leading edge of 1964 Alaska tsunami beginning with inception around 7:36 p.m. 27 Mar. Star marks earthquake epicenter in Unakwik Inlet at northern end of Prince William Sound. Shaded area delineates area of crustal uplift (+) and subsidence (−) that accompanied the earthquake. Maximum uplift was about 9 m, maximum subsidence about 1½ m. (Adapted from Spaeth and Berkman 1967)

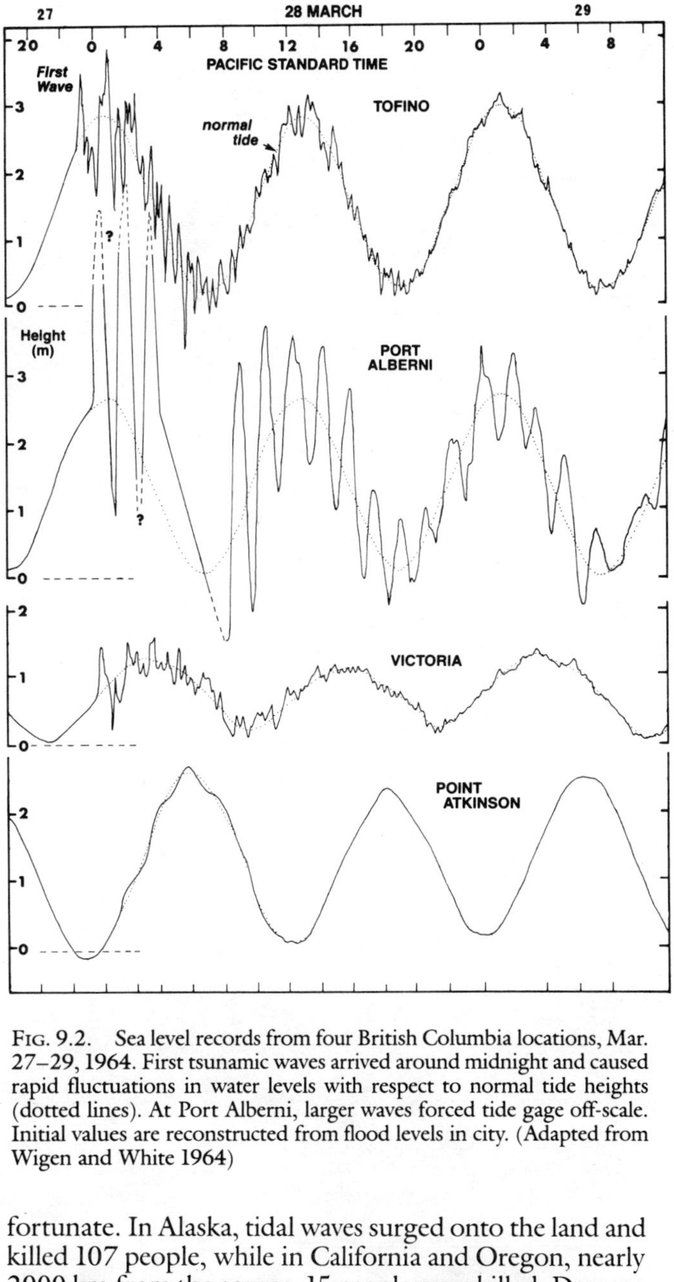

FIG. 9.2. Sea level records from four British Columbia locations, Mar. 27–29, 1964. First tsunamic waves arrived around midnight and caused rapid fluctuations in water levels with respect to normal tide heights (dotted lines). At Port Alberni, larger waves forced tide gage off-scale. Initial values are reconstructed from flood levels in city. (Adapted from Wigen and White 1964)

Waves of the magnitude generated by the Alaska earthquake are rare. Of the 176 instances recorded between 1900 and 1970 in the Pacific, 35 caused damage near their source but only 9 created widespread destruction. Nevertheless, it is important to be familiar with the nature of such waves. How are they generated and what factors determine their destructiveness? Is the first wave the highest? What effect would a tidal wave have on a ship at sea?

[1] Richter Scale: A logarithmic scale that expresses the intensity of a seismic disturbance in terms of the energy it dissipates. A unit (1) increase in the scale implies a 10-fold increase in energy. Thus, an earthquake that measures 4 on the Richter Scale is 10 times more energetic than an earthquake that measures 3, and 100 times more energetic than one measuring 2. The smallest disturbance that can be felt on land has a scale reading of 1.5; an earthquake of 4.5 causes slight damage; a reading of 8.5 means a devastating earthquake. A seismic disturbance of 8 on the Richter Scale has the equivalent energy of a 250-megaton thermonuclear bomb. In comparison, the atom bomb dropped on Hiroshima in 1945 was only 0.02 megatons, a factor of 12,500 less powerful.

FIG. 9.3. Property damage in Port Alberni from 1964 Alaska tsunami. (*Vancouver Sun* photo)

FIG. 9.4. Property damage in Port Alberni by 1964 Alaska tsunami. (*Vancouver Sun* photo)

Generation

The formation of so-called tidal waves has absolutely nothing to do with the tide, the vertical rise and fall of sea level produced by the sun and moon. For this reason, a tidal wave is now more properly referred to as a tsunami. Meaning "harbor wave" in Japanese, it is a recognition of that country's pioneering work on the subject, and the unparalleled destruction such events have wreaked on its inhabitants throughout history. Universal acceptance of the name tsunami has also replaced terms like seismic sea wave and seaquake, which were in common use until recently. Seaquake now has a specialized meaning related to earthquake shock vibrations transmitted through the water. (Two B.C. mariners reported noticing such brief vibrations when anchored in the Gulf Islands during a calm day, at the time of a small earthquake in the area.)

Essentially, tsunamis appear to be associated with earthquakes generated beneath the seafloor with magnitudes greater than 6.5 on the Richter Scale and centers not deeper than about 100 km below the seabed. However, even if such a tremor occurs it rarely generates a tsunami. This is because the great majority of earthquakes in the Pacific cause a sideways slipping of the seafloor rather than an upward or downward displacement, which is needed to cause distortion of the ocean surface (Fig. 9.5). Nevertheless, such faults do form across large areas of the seafloor. When they do, a tsunami is created and spreads outward as a series of waves, somewhat analogous to the pattern orite crashing into the ocean would produce a tsunami, although there are no records of such an event.

It is postulated that about 1% of the seismic energy of submarine earthquakes goes into the generation of tsunamic waves, and that heights of the waves at a coast within a range of 800 km of the earthquake's origin are proportional to the square root of the earthquake energy. On this basis, a submarine earthquake of magnitude 7 on the Richter Scale would cause wave heights of around 2 m at the coast and an earthquake of magnitude 8 would cause wave heights of 10 m (Fig. 9.6). Within the generat-

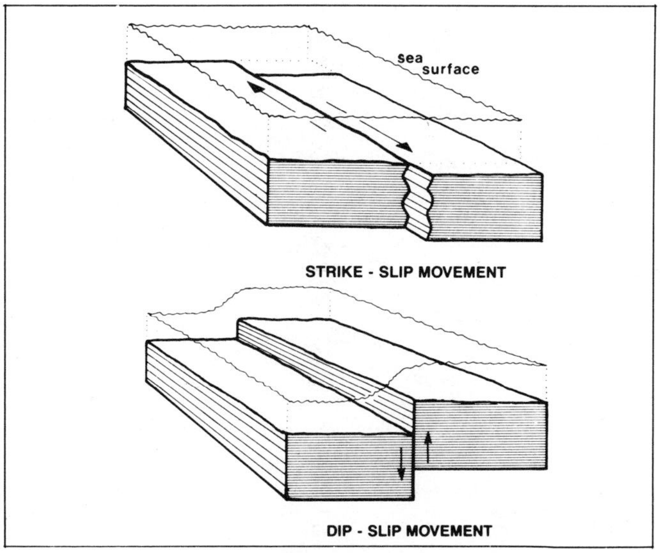

FIG. 9.5. Two types of seafloor crustal faults (highly exaggerated in scale). Tsunamis are generated by vertical dip-slip dislocations that lead to rapid deformation of sea level.

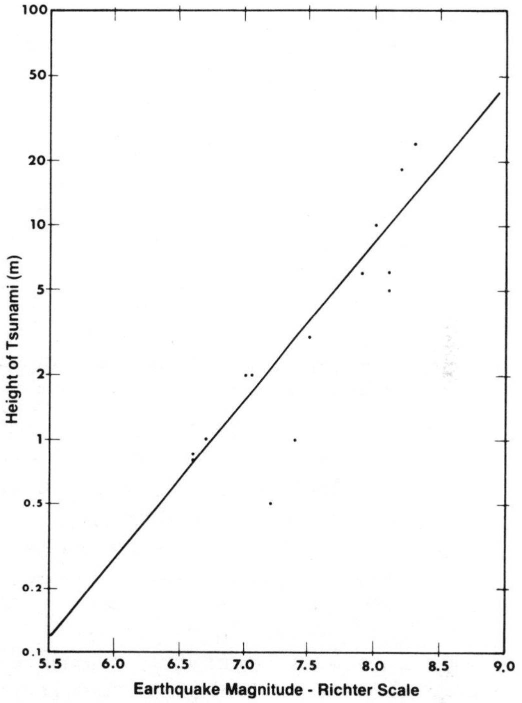

FIG. 9.6. Heights of tsunamis expected at coasts within 900 km of submarine earthquake. Based on data collected near Japan 1923–1957. Dots give measured wave heights for various earthquake magnitudes. (Modified after Wilson 1964)

formed by a stone cast into a pond. For the most part, tsunami generation is confined to the rim of the Pacific Ocean where shifts of the oceanic crust relative to the continental land masses have created highly active earthquake zones (Chapter 1). Statistically, 62% of all tsunamis occur in the Pacific Ocean, 20% in the Indian Ocean, and the remaining 18% in the Mediterranean Sea and the North Atlantic Ocean. The South Atlantic is relatively aseismic and tsunami-free.

Not all tsunamis are directly attributable to seafloor motions. Most damage along the Alaska coast in 1964 was, in fact, due to locally derived waves formed by underwater landslides and shoreline slumps that were triggered by the land-centered earthquake. Because tsunamis of this type are usually generated in confined areas like inlets, they can be highly destructive. Near Valdez, Alaska, an underwater slide created a tsunami that deposited driftwood 52 m above the low water mark and splashed sand to a height of 67 m. A massive rockfall at the head of Lituya Bay, Alaska, on July 9, 1958, precipitated a giant wave in narrow Gilbert Inlet, which ascended 518 m up the opposite headland. A similar but much less dramatic event occurred in British Columbia in 1975. On April 27, a large underwater slide down-inlet from Kitimat sent waves "boiling ashore." After an initial fall of 4.6 m below low tide, the sea level near the Kitimat waterfront rose by 7.6 m in only a matter of minutes and then surged back and forth for about an hour before returning to its normal state. Damage from these events and others in the same week was light, however, as they occurred near the time of low tide. Tsunamis may also be generated by the explosion of submarine volcanoes. When Krakatoa erupted on Aug. 27, 1883, in the East Indies, it created waves 30 m high that crashed over neighboring islands and drowned more than 36,500 people in nearby Java and Sumatra. The atmospheric shock waves from the volcanic blasts induced sea level oscillations that were detected by tide gages as far away as the English Channel. Conceivably a large mete-

ing region itself, tsunamis are thought to begin as a complex system of waves with a broad spectrum of periods, heights, and wavelengths. Eventually the waves sort themselves out through dispersive effects with longer waves leading the way out of the generating region, though present knowledge regarding the types of initial waves and the transformations they undergo is still lacking. In reality, there is often more than one group of waves because of aftershocks and readjustments of the seabed following the initial upheaval.

Away from the generating region, tsunamis, as other wave phenomena, are affected by refraction, shoaling, dissipation, diffraction, resonance, and reflection. These factors, together with those that produce the waves in the first place, lead to a convoluted wave pattern whose behavior at a coast is extremely difficult to predict or even to reconstruct. Table 9.1 gives some indication of the complexity of the problem by illustrating the considerable variability in maximum wave heights at, and between, various locations in the Pacific for the five most recent ocean-wide tsunamis.

TABLE 9.1. Maximum waves (rise or fall) recorded for five recent tsunamis within the Pacific Ocean (heights in metres). Where maximum wave exceeded gage limit (+ sign), values may be misleading. Hilo, Hawaii, for example was devastated by the 1960 tsunami yet only suffered minor flooding from the 1964 tsunami despite its greater recorded height. Heading gives year and origin of tsunamis. (Adapted from Spaeth and Berkman 1967)

Location	1946 Aleutian Islands	1952 Kamchatka Peninsula	1957 Aleutian Islands	1960 Chile	1964 Alaska Peninsula
Sitka, Alaska	0.79	0.46	0.79	0.91	4.36
Tofino, B.C.	0.60	0.61	–	1.40	2.47
Neah Bay, Wash.	0.37	0.46	0.30	0.73	1.43
Crescent City, Calif.	1.80	2.07	1.31	3.32	3.96+
Talcahuano, Chile	–	3.66+	1.40	5.06	1.65
Hilo, Hawaii	–	2.41	2.71	2.93+	3.81+
Honolulu, Hawaii	1.25	1.34	0.98	1.68+	0.82
Wake Island	–	0.52	0.73	1.01	0.15
Kushimoto, Japan	–	–	–	3.20	0.79

Wave Travel

Once generated, a tsunami consists of a series of waves that spread rapidly away from their source. In the deep ocean, waves attain speeds of over 900 km/h. The crest-to-crest separation between each wave in the series is typically around 100–400 km so that even at their fantastic speeds it usually takes 10–60 min for successive crests to pass a given point. Because their wavelength is appreciably greater than the depth of the ocean, tsunamis fall into the category of shallow-water waves (Chapter 6, 8). As a consequence, they behave in midocean somewhat like ordinary wind waves that propagate through knee-high water, in that their speed, C, depends only on the local water depth, D, through the simple formula $C = \sqrt{gD}$. According to this equation, for example, tsunamis would travel at a speed of 620 km/h or 335 kn in water 3050 m deep.

Because tsunamis continually feel the presence of the bottom, they must always adjust their speed according to the depth. They decelerate in shoaling regions and accelerate in deepening regions. As a result, the concentric wavefronts begin to distort as different parts of the front move across areas of different depth in the ocean. The part of a wavefront over the shallow continental shelf will move appreciably slower than the part of the wavefront over an ocean abyss, and, therefore, lag behind (Fig. 9.1). For this reason, it took the first wave of the 1964 tsunami 3 h, 24 min to travel the 1800 km (973 nm) over relatively shallow water from Alaska to Tofino, for an average speed of 530 km/h (286 kn). On the other hand, it took only 5 h, 17 min for this wave to cross the 4386 km (2368 nm) to Honolulu, for an average speed of 830 km/h (448 kn). Obviously, a detailed knowledge of the ocean depths is essential for accurate predictions of arrival times around the Pacific rim.

Height at Sea

Contrary to what one might think, the height of a tsunami at sea is always small, rarely exceeding 1 m from crest to trough. And, because the crest-to-crest separation is so large, the profile of these waves at sea is extremely low. A ship would not detect their passage as any possible effect would be lost in the confusion of sea and swell. During the April 1946 tsunami at Hawaii, ships standing off the coast observed tremendous waves breaking on the shore but detected no unusual motion at their offshore locations.

Height Near Shore

When it arrives at a shore, a tsunami may take several forms. It may cause nothing more than a series of gentle rises and falls of the water level over periods of 10–60 min, somewhat like a time-lapse version of the tide. Alternatively, the sea level may rise and fall more quickly, and produce swift floodlike currents. Withdrawals may be rapid and destructive and sweep all before them. Or if conditions are right, a tsunami may manifest itself as a cresting wall of water tens of metres high that crashes ashore with devastating force.

The first wave of a tsunami is usually not the highest. In fact, the approach of a tsunami is often heralded by a slight rise in the water level as the first small crest arrives, followed some time later by a much larger ebb associated with the first trough. At Kodiak, Alaska, the first wave from the 1964 tremor produced nothing more than a gentle rise in sea level followed by a gradual fall. The second wave, however, was a 10-m high monster that pushed 100 t boats over a breakwater and then carried them three blocks into the city.

Tsunamis do not affect all areas equally. Figure 9.2 illustrates that Tofino, on the west coast of Vancouver Island, is only moderately susceptible to these waves, whereas Port Alberni, some 65 km inland, is highly susceptible. In Crescent City, Calif., the 1964 tsunami killed 11 people but in Hilo, Hawaii, no deaths were reported, despite the fact that the waves at both places reached heights of 4 m. Clearly, there are other factors that determine the overall effect of a tsunami aside from the original amount of energy put into it at the generating region.

Proximity

The force of a tsunami is most likely to be extreme if the waves are generated nearby. Japan suffers considerably in this respect and coastal communities in Alaska were hard-hit in 1964 because of their proximity to the wave-generation area.

Tsunamis lose energy as they propagate. A fraction of this energy is lost directly through friction as the waves "rub against the bottom." In addition, there is a decrease in the wave energy as the initial impulse given to the tsunami in the generation region becomes spread over an ever-increasing area of the ocean. The latter is similar to the way sound diminishes as it spreads away from its source, but with an important difference. Whereas sound intensity diminishes inversely as the square of the distance, the energy in tsunamis diminishes inversely as the distance alone. Therefore, tsunamis weaken very slowly and distance alone is not a guarantee of immunity. Often it is more important to be out of the direct "line of sight" to the generation region as most tsunamis tend to be highly directional. But, the best protection is an obstacle course of straits, passages, islands, shoals, and sharp corners through which the waves effectively dissipate their energy. At Point Atkinson in the well-protected Strait of Georgia region, the 1964 tsunami caused only a 25-cm (10 in.) rise in sea level, while at the more exposed Argentine Islands in Antarctica, 13,000 km and 16 h later, a 60-cm (24 in.) rise was recorded.

Shoaling

As with other sea waves, tsunamis steepen as they move from the deep ocean into shallow water. The greatest amplification of the wave takes place near shore. If the slope of the shore or beach is gradual over many kilometres, much of the wave power is dissipated before it reaches the land. Reefs like those around the Fijian Islands are particularly helpful in this regard. Where the beach is small compared to the distance between crests, however, the waves may grow to a large swell that breaks at its crest. The shoreward run-up of the waves can then carry them over low-lying areas as a flood whose turbulent crest will quickly subside once it is ashore. On the other hand, if a tsunami encounters a steep, abrupt shore, the wave will slosh up the beach but its run-up height will be no higher than the height at which it is breaking. Under such circumstances, most damage will be due to flooding, not direct wave force. Underwater ridges near the shore will tend to focus the wave energy, and make it significantly higher than normally expected. There is some evidence that Crescent City, Calif., received large waves in 1964 partly because Cobb Seamount, 740 km northwest, acted like a lens and focused the tsunami toward the city. (Submarine canyons, of course, have the opposite effect and tend to lower the wave height.)

If a tsunami propagates into a narrowing embayment, it may be funneled to much greater heights than if it encounters a straight shoreline. The tsunami from the Sanriku earthquake in Japan, June 15, 1896, amplified into devastating 30-m waves at the head of Kamaishi Bay where it claimed over 27,000 lives. But every tsunami appears to be different, and quite often embayments are no more affected than other localities as the structure of the bottom outside of the bay apparently also affects the behavior of waves inside the bay.

At abrupt continental margins, a considerable portion of the energy associated with an approaching tsunami may be reflected back to sea, and effectively reduce the impact of the waves at the coast. Alternatively, the waves may be guided by the bathymetry in such a way that they become trapped near the coast. Trapping of this kind can take place around islands, along coastlines, or over undersea ridges and escarpments with the waves slowly "radiating" their energy back to sea over a period of time (see Fig 8.7).

State of the Tide

The higher the tide at the time a tsunami arrives ashore, the greater the likelihood of flooding. The worst set of circumstances in this regard is a high spring tide during a period of strong onshore winds. Waves of the 1964 Alaska earthquake first arrived at the shores of Vancouver Island about the time of a normal high water so, although sea levels rose above higher high water in many locations (Fig. 9.7), conditions were not extreme given the possible heights tides can attain in these areas.

Natural Oscillations

Despite a distance of 65 km from the coast, the twin cities of Alberni–Port Alberni suffered widespread damage from the Alaska tsunami. Its arrival at these cities, moreover, was atypical in that the 97-min delay between the first and second crests was much longer than usual. (This, and the fact that the first wave was smaller, allowed people time to evacuate the area.) Why these peculiarities? It seems that the combination of Trevor Channel and Alberni Inlet, which leads to the cities, had a natural resonant frequency that matched that of the tsunami, and allowed the waves to amplify (Fig. 9.2). In other words, the series of oceanic waves was able to make the water in the inlet slosh back and forth near its natural "sloshing frequency" just as water in a bathtub can be made to surge higher and higher by the appropriate rhythmic motion of one's hand. The relatively large tsunamic waves at Crescent City also appear mainly to have been the result of resonant conditions brought about by the bowl-like bathymetry of the adjacent continental shelf.

Ignorance

In San Francisco, 10,000 people jammed the beach areas to watch the Alaska tsunami wash ashore. Fortunately, for them, the waves were small! In contrast, the beaches in Chile, Japan, and Hawaii were deserted after the warning was issued; people had learned from previous experience the power of tsunamis. Most deaths in Crescent City from these waves were due to ignorance. The owners of a tavern that had suffered damage from the first two waves, returned with friends to remove money from the building. As everything appeared normal, they stayed to have a beer and were caught by the third wave, the largest of the series, and five were drowned. In 1960, the Chile tsunami killed 61 people in Hawaii who had failed to heed the warning to get to higher ground.

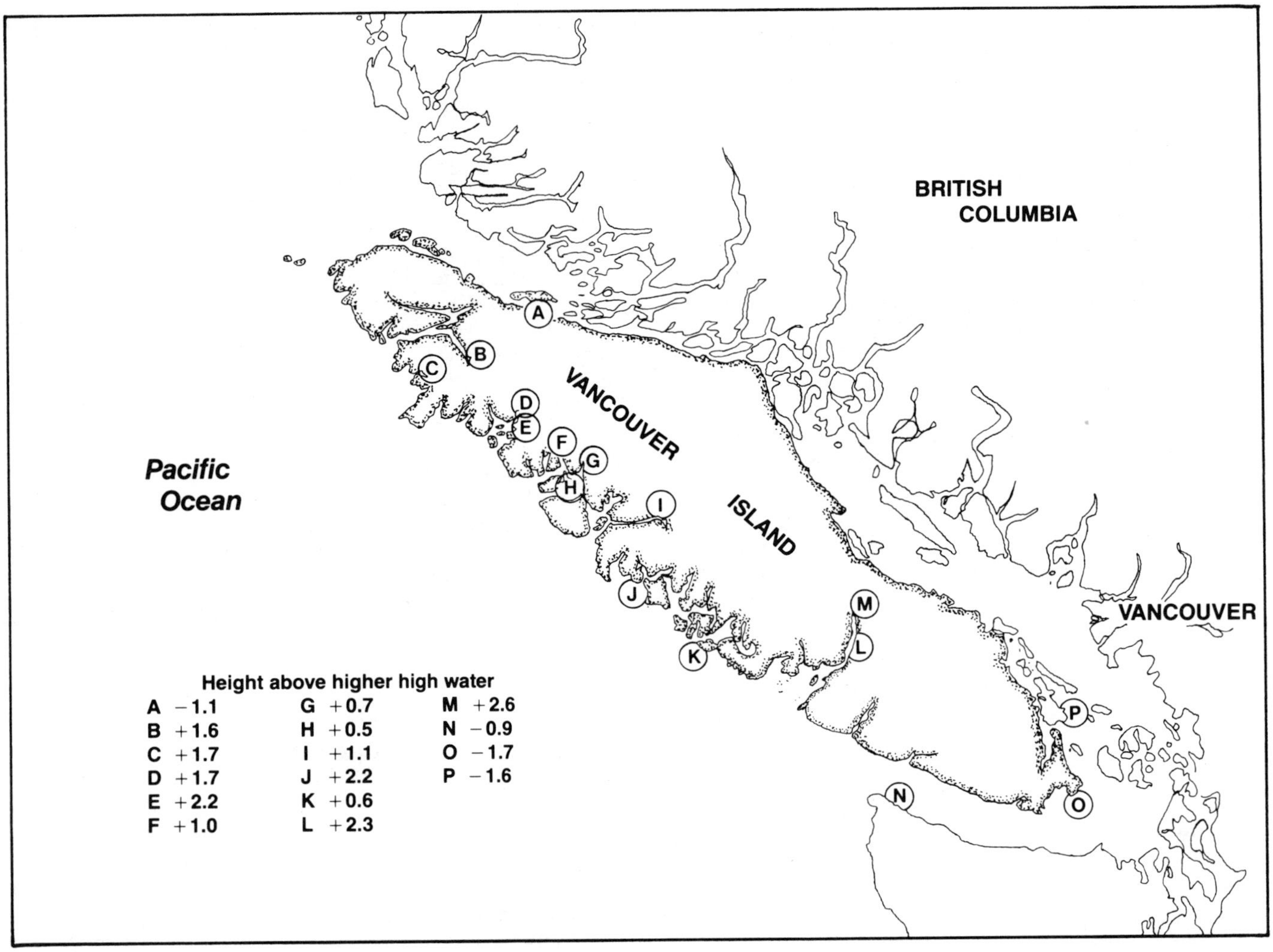

FIG. 9.7. Locations in southwestern British Columbia that reported significant tsunami activity Mar. 27–29, 1964. Numbers give height in metres of maximum wave crest above HHW, large tide (negative values indicate less than HHW). A, Alert Bay; B, Port Alice; C, Klaskino; D, Fair Harbour; E, Amai Inlet; F, Zeballos; G, Esperanza; H, Tahsis; I, Gold River; J, Hot Springs Cove; K, Tofino; L, Franklin River; M, Port Alberni; N, Neah Bay; O, Victoria; P, Fulford Harbour. (Adapted from Wigen and White 1964)

Population

The population of an area has a lot to do with the destructiveness of a tsunami. The highest waves reported in British Columbia in 1964, for example, were at Shields Bay on the west coast of the Queen Charlotte Islands. One wave was estimated to be over 9 m above the low-tide level. Damage was light compared to Alberni–Port Alberni as only a logging camp was there at the time.

Tsunami Warning System

After the devastating Aleutian tsunami of April 1, 1946, and its widespread damage to Hawaii, officials in the United States began to push for the formation of an early warning system in the Pacific Ocean. To do this, it was necessary to design tide gage stations capable of detecting abnormal changes in sea level and to have a communication network linked via a central dissemination center. With the engineering problems overcome, the system of tide gages (and seismic gages) in U.S. territory went into operation 2 yr later. Japan added gages in 1949. Canada waited until after the disastrous Chile tsunami of May 1960 to join the network, as did other nations like Chile, Taiwan, New Zealand, and the Philippines. No major tsunami occurred in the next few years so Canada withdrew in 1963 and, therefore, received no prior warning of the 1964 Alaska tsunami. Needless to say, Canada is a member again!

To date, Canada maintains two special tide gage stations within the Pacific network of 30 odd stations. Located at Tofino and Langara Island on the northwest tip of the Queen Charlottes, these stations are capable of direct telephone contact with the warning center in Honolulu. If an abnormal change in the water elevation occurs at these stations over a period of 5 min, the machine immediately switches to "warning mode," then dials Honolulu and repeats the message . . . *Warning Tsunami*. If such gages determine that a "tidal wave" is headed toward B.C., the Honolulu Observatory first contacts the Civil Defense Preparedness Agency in Bothwell, Wash., which in turn telephones (collect) to RCMP "E" Division Headquarters in Victoria, B.C. They then contact the Provincial Emergency Programme Co-ordinator (PEPC), the RCMP Duty Officer, and the Department of National

Defense. Based on the known epicenter of the earthquake and other gages in the network, a time of arrival is determined from specially prepared charts. Provided there is more than 2½ h, the above group, together with the Regional Tidal Superintendent, evaluate the information and determine the likely damage to low-lying areas. If there is a possibility the waves will reach the coast, people in these areas are evacuated by local civil defense authorities. On the other hand, if there is less than 2½ h notice because of the proximity of the earthquake, the PEPC contacts Broadcast News in Vancouver, which then issues a warning via special broadcasts over TV and radio.

The probability that a large tsunami from the ocean will propagate far into Juan de Fuca Strait or the Strait of Georgia is very low. It is also unlikely that the occurrence of a major earthquake off the coast will generate a tsunami because faulting in this geographical location is mostly of the sideways slip type. "Unlikely," however, does not mean impossible. In fact, there is a documented case of tsunamic waves generated by just such an occurrence. According to Murty (1977), the earthquake of June 23, 1946, that originated on the east coast of Vancouver Island was sufficiently strong to break telegraph cables on the seafloor and to cause the water level at Franklin River in Alberni Inlet to rise 6.1–9.1 m above average. At Sisters Rock near Texada Island in the Strait of Georgia, the tsunami attained an amplitude of over 2 m. A second wave in the Strait had an amplitude of 1.2–1.5 m and was responsible for the death of one person near Mapleguard Point opposite the southern end of Denman Island.

PART IV
OCEANOGRAPHY OF INSHORE WATERS

Chapter 10. Strait of Georgia

The Strait of Georgia is by far the most important marine region of British Columbia. More than 70% of the population of the province is located on its periphery and its shores provide a foundation for expanding development and industrialization. The Strait is a waterway for a variety of commercial traffic and serves as a receptacle for industrial and domestic wastes from the burgeoning urban centers of greater Vancouver. Salmon runs to the rivers that enter the Strait of Georgia are the basis for one of the world's largest commercial salmon fisheries; its resident coho and chinook salmon form an important and ever-increasing recreational fishery. The Strait also provides an area for the spawning and growth of herring and is the largest overwintering location for waterfowl in Canada. Widespread recreational use of the Strait by boaters, sports fishermen, bathers, and campers make tourism in British Columbia a major industry. In short, the Strait of Georgia constitutes a multiple-use aquatic environment that must be considered a national asset worthy of utmost consideration and protection.

Physiography

The Strait of Georgia occupies an inundated portion of the northwest–southeast trending Georgia Depression that lies between the intrusive rocks of the Coast Mountains and the intrusive, metamorphic, and sedimentary rocks of Vancouver Island (Fig. 10.1). On the average, it is about 222 km (120 nm) long and 28 km (15 nm) wide; islands occupy roughly 7% of its total surface area of 6800 km^2 (200 nm^2). The average depth within the Strait is around 155 m, and only 5% of the total area has depths in excess of 360 m. The maximum recorded depth of 420 m is immediately south of the largest island in the Strait, Texada Island, and is rather shallow compared with soundings obtained in some of the adjoining inlets (depths in Jervis Inlet reach 730 m).

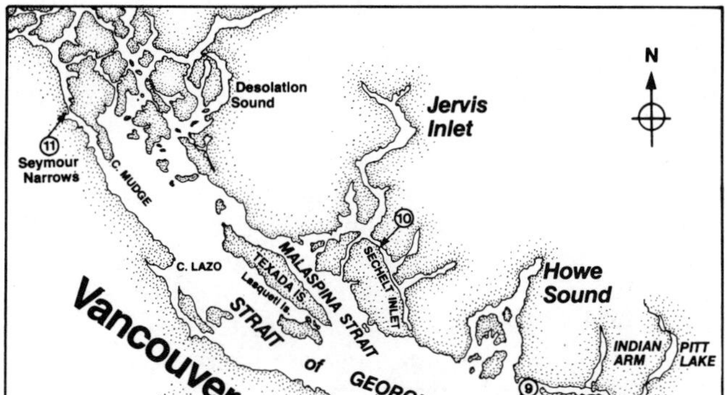

FIG. 10.1. Major geographical features of Strait of Georgia. Circled numbers: 1, Admiralty Inlet; 2, Deception Pass; 3, Rosario Strait; 4, Boundary Passage; 5, Active Pass; 6, Porlier Pass; 7, Gabriola Passage; 8, Dodd Narrows; 9, Burrard Inlet; 10, Skookumchuck Narrows; 11, Discovery Passage.

To the north, the Strait of Georgia is linked to the Pacific Ocean via several narrow but relatively long channels, notably Discovery Passage and Johnstone Strait, and by the broader Queen Charlotte Strait. To the south it is linked to the ocean via Juan de Fuca Strait and a few comparatively wide channels between the San Juan and Gulf Islands (Fig. 10.2). Of these, Haro Strait and

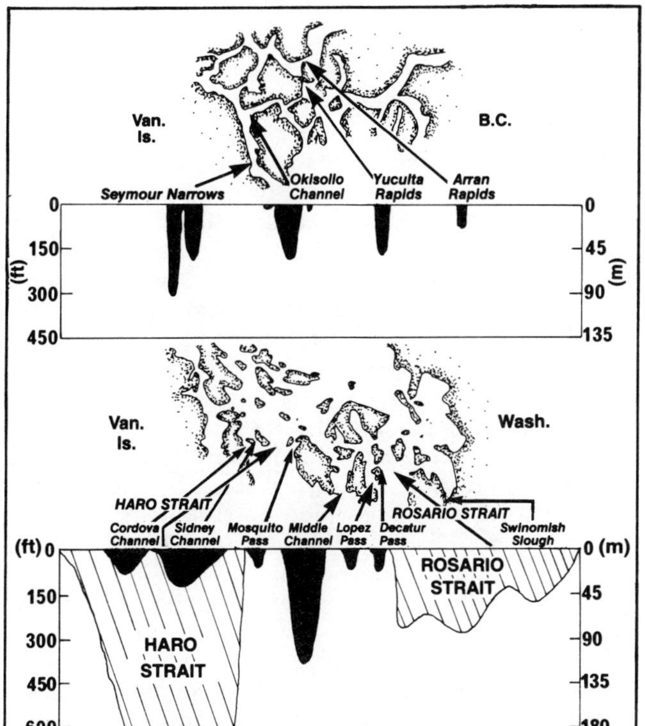

FIG. 10.2. Cross-sections of channels leading into Strait of Georgia, (*top*) northern approaches, (*bottom*) southern approaches. (From Waldichuk 1957)

Rosario Strait have a cross-sectional area appreciably larger than all other channels combined. Along the eastern margin of the Strait of Georgia, erosion by ice moving down former rivers during repeated periods of glaciation created long, steep-sided valleys that were later flooded to form fiords as sea level rose following the last glacier retreat. The longest of these, Jervis Inlet, cuts back more than 61 km (33 nm) into the Coast Range. The combined effect of glacial scouring and subsequent flooding has further created a complex of islands, sounds, and passages along the eastern coastline of the basin. By contrast, the western side of the Strait has few inlets and a more regular coastline. The distinctive Gulf Islands in the southwest corner of the region were originally part of Vancouver Island but have since been detached through glacial erosion of the poorly resistant sedimentary rocks. Freshwater discharge into the Strait comes mainly from the Fraser River, which empties directly into the basin near Vancouver, with an important secondary contribution from the Squamish River that enters the Strait via Howe Sound. Appreciably smaller freshwater discharges come from the rivers of Vancouver Island, such as the Cowichan, Chemainus, Nanaimo, and Courtenay and from the rivers that empty into the inlets on the eastern coast.

Temperature and Salinity Distributions

The temperature of the water within the Strait of Georgia varies with depth, proximity to the Fraser River delta, and season. For simplicity the water column in the Strait can be divided into two layers; the upper layer, shallower than about 50 m, and the lower layer below this level to maximum depths of around 400 m. In the lower layer observed temperatures are nearly uniform throughout the year, typically 8–10°C over the entire basin from Cape Mudge to Haro Strait. Winter values tend to be higher than summer values but only by about 1C°. There is also a slight tendency for temperatures to decrease northward along the channel at depth.

It is in the upper layer that most seasonal and along-the-strait variation in water temperatures takes place, and this can be quite pronounced (Fig. 10.3a, b). Beginning in late fall, cool air begins to rapidly lower the temperature of surface waters. Due to storm winds and cooled water sinking as it becomes more dense than the water beneath, temperatures also begin to decrease throughout the upper 50 m or so of the Strait. By late winter (February–March), near-surface waters are coldest, sometimes falling to as low as 5–6°C, but below 50 m there is a gradual increase in temperature to around 9°C at the bottom. The coldest water in the Strait of Georgia at this time is frequently associated with Fraser River runoff, which loses heat as it flows through the wintery British Columbia interior.

With the advent of spring, air temperatures rise and storm activity abates to permit increased retention of solar energy in the upper waters of the Strait. By the middle of May sufficient warming has taken place in the top 10–20 m for near-surface temperatures to reach around 15°C in certain areas. The Fraser River freshet starts around this time and forms a brackish layer over large portions of the Strait. Increased stability of the top few metres of the water column allows further warming to occur, aided by longer periods of more intense solar radiation and warmer air temperatures. In July it is not unusual to find patches of water in mid-Strait that exceed 20°C. Similar water temperatures occur in protected areas, such as Departure Bay near Nanaimo and Burrard Inlet off Vancouver, and in numerous small coves and bays along both sides of the Strait. Only near tidally mixed regions at the northern and southern approaches to the Strait of Georgia do water temperatures remain consistently low (around 10°C) throughout the summer. However, when winds are offshore even areas like Departure Bay can have cool surface waters as the warmer water is driven offshore and colder deeper water upwells to replace it (see Chapter 5).

Maximum surface temperatures in the Strait of Georgia generally occur around the first week of August. By the end of the month substantial cooling sets in and surface water temperatures again fall to around 15°C over most of the region. Near the end of autumn, enhanced storm activity, combined with decreased solar heating, relatively low freshwater runoff, and cool wintery air, creates nearly uniform surface temperatures of about 9°C within the entire Strait. The upper layer continues to cool until late winter and the cycle is repeated, with slight variations from year to year due to annual fluctuations in local climate.

Like temperature, the salinity distribution in the Strait of Georgia has a marked two-layer structure (Fig. 10.4a, b). The top of the lower layer lies approximately 50 m deep and is delineated by a salinity of 29.5‰; below this, salinities gradually increase to near-bottom values of 30.5‰ in summer and 31.0‰ in winter. Above 50 m, on the other hand, salt content varies considerably with season and distance from the mouth of the Fraser River estuary, where salinities are always comparatively low. Very localized regions of low-salinity surface water are also often associated with the smaller rivers that empty into the Strait.

During the period of consistently low Fraser River discharge and strong wind mixing from early December to early April, freshwater runoff in the upper layer is generally confined to a region well south of Texada and Lasqueti islands. In this portion of the Strait of Georgia, regions with nearly uniform salt content form in the upper layer, where salinities increase from surface values of 27–28‰ to 29.5‰ near 50 m depth; other areas under more direct influence of the Fraser River may at the same time have salinities of 25‰ or less at the surface. Within the northern Strait, the overall range of salinity is consistently small in winter and near uniform salinity conditions always prevail.

With the Fraser River freshet in late May, a layer of brackish water with salinities of less than 15‰ forms the top few metres over most of the central and southern sectors of the Strait of Georgia. The surface water during this period often has a sweet taste and is drinkable. However, in the northern portion of the Strait the surface water is typically saltier than 25‰. By August, the peak runoff from the Fraser River is over and salinities in the upper 50 m begin a gradual increase in value. Localized regions of nearly brackish surface water still occur, but their extent and persistence continue to diminish with the approach of winter.

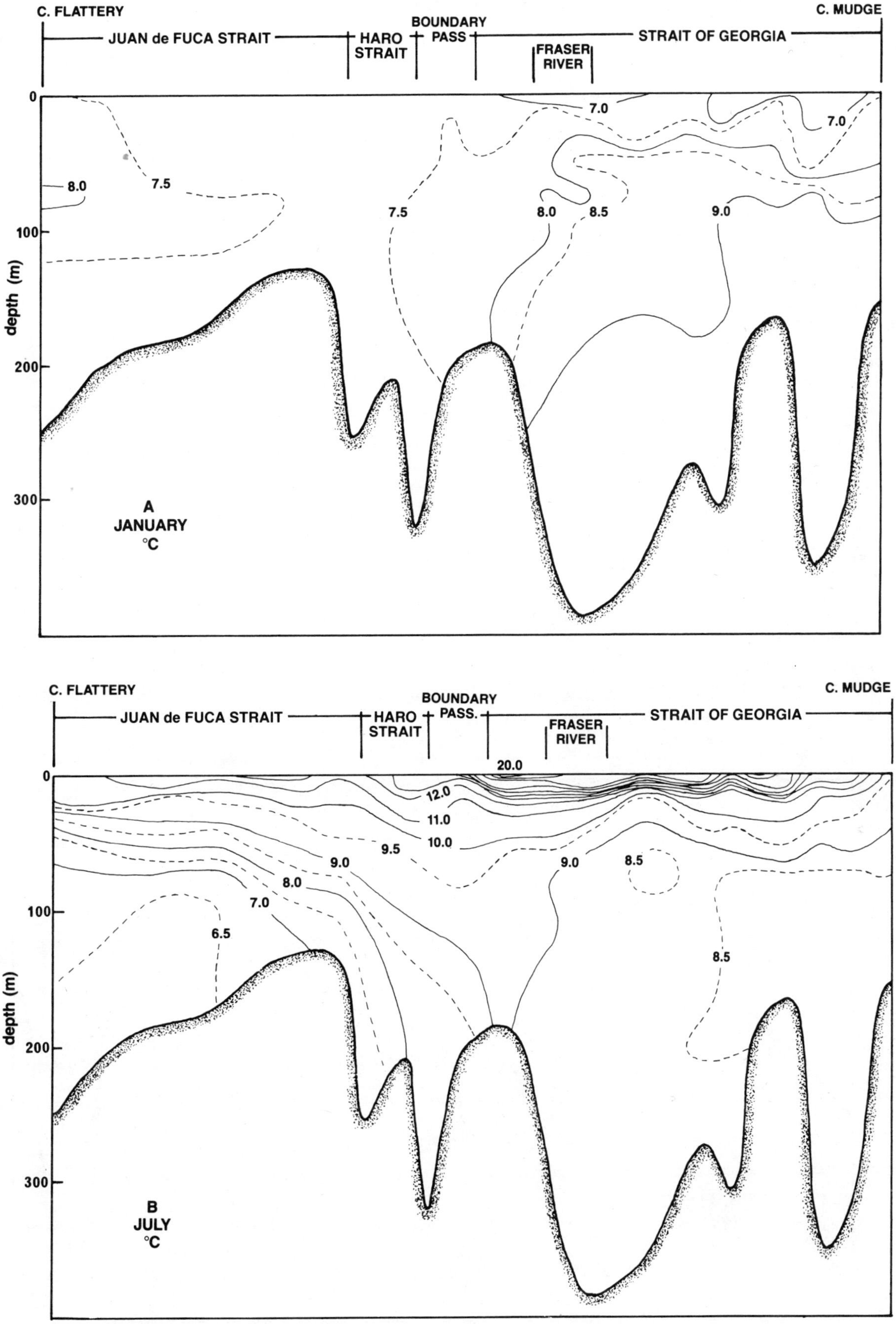

FIG. 10.3. Along-channel sections of water temperature from western end Juan de Fuca Strait to northern end Strait of Georgia (A) January 1968, (B) July 1968. (From Crean and Ages 1971)

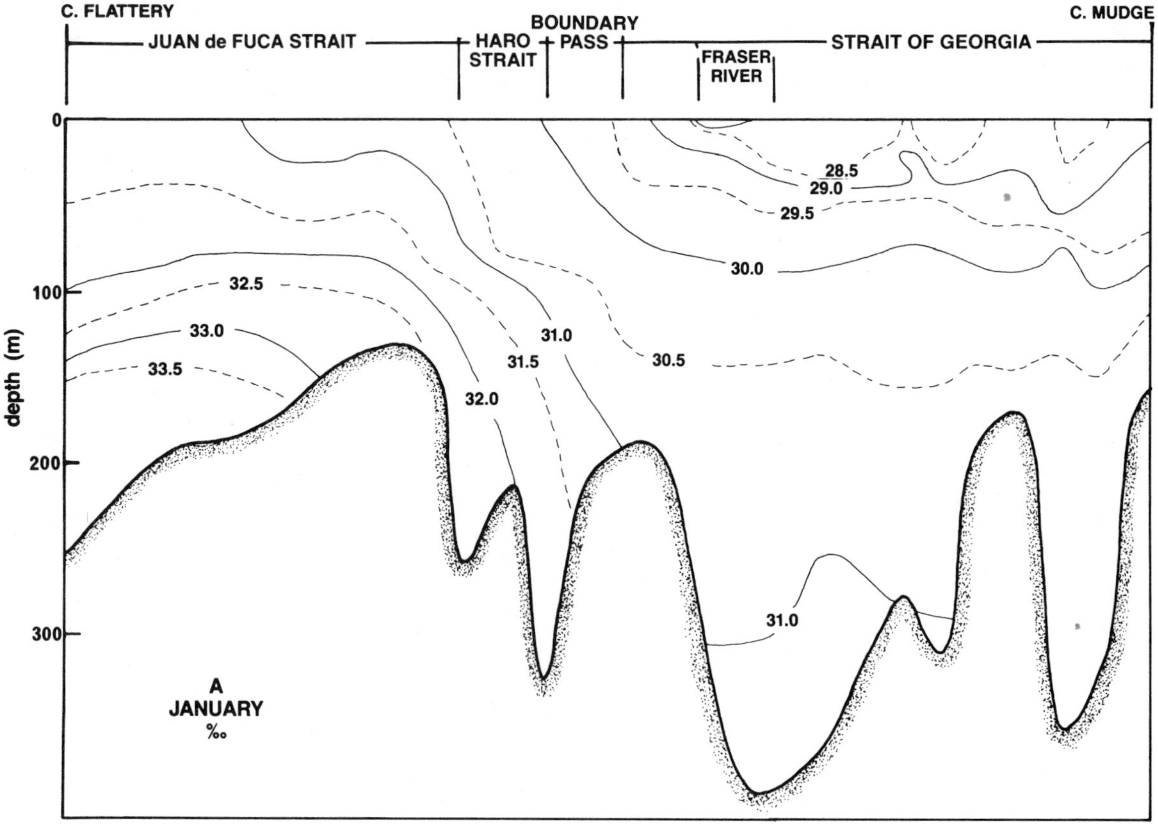

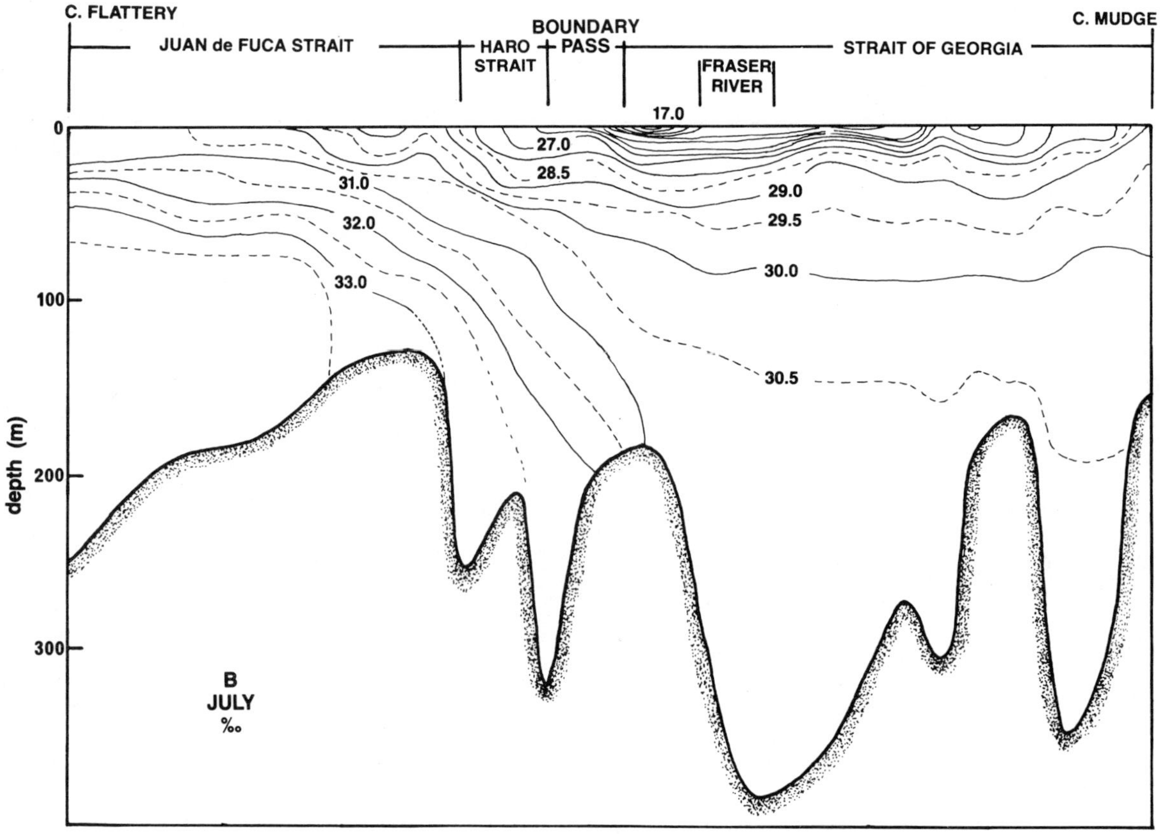

FIG. 10.4. Along-channel sections of salinity from western end Juan de Fuca Strait to northern end Strait of Georgia (A) January 1968, (B) July 1968. (From Crean and Ages 1971)

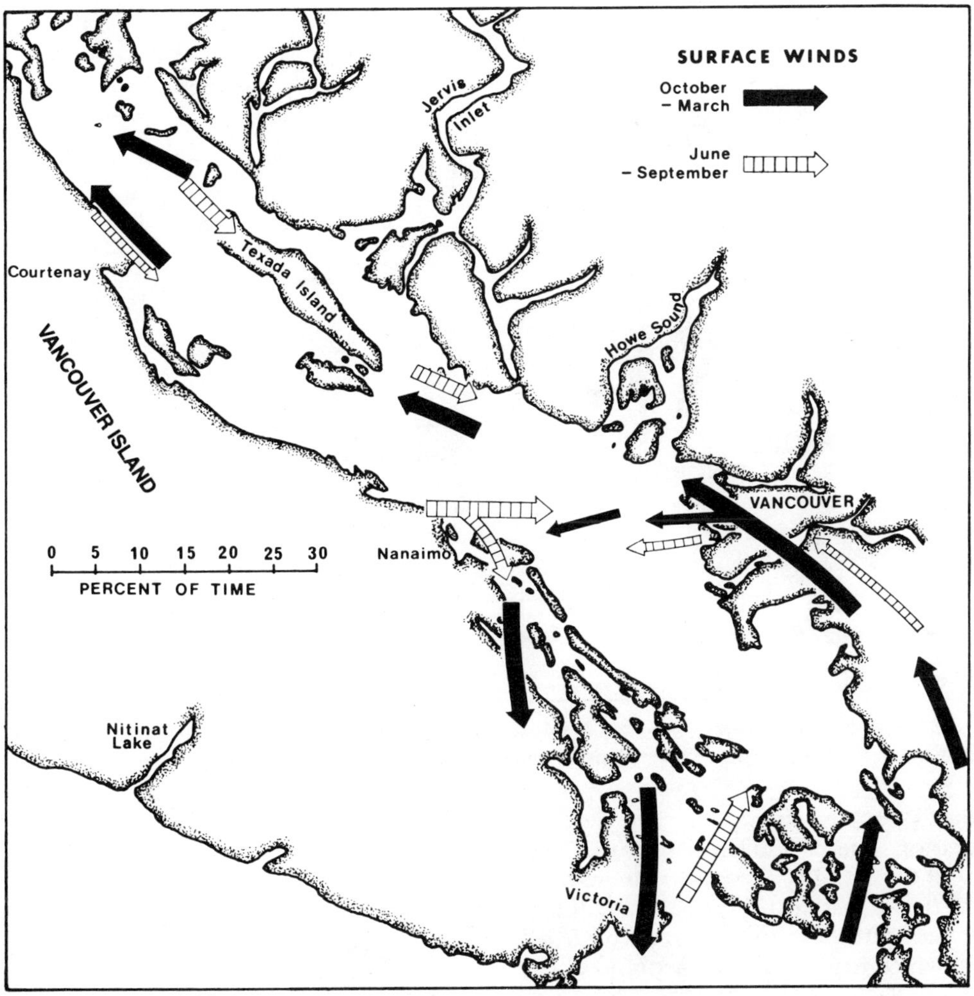

FIG. 10.5. Prevailing surface wind patterns over Strait of Georgia in winter (solid arrows) and summer (hatched arrows). Thick arrows correspond to speeds 4.5–9 m/s (8.7–17.5 kn); thin arrows less than 4.5 m/s. Comparison of length of arrow to scale on left yields frequency of occurrence of particular wind. (Modified after Barker 1974)

Wind Patterns

In the exposed areas of the Strait of Georgia, prevailing winds are predominantly from the northwest in summer and the southeast in winter. The northwesterlies are associated with the clockwise motion of air around the North Pacific High, which centers west of California in midsummer; the southeasterlies are associated with the strong anticlockwise flow of air around the Aleutian Low, which develops just south of Alaska in winter (see Fig. 2.14). Significant modifications of this general pattern are produced by the complicated topography of the surrounding land. The funneling effects of Juan de Fuca Strait, Puget Sound, and the Fraser Valley play important roles in governing the overall wind pattern in the Strait of Georgia. In the southern Strait, for instance, there appears to be a closed anticlockwise wind pattern from October to March (Fig. 10.5), and a turn from southeasterlies to easterlies of winds off the Fraser River, because of the influence of the Fraser Valley. By spring, the winds are predominantly southeasterly to easterly. In summer, winds are typically lighter and more confused than in other seasons. Southeasterlies and southwesterlies dominate the southern Strait but northwesterlies dominate the northern Strait, where they conform more to the oceanic wind pattern.

The winter wind pattern is altered further by the invasion of polar continental air funneled onto the Strait through the inlets and valleys that lead from the B.C. interior. As this air descends toward the coast, its speed is often sharply increased to produce the gale strength Squamishes that howl forcefully down inlets like Howe Sound. These winds have some important effects on the wind structure in the Strait of Georgia. For example, if the Squamishes occur when there is a low-pressure area over northwestern Washington, northwest winds instead of southeast winds occur over the northern and central regions of the Strait (Fig. 10.6). An interesting feature for Vancouver sailors is that, although Howe Sound Squamishes are usually associated with strong easterly winds through the Fraser Valley, the accompanying winds in Burrard Inlet are often light and variable. In other words, it is possible to have gale force winds in Howe Sound and at Abbotsford, but nearly becalmed conditions in the

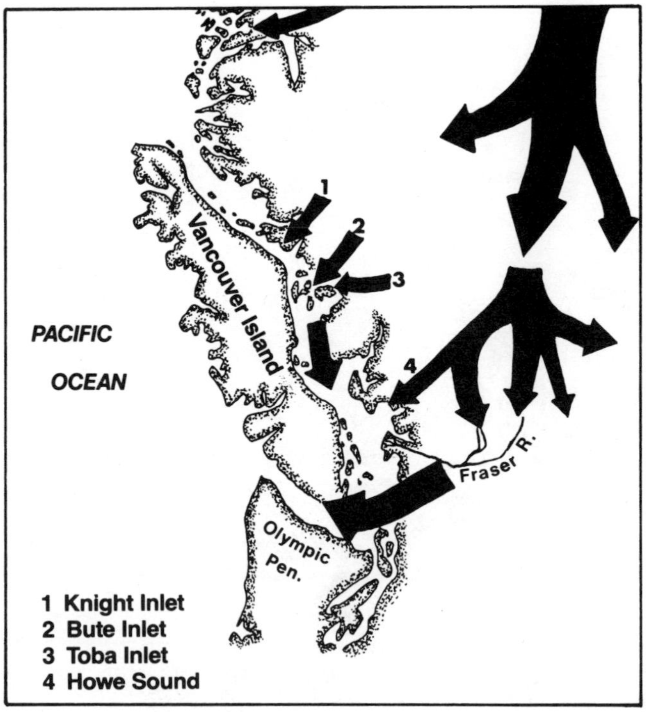

FIG. 10.6. The main paths of cold air in polar outbreaks. Seaward flow along mainland inlets is responsible for Squamish-type winds. (From Tyner 1951)

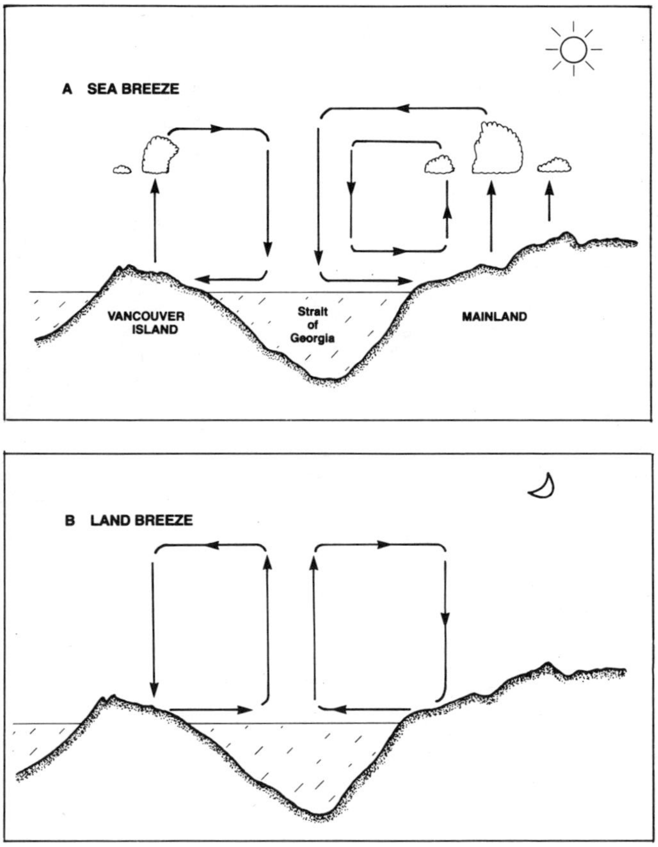

FIG. 10.7. Day sea breeze and night land breeze over Strait of Georgia region in fine weather.

sailing areas around Vancouver. Squamishes were reported on one occasion to reach 130 km/h in Dean Channel in the vicinity of Ocean Falls. Strong winds are most often associated with the passage of active frontal disturbances, however. The more vigorous storms are preceded by southeast gales, which tend to follow the trend of the Strait of Georgia and the Coast Mountains. Moreover, strong southeasterly winds that precede southwest–northeast oriented cold fronts frequently veer to strong northwesterly winds as a ridge of high pressure strengthens behind the advancing front.

In summer, the differential heating of the land and water generates breezes that significantly alter the prevailing wind regime during periods of fine weather. There are two types: (a) the sea breeze from the water to the land, produced by the greater heating of the land than the water during the day (Fig. 10.7a), and (b) the land breeze from the land to the water, produced by the greater cooling of the land than the water at night (Fig. 10.7b). Over the eastern side of the Strait of Georgia, the sea breeze generally sets eastward onto the mainland around 10 a.m., strengthens to about 4–7 m/s (8–14 kn) at midafternoon and then dies away before sunset. Consequently, the time of least chop begins about 2 h before sunset, and the smoothest crossings of the Strait begin in the late afternoon to early evening. (See *Small Craft Guide*, Vol. 1, 1979, p. 28–29, published by the Canadian Hydrographic Service.) Often in summer, the combination of the sea breeze and prevailing northwest winds (associated with large-scale westerly flow of air over the coast) results in strong (over 8 m/s) onshore winds on the eastern side of the Strait. Near gale force winds in summer can occur in Burrard Inlet if each of these two wind systems is intense enough. On the western side of the Strait the sea breeze will partially counterbalance any pattern of prevailing northwest winds, and produce weaker, more northerly, winds.

During the evening, the land breeze develops and winds blow onto the eastern half of the Strait from the mainland at speeds that peak near midnight to less than 4 m/s. On the western side of the Strait, a similar system of onshore and offshore breezes develops as a consequence of the daily heating and cooling of Vancouver Island. These winds, of course, are in the opposite direction to their counterparts on the mainland side.

Waves

Wave heights in the Strait of Georgia are limited by the fetch of the wind and to a lesser degree by its strength and duration. The total fetch is further limited by obstructions like Texada and Lasqueti islands, which make it almost impossible for wind waves to propagate and grow unimpeded along the entire length of the basin. No one who has been out in the Strait during a strong blow would argue that wave heights cannot at times be quite appreciable. But claims such as one boater made that he had "just survived 30-foot rollers off Nanaimo" during a gale are exaggerations (5–10 ft waves were seen in the same area from the research vessel *Richardson*). Had he claimed to have endured such conditions south of Cape Mudge or off the main arm of the Fraser River where large rips can develop, his story would have been more believable.

Some of the first scientifically recorded wave measurements in the Strait of Georgia were made in 1968–69 by the B.C. Research Council, to assess wave effects on the first "stretched" B.C. ferry, MV *Queen of Esquimalt*, between Swartz Bay and Tsawwassen. These were followed by similar studies at Halibut Bank prior to the first stretched ferry sailings between Nanaimo and Horseshoe Bay. More recent investigations of the wave climate of the Strait have been undertaken by Environment Canada and the Department of Public Works using bottom-moored waverider accelerometer buoys. Wave records from these buoys have been obtained off West Vancouver in Burrard Inlet in 40 m of water, off Sturgeon Bank near the Iona Sewage Outfall in 139 m, and off Roberts Banks near the coal port in 110 m (Fig. 10.8a). Observations were made for 17 mo off West Vancouver and 26 mo at the other localities, with continuous readings over a span of 20 min, each 3 h. Sturgeon and Roberts Bank are especially well exposed to waves from the northwest generated over a possible wind fetch up to 120 km (63 nm) and, therefore, should yield extreme wave data indicative of other exposed areas in the Strait.

Heights

Wave statistics from the three locations are presented in Fig. 10.8a, b; 10.9. During the entire period of observations in the Strait, significant wave heights never exceeded 2.1 m off Roberts Bank or 2.7 m off Sturgeon Bank, while corresponding maximum heights were always less than 3.3 and 4.0 m. Only 10% of the time did average wave heights at both localities exceed 0.8 m and maximum wave heights exceed 1.2 m; however, 60% of the time the maximum waves exceeded 0.3 m. Calm conditions prevailed 31% of the time at Roberts Bank and 27% at Sturgeon Bank. The curves for West Vancouver reflect the protected nature of Burrard Inlet where maximum probable waves rarely exceeded 1.8 m and maximum wave heights were greater than 0.6 m only 10% of the time. Fewer than 30% of the latter records showed waves higher than 0.3 m.

Periods

A comparison of wave heights and wave periods reveals other facets of the wave climate in the Strait of Georgia. At the two exposed locations, for example, there is a tendency for significant wave heights to increase as the period increases, at least up to periods of around 5 s. A typical 5-s wave will have a significant wave height of 0.5–1.0 m or a maximum wave height of 0.8–1.6 m, suggesting that waves with periods of 5 s or less are seas actively worked on by the winds. Wave periods of 5–6 s represent a transition from sea to swell-type waves, when both large and small wave heights occur with almost equal regularity. At periods exceeding 6 s, the observed wave heights are generally small and indicative of long, low, swell-like waves, which for a period of 6 s would have a wavelength of approximately 55 m based on Table 6.3.

FIG. 10.8. Measured wave-height distributions for three locations (*) Strait of Georgia. Upper curves (A) (B), percentage of time maximum probable wave, H_{max}, exceeds given height; lower curves percentage exceedance for significant wave, $H_{1/3}$. Statistics based on observations March 1973–May 1974 West Vancouver, and February 1974–April 1976 Sturgeon and Roberts Bank.

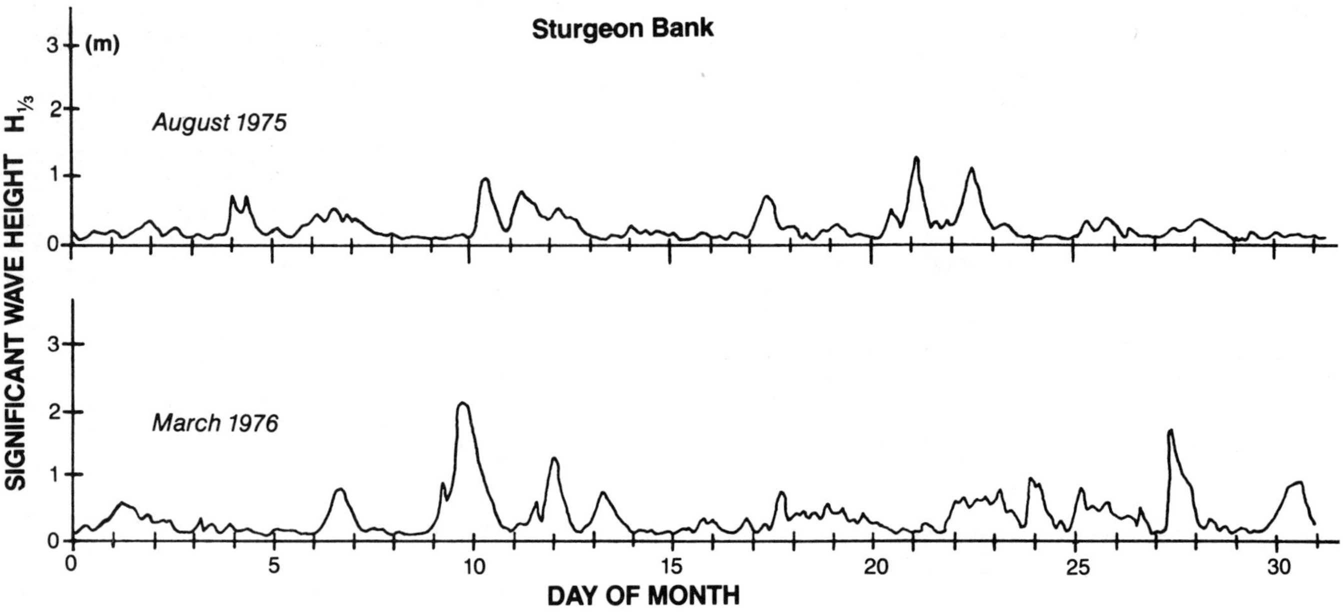

FIG. 10.9. Significant wave height versus time at Sturgeon Bank for 2 mo. Height in metres based on measurements taken every 3 h. (Adapted from Canadian Marine Environmental Data Service reports)

Nevertheless, seas with significant wave heights up to 2.5 m and periods of 7–8 s occur on occasions. Such sea conditions are more likely to occur at Sturgeon Bank than at Roberts Bank because of a steepening effect by opposing currents from the North Arm of the Fraser River.

The vast majority of waves in the open Strait off the Fraser River estuary have periods clustered in the range of 2–4 s and maximum probable heights below 1.5 m. The most probable maximum wave period according to the observations off Sturgeon and Roberts Bank is 9 s which, for low sinusoidal swell in deep water, would correspond to a maximum wavelength of about 125 m. Periods of 5–6 s occur as much as 30% of the time off Sturgeon and Robert Banks.

Wave records off West Vancouver are noticeably different from those within the Strait. Significant wave heights off West Vancouver are always less than 1.0 m, but span a surprisingly wide range of wave periods of up to 7 s, indicating that both sea and swell have comparable heights within Burrard Inlet.

Of course not all wave conditions in the Strait of Georgia are represented by the above statistical results. On one ferry crossing the author took between Swartz Bay and Tsawwassen on a blustery spring day, northwest winds were gusting to 25 m/s (50 kn) and huge breaking seas seemed to engulf the Strait in foaming white water. Onboard the spray-drenched MV *Queen of Sidney*, babies cried and many passengers ended up on the deck. The cafeteria was closed soon after the boat left Active Pass, though many patrons were too seasick to notice. A few later sailings that day were canceled, but not because of the 3–4 m high waves. As attested to by the minimal amount of wave shelter provided by the small breakwaters at Tsawwassen, the occasions when sailings are canceled are usually due to adverse winds when docking rather than the sea state.

Rips

The highest and steepest waves in the Strait of Georgia occur in rips where strong currents oppose wind waves generated over long fetches during gale force winds. Waves in these regions can be devastating and small craft should make every effort to avoid them.

Particularly dangerous rips occur at three locations in the Strait (see Fig. 10.1): (1) in the vicinity of Boundary Passage during periods of strong northwest winds and large flood streams. The sea state is further confused in this area by intense upwellings and eddies associated with tidal mixing in the channel. In the large backeddy that forms north of Tumbo Island on the flood, the seas will be more subdued; (2) seaward of Steveston Jetty and North Arm Jetty of the Fraser River delta during times of west to northwest gales. Extremely dangerous rips for such winds are created during periods of substantial river discharge (summer freshet) or near times of low tides when entrant speeds of the river can reach 2.5 m/s (5 kn) in the main channel. Near the river mouth, shoaling of the bottom causes a further amplification of the wave heights and invariably the water is rougher than in adjacent regions of the Strait. Numerous boats have capsized and their occupants drowned when attempting to enter the river in a northwest blow, and the number will undoubtedly grow unless the increasing number of boaters on the lower mainland are made more fully aware of the risks; and (3) south of Cape Mudge during strong southeast winds and a flood stream. The region from Cape Mudge to about Willow Point is locally renowned for some of the most awesome rip conditions in the Strait of Georgia, and owes this distinction to the relatively long wind fetch and strong tidal currents that surge southward out of Discovery Passage. The author can personally attest to the validity of this reputation. On the homeward leg of a March 1977 cruise, the CSS *Parizeau* encountered an area of high seas roughly 1.5 km south of Cape Mudge at the time of a 3 m/s (6 kn) flood and a southeast gale with winds gusting to 30 m/s. The currents had literally stopped the waves at the leading edge of the advancing tidal stream. The waves steepened into sharp, white-water peaks over 5 m high, churning and foaming in their effort to propagate against the flow. With the passage of each standing crest, the bow of the *Parizeau* would drop suddenly and plow almost immediately into the next wave, sending a blinding spray of water over the bridge and leaving stomachs suspended in air. After a few kilometres the ride ended, and the 2- to 3-m seas south of the tide rip seemed like the proverbial millpond.

From the air, flood waters that intrude into the Strait of Georgia via Discovery Passage appear as a slowly widening jet, whose axis lies roughly parallel to the 90-m bottom contour (Fig. 10.10). As the intense rips occur near the leading edge of this intruding jet where its currents first begin to oppose waves from the southeast, they can be avoided by keeping well to the east of midchannel or by waiting for a turn to ebb.

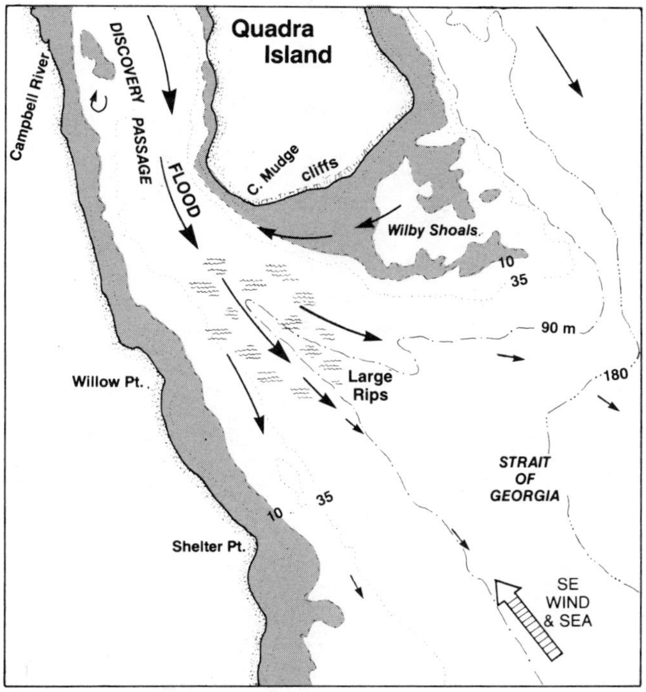

FIG. 10.10. Approximate extent of large rips south of Cape Mudge during a flood current and moderate-to-high seas from southeast. Area covered by rips varies according to velocity of current and strength and duration of wind. Depths in metres (for location see Fig. 10.1.)

Substantial rips can also form at the Strait of Georgia entrances to Porlier Pass and Active Pass during a flood stream and northwest winds. If the flooding jet of water curves northward when it leaves the pass, as it has been observed to do on occasions at Porlier Pass, there can be a further steepening of the waves in the rip. Southeast winds generally produce less intense rips due to the comparatively short wind fetch, especially at Active Pass.

Tides

As noted in Chapter 3, the inward progressing ocean tide enters the Strait of Georgia via Juan de Fuca Strait and is reflected from the constricted northern end of the channel. This southward reflected portion of the tidal wave then combines with the northward advancing portion of the tide to produce a standing wave in the Strait of Georgia, with distinct features from the more progressive-type tidal wave traveling along Juan de Fuca Strait. The fact that the reflected portion of the semidiurnal tide is partly attenuated by frictional effects before it can reenter Juan de Fuca Strait confines the pure standing wave pattern to the Strait of Georgia.

The standing wave nature of the tides causes the entire water level along the Strait of Georgia to move up and down in unison roughly every 12 h, 25 min, as if it were pivoted at an imaginary line drawn eastward through a point south of Saturna Island (see Fig. 3.25). As a consequence of this teeter-totter arrangement, the tidal ranges of both the main semidiurnal constituent (the M_2 tide) and the main diurnal constituent (the K_1 tide) increase from south to north (Fig. 10.11a, b). Moreover,

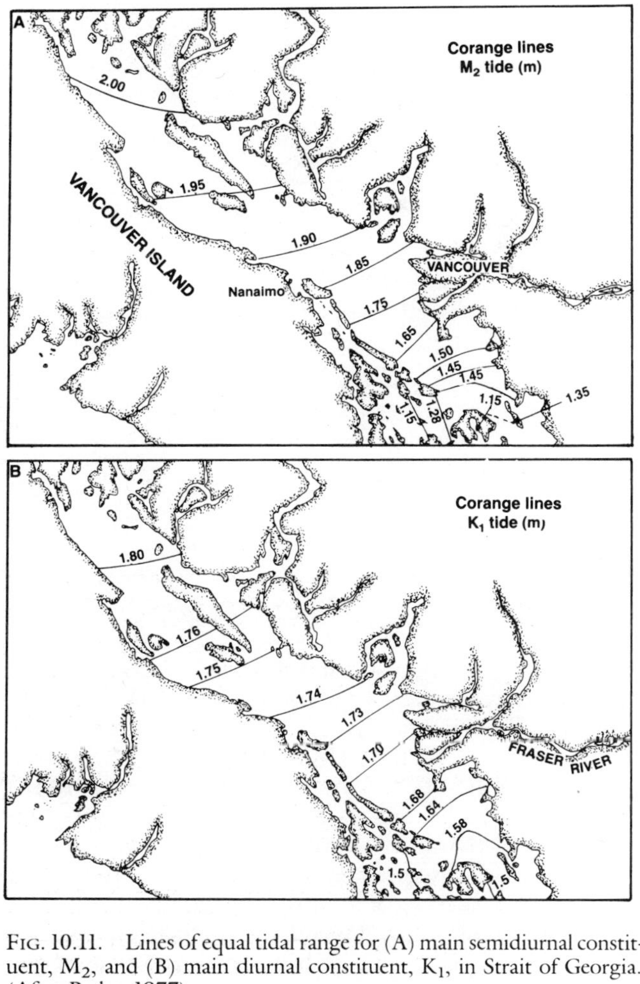

FIG. 10.11. Lines of equal tidal range for (A) main semidiurnal constituent, M_2, and (B) main diurnal constituent, K_1, in Strait of Georgia. (After Parker 1977)

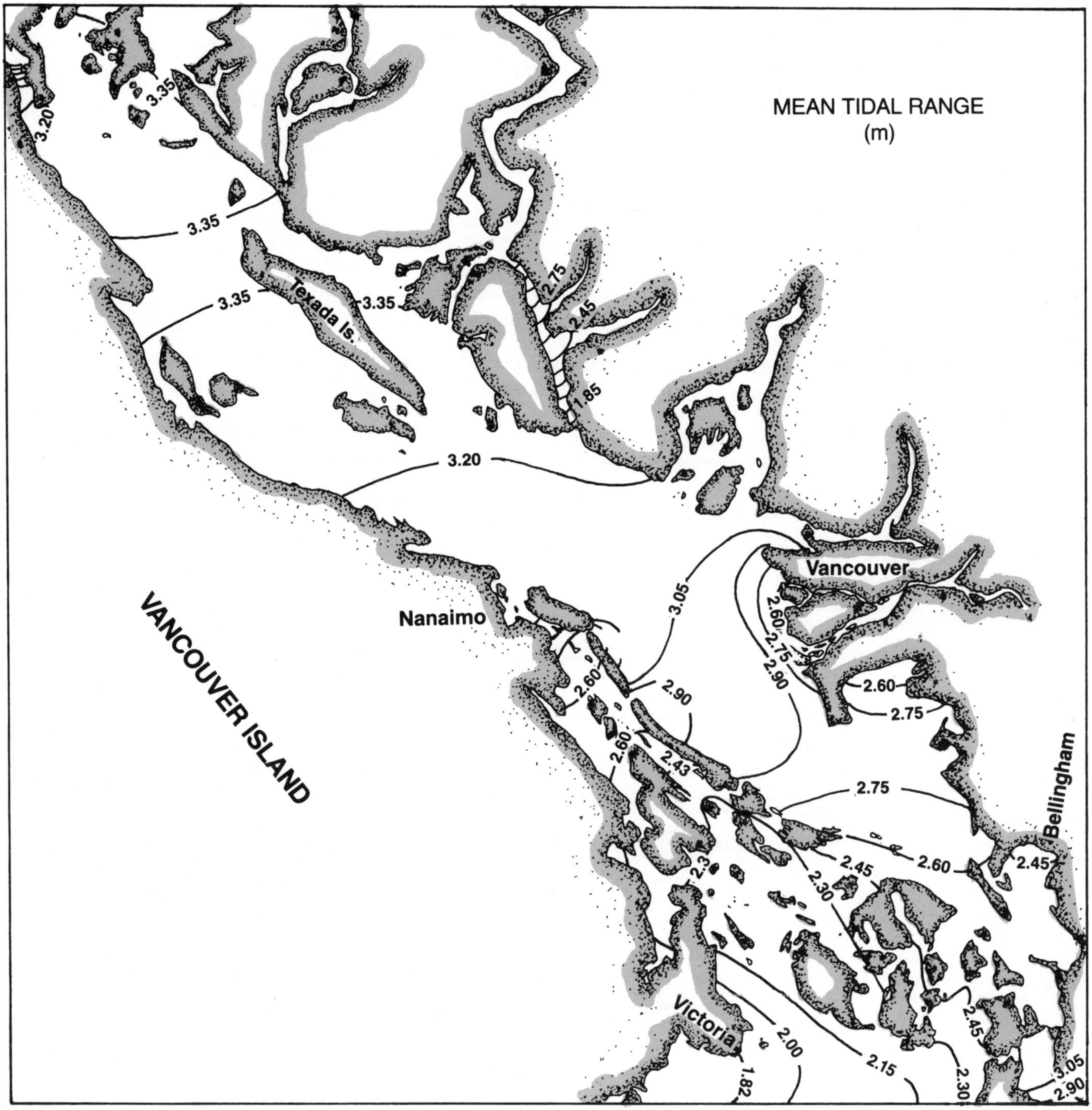

FIG. 10.12. Lines of equal mean tidal range in Strait of Georgia. (From Barker 1974)

between any two places within the Strait, the difference in time between identical stages of the tide never exceeds 30 min, quite unlike the situation in Juan de Fuca Strait. The actual tide is largely a combination of the M_2 and K_1 constituents and behaves in the same manner (Fig. 10.12). As semidiurnal elevations are slightly more important than diurnal ones, tides in the Strait are mixed, mainly semidiurnal. Consequently, there are differences in elevation between successive high waters and between successive low waters, the diurnal inequality. The diurnal inequality for successive high waters is always less than that for successive low waters. Moreover, the sequence of the tide always follows the pattern of Higher High Water, Higher Low Water, Lower High Water, Lower Low Water illustrated in Fig. 3.5 for Point Atkinson near Vancouver.

In addition to its daily inequality in height, the tidal range in the Strait of Georgia undergoes a biweekly variation due to cyclic changes in the declination and phase of the moon, with the declinational influence decreasing northward along the channel. Spring tides each 15-day period occur almost exactly 26 h following a new or full moon, which compares to a corresponding delay of about 15 h in the eastern portion of Juan de Fuca Strait.

Each year maximum tidal ranges occur during the summer and winter solstices when both the sun and moon attain their greatest north–south declinations at the same time. Minimum ranges occur during the spring and autumn equinoxes. Therefore, beginning in March, when Lower Low Water in the Strait is in the late afternoon, the low water begins to occur earlier each day and to become more extreme. By late June, lowest daylight tides take place at midday when the sun is overhead and a new moon coincides with the maximum lunar declination south of the equator. If one wishes to swim at Spanish Bank in Vancouver in the summer it is best to wait until evening after the water has been warmed as it crosses the broad sand flats on the rising tide.

The Coriolis effect produces only slightly greater tidal ranges at the boundaries of the Strait compared to those at interior regions. This is expected because, except in the southern portion, currents in the main body of the Strait of Georgia are weak and, therefore, subject to negligible deflection as a result of the earth's rotation. On the contrary, the across-strait difference in the tidal range in Juan de Fuca Strait is much more pronounced because of appreciably stronger tidal currents in that channel. Finally, the "centrifugal" force on the tidal wave that turns the corner into the Strait of Georgia from Haro Strait causes the tides to occur a few minutes earlier on the American side south of Boundary Bay than elsewhere in the southern portion of the Strait. Generally speaking, however, the large-scale curvature of the channel has a minor effect on the flow compared to that of local bends and points of land.

Tidal Streams

As with other coastal regions, currents in the Strait of Georgia are affected by tides, winds, river discharge, the Coriolis force, centrifugal forces, and channel bathymetry. Because of along-channel changes in the relative importance of these factors, the Strait actually possesses a diversity of circulation patterns superimposed on the tidal flow pattern. The Fraser River runoff, for instance, is of more direct importance to the flow structure in the surface layer of the central and southern portions of the Strait than to the northern sector. Although tidal streams are the predominant form of motion, their strength varies considerably over the length and depth of the Strait to further complicate an already complex oceanographic regime.

Measured tidal streams within the Strait of Georgia exhibit the characteristics of a standing wave where maximum flood occurs roughly 3 h before high water and maximum ebb around 3 h before low water (Fig. 10.13).

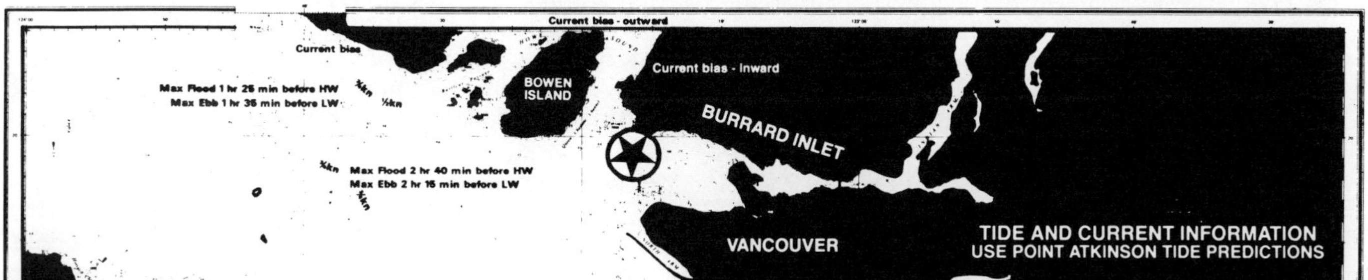

FIG. 10.13. Tidal streams and times of maximum ebb and flood relative to tide heights at Point Atkinson (star). Except near Fraser River delta, maximum surface tidal currents occur approximately midway through a semidiurnal tidal cycle, 2–4 h prior to high or low water. (Courtesy S. Huggett and P. Crean)

This correspondence between tide and tidal currents is best established within the northern part of the Strait and weakens slightly southward where the progressive nature of the ocean tide becomes more prevalent. Like the tides, tidal streams in the Strait can be classified as mixed, mainly semidiurnal with a diurnal inequality in the strengths of successive floods and successive ebbs. The sequence of tidal streams is stronger maximum ebb, stronger maximum flood, weaker maximum ebb, weaker maximum flood, which is consistent with the tide height sequence described in the previous section. In terms of constituents, the semidiurnal (M_2) tidal streams are generally much stronger than the diurnal (K_1) tidal streams. The main reason is that, although both the semidiurnal and diurnal streams transport roughly the same volume of water, the more rapidly varying semidiurnal streams must transport the water in half the time of the diurnal streams.

The following general features of the tidal flow are found during a flood stream. As the flow turns the corner toward the San Juan Islands, the water is deflected slightly eastward by centrifugal forces that produce a high water that is higher and earlier on the U.S. side than the Canadian side. A large portion of the flow then floods into Haro Strait while the rest moves northward through Middle Channel and Rosario Strait. Increased friction in these relatively constricted passageways greatly retards the flow and forces a partial backing up of water into Puget Sound (see Tidal Streams, Chapter 3). Although the transport of water in and out of the Strait of Georgia by the tides takes place at either end, the exchange through the southern channels is about 15 times that through the northern ones. For the most part then, the tides enter and leave the Strait of Georgia via Haro Strait and Rosario Strait, the volume through the former about 3 times that through the latter.

Most water that floods into Haro Strait continues along the main channel and eventually enters the Strait of Georgia through Boundary Passage. However, a significant portion surges northwestward via Swanson Channel, then into Trincomali Channel, where it piles up between the Gulf Islands as a sort of head box, an engineering term for a partially enclosed container used to maintain a hydraulic head. As a consequence, the water streams through the passes into the Strait of Georgia to produce the strong flood currents associated with Gabriola Pass, Porlier Pass, and Active Pass (Fig. 10.1). Dangerous rips often appear in the eastern ends of these passes at this time.

Within the main portion of the Strait of Georgia, the flood streams set to the northwest more or less parallel to the orientation of the shore. To the north of Point Roberts, the tidal flow undergoes a marked decrease in speed due to an appreciable increase in the cross-sectional area of the channel. Because the Strait is fairly wide, currents are also influenced by the Coriolis force, which deflects them very slightly to the right. This produces a small tilt of around 10 cm across the Strait toward the mainland side, enough to generate slightly stronger currents on that side than on the Vancouver Island side. Somewhere south of Quadra and Cortes islands the northward propagating tide encounters the southward propagating tide through the northern channels, so the tidal patterns in the northern part of the Strait are confused and highly variable. Within the northeastern sector, the tidal motions apparently swirl round and meet near Lund, Desolation Sound.

Consistent with the standing wave nature of the tide, there is slack water at high tide within the Strait of Georgia and a reversal in flow direction with the ensuing ebb. In contrast to the flood, the ebb is to the southeast and deflected slightly toward the Vancouver Island side, whereby tidal streams on the western side of the Strait are a few centimetres per second stronger and have a slightly longer duration than those on the mainland side. Tidal currents in the passes and inlets that adjoin the Strait are also reversed on the ebb and rips are infrequent. There is an acceleration of the ebb in the narrow portion of the Strait south of Point Roberts.

Except in passes and narrows, tidal streams in the Strait of Georgia are typically weak, their presence in the surface waters often masked by wind-generated currents and river runoff. For this reason, it is sometimes impossible to know what fraction of the observed current is actually due to the tides. To accurately determine the tidal flow throughout the entire Strait, current measurements over a broad area at many different depths and stages of the tide would be needed. This would be expensive, difficult, and time-consuming. Fortunately, there is a way around the problem, and that is to develop a computer-simulation of the tides and associated tidal streams. Pat Crean, of the Institute of Ocean Sciences, Sidney, B.C., has developed such a model for the waters of the Strait of Georgia–Juan de Fuca Strait system.

More than a decade of full-time effort was required to bring Crean's model to its present high level of sophistication. During that time, the model has undergone considerable modification to include increasing numbers of physical mechanisms that can affect the tides, and to make use of improved computer technology and numerical analysis techniques. Moreover, it has been necessary to measure tides and currents at selected locations in the two Straits to check the validity of computer-simulated results. Because results from the present model consistently reproduce actual measurements for a wide variety of tides, the ability to accurately predict the tides and tidal streams has improved immeasurably. It is now possible, for example, to simulate the tidal flow over the entire seaway at a given time and to estimate the long-term, tidally induced movement of pollutants, such as oil. The computer model further assists scientists to establish which physical processes most affect the oceanic variability of the inside coastal waters. Ultimately, it is hoped that the model will be refined to the point where it will be possible to simulate the influence of winds, Fraser River runoff, and conditions in the open Pacific on the circulations in the Strait of Georgia and Juan de Fuca Strait.

Selected tidal flow charts from the model have been reproduced in Fig. 10.14, 10.15. The first set of charts (Fig. 10.14a, b) gives instantaneous pictures, or "snapshots," of the surface tidal streams for maximum ebb and flood throughout the entire seaway from Cape Flattery to Cape Mudge. These so-called "coarse grid" charts are derived from an earlier version of the model. Figure

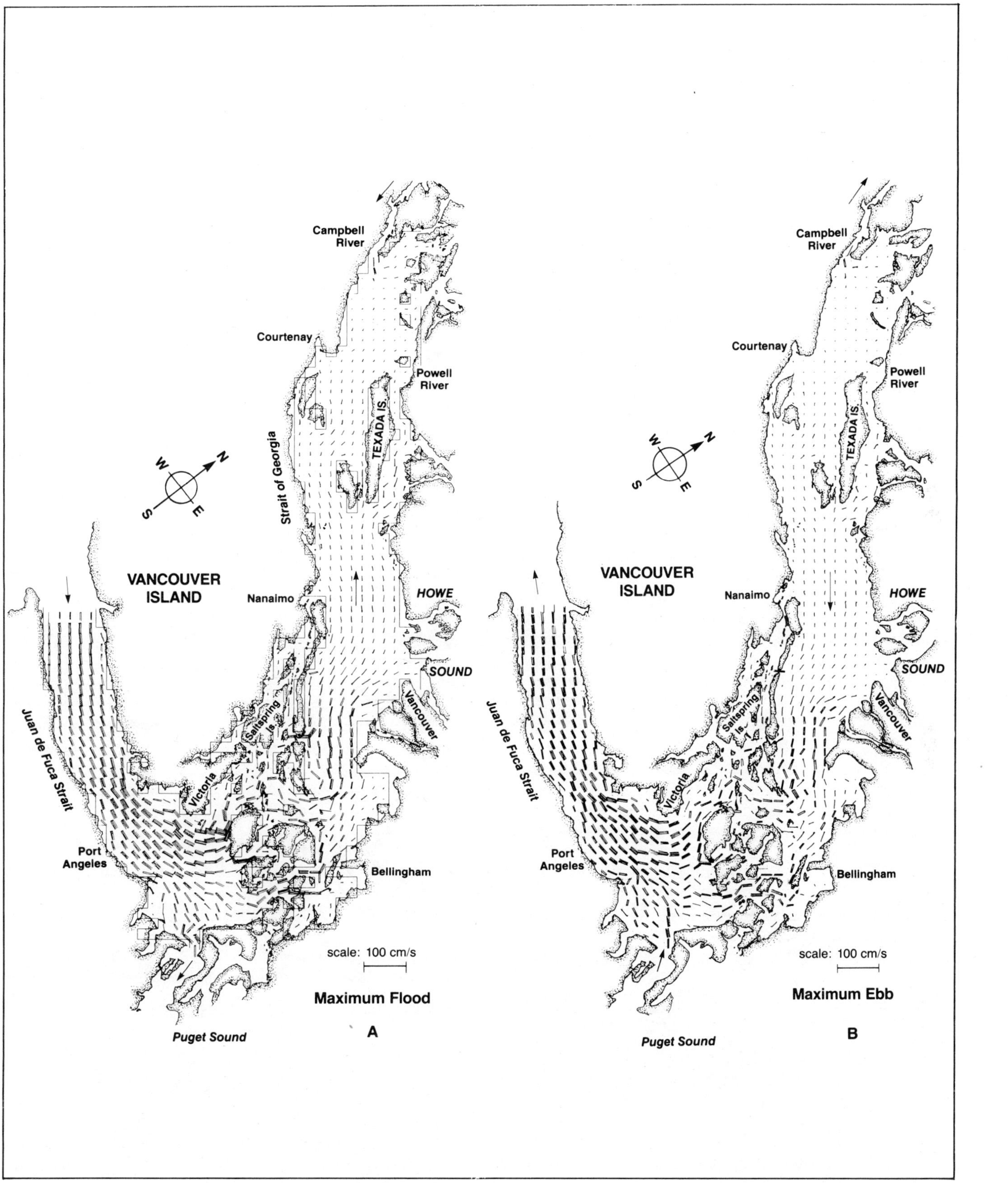

FIG. 10.14. Tidal streams in Strait of Georgia–Juan de Fuca Strait system calculated from Crean's coarse grid computer simulation model for large semidiurnal tide. Plots correspond to (A) maximum flood and (B) maximum ebb. Model calibrated by observed current and tide height data from the two straits. Direction of flow at center of each 4 × 4 km grid square is away from dot, total length of each line (vector) emanates from dot proportional to current speed (actual speed obtained from scale 100 cm/s or 2 kn long). Where current is strong and line too long to fit neatly on plot, it has been broken into segments and placed side by side. Actual coastline has been drawn over computer shoreline. (Courtesy P. Crean)

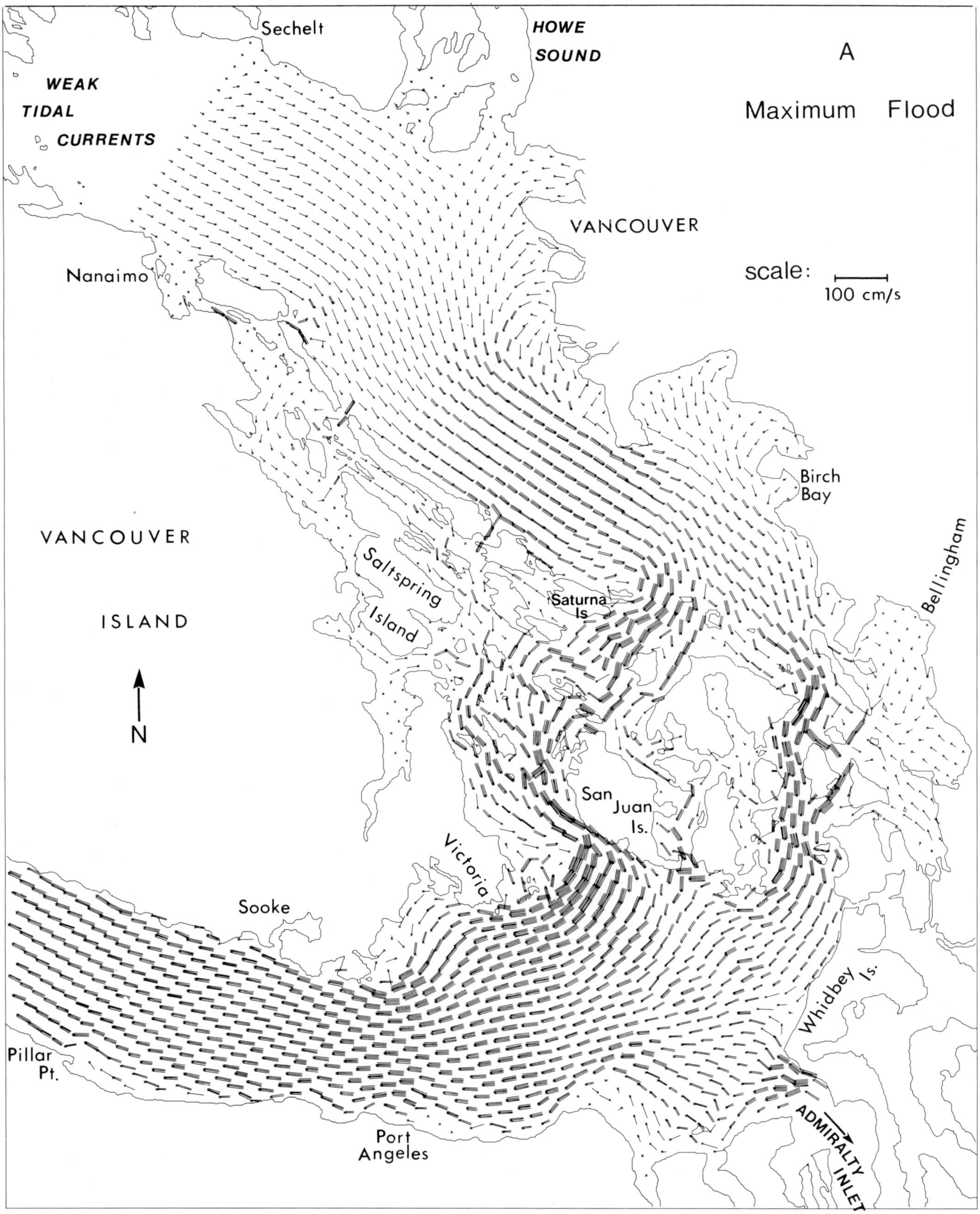

FIG. 10.15. Tidal streams in Juan de Fuca Strait and southern portion of Strait of Georgia calculated by Crean's fine grid computer simulation model. Plots correspond to three stages of mixed tide at Point Atkinson (A) maximum flood, (B) maximum flood plus 2 h, and (C) maximum ebb. Velocity vectors emanate from crosses and represent average surface currents over 2 × 2 km grid squares. Total length of vector at each location proportional to flow speed. Development of large counterclockwise eddy east of Race Rocks follows maximum flood; large eddies form east of Saturna Island and at southern end Haro Strait. (Courtesy P. Crean and A. Douglas)

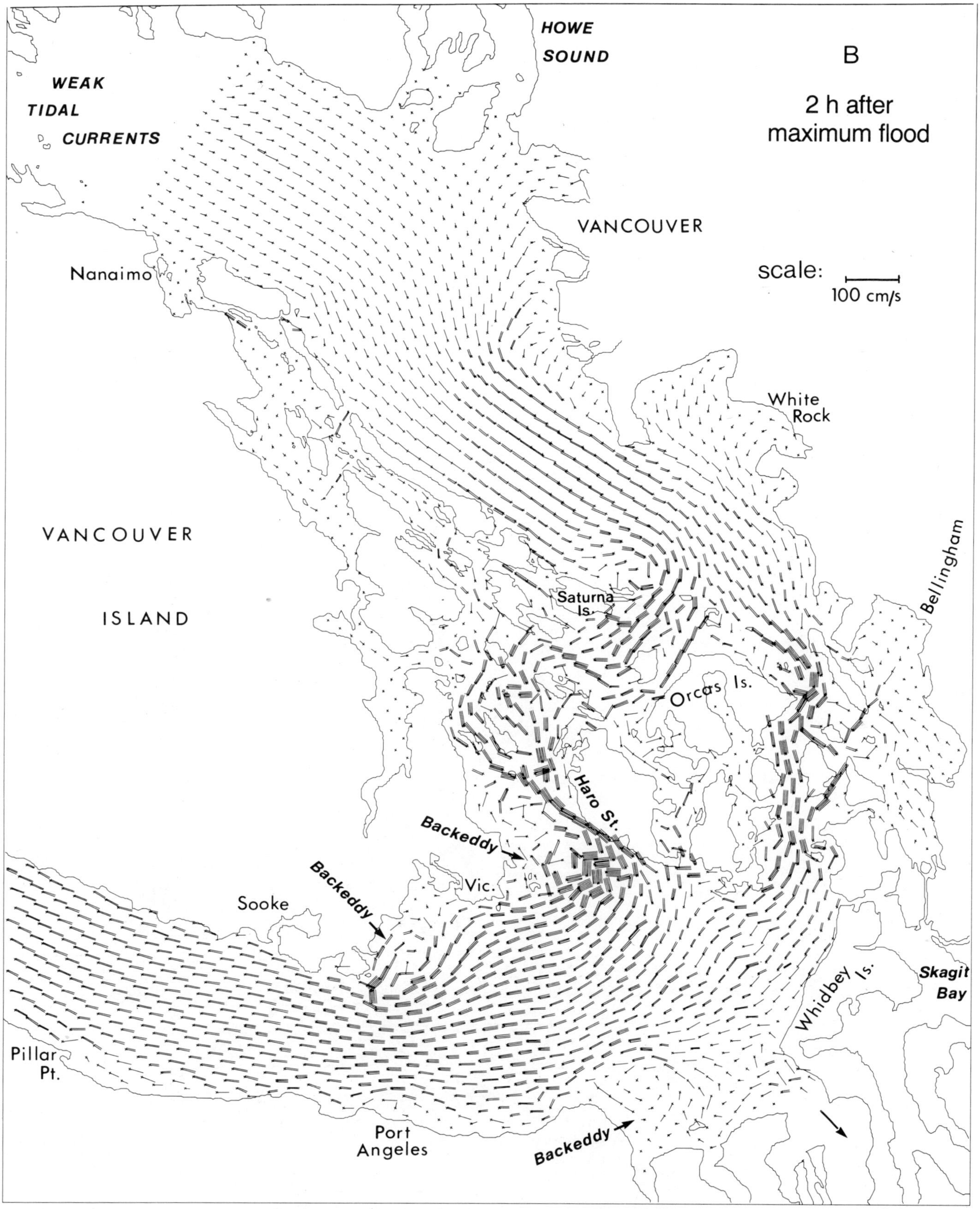

FIG. 10.15. Tidal streams in Juan de Fuca Strait and southern portion of Strait of Georgia calculated by Crean's fine grid computer simulation model. Plots correspond to three stages of mixed tide at Point Atkinson (A) maximum flood, (B) maximum flood plus 2 h, and (C) maximum ebb. Velocity vectors emanate from crosses and represent average surface currents over 2 × 2 km grid squares. Total length of vector at each location proportional to flow speed. Development of large counterclockwise eddy east of Race Rocks follows maximum flood; large eddies form east of Saturna Island and at southern end Haro Strait. (Courtesy P. Crean and A. Douglas)

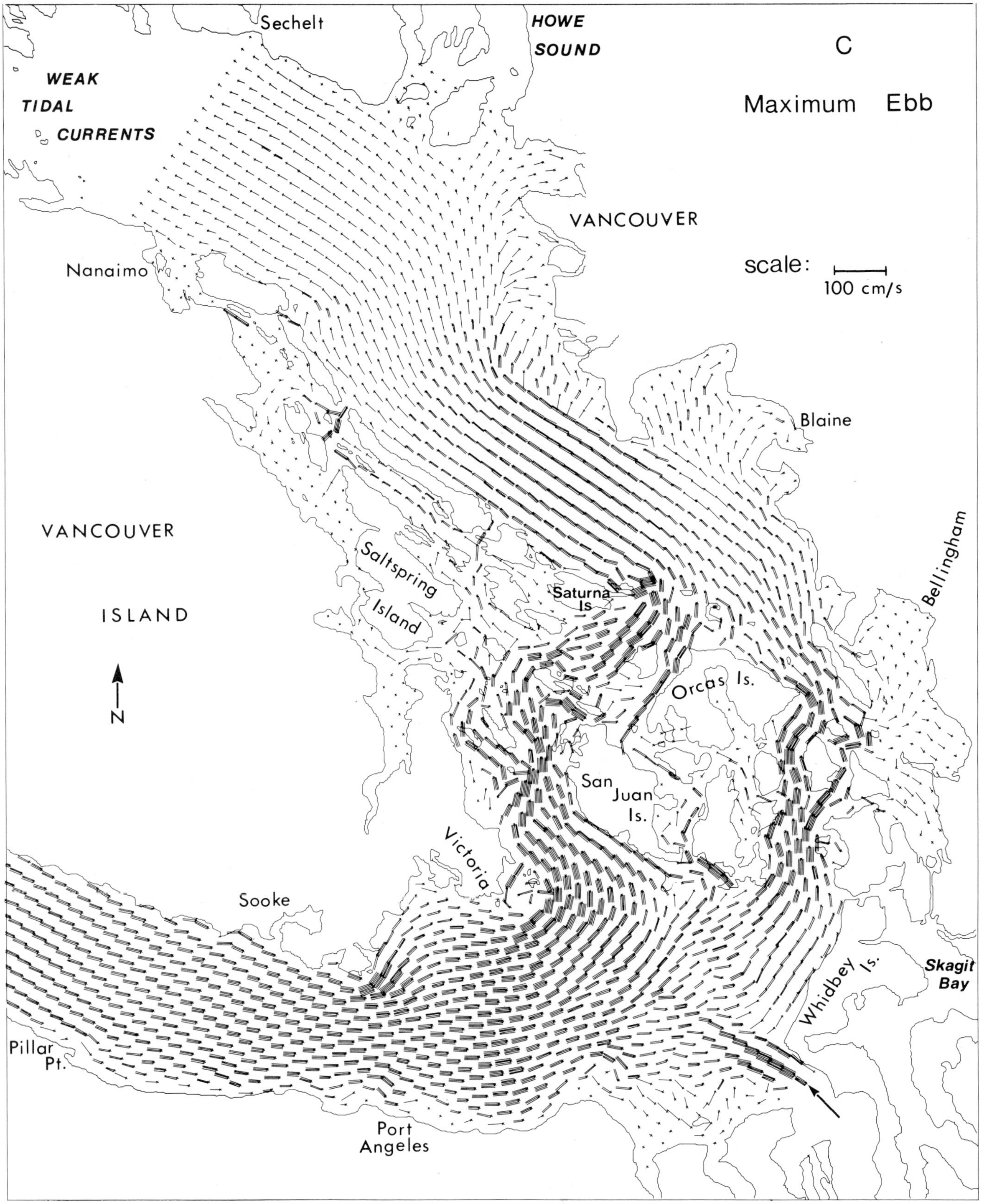

FIG. 10.15. Tidal streams in Juan de Fuca Strait and southern portion of Strait of Georgia calculated by Crean's fine grid computer simulation model. Plots correspond to three stages of mixed tide at Point Atkinson (A) maximum flood, (B) maximum flood plus 2 h, and (C) maximum ebb. Velocity vectors emanate from crosses and represent average surface currents over 2 × 2 km grid squares. Total length of vector at each location proportional to flow speed. Development of large counterclockwise eddy east of Race Rocks follows maximum flood; large eddies form east of Saturna Island and at southern end Haro Strait. (Courtesy P. Crean and A. Douglas)

10.15a–c, the more recent "fine grid" charts, give a much more detailed picture of the currents but are limited to the southern Strait of Georgia and eastern Juan de Fuca Strait. (Results from the coarse grid model are used as "input" to the fine grid model.) The times of the three tidal stream patterns correspond to maximum flood, 2 h after maximum flood and maximum ebb. In each case, the arrows (or vectors) denote the average velocity of the tidal streams from sea surface to seafloor at a particular locality for the specified stage of the tide. For the coarse grid charts, a velocity vector represents the average tidal stream at the center of a square grid of the water surface 4 km long by 4 km wide; for the fine grid charts, the average is over a smaller grid surface 2 km long by 2 km wide. A similar averaging procedure applies to the depth of water at each grid. (The expense of running programs for the computer model grows rapidly with decreasing grid spacing. Therefore, cost, aside from scientific limitations, becomes a major prohibiting factor to enhancing the detail resolvable by the tidal model.) The model takes into account the variable water depth and shoreline configuration, the earth's rotation (the Coriolis effect), friction effects, and the earth's gravity. It does not include the influence on the circulation of the winds, atmospheric pressure, river runoff, and variable water density.

The computer simulation deals only with the "pure" tidal streams so there can be considerable difference between observed and modeled surface currents currents due to the effects of runoff and winds. This will be particularly true at times of large Fraser River discharge or strong winds. The prudent reader should consider the current charts a guide to typical ebb and flood conditions but not assume them to be gospel. Furthermore, narrow channels like Dodd and False narrows are crudely approximated in the simulation. (Because the actual volume of water that moves through them is small, the passes are not critical to proper simulation of the tidal streams in the main channels, although they do possess considerable local importance.) Nor can localized backeddies be delineated by the model if their size is comparable to, or smaller than, the size of the grid squares over which the flow and depths are averaged. Also, the actual shoreline can be approximated, at best, by a series of short, straight lines corresponding to sides of the square computer grids along the land boundaries in the model. As a consequence, tidal streams near the shore may differ appreciably from reality. Finally, the model is not capable yet of simulating the strong onshore–offshore flow associated with drying tidal flats such as the Fraser River delta. In Fig. 10.15, the depth of these banks has been set arbitrarily at 5 m, so the currents only crudely model the true flow. Future models, however, will accurately include flow onto the banks as it can be a navigational hazard to vessels near the delta.

Despite its shortcomings, the computer simulation reveals numerous aspects of the tidal streams within the Strait of Georgia and Juan de Fuca Strait. (The latter region will be discussed in the next Chapter.) For example, the coarse grid charts (Fig. 10.14), show that tidal streams in the Strait of Georgia become progressively weaker to the north and eventually reverse where the flow from the south meets that entering the Strait via the northern channels. Although the currents tend to be in the same general direction throughout most of the Strait, those in the southern portion of Malaspina Strait to the east of Texada Island circulate in the clockwise direction during the flood. The fine grid charts (Fig. 10.15) reveal numerous large-scale eddy patterns in the Strait of Georgia, especially during the flood. The large counterclockwise eddy that forms north of Tumbo and Saturna islands during the flood appears to be one of the major circulation systems in the Strait.

The computer model is capable of simulating other facets of the tidal regime in the Strait of Georgia and Juan de Fuca Strait in addition to the currents. Figure 10.16 shows it is possible to obtain three-dimensional plots of the sea surface elevation relative to an arbitrary reference plane. The large drop in water level between the two Straits clearly shows the delaying effect the narrow channels between the San Juan and Gulf islands have on the passage of the inward propagating oceanic tidal wave. It is differences of this kind that account for the rapid tidal currents through the passages that connect the Strait of Georgia to Juan de Fuca Strait. The times of high and low water among the San Juan and Gulf islands coincide with those in Haro Strait, not those in the Strait of Georgia.

Measured Currents

Although there were a few observations of currents in the navigable tidal passages of southwestern B.C. as early as 1895, the measurement of currents in the Strait of Georgia proper only began fairly recently. This was perhaps due as much to a lack of effort and interest as to a lack of technology. In fairness, however, it should be pointed out that current measurements are neither easy nor inexpensive to obtain.

As any mariner knows, determination of the direction and speed of the surface current is not a simple task. So how is it done? One of the oldest and least expensive methods is to employ the "drift bottle" technique made famous by shipwrecked sailors in Hollywood films. For scientific purposes, sealed, nearly submerged bottles with an official card are used and offer a small reward to the finder, if the card is returned with the time and location of the find. A rough estimate of the surface drift can then be determined from a knowledge of the time it took the bottle to travel the distance between its release and recovery. During the summers of 1926 and 1929, 1636 such bottles were released throughout the Strait of Georgia. The 41% recovered indicated there was a net northwestward drift along the eastern side of the Strait and a southeastward drift along the western side over many tidal cycles. On the basis of this, Waldichuk (1958) suggested there was a general counterclockwise circulation in the Strait superimposed on the ebb and flood of the tidal streams. A smaller gyre that circulated in the same direction also seemed to exist to the south of a line drawn from Sand Heads to Active Pass.

The difficulty with this technique is that the bottles are not really an accurate way to measure currents, as winds and waves affect them in unpredictable ways, and

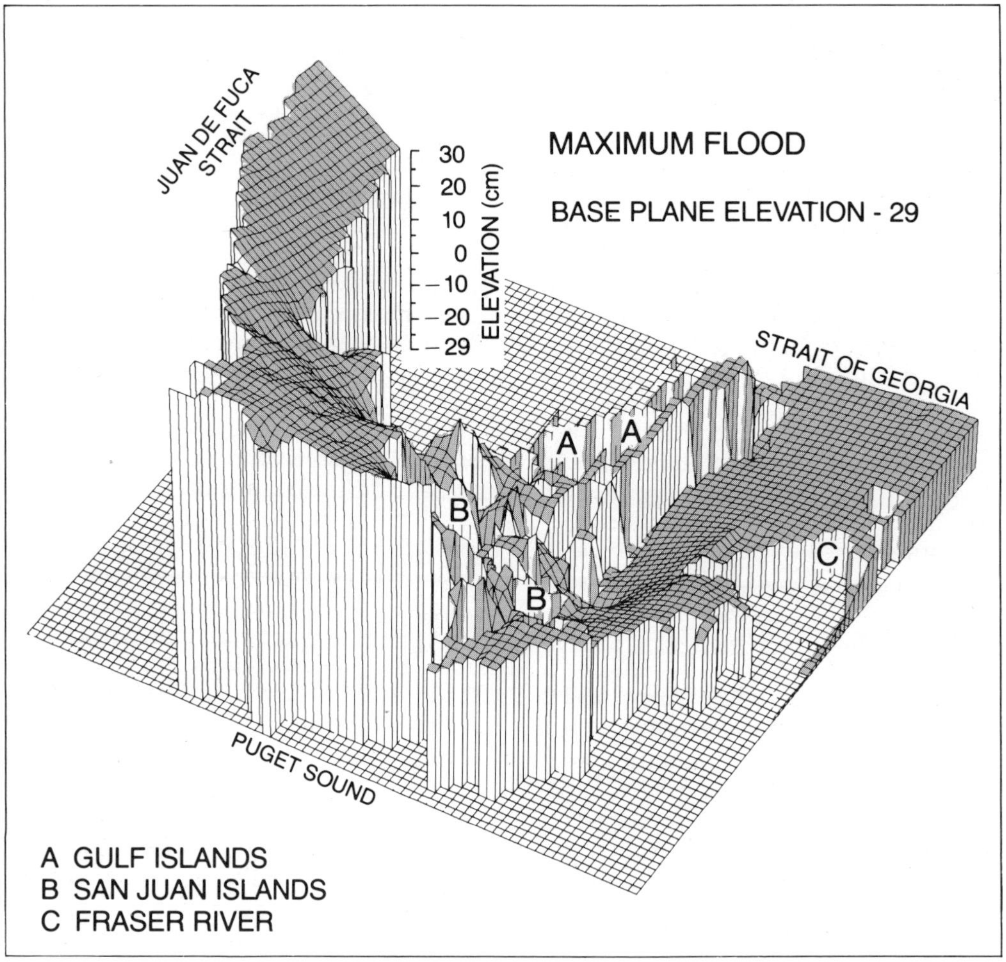

FIG. 10.16. Computer simulation of sea surface elevations during maximum flood, from seaward entance to Juan de Fuca Strait to northern end Strait of Georgia. Base plane is arbitrarily −29 cm; heights of water surface above this level in cm. Note slightly higher elevations on U.S. side of Juan de Fuca Strait due to Coriolis effect on flood streams and steep water slopes through islands. (Courtesy P. Crean)

they may lie on the shore a long time before being recovered. A more sophisticated method is to attach current meters (instruments that record the direction and speed of the current at a fixed position) at various depths on an anchored line. This works fine except near the surface, where logs and marine traffic can do considerable damage and the bouncing action of waves can cause the instrument to give meaningless readings. Consequently, observations of this type are usually not taken shallower than 5 m below the surface. Because there can be considerable difference in the speeds and directions of currents at different depths, especially in regions affected by river discharge, other methods are needed to further determine the behavior of surface currents.

Fortunately, there are two other methods especially applicable to measuring currents close to the surface. The first uses small drifting buoys called drift drogues, whose motion can be tracked day and night by ship or land-based radar. (In offshore regions, satellites are used to track drifting buoys.) An essential part of their construction is a structure with a relatively large surface area that hangs below the surface float to minimize the effect of winds and waves. Called a drogue, it ensures that the system as a whole doesn't move relative to the water, just as a sailboat's keel is meant to prevent leeway. Provided they don't run aground, the buoys will give a measure of the average speed and direction of the current over a depth comparable to the keel depths of most yachts (roughly 2 m). Some of the first measurements by tracked drifters in the Strait of Georgia were taken by Pickard in the summer of 1954. (He actually used "drift poles" that don't have a drogue, but the idea is the same.) Observations lasted 25 h (1 tidal cycle) and were taken at six positions between Point Roberts and Galiano Island, and at two positions in Trincomali Channel. Surface currents averaged 23 cm/s (0.45 kn) on the flood and 43 cm/s (0.85 kn) on the ebb, with maxima to 100 cm/s (2 kn) in each case. In the Strait, these flows were attributed to the tides and to the large Fraser River runoff, as the winds were generally less than 2.5 m/s (5 kn).

The second method uses the silty water from land drainage as a natural tracer of the surface water movement and is applicable to the regions of the Strait of Georgia directly affected by the Fraser River runoff. Aerial and satellite photographs are valuable in this regard although cloud cover can often be a nuisance (Fig. 10.17, Pl. 12).

FIG. 10.17. Satellite (LANDSAT) image shows southeasterly movement of Fraser River sediments into southern Strait of Georgia, July 20, 1974. Sediment dispersal pattern (midway between maximum ebb and low water slack) suggests movement of surface water along Gulf Islands into Porlier Pass, with a cyclonic eddy to east of Gabriola Island (A). Suspended sediments from north arm of Fraser River flow into Burrard Inlet (B); sediment discharge of Skagit River into Skagit Bay (C), and Nooksack River into Bellingham Bay (D). (Courtesy R.A. Feely)

Because the surface can be expected to move primarily under the influence of tides, winds, and the Fraser River runoff, the question is, "How closely does the behavior of the observed surface currents resemble that of the currents expected to be produced by the local winds and tides?" To answer this question it is expedient to divide the Strait into its three somewhat different regions (Fig. 10.18).

The Northern Strait

This portion of the Strait is typified by weak and variable tidal currents (see for example Fig. 10.14). Over most of the region these currents attain speeds of only around 10 cm/s, with the most marked exception at the southern approaches to Discovery Passage. Tidal streams through Sabine Channel between Texada Island and Lasqueti Island, and in Malaspina Strait, can reach 50 cm/s or

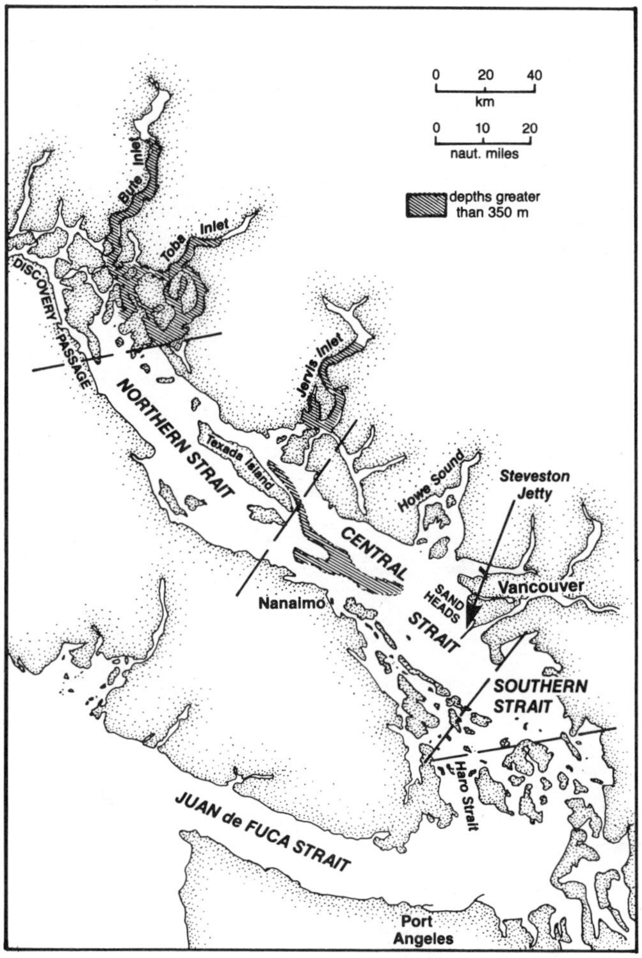

FIG. 10.18. Three oceanic regions of Strait of Georgia. Depths greater than 350 m (200 fa) indicated by hatched areas.

more on occasion, but in general are of the order of half this value.

Aside from a general knowledge of the tidal motions, little else is known about the surface currents in the northern Strait as so few observations have been reported. Although drift bottle studies in the 1920s indicated a general counterclockwise circulation in this area, with westward drift at the northern end and southward drift on the Vancouver Island side, this has yet to be confirmed. With the information in Chapter 4, however, the importance of wind currents in this region can be speculated as generally weak, despite the persistence and strength of the local winds. The reason is that the water does not normally have a distinct layering over most of the year. Consequently, the maximum speed of surface water under a steady, along-the-strait wind is only about 3% of the wind speed. Even for a persistently strong 10 m/s (20 kn) wind, this is only a current speed of 30 cm/s (0.6 kn) at best. For winds across the strait, or of short duration (less than about 12 h), the speed will be somewhat less.

In summer, the overall situation may be altered slightly when the sun warms the upper metre or so of water thereby making it lighter than the cold water beneath. This creates the "slippery water" situation discussed in Chapter 4 and permits the surface layer to slide downwind with greater ease. Provided the surface layer is not so churned up by wave action that it loses its identity, a 10 m/s wind could now generate surface currents that exceed 50 cm/s (see Fig. 4.3).

The Central Strait

This portion of the Strait is characterized by moderately strong tidal streams and by the influence of the Fraser River runoff. Its proximity to the densely populated areas of greater Vancouver makes it the most used marine passageway in British Columbia. As it receives most domestic and industrial waste from the lower mainland and is an important feeding area for migrating salmon, it has been the most studied region of the Strait. Hence, it is possible to see how the observed currents resemble those a mariner might expect according to the winds and tides. For this purpose, the findings of the drifting drogue studies made between 1966 and 1968 will be used. Reference will also be made to some recent aerial photographic studies.

To begin, it is necessary to appreciate the importance of the Fraser River runoff to water circulation in the Central Strait. Approximately 75% of freshwater runoff into the Strait of Georgia is from the Fraser River, which itself drains about 25% of the total land area of British Columbia. All but 30% of this discharge empties into the Strait from the main channel near Steveston. During the snow-melt period of late spring to summer the lighter water pouring from the river spreads over a portion of the Strait as a brackish, silt-laden layer with thickness from 1 to 10 m. Because the discharge from June to mid-July can exceed 11,000 m^3/s (390,000 ft^3/s) (Fig. 10.19), the total surface area covered by this murky water may be quite large, particularly when the numerous patches and fingers of fresh surface water that have broken from the main pattern are considered. In winter, on the other hand, the river outflow tends to be more confined to the nearshore region of the Fraser River delta; by mid-March the discharge has typically diminished to around 1000 m^3/s (35,300 ft^3/s) and is less able to spread over the central

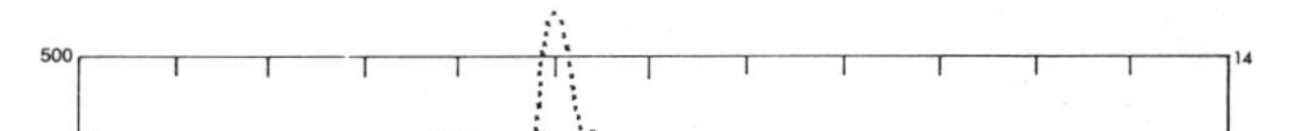

FIG. 10.19. Volume of water per second through cross-section of Fraser River in 4 different years. (Measurements taken at Hope, head of Fraser Valley.) Only portion of 1948 runoff curve shown. (Large runoff that year caused severe spring floods in Fraser Valley.)

part of the Strait. Except under very rough wave conditions, the boundaries between the light colored brackish water and the darker salty water of more oceanic origin are usually very distinct. The runoff's importance is two-fold: first, the momentum of the river water flowing from the main arm causes it initially to be directed southwesterly toward the Gulf Islands; and, secondly, the river is a source for light surface water that produces the slippery water situation necessary for enhanced wind-generated currents. Superimposed on the currents produced by winds and river runoff is the more regular tidal streams that ebb to the southeast and flood to the northwest at speeds of about 50 cm/s (1 kn) half way through the normal tidal range of 3.2 m.

In summary, observations of the interplay of the various mechanisms that produce surface currents are:

1) Drift drogues released near the mouth of the Fraser River indicate that the speed at which the river water first overflows the Strait depends on the volume of runoff and the stage of the tide. During the summer runoff, outflow speeds to 2.5 m/s (5 kn) can occur near low water during large tides, and speeds of 1.0–1.5 m/s (2–3 kn) are not uncommon at less extreme low tides. These reduce to around 0.5 m/s within about 5 km of the river mouth as the buoyant plume of brackish water spreads laterally over the Strait.

At high tide the outflow speeds are typically below 50 cm/s. During winter when the runoff is low, currents are up-river as far as New Westminster during the later stages of a rising tide. This is normal so anyone who uses the river is advised to pick the tides to his advantage.

2) In the presence of light winds, the plume of fresh water leaving the river does one of two things, depending on the state of the tide. During an ebb, the plume will maintain its southwest direction despite the fact that the tidal currents are to the southeast (Fig. 10.20a). During a flood, on the other hand, the surface plume will turn sharply to the north at a greater rate than would be predicted from the tidal currents alone (Fig. 10.20b). Why these unexpected results? One possible explanation is that the Coriolis force, which attempts to turn the plume to its right, is able to balance the effort of the ebb currents to move the plume to its left. During a falling tide the ebb currents increase in speed but so does the speed the river runoff enters the Strait. As the Coriolis force on the runoff also increases with its speed, the effect of the tidal current is kept in check. Thus, even when the tide tables indicate an ebb current, a boat within the plume emanating directly from the river could drift to the southwest. If the ensuing flood is weak, this drift may extend all the way to the Gulf Islands. Usually, the Coriolis force begins to prevail over the weakening ebb and the plume turns northward. During the flood, the combined effect of the tidal currents and the Coriolis force turns the plume directly northward within about 15 km of the river mouth. It may then drift toward Burrard Inlet at speeds of 50–100 cm/s (1–2 kn) on a large flood.

3) Away from the direct influence of the river's discharge the surface currents are expected to be more dependent on local winds and tides. But this is not always the case. The drift drogue observations of 1966 and 1967 showed that a section of the plume that had been turned northward on a flood could continue to move northward at over 50 cm/s, even when the tidal currents were ebbing strongly to the southeast. This motion inevitably carried the surface water toward the western side of Howe Sound or toward the vicinity of Wilson Creek on the Sechelt Peninsula. The water then moved parallel to the shore or turned immediately westward toward Texada and Lasqueti islands. Although the brackish surface layer moved against the ebb, observations showed it did slow significantly at the time of maximum ebb, as would be anticipated. Similarly, maximum northward speeds occurred at times of maximum flood.

Because the winds were generally around 5.0–7.5 m/s (10–15 kn) from the southeast at the time of these particular drift drogue studies, it is natural to assume that the persistent northward drift of this particular portion of the plume was due solely to the influence of the wind. But the observed depth of about 3 m for the slippery brackish layer and Fig. 4.3 show that, at most, these winds would drag the surface layer along at 25–30 cm/s (0.5–1 kn). Obviously, this is not rapid enough to account completely for the movement of the top layer against the ebb, which is also about 50 cm/s. In other words, there are fairly strong currents in the Strait of Georgia that aren't directly relatable to local winds and tides. For want of a better term, oceanographers call these unpredictable motions residual currents. Although of little consolation to anyone who attempts to determine the surface currents, in retrospect the residual flow was to the north at about 50–100 cm/s in the present example. However, until there are more measurements and more concrete explanations for what produces such currents there is little to be said about them. One thing is fairly certain. The initially northward moving surface layer will start flowing nearly parallel to the shoreline once it reaches the vicinity of the Sechelt Peninsula, regardless of what mechanism is pushing it along. Moreover, it appears that surface currents generated by moderate winds, although at times significant, are not an over-riding factor in determining the speed and direction of the surface waters in the Central Strait.

4) Through the action of winds, tides, and residual flow, the silty river outflow has an irregular distribution over large areas of the central and southern portions of the Strait; the greatest area is covered in summer when runoff is greatest. The result is a surface patchwork of dark- to light-colored water; the darker areas correspond to salty oceanic water and lighter areas to recently added Fraser River outflow. Less recent Fraser runoff, which has become mixed to varying degrees with the oceanic water, appears in various shades of grey (Fig. 10.21). A sharp boundary between the dark- and light-colored water often marks the location of a rapid change in the speed of the surface current. Currents that flow parallel to the boundary on the darker side may, in fact, differ by as much as 1

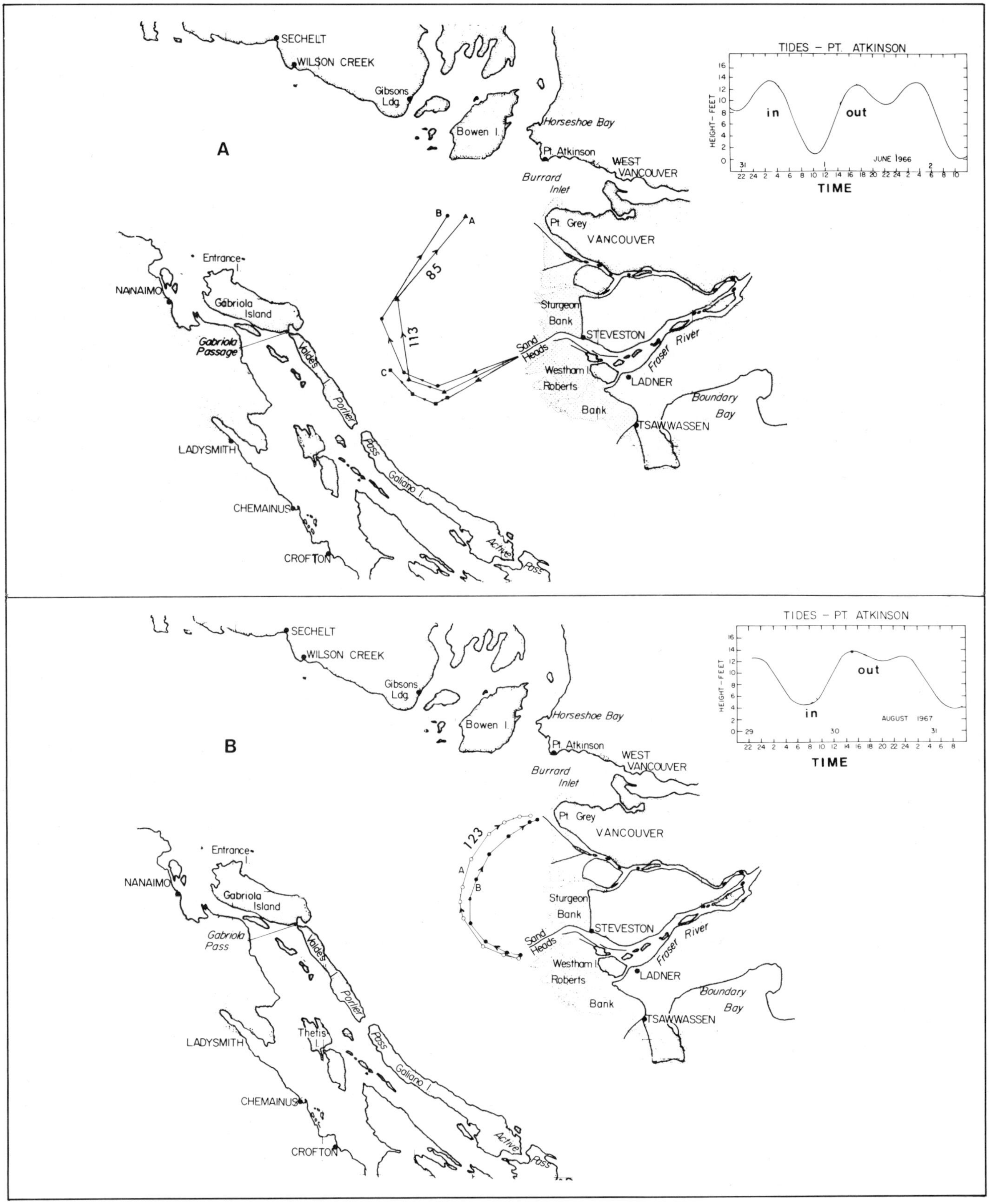

Fig. 10.20. Paths taken by surface drogues released near Sand Heads at mouth of main arm of Fraser River. Top inset gives times of release and recovery with stage of tide at Point Atkinson. (Point Atkinson tides lag those at Sand Heads by about 18 min.) (A) Three drogues released on ebb, June 1, 1966. Light winds. On flood, drogues moved northward attained speeds 85–113 cm/s. (B) Two drogues released at low water, Aug. 29, 1967. Light winds. Each dot (or circle) represents position of drogue on hour with 1 h between each position. Example shows drogue speed 123 cm/s on one leg of path. (Adapted from Giovando and Tabata 1970)

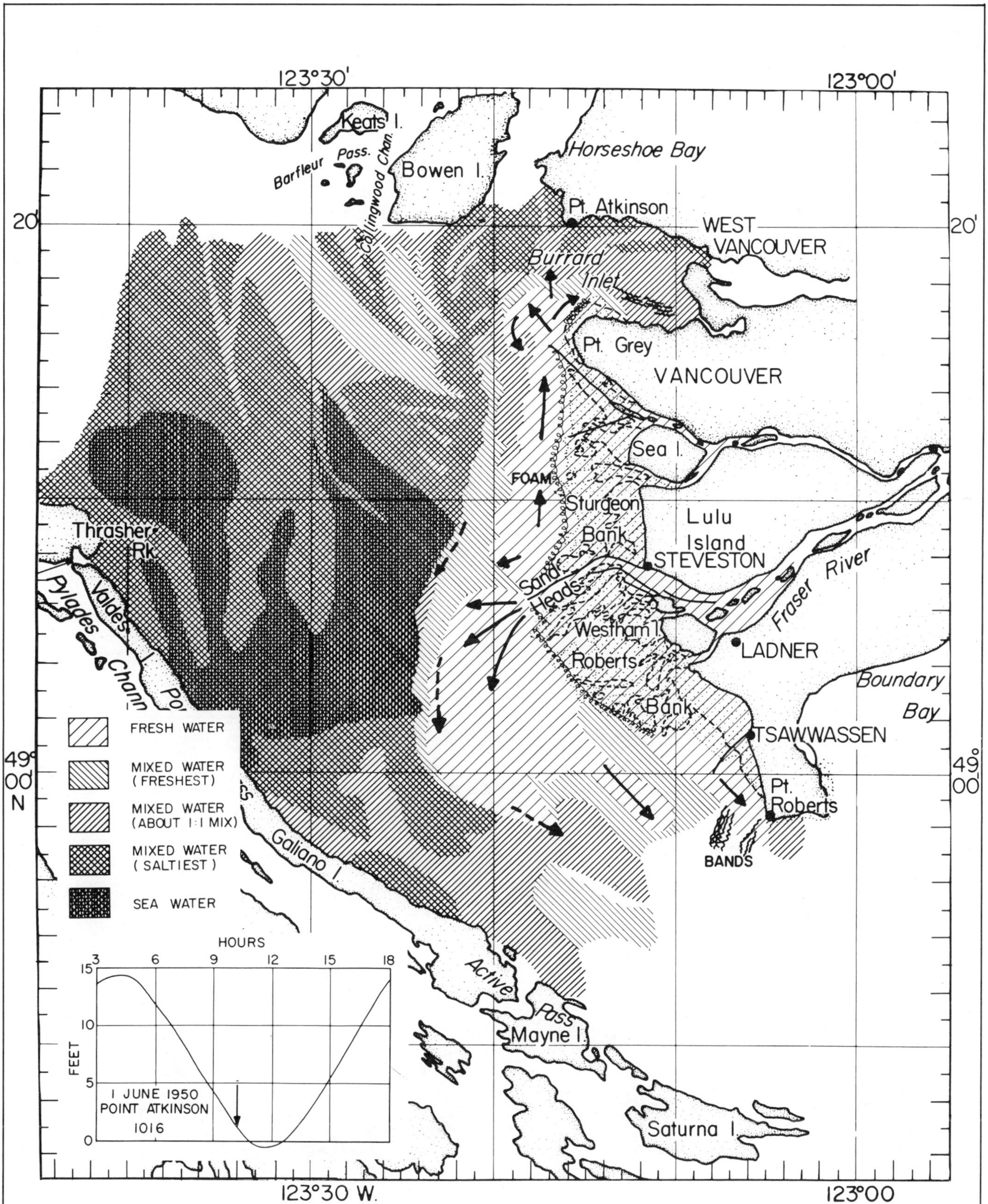

FIG. 10.21. Surface distribution of brackish and oceanic waters in central Strait of Georgia in final stage of large ebb, June 1, 1950 (deduced from series of aerial photographs). Note northward current at edge of Sturgeon Bank despite southeast ebb. Bands due to internal waves; spiral lines on delta represent foam associated with breaking wind waves. (From Tabata 1972)

m/s (2 kn) from their counterparts on the lighter side within a boat-length of 10 m or so (see Fig. 3.33).

5) An obvious example of a wind-driven current occurred in early June 1967 during 10 m/s northwesterlies. As shown in Fig. 10.22, the surface layer was driven to the southeast at speeds to 1 m/s (2 kn) at a time when the tidal streams were almost nonexistent. On this occasion, the drift drogues were tracked as far as Boundary Passage where they were then deflected rapidly to the east by the flood that began to develop near the end of the tracking session. Therefore, in a well-defined brackish layer, winds greater than 10 m/s generate surface currents that dominate the motion of the surface waters.

6) One of the more curious features of the residual flow in the central Strait was uncovered by studies in 1968. In attempting to discover the fate of the treated sewage that enters the Strait from the Iona outfall, oceanographers found that the current near the Fraser River delta north of the Steveston Jetty was often to the north, even during strong ebbs. However, about 2 km to the west of the edge of the delta this current disappeared. Aerial photographs have verified these findings and show that the silty water near the delta often moves in a different manner from that offshore (Fig. 10.21). One explanation for many features of this "nearshore" current is that it is generated by internal gravity waves as they break against the delta. Alternatively, the current may be driven by the along-shore hydraulic head due to the higher freshwater level at the entrance to the main arm relative to the north arm. As the speed of this current can be as much as 1 m/s (2 kn) about a kilometre from the edge of Sturgeon Bank, the mariner should choose his route accordingly. This is particularly true between May and September when the nearshore current is most likely to exist (Pl. 12).

The Southern Strait

As indicated by the computer-modeled charts in Fig. 10.14, 10.15, this region is characterized by relatively strong tidal currents that typically attain speeds of over 50 cm/s on normal tides. The influence of the Fraser River runoff is also important as most of the river water that accumulates in the Strait must eventually work its way toward the Pacific Ocean via the southern passes. Depending on the degree of mixing it has undergone, there is the possibility a well-defined brackish layer will extend into parts of the southern Strait, particularly in summer. Consequently, the southwesterly to southeasterly winds that dominate these regions could augment the flood currents and weaken the ebb currents. Northwesterlies would have the opposite effect. Near the extreme southern end of the Strait, on the other hand, the flood currents are so strong and the tidal mixing so intense that any surface wind effects should be completely overshadowed.

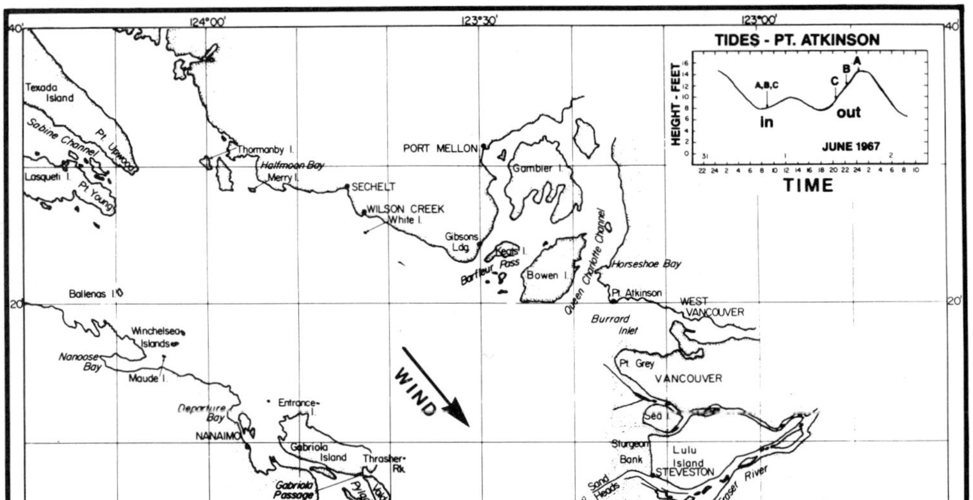

FIG. 10.22. Paths taken by three surface drogues released near Sand Heads during small tides but strong 10 m/s northwest winds (see inset, lower left). Drogue positions given for each hour. Maximum 175 cm/s (3.4 kn) for one drogue just north of Mayne Island. Near end of tracking session, flood entering via Boundary Passage appeared to deflect brackish water to east. (Adapted from Giovando and Tabata 1970)

Figure 10.23 is a crude schematic picture of the "average" surface currents in the Strait of Georgia based on the information just presented. It can be thought of as the resultant drift to be expected if the effects of tides, winds, and runoff were averaged over many months.

Water Renewal

Dilution of the Strait of Georgia waters by Fraser River runoff, together with intense tidal mixing in Haro and Rosario straits and inward movement of oceanic water along Juan de Fuca Strait, lead to water renewal within the deeper portions of the basin. This frequent flushing of the deeper basins is responsible for maintaining the present water quality throughout the Strait.

To understand this process, it is necessary to begin at the surface where fresh water forms a relatively thin brackish layer over much of the southern Strait. Because this fresh water cannot accumulate within the basin, most of it (typically 70–80%) slowly makes its way seaward toward the Pacific Ocean via the southern passes, where vigorous tidal action over the sills mixes it with the cold, salty oceanic water that moves inward along Juan de Fuca Strait (Fig. 10.24). Part of this newly formed mixture

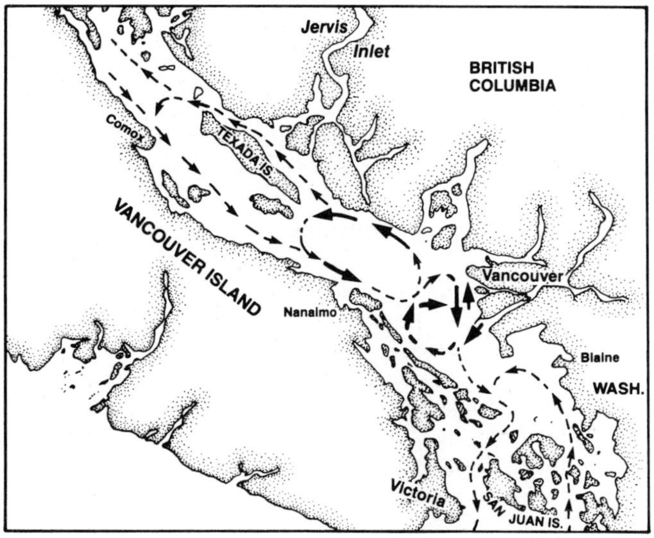

FIG. 10.23. Average (or net) surface circulation in spring and summer in Strait of Georgia as interpreted by author by a variety of oceanographic sources. Large arrows indicate currents measured by current meters and drift drogues; small arrows indicate currents measured by drift bottles.

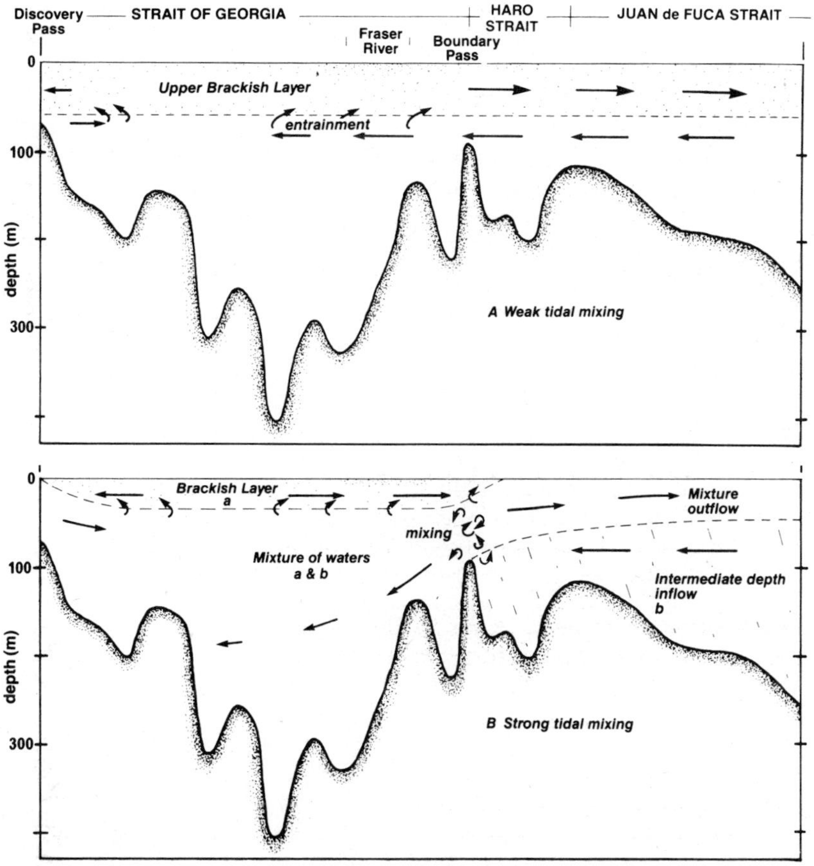

FIG. 10.24. Average circulation in the Strait of Georgia–Juan de Fuca Strait system. (A) Hypothetical circulation that would arise in the absence of strong tidal mixing in the southern and northern approaches to the Strait of Georgia. (B) Actual circulation from vigorous tidal mixing in passes. Brackish Fraser River water (a) mixes with saline oceanic water (b) entering at depth from Juan de Fuca Strait and Discovery Passage. Part of water mixture types (a & b) sinks in Strait of Georgia, rest moves seaward within upper layers of adjoining channels. (Modified after Waldichuk 1957)

(now less dense than the original oceanic water) continues its seaward advance through Juan de Fuca Strait as a more saline surface layer about 100 m thick. The remainder (now more dense than the brackish water flowing southward into the passes) sinks back into the Strait of Georgia at subsurface depths and flows slowly northward aided by pumping action of the tides.

It should be fairly obvious from the above description that the properties of water that flows back at depth into the Strait of Georgia are strongly determined by the properties of water present near the sills in Juan de Fuca and Haro straits. Observations show that relatively low-salinity, oxygen-rich water enters Juan de Fuca Strait from the Pacific in midwinter (December) and begins to move up-channel. About 2 mo later it has worked its way into Haro and Rosario straits. By early spring this oxygen-rich water mass has crossed the sills that tend to isolate the Strait of Georgia and has begun to sink into deeper basins, and gradually replaces older water whose oxygen levels have been depleted by biological activity and decay processes. Through the summer, oxygen levels begin to diminish at depth in the Strait because of the combined effects of internal consumption and lower dissolved oxygen values in Juan de Fuca Strait. Minimum values occur in early fall approximately 3 mo after the midsummer minimum at the seaward entrance to Juan de Fuca Strait. Dissolved oxygen levels then begin to increase again in the lower layer of the Strait of Georgia until the following spring. This process further demonstrates that oceanographic conditions on the continental shelf play a vital role in the replacement of bottom water in the Strait of Georgia. The same can be said of Puget Sound, Saanich Inlet, Bute Inlet, and other estuarine regions within the inshore waters. Similarly, the replacement of bottom waters in basins on the west coast of Vancouver Island, such as Alberni Inlet and Nootka Sound, depend on conditions on the shelf (see Chapter 13).

The situation in the northern Strait is slightly different. There, the waters are more homogeneous and overall tidal mixing is less intense. As a consequence, water renewal results mainly from convective overturning, when the surface water sinks after intense cooling by the air during unusually cold winters. Even then, the cooled surface water can only sink to intermediate depths. Therefore, water near the bottom of the northern Strait can only be replaced by the slow northward movement of the deep water from the southern Strait. (It is conceivable that relatively high density flood currents from Discovery Passage could, at certain times of the year, lead to bottom water formation in the northern Strait, although this has yet to be substantiated.)

In summary then, if there were no tides, no Fraser River, and a more restricted connection to the ocean, density currents in the Strait of Georgia would be far less intense and dissolved oxygen needed for marine life within the basin might soon be depleted below depths of around 30 m; above this depth, near surface mixing and convective overturning in winter would keep the water well supplied with air. In certain inlets this has already happened, sometimes because of natural causes, sometimes not. A case in point is Saanich Inlet north of Victoria. Because of a weak estuarine circulation (a combination of very little river inflow and weak tidal motions) and a sill separating the inlet from Satellite Channel (Fig. 10.25), there is no consistent replacement of the water below the sill depth of about 70 m. As a consequence, water below 200 m is usually deficient in oxygen and often contains hydrogen sulphide, "rotten egg" gas. Under such conditions, the deeper water will become devoid of higher forms of marine life, such as fish and squat lobsters, which can move to more suitable waters. Nonmobile animals such as anemones, sponges, and sea-squirts will die or, if the period of anoxia is brief, become seriously unhealthy. Such a situation does not exist in the Strait of Georgia because deeper waters are completely replaced by more oxygen-rich water over a relatively short period of a year or so, and water above 30 m is completely replaced roughly every month.

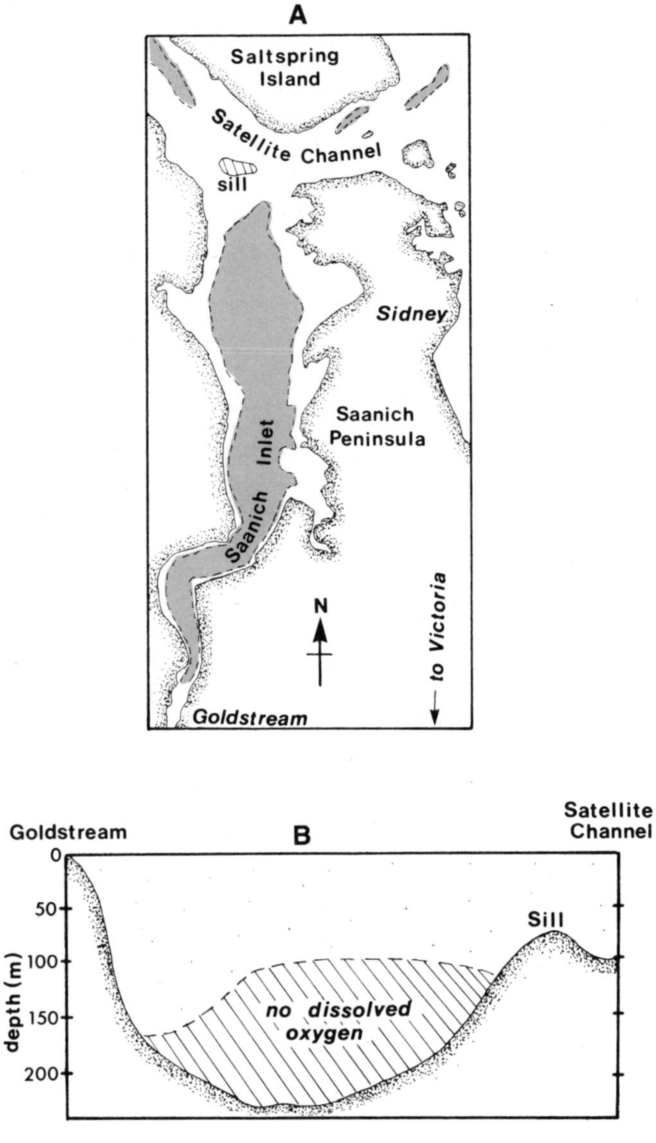

FIG. 10.25. Saanich Inlet. (A) Location of sill and regions with depths exceeding 100 m (shaded area). (B) Hatched area in north–south profile illustrates the extent of oxygen-depleted waters at certain times of year. (From Herlinveaux 1962)

Fraser River Estuary

The Fraser is one of the world's major rivers and no description of the Strait of Georgia region would be complete without a few specifics about the oceanographic setting of its estuary. Volumes could be written on this subject alone, but only some highlights will be touched on here.

Through an intricate network of tributaries, the Fraser River drains approximately 230,000 km^2 (90,000 mi^2) of British Columbia or about ¼ of the province. The river originates in the Rocky Mountains near Jasper, Alta., 1370 km from its mouth, descends rapidly until it reaches Hope, then spreads to a flat alluvial valley to begin a 160 km journey to the Strait of Georgia. In addition to the largest salmon runs of any river in North America, its silt-laden waters have formed the largest estuary on the Pacific Coast of Canada and support important fish and wildlife populations. The delta provides fertile agricultural land and building lots for the sprawling population of the lower mainland. Deep-sea vessels can navigate the main channel to New Westminster, 30 km upstream from the Strait (Fig. 10.26). (Millions of tonnes of sediment must be dredged annually to maintain a 9.8-m draught on a 3.7-m tide.) A northern arm separates from the main channel at New Westminster and is navigable to ships with a 3.7-m draught. The mouth of the Fraser River adjoins the Strait of Georgia along a 37-km delta-front from Point Grey to Point Roberts Peninsula. An abandoned portion of the delta-front extends 13 km east from Point Roberts and faces into Boundary Bay. Roughly 75% of the total river outflow into the Strait is discharged through the South (main) Arm, with the remaining outflow divided among the North Arm (15%), the Middle Arm (5%), and Canoe Pass (5%).

For geographical purposes, New Westminster can be considered the head of the estuary. At the end of the last ice age 8000 yr ago, it was there the river began to fan out as a deltaic plane. Since then, the delta has been adding sediments at a rate of 12 million m^3/yr and has formed deposits 100–200 m thick over glacial deposits. Estimates presently indicate that the front of the delta off Sturgeon Bank is advancing seaward at 2.3 m/yr at the low water mark and 4.6 m/yr at a depth of 30 m. By comparison, the southern end of Roberts Bank is thought to be retreating inland at around 12 m/yr at a depth of 30 m.

Winds

As with the Strait of Georgia proper, the Fraser River estuary is influenced by the large-scale pressure systems that develop over the coast. However, the valley and local mountains somewhat alter the winds associated with these systems. Thus, although maximum hourly winds blow from the northwest and occasionally from the southwest, the prevailing direction is easterly. During the summer, local winds are dominated by the sea–land breeze circulation; the eastward blowing sea breeze sets up around 10 A.M., strengthens until midafternoon to about 5.0–7.5 m/s (10–15 kn) and then dies away before sunset. The weaker land breeze blows onto the Strait until early morning. In winter, strong outflows of cold arctic air pour down the Harrison and main Fraser valleys to augment the comparatively light land breezes associated with nighttime cooling. Though the northern portions of the estuary, such as Sea Island, are partially protected from the full strength of these winds by local topography, northeast

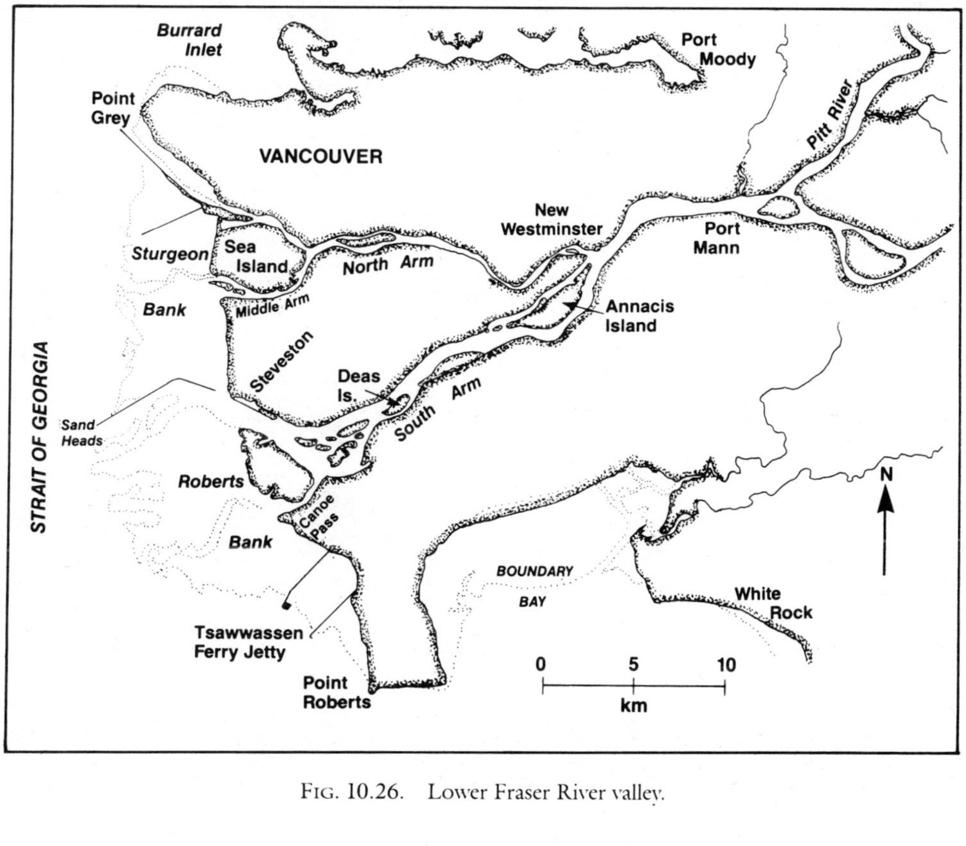

FIG. 10.26. Lower Fraser River valley.

winds in excess of 15 m/s (30 kn) often blow across the more exposed areas of the estuary to the south. The Tsawwassen Ferry causeway can be an especially chilly place in winter when cold Arctic air is sweeping the Fraser Valley.

Periods of calm airs occur less than 10% of the time over the estuary; they are twice as likely to occur in fall as in spring and are most frequent at night and early morning. Episodes of light winds (0–3 m/s) persist for days, and are most frequent in fall and winter (50%) and least frequent in spring (20%).

Fog

Fog is infrequent in March through August but begins in September and is most likely to develop in October through February; 80% of all foggy days occur in these 5 mo. For the most part, fog is the radiation-type, formed by the ground cooling on clear, calm nights and, therefore, is usually most dense in early morning just before sunrise. The frequency of fog by industrial pollutants and domestic fuel consumption has decreased significantly in recent years. Periods of fog visibilities less than 1 km dropped from 104 days in 1943 to 35 days in the late 1960s, due perhaps to the changeover from sawdust to natural gas as the main source of heat for residential homes. There are fewer fine solid particles in the air for water drops to condense on, so "pea-soup" fogs are less common throughout the entire lower mainland.

River Discharge

As noted earlier, the volume of fresh water carried by the Fraser River varies considerably from year to year. It also changes with season and has marked fluctuations over periods of a few days. Snow-melt, which constitutes about two-thirds the total runoff, begins in April and runoff increases to a maximum in late May and early June (Fig. 10.19). By late August, the river discharge has significantly diminished and by early December has reached the low level it will maintain until the following spring. The average daily discharge measured at Hope between 1912 and 1956 ranged from 570 m^3/s (20,000 ft^3/s) in winter to 8800 m^3/s (310,000 ft^3/s) in summer. Discharge rates increase downstream of Hope; are approximately 20% greater at Mission, and 30% greater at Port Mann. The largest daily flow on record 15,200 m^3/s (536,000 ft^3/s) was on May 31, 1948, when over 220 km^2 of the lower Fraser Valley were flooded; the lowest daily flow of 340 m^3/s (12,000 ft^3/s) was Jan. 8, 1976.

Tides

As with any channel that opens onto a coastal basin, the lower portion of the Fraser River is influenced by the tide at its entrance. As it is moving into a river channel, however, the nature of the tide is affected by the volume of water carried downstream and by the relatively shallow depths of the channel. During the winter period of low discharge, for example, the tidal influence reaches as far as Chilliwack, 120 km upstream of Sand Heads, but during the high-discharge freshet it reaches only to Mission, 75 km upstream. The river discharge also affects the daily range of the tide along the river while the shallowness of the river causes a particular stage of the tide to be delayed compared with the tide in the Strait of Georgia. These features can be seen in *Canadian Tide and Current Tables*, Vol. 5, for the Fraser River, where the predicted tide heights at three river locations, Steveston, Deas Island, and New Westminster were determined on the basis of the corresponding tide height at Point Atkinson at four different runoff stages.

The main effects of river discharge on Fraser River tides can be summarized as follows: (1) like tides in the Strait of Georgia, the tide in the river is mixed, mainly semidiurnal, but there is a marked up-river decrease in the diurnal inequality associated with the up-river decrease in the daily range of the tide (Fig. 10.27). The greatest tidal

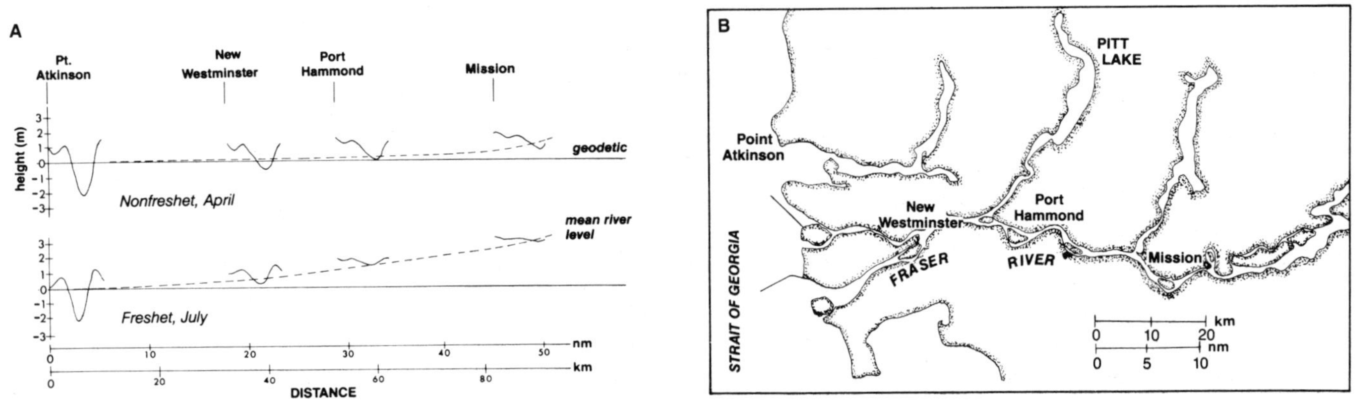

FIG.10.27 (A) Upstream modification of tide in main arm of Fraser River. *(left)* tidal curve over a 24-h period at Point Atkinson, height measured in metres relative to geodetic (local mean sea level). Remaining curves show upstream alteration of tidal cycle as measured relative to mean river level (broken line) for both nonfreshet and freshet conditions. (B) Map of area. (Modified after Ages and Woollard 1976)

range at any one time is found near the river mouth at Sand Heads, which has a mean range of 3.1 m and a large (spring) tide range of 4.8 m; (2) the disparity between the tide range in the Strait and in the river is *least* at times of low runoff. During large spring tides of 4.9 m at Point Atkinson, for instance, the range at low runoff periods varies from 3.9 m at Steveston to 3.1 m at Deas Island, to 2.3 m at New Westminster; and, (3) the disparity between the tide range in the Strait and in the river is *greatest* at times of large runoff. For the above spring tide at Point Atkinson and a typically large discharge rate, the tidal range varies from 3.3 m at Steveston to 2.1 m at Deas Island, to 0.8 m at New Westminster.

Tides in the river are always delayed relative to the Strait of Georgia. Starting at Sand Heads, a particular stage of the tide occurs progressively later as the tidal wave progresses inland. Although it may seem strange at first, the magnitude of the river discharge has a negligible effect on this delay below New Westminster. The height of the tide at the river mouth is important. For example, a large spring tide of 5.0 m in the Strait will be delayed by only 5 min at Steveston, 10 min at Deas Island, and 50 min at New Westminster; on the other hand, a small neap tide of 0.5 m will be delayed 25 min at Steveston, 70 min at Deas Island, and 180 min at New Westminster. Curves in Fig. 10.28 give the time delay of the tide at three river locations

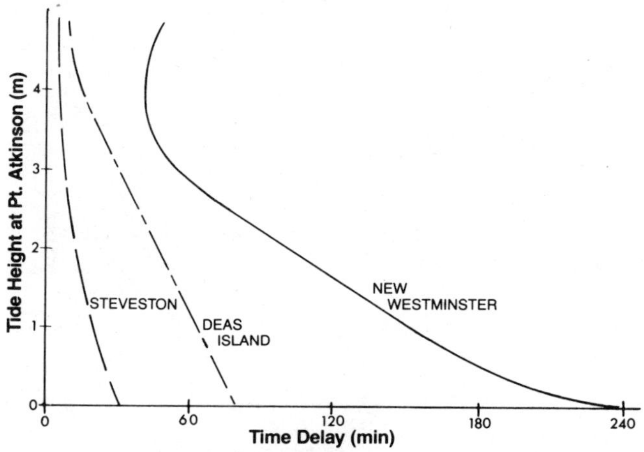

FIG. 10.28. Tidal heights and time differences in main arm Fraser River. Curves give time in minutes that low or high water at each location lags that at Point Atkinson for given tide height. In general, added time for tide to propagate up river from Strait of Georgia decreases as height of tide increases. (Adapted from *Canadian Tide and Current Tables* Vol. 5, 1979)

referred to Point Atkinson. In general, high waters at New Westminster occur around 1 h later than those in the Strait of Georgia, and low waters about 2 h later. A portion of the increasing tides also propagates into Pitt Lake via Pitt River, a few kilometres upstream of New Westminster. Delays in high and low water are typically around 1–2 h in the lake relative to New Westminster; the range of the tide is commonly around 1 m.

The previous features raise two important questions. Why are low tides delayed more than high tides? And, why doesn't the river rate affect the delay of the tides? Because the average depth, D, of the river channel is 10 m it should theoretically take slightly less than 1 h for the tidal wave to travel from Sand Heads to New Westminster at normal propagation speed, $C = \sqrt{gD}$. This alone accounts for delay of high water along the river; it is simply due to the finite time it takes the tide to propagate along the river channel. However, because adding or subtracting a few metres of tide to the river depth, D, doesn't significantly alter this speed, the longer delays of low waters are not directly due to changes in river depth. Instead, as the tide begins to fall below its mean level, bottom friction becomes increasingly important and acts to retard the seaward retreat of the water; the lower the tide the shallower the river and the greater the frictional drag. As a consequence, tides in the river rise much more quickly than they fall, though the whole process still takes just over 12 h (Fig. 10.29). Moreover, the tidal wave is unaffected by the runoff because associated tidal streams are independent of the river current. Flood currents associated with the rising tide at the river mouth oppose the river flow to cause a net slowdown in the downstream flow and a subsequent backing-up of water that travels upstream as a high-water bulge. Ebb currents formed at the river mouth on a falling tide augment the river flow to cause a net speed-up in the seaward flow and a subsequent lowering of the water level that moves upstream as a low water depression. The amount of runoff determines the average slope of the river surface but doesn't affect the strength of the tidal streams. In effect, the tide "diffuses" up-river as if the river flow were nonexistent.

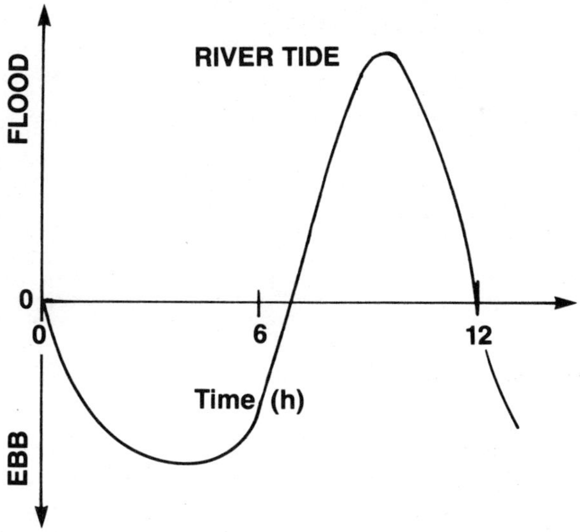

FIG. 10.29. Tidal streams in river. Ebb is weaker but longer duration than flood.

Flow Reversals

During the winter low-runoff period, moderate to strong flood streams can reverse the river flow as far as Mission; the strength of the reversal diminishes upstream. At times of large freshet, the flow is outward at all stages of the tide all the way to the mouth.

A unique feature of the river system is the negative delta that has formed at the southern end of Pitt Lake.

FIG. 10.30. Aerial photograph of lower Pitt system. Delta formed at south end of Pitt Lake by upstream movement of Fraser River sediments along Pitt River to lake. Pitt tidal delta has advanced 6 km into lake in last 4700 yr, average rate 1.28 m/yr. (Ashley 1977; British Columbia Government Photo 1954)

Sediment is carried into the lake at its seaward end contrary to the direction of the Pitt River (Fig. 10.30). The reason for this apparent paradox is that flood streams associated with the tide that enters the lake create stronger currents than the combined ebb stream and river flow, though the latter are of longer duration. As the stronger flood currents can move the sediments more readily than the weaker ebb currents, sand moves upstream into the lake. Maximum floods occur in winter and can attain speeds of 50 cm/s (1 kn) at the entrance to the lake; in summer the larger flood streams are only a fraction of this speed.

Salt Wedge

Currents in the tidally influenced region of the Fraser River below New Westminster regularly have characteristics of an intensified version of the estuarine-type circulation in fiords. On a large flood, a wedge of clear, salty water from the Strait of Georgia will work its way up-river along the bottom, despite the fact that the fresh silty water above is flowing swiftly downstream (Fig. 10.31). On occasion, this salt wedge may penetrate as far as Annacis Island, 22 km upstream of Steveston, before it is swept out with the ebb. The time of maximum intrusion appears to lag behind high water at the river mouth by 60–80 min.

Reversals in the direction of the currents with depth at the seaward end of the estuary may be quite abrupt when the salt wedge is in the river. Within the lower reaches of the Main Arm, 100 cm/s (2 kn) down-river flow at the surface is often accompanied by a 50 cm/s (1 kn) up-river flow at a depth of only 6 m or so (Fig. 10.32).

The maximum upstream distance of penetration of

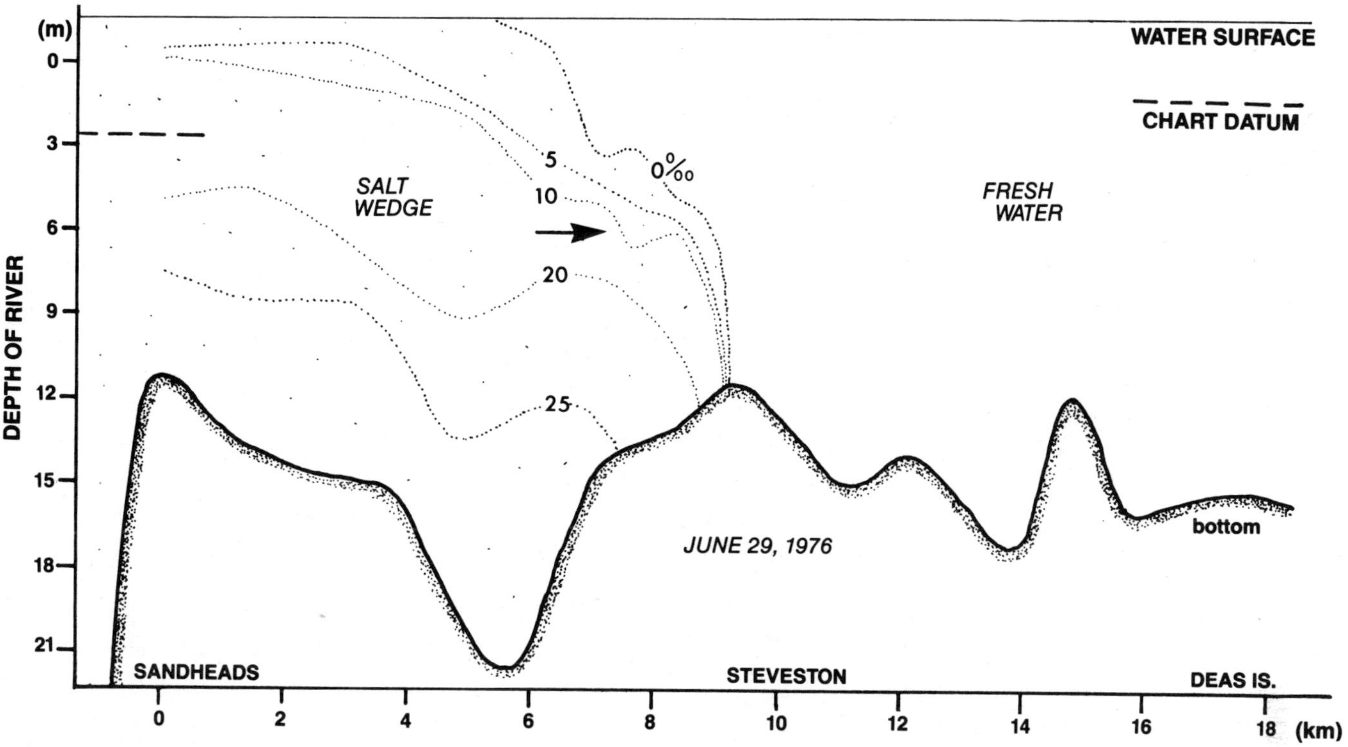

FIG. 10.31. Position of salt wedge in main arm Fraser River near high tide in freshet runoff conditions, June 1976; River depth (m), salinity (‰). During nonfreshet (winter) runoff, same salt wedge could penetrate upstream past Deas Island. (Adapted from Ages 1979)

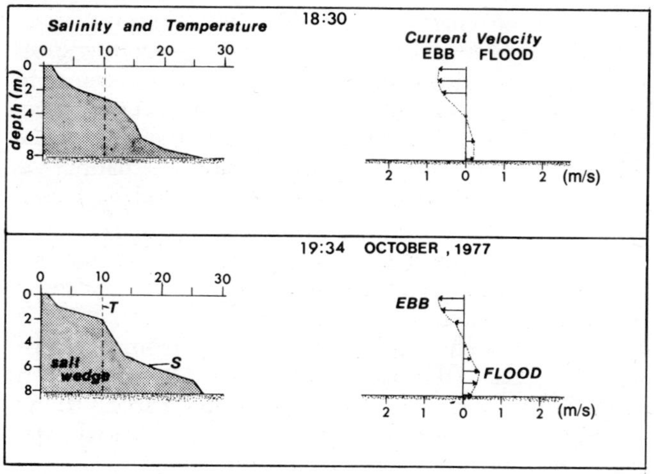

FIG. 10.32. Profiles of temperature (°C), salinity (‰), and current velocity (m/s) in Fraser River near mid-channel south of Steveston, Oct. 17, 1977, at 18:30 and 19:34 (PST). Water temperature uniform from top to bottom but salt wedge present. Currents reverse from outflow at surface to inflow at depth. (Adapted from Ages 1979)

the salt wedge depends on the strength of the river discharge. At low discharges, it may reach Annacis Island in the main channel and penetrate halfway along the shallower North Arm; at intermediate discharges the upstream penetration is limited to Deas Island tunnel in the Main Arm and the Oak Street Bridge in the North Arm. During large runoff peaks in summer, the salt wedge can penetrate only a short distance into the river.

Provided that river discharge is not so great nor the tide so low to prevent the salt wedge from entering, a deeply keeled vessel can actually drift upstream against the surface current of the river. To quote Dawson (1920) ". . . a deep-draught vessel which was being towed in, at a certain stage of the tide, was carried forward by the undercurrent faster than the tug with half the draught could make against the swift-running surface water. The tug had thus difficulty to avoid being over-run by the vessel it was towing."

Buoyancy

Ocean water is a few percent more dense than fresh water and, consequently, more buoyant. A swimmer floats higher in the "salt chuck" than in a lake because his body needs to displace a smaller volume of water to support his weight. By the same token, a boat proceeding into the Fraser River from the Strait of Georgia will experience a decrease in buoyancy when it encounters the river water and, if loaded to the gunwhales, may suddenly find its deck awash as its hull rides deeper in the water.

Burrard Inlet–Indian Arm

Unlike most west coast inlets, Burrard Inlet lacks a sill at the seaward entrance, is relatively shallow, is not bounded by steep, precipitous cliffs, and receives considerable fresh water from an external source, the Fraser River. Only at its eastern extremity does this marine indentation transform into a truly fiordlike setting (Indian Arm) nestled in the heart of the Coast Mountains (Fig. 10.33). But, what the inlet lacks in rugged beauty is compensated for by its socio-economic importance to British Columbia.

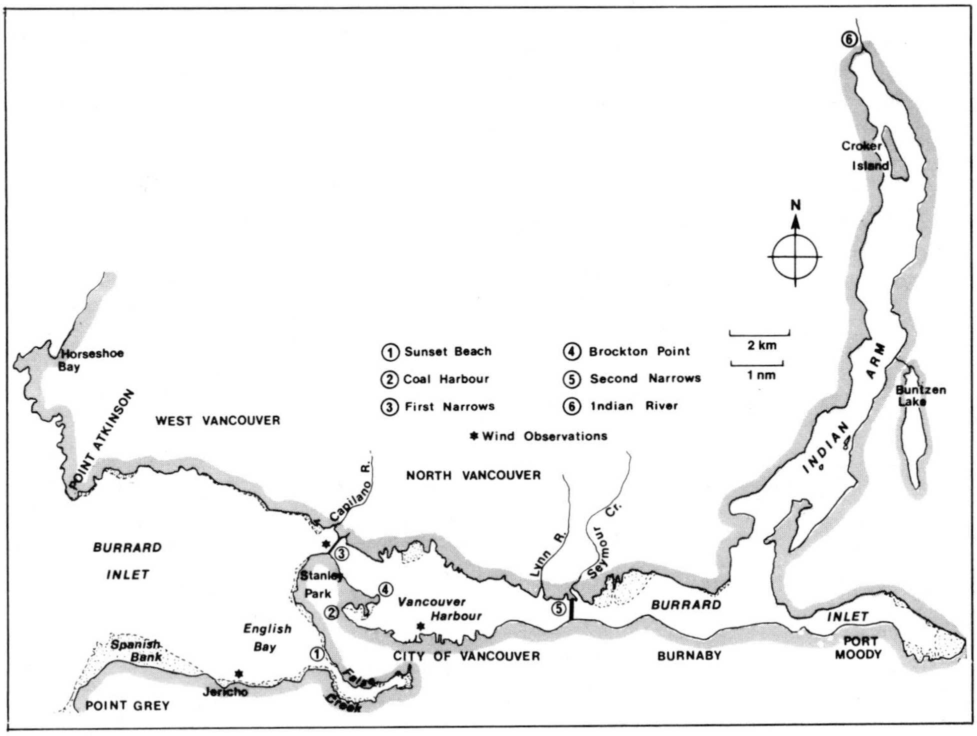

FIG. 10.33. Map of Burrard Inlet-Indian Arm. Also shown are wind observation locations for Table 10.1–10.3.

Burrard Inlet is divisible into an "inner" and "outer" basin. The relatively wide outer basin has a maximum breadth of 7 km and extends about 9 km from Point Atkinson to the First Narrows (Pl. 12) through one of the more densely populated areas of the province. The narrower, 3.5-km wide inner basin, Vancouver Harbour, extends a further 8.5 km eastward to the Second Narrows through the highly commercialized and industrialized sectors of Vancouver. Sandy beaches along much of the perimeter of the outer basin, against the backdrop of the local mountains and the skyline of the city, create a scenic recreational shoreline that is actively used throughout the year. This portion of the inlet is also heavily used by pleasure craft and provides an essential marine highway for commercial vessels. English Bay at the eastern extremity of the basin provides anchorage for as many as 16 deep-sea ships waiting to load or unload cargo at the ports of greater Vancouver. Within the inner basin, the foreshore has been greatly modified by landfill and dredging, and the few recreational areas that remain haven't been improved by industrialization.

Bathymetry

Water depths in the outer basin of Burrard Inlet decrease steadily eastward from a maximum of around 100 m in midchannel south of Point Atkinson to about 15 m over the sill at the First Narrows under the Lions Gate Bridge. There are extensive shallows less than 10 m deep along the southern shore at Spanish Bank and English Bay from Jericho Beach to the vicinity of Second Beach in Stanley Park. The deepest parts of the basin are within a broad trough that begins well into the Strait of Georgia and cuts eastward to the vicinity of First Narrows. An isolated, 30-m deep depression is located immediately outside the Narrows.

There are few dangerous shoals in the outer inlet and these are well charted. This does not, however, deter groundings. One of the most frequented grounding spots is the small shoal off Sunset Beach near the entrance to False Creek (Fig. 10.33). More than once, cutters from the nearby Coast Guard Station have had to free stranded sailboats in this area; on one occasion a "mariner" is reported to have navigated himself aground by a road map!

Within Vancouver Harbour, extensive shoals fan out from Stanley Park, and there is an isolated offshore shoal area near the eastern end. The maximum depth of 66 m is in the center of the basin. A wide swath from south of Capilano Creek to the vicinity of Brockton Point has been dredged to 15 m, to allow deep-sea shipping access to the harbor. Another 15-m deep dredged area is located adjacent to Neptune Terminals west of the mouth of Lynn Creek, and two lesser dredged areas near the mouth of Mackay Creek. Within Indian Arm, depths are more typical of coastal fiords, and average 120 m with a maximum of 245 m to the south of Croker Island. The Indian River at the head of the fiord provides the main source of fresh water and the broad sill-like shallows at the southern end restrict exchange of salt water with Burrard Inlet. As with most fiords, the total length of Indian Arm (22 km) greatly exceeds its average width (1.3 km).

Winds

Because of the funneling effect of the northshore mountains and the ridges of the southern shore, winds in Burrard Inlet are predominantly east–west. Recorded winds from Point Atkinson, First Narrows Bridge, Jericho, and Centennial Pier in Vancouver Harbour show that easterlies occur most often and that the wind strength, regardless of direction, is greatest in winter and least in summer. Somewhat surprisingly, periods of persistently light winds longer than 24 h are most prevalent in the autumn and winter and least prevalent in the spring.

Table 10.1 shows the speeds and percentage occurrence of winds from the prevailing east–west directions, averaged over each month, at the First Narrows Bridge. Roughly speaking, easterly winds occur about 40% of the time from November to January, 35% of the time from August to October and February to March, and 30% of the time from April to July. Easterly winds occur most frequently in winter and least frequently in summer in Burrard Inlet. If northeast and southeast winds are included in the above totals, winds from easterly quadrants occur about 70% of the time in December and January, 60% of the time during October, November, February, and March, and about 50% of the time from March to September.

Westerly winds occur most often from late spring to summer (20–30% of the time) and least often in winter. Winds from the north, south, southwest, and northwest, on the other hand, occur less than 5% of the time throughout the year. The average speeds of westerlies exceed those of easterlies for all months, and are strongest in January and weakest in August. Strong westerlies in winter are associated with storm fronts that pass across southwestern British Columbia on their way inland and last 24–36 h. In spring and summer, westerly winds are associated with the sea-breeze effect set up by solar heating of the land. On more than 60% of the days during these months there is a shift to westerly winds about 3–5 h after sunrise, followed by a return to predominantly easterly winds (land breeze) near sunset. Calm periods and light winds occur generally a few hours before sunrise and a few hours after sunset.

Because of the local topography, prevailing winds along the southern side of the outer basin are from the southeast and northwest, a slight modification from conditions on the northern shore. The land-sea breeze effect in summer (June, July, and August) is the reason that winds from the eastern quadrant (and calms) are most

TABLE 10.1. Winds at First Narrows (Lion's Gate) Bridge, Vancouver. Percentage occurrence of winds from 8 directions for each month (e.g. June westerly winds occur 25% of time). Also percentage of calms and mean wind speed for each month regardless of direciton. Data from January 1969 to June 1974.

DIRECTION		MONTH											
		J	F	M	A	M	J	J	A	S	O	N	D
N		2	1	1	1	2	1	1	0	1	1	2	2
NE		17	15	16	14	9	8	5	6	10	12	17	16
E		38	36	35	31	28	33	30	33	33	35	39	43
SE		13	10	8	9	9	12	11	11	9	9	8	12
S		3	2	2	3	1	1	1	1	1	1	2	2
SW		2	2	2	3	4	3	3	2	3	2	2	2
W		11	12	19	22	27	25	29	26	20	17	10	7
NW		1	1	1	2	1	1	1	1	1	1	1	1
Calm		13	21	16	15	19	16	19	20	22	22	19	15
Mean speed	(*m/s*)	3.5	3.5	3.6	3.6	3.2	3.1	2.9	2.9	3.1	3.1	3.4	3.6
	(*kn*)	6.8	6.8	7.1	7.1	6.3	6.1	5.7	5.7	6.1	6.1	6.7	7.0

prevalent in the morning hours, whereas western quadrant winds prevail in the afternoon (Table 10.2). Easterlies and southeasterlies, for example, occur on 59% of the days at 7 A.M.; westerlies and northwesterlies on 55% of the days at 3 P.M. Aside from those winds associated with passing frontal systems, summer winds on the average are strongest in the early afternoon and weakest around midnight. Moreover, Table 10.2 shows that 41% of the wind shifts from morning easterlies to afternoon westerlies occur between 10 and 11 A.M. During periods of polar outbreaks, however, strong and easterly outflow conditions will be maintained along the south shore despite the clear skies. Easterly winds also can be expected to prevail in winter during periods of inclement weather prior to the passage of a front or low-pressure system.

The wind pattern in Vancouver Harbour is similar to that in the outer basin with a few modifications (Table 10.3). Easterly winds, for instance, occur more frequently throughout the year in the inner basin and summer westerlies are turned more to the northwest. The sea-breeze effect is again quite noticeable and strongest westerlies are linked to passing winter storms. Calms are less frequent within the inner basin; winds from the west, southwest, and southeast blow less than 10% of the time each month.

There is no detailed wind information for Indian Arm, though it is expected that funneling of air by the rugged surrounding land leads to intensified winds. Northerly winds in Indian Arm presumably accompany easterlies in Burrard Inlet and southerly winds accompany westerlies.

The air lifted by the northshore mountains and the convergence of winds that flow up the local valleys lead to significantly greater annual precipitation on the northern side of the inlet than the southern side. The first effect causes approximately a 4½-cm increase in precipitation for every 300-m rise in elevation above sea level on the north shore. Along the axes of the two main valleys, Capilano and Seymour, the added effect of the wind convergence causes a 6-cm increase in annual precipitation for every 30-m rise in elevation; this is equivalent to an increase in annual precipitation of 2 cm/km for Seymour Valley and 3 cm/km for Capilano Valley.

Tides

From Point Atkinson to the head of Indian Arm the tides are mixed, mainly semidiurnal with a strong declinational variation over 2 wk. There is a slight increase in the tidal range eastward of Second Narrows to a maximum in Port Moody and an accompanying delay of Higher High Water by approximately 30 min relative to Vancouver Harbour. Lower Low Water, on the other hand, occurs almost simultaneously throughout the entire inlet system.

Within Burrard Inlet, mean Higher High Water is 4.4 m and mean Lower Low Water 1.1 m, a mean tidal range of 3.3 m. Large tides by comparison have a range of 5.0 m, and vary from a high-water level of 5.0 m near midnight in late December to a low-water level of 0.0 m (chart datum) near noon in late June. Extreme tides in Burrard Inlet have attained high-water marks of 5.6 m and

TABLE 10.2. Winds at RVYC, Jericho (See Fig 10.33). Percentage occurrence of winds from 8 directions averaged over each hour during June, July, and August (1975–78), with percentage calms and mean wind speed. Bottom line gives hour at which wind shifted from land breeze (easterly) to sea breeze (westerly) expressed as percentage of total wind shifts. (From Emslie 1979)

	TIME (PDT) (h)																							
	00	01	02	03	04	05	06	07	08	09	10	11	12	13	14	15	16	17	18	19	20	21	22	23
DIRECTION N	1	1	0	0	0	0	0	0	2	3	2	2	3	3	3	5	6	5	5	6	4	3	0	0
NE	4	5	1	3	3	2	3	3	5	5	8	6	3	5	5	4	5	4	4	2	4	3	3	3
E	19	19	18	18	16	16	15	19	20	22	18	13	17	11	10	8	9	8	7	9	9	13	16	19
SE	33	32	38	37	39	39	41	40	36	30	27	29	19	19	19	19	16	17	17	19	20	21	25	28
S	6	8	5	9	9	7	8	5	2	2	2	2	5	4	2	2	5	5	6	5	5	6	7	6
SW	5	4	5	4	3	5	4	4	3	3	3	4	4	6	6	6	9	12	14	13	13	10	7	7
W	6	5	5	4	5	5	5	6	8	10	14	16	19	17	17	19	18	20	21	23	20	17	3	7
NW	1	1	2	2	2	2	2	4	4	11	17	25	28	34	37	36	31	28	24	18	17	10	4	3
Calm	25	25	26	23	23	24	22	19	20	14	9	3	2	1	1	1	1	1	2	5	8	17	25	27
Mean speed (*m/s*)	1.7	1.7	1.7	1.9	1.9	2.0	2.0	2.1	2.2	2.5	2.7	3.0	3.3	3.5	3.7	3.6	3.4	3.3	3.2	2.9	2.5	2.1	1.7	1.7
Mean speed (*kn*)	3.3	3.4	3.3	3.6	3.6	3.7	3.8	4.0	4.3	4.9	5.3	5.9	6.5	6.9	7.1	7.0	6.7	6.5	6.3	5.6	4.8	4.0	3.4	3.3
% wind shift	–	–	–	–	–	–	1	3	10	15	22	19	14	10	3	3	–	–	–	–	–	–	–	–

TABLE 10.3. Winds in Vancouver Harbour as measured at Centennial Pier (March 1969 to February 1974). Values are percentage occurrence of winds from a particular direction for each month. (See caption Table 10.1.)

DIRECTION		MONTH J	F	M	A	M	J	J	A	S	O	N	D
N		1	2	2	3	5	3	5	3	3	3	3	3
NE		5	4	3	4	4	3	2	3	5	4	4	6
E		52	55	59	48	40	48	38	42	41	50	65	60
SE		15	7	8	8	8	10	10	13	14	10	8	13
S		7	5	4	6	6	6	7	7	4	4	4	6
SW		5	2	3	5	3	1	2	1	3	1	2	2
W		9	10	11	14	17	12	13	11	12	12	8	6
NW		3	5	7	10	15	15	21	17	12	9	3	2
Calm		3	9	3	2	2	2	2	2	6	7	3	2
Mean speed	(*m/s*)	2.9	2.8	2.6	3.0	2.8	2.7	2.5	2.6	2.5	2.6	2.5	2.7
	(*kn*)	5.6	5.5	5.0	5.8	5.4	5.2	4.9	5.1	4.8	5.0	4.9	5.3

low-water marks of −0.4 m. Therefore, tides within the inlet system only rarely fall below chart datum though such conditions might occur during strong easterly blows in midwinter at low spring tide. Within Indian Arm the mean tidal range is 3.3 m and the spring tide range 4.9 m.

Oceanography

As with other inlets along the coast, temperature and salinity distributions within Burrard Inlet and Indian Arm vary considerably with season, particularly in the upper portion of the water column. Factors that affect these distributions include conditions in the Strait of Georgia, the amount of local land drainage, the influx of Fraser River water, the tides, and winds.

Throughout most of the year the inlet system has a two-layer structure, with relatively warm, low-salinity water in about the top 5 m overlying colder, more saline water beneath. This configuration is maintained by freshwater discharge from the various rivers, land seepage, and sewer discharge, which together move seaward near the surface with a compensating inflow of saltier water at depth from the Strait of Georgia. The Indian River provides most of the freshwater input to Indian Arm, though as much as 40% can come from Buntzen Lake via the Buntzen Power Plant. Seymour River provides most of the direct freshwater inflow to Vancouver Harbour, and the Capilano River is the main direct source of fresh water into the outer basin. The amount of flow in these rivers is closely tied to local precipitation and tends to be greatest during the rainy season of autumn to winter. Only during a few days each year does the discharge from the north-shore rivers exceed 150 m^3/s (roughly 1% of the discharge rate of the Fraser River) and these are usually periods of flooding associated with heavy rainfall; the record measured discharge for the largest of these rivers, the Capilano, was 408 m^3/s on Nov. 26, 1949. During exceptionally cold winters, however, the precipitation may be held in the snowfields and not released until spring.

The North Arm of the Fraser River and not local rivers is the main source of brackish water in the outer portion of Burrard Inlet. This river water is especially prevalent during the peak runoff period of late spring and summer. The combination of flood currents and the hydraulic head of the river drive a silty layer of fresh water northward along the edge of Point Grey, then eastward along the southern shore of the inlet where it has a murky, green color, a contrast to the darker green of the more salty oceanic water. During certain conditions of wind and tide, Fraser River water can also penetrate into Vancouver Harbour, although in a considerably modified form as a result of tidal mixing at the First Narrows. This brackish layer accounts for the comparatively lower salinities and higher temperatures of the surface waters along the southern shore of the inlet in summer. It also impedes the natural seaward flow of fresh water that enters the inlet system from the indigenous rivers, thereby disrupting the normal estuarine circulation.

Near surface waters are warmest in late July to early August throughout the entire inlet (Fig. 10.34a). Temperatures in the top metre may reach 20°C or more along the southern shore of the outer basin, over the shallow region that extends eastward from the Second Narrows into Port Moody, and over most of Indian Arm. If winds are light in these regions, the surface water temperature will begin to approach air temperature. Unfortunately for swimmers, the thermocline is shallow and water temperatures invariably decrease by 5–10C° within 5 m of the surface. Below 20 m, water temperatures become uniformly cold at around 10°C. The coldest surface water in the inlet during the summer is always in Vancouver Harbour because of the intense tidal mixing within First and Second Narrows. Maximum surface temperatures in this basin are typically less than 15°C, except in sheltered embayments such as Coal Harbour.

In winter, the waters of the inlet system are almost uniformly cold from top to bottom (Fig. 10.34b). Though there is some variation from winter to winter,

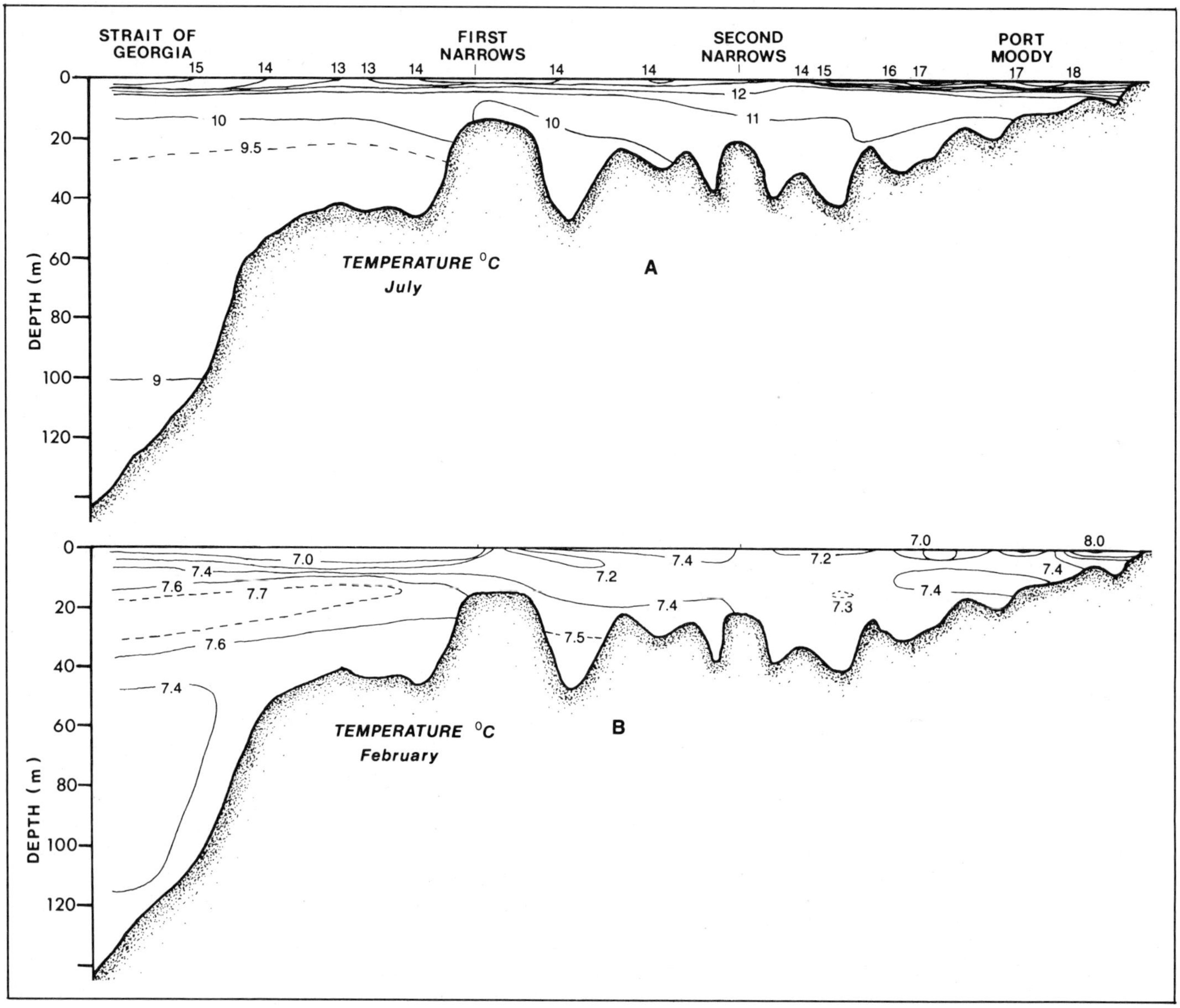

FIG. 10.34. Temperature (°C) distributions in mid-channel sections through Burrard Inlet. (A) July 7–9, 1966; (B) Feb. 15–18, 1962. (Adapted from Anon 1973)

temperatures of between 6–8°C are common, with slightly colder water near the surface due to heat loss to the atmosphere. As at other times of the year, waters within the inner basin of Burrard Inlet exhibit the smallest temperature differences over depth of the entire system.

The Fraser River water in the outer basin creates a complex surface salinity distribution that varies greatly with tide, wind, and runoff conditions. Below 10 m, the salinity distribution becomes more uniform with values of 29–30‰ throughout the year (Fig. 10.35). Lowest surface salinities occur in the summer. The general distribution at this time consists of a tongue of low-salinity water from the Fraser River of 10‰ or less that enters from the vicinity of Point Grey and moves toward the First Narrows (Fig. 10.36). On the average, saltiest water lies over the northern half of the outer basin with maximum values of around 20‰ in the vicinity of Point Atkinson. The presence of relatively high-salinity surface water east of Jericho Beach to False Creek suggests that surface flood currents in the outer basin favor a northeasterly set rather than an easterly set. However, the average distribution in Fig. 10.36 is subject to considerable variation. For example, it is possible to find detached pools or "lenses" of low-salinity water of around 10‰ at about any location in the outer basin. Low-salinity pools can also be prevalent seaward of the mouth of the Capilano River. Typically, these pools persist through only part of the tidal cycle before they are destroyed by tides and winds. In winter, salinities over most of the outer basin are around 25‰ at the surface and increase to about 30‰ below 20 m (Fig. 10.35b).

East of the First Narrows, there is a gradual decrease in the surface salinity to the head of Indian Arm, regardless of season. Due to the shallowness of the sills separating these inner basins, salinities remain below 30‰. Highest values are attained just inside Vancouver Harbour where density flow from the First Narrows penetrates to

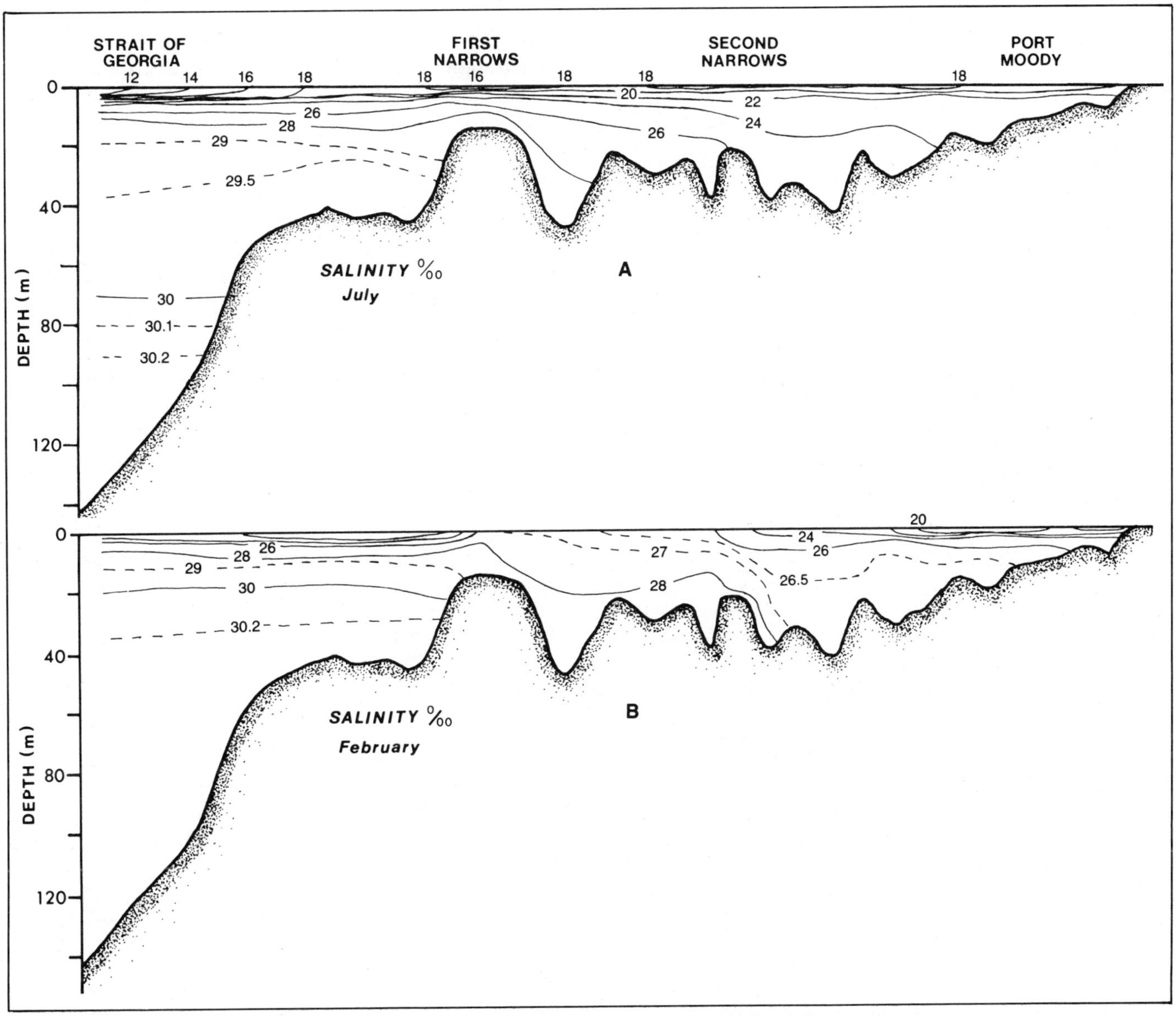

FIG. 10.35. Salinity (‰) distributions in mid-channel sections through Burrard Inlet. (A) July 7–9, 1966; (B) Feb. 15–18, 1962. (Adapted from Anon 1973)

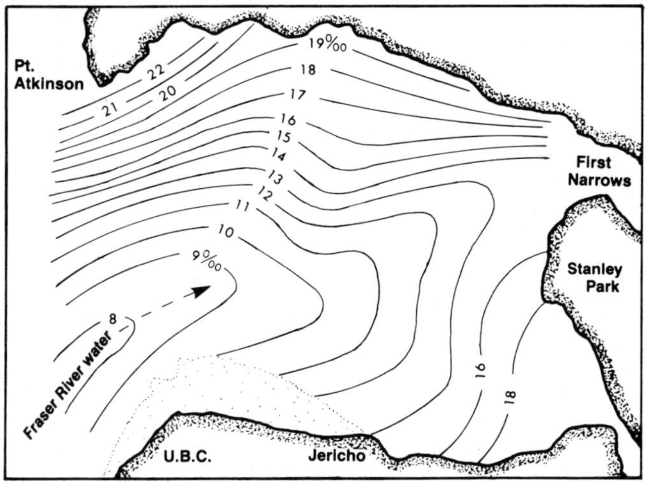

FIG. 10.36. Average surface distribution salinity (‰) in Burrard Inlet during large Fraser River runoff (summer). (From Campbell 1954)

the bottom. Within Indian Arm, surface salinities reach minimum values twice a year: once during summer (July–September) presumably at the time of greatest snow melt; and, again in late winter (January–March) during periods of heavy rainfall. Minimums are around a few parts per thousand at the head of the fiord, and increase to about 10‰ midway along the channel, to about 15‰ at the southern end. Maximum salinities occur in spring and early winter.

Currents

In this section attention will be mostly confined to the flow structure in the outer basin of Burrard Inlet. Those interested in currents in the inner basin are referred to *Vancouver Harbour Tidal Current Atlas 1981*, published by the Canadian Hydrographic Service, which shows the speed and direction of surface tidal streams at hourly stages of the tide. Only a cursory description of the currents in Indian Arm will be presented.

The Burrard Inlet current patterns described below need to be put in perspective. First, they are based on current meter measurements taken over 25 yr ago, when oceanographic technology was in its infancy. Moreover, only two short observational periods of 25 h each were obtained, and these were collected during summer at periods of weak and variable winds. Consequently, the current patterns should be viewed as general representations of the actual flow at various stages of the tide and from which significant departures can be expected.

On a large flood, the northward flowing tidal streams in the Strait of Georgia turn into Burrard Inlet with a northeasterly set in the vicinity of Point Grey (Fig. 10.37a). During the time these measurements were taken there was an accompanying southeasterly set off Point Atkinson where flood streams entered the inlet from Queen Charlotte Channel. Because of the constricted nature of this channel, the latter feature may indeed typify the flow at large floods, though normally a more westerly set off Point Atkinson is expected. Over most of the inlet, the surface currents are then directed toward the First Narrows, and attain maximum speeds of around 25–50 cm/s. Modifications of this simple picture are subsequently brought about through the influence of winds and shoreline geometry. Westerly winds, for example, would be expected to augment the flood streams in the top few metres of water by approximately 3% of the wind speed and to turn them more to an easterly set; by the same token, easterly winds would reduce the flood and cause it to set more toward the northeast. The wide shoal areas on the southern shore and the protrusion of Point Atkinson into the inlet result in downstream backeddies. East of Spanish Bank, a weak nearshore counterflow of 10–20 cm/s was found in the earlier study, although westerly winds would presumably reverse this tidal flow, particularly in summer when brackish Fraser River water overlies most of the south shore. The flood into False Creek is accelerated, but even under the Burrard Street Bridge it is typically less than 50 cm/s. Funneling of the flood streams through the First Narrows leads to a strong flow into Vancouver Harbour at speeds to 300 cm/s (6 kn) during spring tides. The flow decelerates as it fans out to the east of Brockton Point, though speeds are sufficient to set up well-defined backeddies in the lee of prominences in the harbor.

During small floods, the flow pattern is similar to that of a large flood except that midchannel currents are weaker and tend to broaden more within the inlet (Fig. 10.37b). In addition, there is a more pronounced northerly set of the tidal streams at the Strait of Georgia entrance to the inlet, and the counterclockwise eddy between Sandy Cove and Point Atkinson extends westward to join the northerly set into Howe Sound. Wind drift currents can be expected to play a dominant role in determining the overall flow structure on small floods during periods of moderate to strong winds.

Surface currents on a large ebb are illustrated in Fig. 10.37c. The most pronounced feature is the strong, narrow current, which can often extend from the First Narrows to Point Atkinson. The core of the flow at such times is offshore, except near Reardon Point and the Narrows, and has maximum velocities of around 100 cm/s. Along most of its path the flow broadens generally, but narrows again off Point Atkinson, where it merges with the southerly flow from Howe Sound. On other occasions, the narrow current appears to be deflected toward mid-channel rather than to hug the northern shore, although why it should do so is not yet understood. Obviously further work in this area is needed before a complete understanding of the flow behavior is achieved.

Because of Stanley Park, an extensive anticlockwise gyre is set up over much of English Bay on the ebb resulting in weak and variable currents along the adjoining recreational beaches. The flow pattern near the southern shore is further complicated by the ebb from False Creek directed westward along Jericho Beach and Spanish Bank. The continued northerly set off Point Grey on the ebb is due to the hydraulic head of the North Arm of the Fraser River, which continues to drive surface waters toward the north. Needless to say, winds will modify the entire flow pattern depending on direction, strength, and duration.

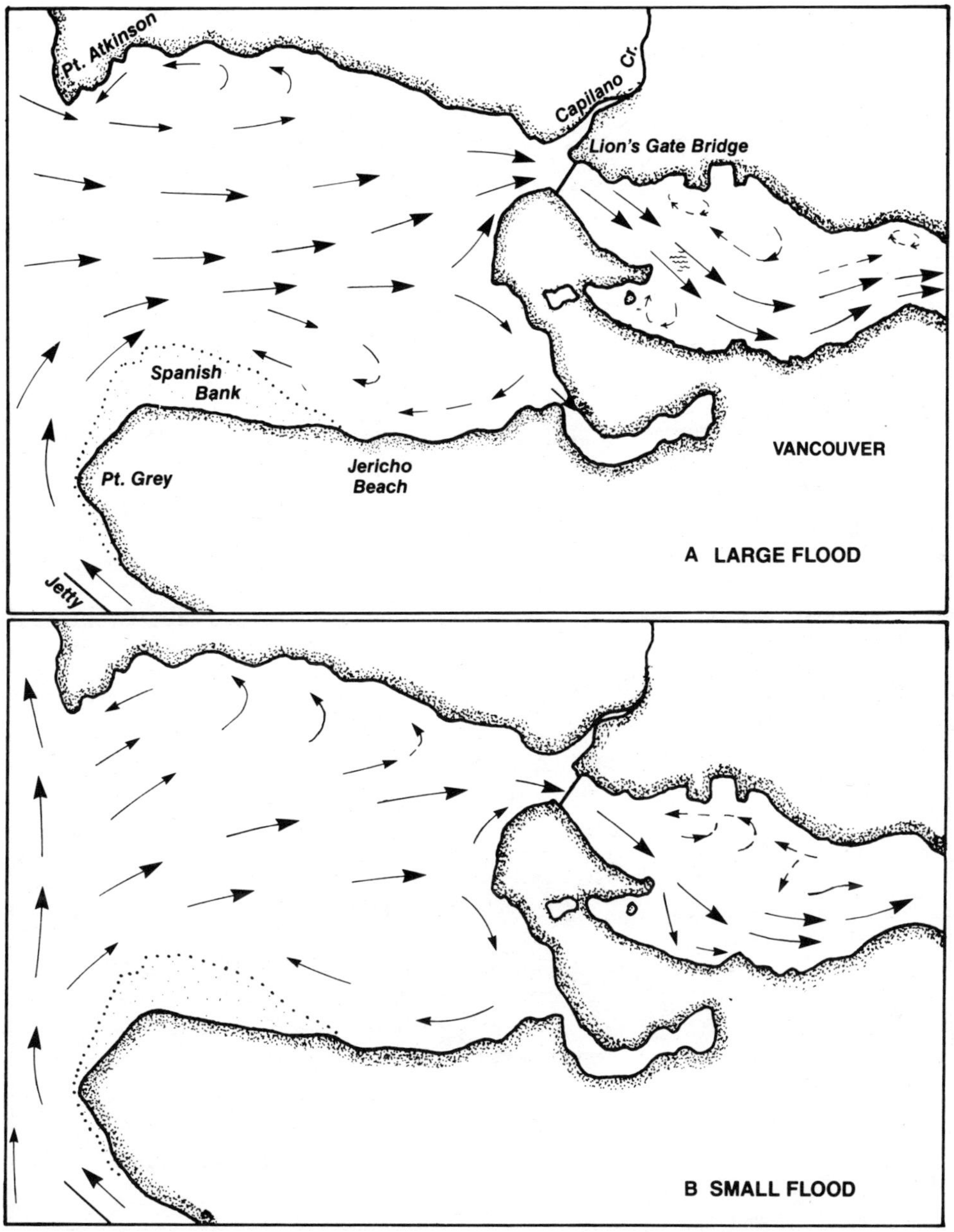

FIG. 10.37. Tidal currents in Burrard Inlet; (A) large flood, (B) small flood, (C) large ebb, (D) small ebb. Larger arrows, 25–50 cm/s (0.5–1.0 kn); small arrows, less than 25 cm/s (except First Narrows where flow is generally over 50 cm/s). (Adapted from Campbell 1954)

During small ebbs, the northshore jet is weaker and less well established so the counterclockwise eddy does not form over the eastern portion of the inlet. However, ebb streams still tend to be directed northward along the beaches of Stanley Park (Fig. 10.37d).

Waves

The limited fetch within Burrard Inlet makes it impossible for local winds to generate waves of any appreciable significance. As a consequence, the comparatively large seas sometimes found at the entrance to the inlet originate with strong winds in the Strait of Georgia. Low swell-like waves propagate into the inlet following the passage of frontal systems, but these rapidly diminish in strength along the axis of the channel. Within Vancouver Harbour, boat-generated waves are probably as important as wind waves when it comes to disturbing the shoreline.

Despite the small wave heights, vigorous rips occur in regions of strong tidal flow. During westerly winds, rips dangerous to small boats often form in the vicinity of Point Atkinson and seaward of the First Narrows on the ebb. The latter may be further increased by the outflow from the Capilano River. Noteworthy rips are also formed over Burnaby Shoal in Vancouver Harbour on larger floods.

Slick bands associated with the up-inlet propagation of internal gravity waves that originate in the Strait of Georgia are common in the outer basin in summer. As

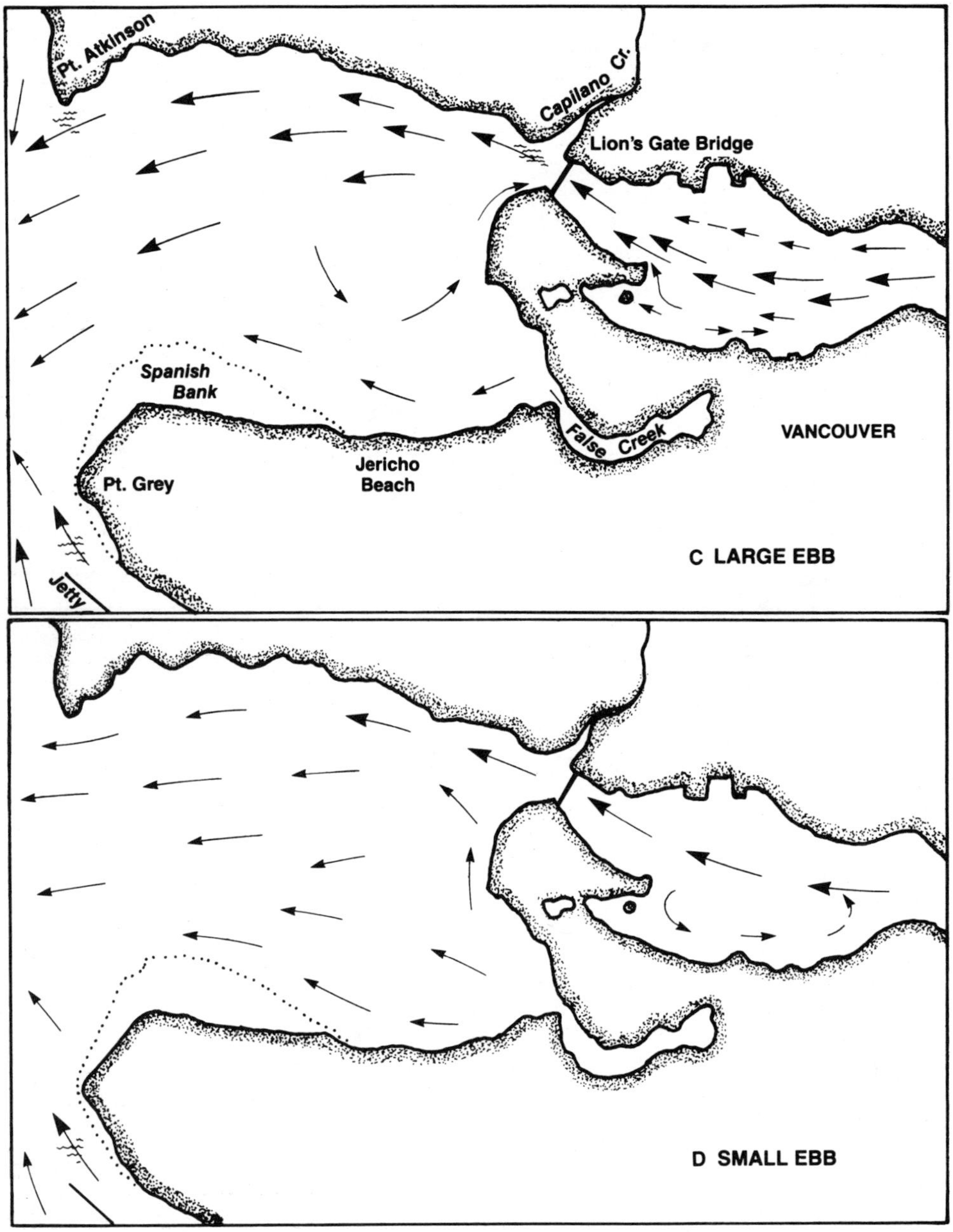

FIG. 10.37. Tidal currents in Burrard Inlet; (A) large flood, (B) small flood, (C) large ebb, (D) small ebb. Larger arrows, 25–50 cm/s (0.5–1.0 kn); small arrows, less than 25 cm/s (except First Narrows where flow is generally over 50 cm/s). (Adapted from Campbell 1954)

with surface waves, the energy of these waves is dissipated by breaking on the shores, a process that goes unnoticed by human eyes.

Indian Arm Circulation

In common with most west coast fiords, the basic circulation in Indian Arm consists of a surface outflow of brackish water driven by freshwater accumulation in the basin, with a compensating inflow of salty water at depth. Superimposed on this structure are the ebb and flood of the tidal currents and the wind-driven currents at 3% of the wind speed. At any particular time, therefore, the speed and direction of the surface waters to a depth of about 5 m depend on the amount of river discharge (which determines the strength of the seaward outflow), the range and stage of the tide, and the wind strength and direction. Maximum tidal streams at the shallow southern end of the fiord vary in velocity from 25 cm/s for a 2-m tide to 50 cm/s for a 4-m tide, but are considerably weaker within the main basin itself. As a consequence, the surface outflow in Indian Arm is only completely stemmed by the flood near the mouth of the inlet during periods of low runoff. Over the bulk of the fiord, the surface currents are almost always southward, but of variable strength, weakest for a flood and/or southerly winds and strongest for an ebb and/or northerly winds. Immediately beneath the outflowing brackish layer an intruding jet is formed at the southern end on the flood, which penetrates part way up-inlet before mixing with the surrounding water.

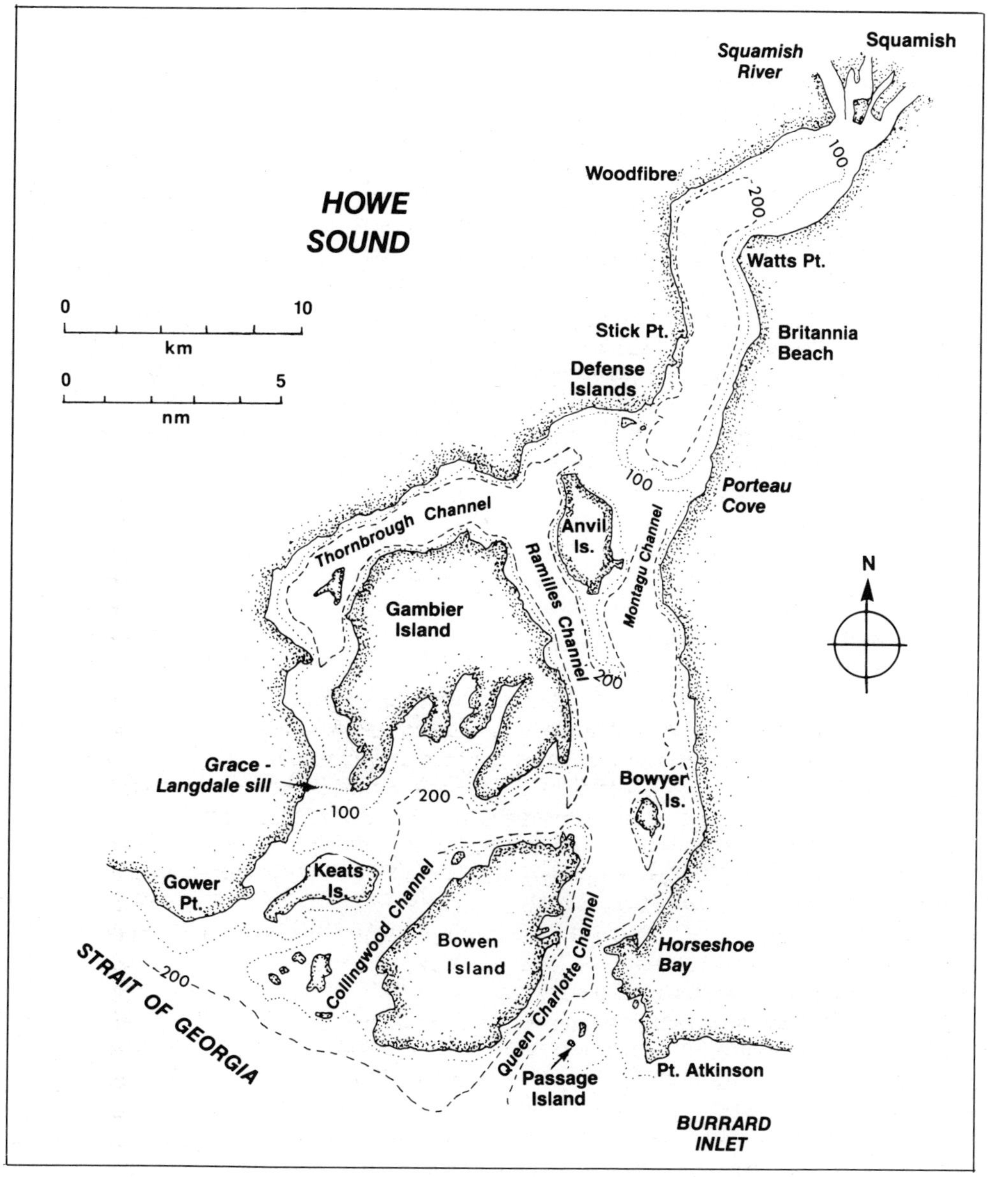

FIG. 10.38. Howe Sound. (Depths in metres.)

Howe Sound

Situated immediately north of Burrard Inlet, Howe Sound is an estuarine-type body of water that cuts 43 km northward into the Coast Mountains to the mouth of the Squamish River (Fig. 10.38). Topographically, it can be subdivided into two separate basins. The island-strewn outer basin is a true "sound," which narrows from 20 km at the entrance between Point Atkinson and Gower Point to about 3.5 km at the Defense Islands, 26 km inland. From this point, the sound gives way to a narrow fiord-like inner basin with steep precipitous cliffs, a large river at its head, and a sill at its entrance (Pl. 2).

Though most visitors to Howe Sound are impressed by its rugged countenance and beauty, the first white men to enter the area were not overly enthusiastic about what they saw. Captain George Vancouver's remarks in June 1792 were anything but flattering:

"Quitting Point Atkinson and proceeding up the Sound . . . we made a rapid progress, by the assistance of a fresh southerly gale, attended with dark gloomy weather that greatly added to the dreary prospect of the surrounding country. The low fertile shores we had been accustomed to see, though lately with some interruption, here no longer existed: their place was now occupied by the base of the stupendous snowy barrier, thinly wooded and rising from the sea abruptly to the clouds; from whose frigid summit, the dissolving snow in foaming torrents rushed down the sides and chasms of its rugged surface, exhibiting altogether a sublime, though gloomy spectacle, which animated nature seemed to have deserted . . ."

For many people today, the "gloomy spectacle" that Vancouver so poetically described is equated to the thick grayish smoke from the Woodfibre pulp mill which, when atmospheric conditions are right, covers much of the Sound with a shroud of unpleasant chemical air.

Bathymetry

Howe Sound began as a river valley that was subsequently gouged and reformed by powerful glacial excavations. At their peak, the ice sheets filled the valley to the present 1950-m level, and when they retreated left behind thick layers of sediment that now overlie the bedrock of the channel. Two major sills were formed as a result. The inner sill extends across the mouth of the fiordlike northern basin between Porteau Cove and the Defense Islands, rises to within 35 m of the water surface, and is thought to be a terminal moraine marking the maximum advance of a recent glacial period. It partially blocks the northward movement of deep oceanic water, and thereby encourages a stagnation of the bottom waters within the inner basin. The comparative shallowness of this area, and the fact that the Squamish Highway descends to sea level at this point, makes it a favorite haunt of scuba divers.

The outer sill lies across the southern end of Howe Sound and consists of glaciomarine deposits that accumulated where the seaward advancing ice sheet from the Sound met the pack ice of the Strait of Georgia. At one time it acted as a barrier to deepwater exchange with the Strait but was later breached by faults. Within Queen Charlotte Channel, erosion by tidal currents has further deepened the breach to the extent there is now a free exchange of water between the Strait and the outer basin of the Sound. Minimum depths over the remaining portion of the sill are around 60 m and average slopes are approximately 1 in 7. A smaller sill (the Grace-Langdale sill) with minimum depths of 30 m extends from the southern tip of Gambier Island to the mainland coast. Depths within the inner basin average 275 m; those in the outer basin average about 240 m. Both basins are flat-bottomed by coastal standards with midchannel reliefs of only a few metres. The bathymetry is appreciably more rugged near the islands and mainland shores where it mirrors the rocky nature of the mountainous terrain. There is also a fairly steep bottom slope seaward of the Squamish River delta; by some estimates, the river sediments are advancing the delta front as much as 7 m/yr.

Winds

Estuarine-type regions like Howe Sound are strongly influenced by winds which, in addition to their direct effect on surface currents, play an indirect role in modifying the oceanographic structure of the deeper waters. This is especially true of the fiordlike inner basin of the Sound where tidal effects appear to be relegated to secondary importance. In Howe Sound as a whole, moderate-to-strong winds are linked with frontal systems associated with Pacific disturbances, arctic outflow, and locally generated sea breezes.

From October to March, frontal disturbances that move across southern British Columbia are commonly preceded by gale-force southeasterly winds that are funneled into strong up-channel winds in Howe Sound. This effect is most pronounced in the northern sector of the Sound where the air is more confined by the local terrain. With the approach of a front, winds can shift from light to storm intensity within a matter of hours, and on one occasion were clocked at average speeds of 25 m/s (50 kn) with gusts to 40 m/s (80 kn) at the head of the Sound. In winter, southerly winds in excess of 7.5 m/s (15 kn) occur approximately 30% of the time and commonly reach speeds of 15 m/s (30 kn) with gusts to 20 m/s (40 kn).

Strong down-channel Squamishes are also fairly common in Howe Sound during the cold winter months. As in other coastal inlets, Squamishes are formed when cold, dense arctic air sitting over the interior of the province surges seaward along river valleys in the manner of a density current. Events of this kind are often triggered by the passage of a low-pressure system down the outer coast and though they tend to occur less frequently than southerly winds, they are generally more persistent once initiated. What few data are available concerning Squamishes in Howe Sound indicate they may persist for 3–5 days and occur on the average about 5–6 days during December and January. Wind speeds frequently reach 15 m/s (30 kn) with gusts of up to 30 m/s (60 kn).

From May through August, the wind pattern in Howe Sound is dominated half the time by a vigorous sea-breeze effect associated with the differential heating of the land and water (see Chapter 2). These diurnal winds tend to be somewhat stronger than their counterparts over much of the Strait of Georgia because of the confined nature of the local terrain, and are most prevalent during clear sunny weather. At such times, the land breeze blows from the north with light winds between midnight and about 9 A.M. With the onset of daytime heating, the winds shift to southerly or southwesterly and by afternoon the sea breeze commonly attains speeds of 10 m/s with gusts to 15 m/s (in one case, gusts of 20 m/s were recorded at Squamish during an intense sea breeze of 15 m/s). The winds die off after sunset and revert to light northerly near midnight. This cyclic summer pattern can, of course, be modified by larger scale atmospheric conditions, such as the passage of weak fronts or a strongly westerly flow of air from the Pacific, and should not be thought to prevail without exception.

Tides

The mixed, predominantly semidiurnal tides in Howe Sound differ little from those in Burrard Inlet, and for most practical purposes Point Atkinson serves as an excellent reference port for both areas. At Squamish, for example, the mean range of 3.2 m and the large tide range of 4.9 m differ by only a few centimetres from the corresponding values of 3.3 m and 4.9 m, respectively, at Point Atkinson. Moreover, the times of high and low water at the head of Howe Sound are usually only a few minutes behind those at Point Atkinson, though this delay can be increased somewhat by Squamish winds. Conversely, southerly winds associated with passing frontal systems will cause the tides to occur a few minutes sooner than predicted and to have slightly greater heights.

Oceanography

Distributions of salinity and temperature in the Sound are similar to other estuarine environments affected by river runoff, winds, air temperatures, and tides. Typically, a relatively low-salinity surface layer 5–10 m thick overrides a much deeper layer of appreciably higher salinity, where conditions are nearly constant throughout the year.

Fresh water to Howe Sound is supplied mainly by the Squamish River which, together with its two major tributaries, the Cheakamus and Mamquam rivers, drains an area of about 3700 km^2, making the Squamish one of the larger rivers of the province. Essentially fed by snow melt, the Squamish River system reaches a maximum discharge rate of around 760 m^3/s in early summer and a minimum discharge rate about 1/10 of this value in late winter. Abnormally large discharges up to 2100 m^3/s have occured in the fall of certain years, due to sudden thaws and heavy rainfall, and have combined with high tides and winds to cause damaging flash floods over the Squamish delta. (The town of Squamish has been inundated with nearly 1½ m of water about once every 16 yr since its founding, and to a somewhat lesser degree about every 7 yr, a problem which hopefully has been alleviated by the construction of a retaining dyke along the river bank.) Accompanying the pronounced summer-to-winter variation in the river discharge is a corresponding seasonal variation in the temperature, salinity, and currents within the surface waters of the Sound.

As the river runoff progresses seaward it entrains saltier oceanic water from below, which results in a discernible down-channel increase in salinity in the upper 10 m of the water column. By the time the brackish layer has reached the outer basin, values are commonly around 15‰ in summer with somewhat higher values in winter, when runoff is lower. There is a further downstream increase in salinity to the mouth of the Sound, where values become comparable to those in the Strait of Georgia. Below 50 m or so, the salinity ranges between 29 and 31‰ throughout the year, with slightly higher values in the outer basin compared to the inner basin, due to the blocking effect of the sill at the entrance to the inner basin. Fortunately for marine organisms that reside below sill-depth in the inner basin, nature provides an approximately biannual refreshening of the deeper waters by an overflow of more oxygen-rich water from the outer basin. Indications are that such overflow events are triggered by a combination of strong Squamish winds and abnormally high river discharge, produced by a heavy rainfall or sudden thaw at higher elevations. Though these events may only last a few days, they are essential to the renewal of the water in the inner basin.

Levings and McDaniel (1980) have shown that, whenever such events are delayed too long, the resulting low oxygen levels lead to widespread depletion of numerous species of benthic organisms. The absence of a major bottom-water inflow in 1977 resulted in the temporary annihilation of shellfish stocks over much of the inner basin.

In the mainstream of Howe Sound, water temperatures in summer remain persistently cool, the only substantial warming occurs in the protected coves of the island-strewn outer basin. It appears that the surface brackish layer formed by the snow-fed Squamish River, whose temperatures hover around 10°C, has little opportunity to warm during its seaward journey to the Strait of Georgia. In winter, it is common for surface temperatures in Howe Sound to drop to 5°C. Below the upper freshet layer, temperatures are consistently around 8–10°C throughout the year, with only minor modifications in the outer basin brought about by incursions of warmer water from the Strait.

Circulation

In contrast to many areas of the coast, currents in Howe Sound have been fairly well documented, especially in the inner basin where effects of the Squamish River are most widely felt. Some excellent work in this regard was conducted by J. Buckley from May to July 1973. Buckley determined the nature of the surface flow by deploying radar trackable drogues and further supplemented his findings with data collected from strings of moored current meters.

Basically, Howe Sound has an estuarine-type circulation structure primarily driven by the freshwater discharge from the Squamish River system. There is a net seaward flow of brackish water in a relatively shallow-surface layer of around 10 m, together with a slow inward drift at depth, which compensates for the loss of salt water to the surface layer (Chapter 2). Observations suggest that currents in the top 3 m are uniform with depth so a ship's keel will lie within a slab of water that moves at the same velocity. However, the simple estuarine flow is modified appreciably by the action of the wind and, to a lesser extent, tides. In addition, there is an ever-present, cross-channel variation in the strength of the seaward surface current in the inner basin, and often a reversal in direction from one side to the other. The strength of the surface current will also vary with season, is weakest during the winter months when runoff is relatively low, and greatest in summer when runoff is high.

The most visible feature of the surface layer is the silty, chalky looking water that originates from the Squamish River and flows seaward along the length of the Sound mixing and widening as it goes (Pl. 2). In many instances, the demarcation between the silt-laden brackish water and the greenish colored salty water of the Sound also marks a distinct change in the currents. At other times the currents are identical on either side of the "line," ruling it out as a reliable indicator of a change in the flow. Although the position of the core of silty water shifts from day to day following shifts in the prevailing winds, and its strength varies with the amount of river runoff, a general picture of the surface currents in Howe Sound is now beginning to emerge. In Fig. 10.39, the core of the river "jet" enters the Sound along the retaining dyke on the Squamish delta at about 50 cm/s. From there it progresses southward until it contacts the shore north of Watts Point, where it is deflected across-channel in the direction of Woodfibre. To the north of the main flow there is a large clockwise backeddy with nearshore up-channel speeds of about 20 cm/s to the vicinity of the river mouth. The core of the brackish layer then turns southward off Woodfibre, often hugging the coastline until "Stick Point" (a local name not on the hydrographic charts) where it separates from the shore to flow down center-channel. (Along the opposite shore there is a weak return flow northward to the vicinity of Watts Point.) By the time the remnant of

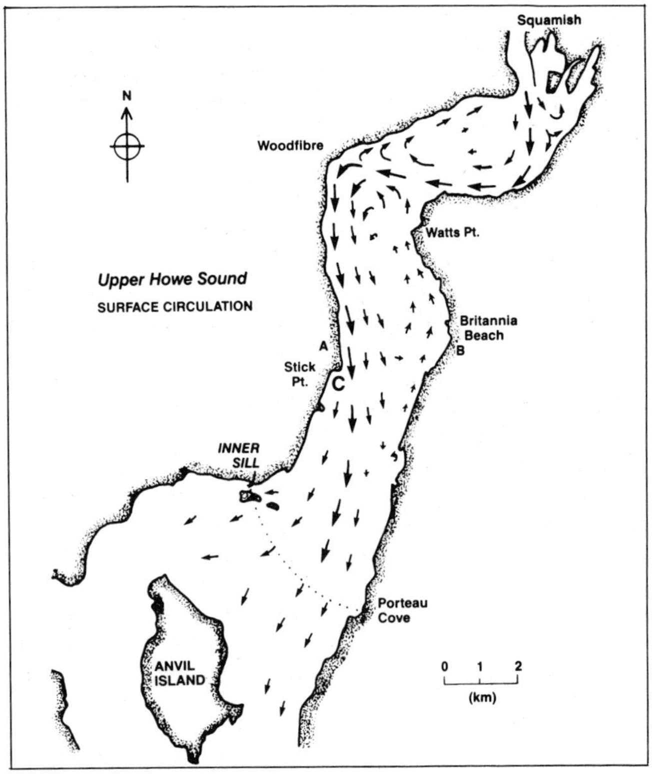

FIG. 10.39. General surface current pattern for upper Howe Sound deduced from surface drogue studies. Arrow size proportional to speed and persistence of flow. (For currents in cross-section between A and B, see Fig. 10.40.) (From Buckley 1977)

the core has reached the sill that divides the two major basins of Howe Sound, its strength has waned and the flow is everywhere seaward, provided there are no strong up-channel winds. Generally speaking, surface currents in the fiordlike inner basin have a strong down-channel component within the core of brackish outflow hugging the shoreline, and a moderately strong up-channel component on the opposite shoreline. Available data further indicates that, under normal circumstances, the down-channel flow along the western side is strongest at the surface, whereas the up-channel current on the east side reaches a maximum at a depth of a few metres (Fig. 10.40).

The abrupt change in the nature of the currents near

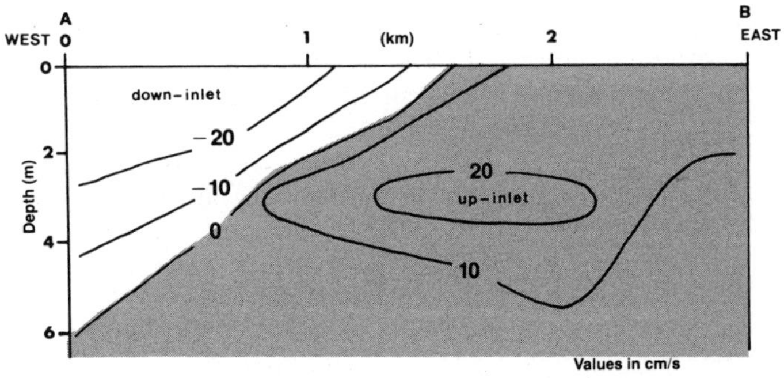

FIG. 10.40. Along-inlet velocity structure in top 6 m of Howe Sound (cm/s), deduced from 3-h averaged drogue motions Sept. 6, 1973. Positive values, up-inlet (landward); negative values down-inlet (seaward). See Fig. 10.39 for locations of A and B. (From Buckley 1977)

Stick Point provides a practical example of the effect of the flow regime on marine traffic in the Sound. Tugboat operators pulling booms or barges up-channel usually work the western shore to take advantage of the weak currents and backeddies, but once they round the point they often become "stuck." Under such conditions it may be more expedient for the operator to head toward mid-channel rather than continue to stem the core of the narrow outflow-jet north of this point.

Winds over the inner basin lead to considerable modification of the generalized flow pattern in Fig. 10.39. In the region between Watts Point and Stick Point, for example, the onset of a fresh-to-strong southerly blow will cause the currents associated with the seaward flowing jet to slow down or stop, whereas from about mid-channel to the opposite shore it will induce up-channel currents with speeds to 50 cm/s (Fig. 10.41). At times, southerly winds in this area may be enough to reverse the flow of the jet, especially during periods of comparatively low runoff. (Except near the northern end of the basin, tidal currents appear to have only a secondary influence compared to winds, and for the most part can be disregarded.) However, there comes a point when, after several hours, a southerly wind loses its ability to counter the down-inlet flow of the surface jet and the currents again begin to speed up, sometimes accompanied by a shift of the core toward mid-channel. This is presumably due to the fact that while the wind is retarding the natural seaward tendency of the brackish layer, a head of water is built up that eventually overcomes the force exerted by the wind, and allows down-inlet currents to regain speeds of 25–50 cm/s. Away from the main core, on the other hand, winds continue to drive the surface layer inland with greatest speeds roughly 1 km from the eastern shore. Between Watts Point and the river delta, the wind and tide have equal influence, though neither is strong enough to appreciably alter the flow regime. In this area, the time of maximum down-inlet velocity closely coincides with the time of high water at Squamish; the core of the jet may be shifted slightly northward by strong southerly winds. Reversals in flow induced by southerly winds at the seaward end of the inner basin occur almost uniformly across the width of the channel with only slightly stronger currents over the eastern half (Fig. 10.42). Up-channel velocities are typically 25 cm/s a few hours after the onset of southerly winds that exceed 15 m/s, though again the build-up of brackish water within the basin can lead to a down-channel set near the western side prior to any slackening of the winds.

Once the strength of the southerly winds begins to fall below about 10 m/s, the circulation in the inner basin returns to its usual pattern shown in Fig. 10.39. Northerly winds such as Squamishes will, of course, augment the velocities of the surface flow in this pattern with the result that surface currents will be everywhere seaward except in the lee of prominences where small backeddies will be formed. Down-channel speeds associated with the core of the jet can be expected to reach 100 cm/s while in other areas maximum speeds of around 50 cm/s will pertain. Finally, under most wind conditions the direct effect of the wind is confined to the top few metres, and even when

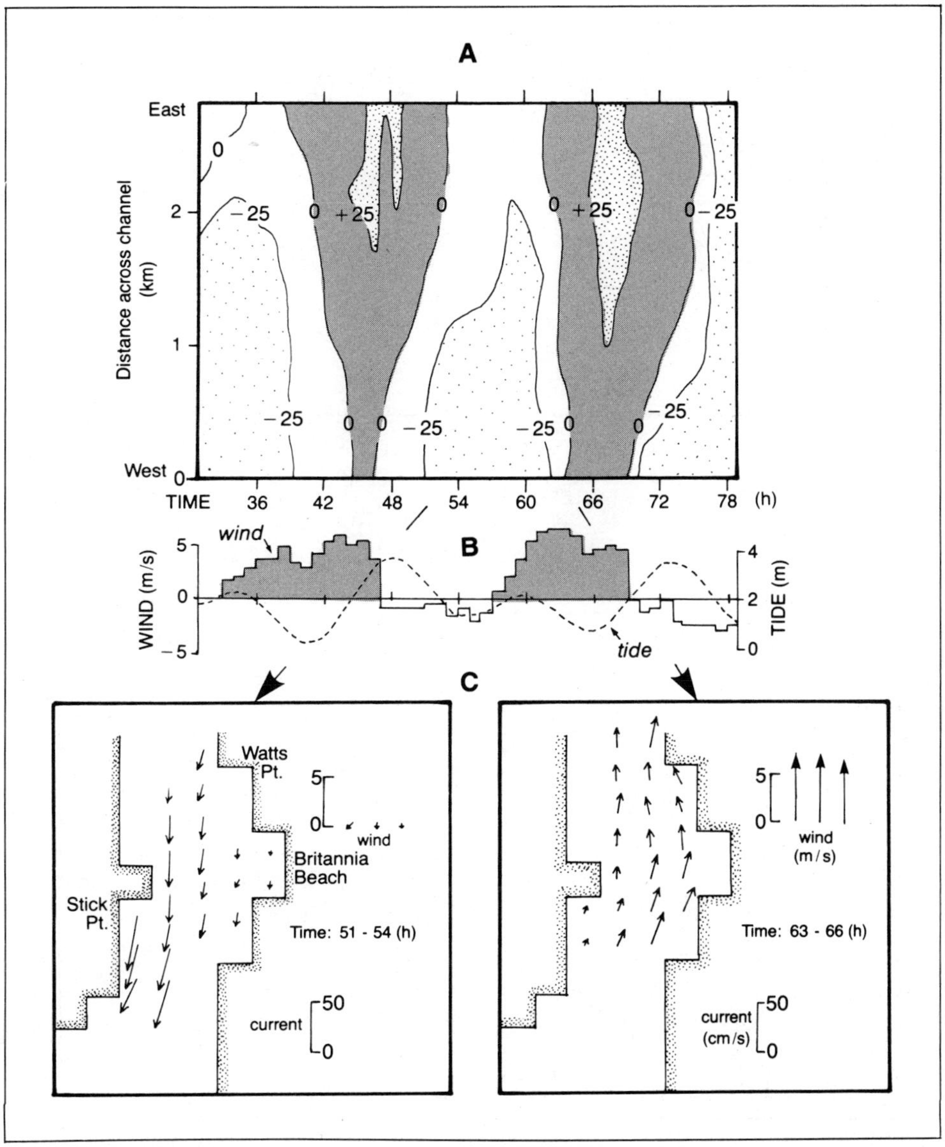

FIG. 10.41. Effect of winds on surface currents in central region upper Howe Sound, May 9–11, 1973. (A) Variation in speed (cm/s) and direction of along-inlet surface current over 48 h between A and B in Fig. 10.39. Positive values indicate up-inlet currents, negative values down-inlet currents. At 72 h for example, flow is down-inlet on west side at over 25 cm/s, but up-inlet on east side at 0–25 cm/s. (B) Wind speed (m/s) and tide height (m) correspond to the times in (A). Positive values (shaded), up-inlet (northward) winds; negative values, down-inlet (southward) winds. At 72 h winds light from north, tide high. (C) Surface current vectors in plan view. Flow averaged over 3 h and surface area of 0.5 nm on side for each of two different time intervals. Compare with (A), (B). Length of current arrows in cm/s, when compared to lower scale; wind speed scale, m/s; and wind vectors shown for each of 3 h over which flow is averaged. (Adapted from Buckley 1977)

the surface outflow has been reversed there is often a persistent seaward current between 5 and 10 m. At greater depths the flow is slowly up-channel to within a few kilometres of the river mouth.

Relatively little is known about the general circulation in the island-strewn outer basin of Howe Sound, though, based on the geometry of the region, surface currents can be expected to be weaker, more variable, and to have considerably more eddylike structure than those in the confined inner basin. Tidal currents play a greater role in this region but their presence will be partially masked by strong winds. Because the tide is delayed by only a few minutes throughout the Sound, maximum floods occur midway through a rising tide, and maximum ebbs, midway through a falling tide. During relatively light to moderate airs, tidal streams attain speeds of around 25 cm/s in the comparatively unobstructed eastern channels with somewhat greater speeds in the narrower passages between the islands and in the vicinity of points of land. Southerly winds in excess of 5 m/s will lead to stronger surface flood currents and weaker ebb currents which, on occasion, may be reversed to up-channel flows. However, northerly winds will accelerate the ebb and decelerate the flood. During the summer months, the added effect of the brackish outflow is to strengthen the ebb and weaken the flood.

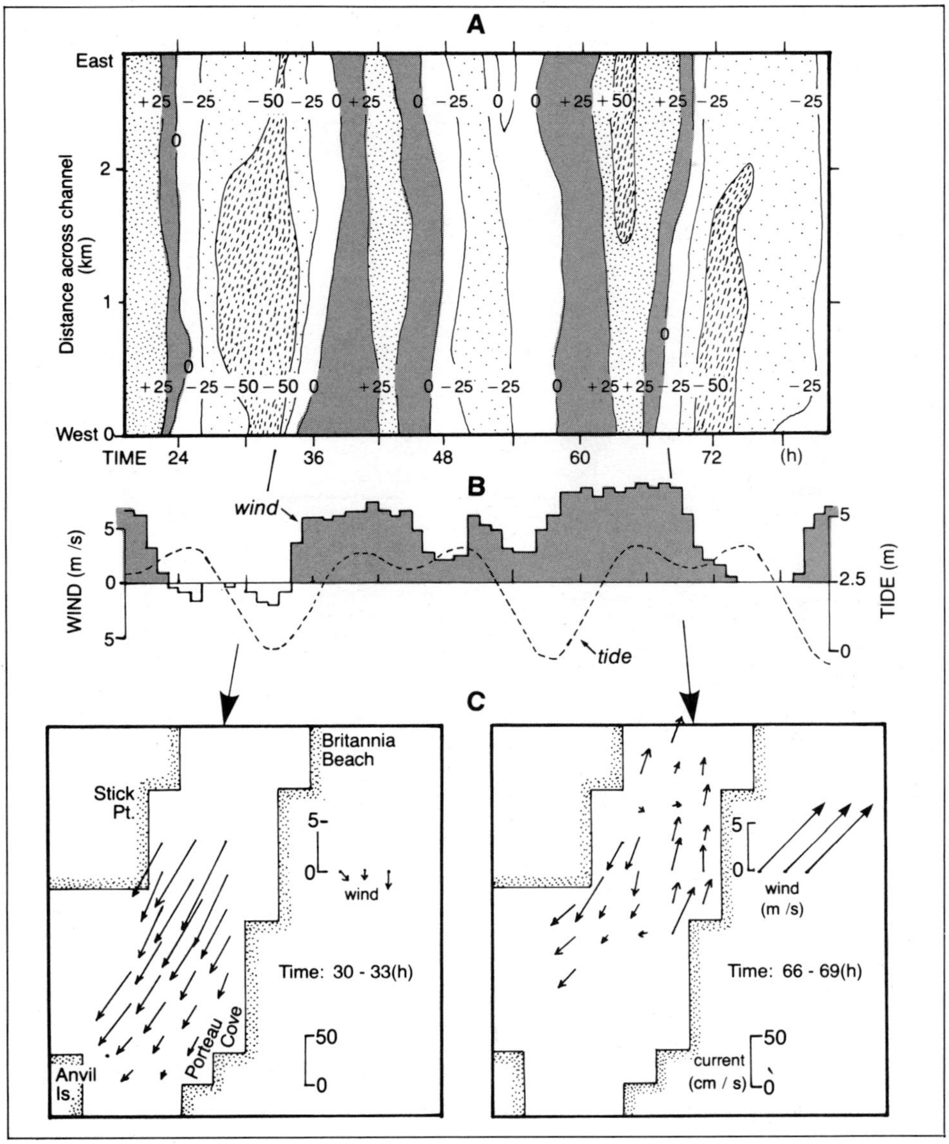

FIG. 10.42. Effect of winds on surface currents at southern end upper Howe Sound, June 27–29, 1973. (A) Variation in speed (cm/s) and direction of along-inlet surface current over 48 h in cross-channel section north of Anvil Island. (B) Wind speed (m/s) and tide elevation (m) at times of current observations. Except for few hours, winds were up-inlet (positive). (C) Plan view of average surface current vectors for two, 3-h periods selected from observational period of (A). Rectangular shoreline consistent with fact that drogue measurements averaged over 0.5-nm grid array. (Adapted from Buckley 1977)

A hydrographic chart might suggest that the silt-laden surface layer that moves seaward from the inner basin would always prefer to enter the outer portion of the Sound via Montagu Channel (Fig. 10.38). Though this is prabably true in general, the few available aerial and current observations suggest there are numerous instances when the opposite occurs. Plate 2, for example, clearly shows that the main core of the brackish jet impinges on Anvil Island, then turns abruptly westward to follow the shoreline into Ramilles and Thornbrough Channels in the "lee" of the island. Nevertheless, a considerable portion of the silty water progresses into Montagu Channel, at least partially supporting the thesis of a surface outflow along the eastern side. Perhaps at the time of the photograph the southerly winds were mainly responsible for forcing the core to the west of Anvil Island, whereas during northerlies or weaker southerlies the brackish layer would more readily flow through Montagu Channel. Logic then dictates that the seaward outflow would continue on a preferred path to the Strait of Georgia via Queen Charlotte Channel. However, it seems that the head of water associated with the North Arm of the Fraser tends to drive brackish Fraser River water northward into this Channel, and forces the brackish Squamish River water to deflect through Collingwood Channel over most of a tidal cycle. Thus, a boater traveling into Queen Charlotte Channel during calmer summer weather will have the advantage of an up-channel set about as far as Bowyer Island (except during the times around maximum ebb) but will subsequently begin to buck a down-channel set north of the island.

Although detailed information concerning the outer

basin is scanty, and probably will remain so for some time yet, the very nature of the convoluted shoreline assures a complex circulation pattern. Backeddies, for instance, undoubtedly prevail downstream of the smaller islands and in the lee of points of land that project into the main channels where nearshore counterflows will be produced, but in the larger embayments such as along the southern end of Gambier Island, currents will be insignificant. Confused and variable currents are also to be expected where the flows from different channels converge, for example north or south of Anvil Island where, depending on the stage of the tide, water moving through Ramilles Channel contacts that moving through Montagu Channel.

Waves

Strong winds can come up suddenly in Howe Sound and, despite the comparatively short fetches, can soon generate short, steep, choppy seas that are particularly hazardous to small craft. Greatest wave heights of 1.5 m are generally produced over the longer fetches available to southerly winds that blow inland from the Strait of Georgia, though waves to 2.5 m have been reported during storm-forced Squamishes. Swell-like waves that originate in the Strait can also penetrate a short distance into the Sound before being dispersed by the topography. Smallest waves at any given time will be found within the sheltered areas between the numerous islands in the outer basin and largest waves will be generated along the exposed eastern channels leading into the northern basin. Especially vigorous seas can be expected to form over the main core of the seaward flowing river jet in the inner basin, following the onset of a southerly gale.

Squamish winds often create an expanse of turbulent breaking waves adjacent to the outside entrance of the Sound. Wave heights diminish rapidly away from the outer islands as the arctic air fans out to the south; nevertheless, the effect is disconcerting to a boater who approaches the region from otherwise flat calm conditions in the Strait. At other times the presence of Squamish winds is not felt until the boater is well within Queen Charlotte Channel.

Chapter 11. Juan de Fuca Strait

Physiography

Juan de Fuca Strait is a long, narrow submarine valley that originates along a depression between the resistant lava flows and metamorphic rocks of southern Vancouver Island and the Olympic Mountains (Fig. 11.1). Over the last 1–2 million yr, the Strait has undergone excavation on at least four separate occasions, as continental ice sheets moved seaward during periods of worldwide cooling.

East of the line between Jordan River and Pillar Point, the region is characterized by a gently sloping U-shaped, cross-channel profile linked to recent glacial processes. A large terminal moraine, the Victoria–Green Point sill, marks the furthest point of advance of an ancient ice sheet that once existed in the Strait. Westward of this area to the Pacific entrance off Cape Flattery, the Strait takes on a V-shaped profile that resembles a mature river valley. Further seaward, the channel turns sharply to the southwest where it becomes irregular and is cut with deep incisions such as the Juan de Fuca Canyon.

From its entrance to about 100 km (55 nm) eastward, the Strait is approximately 22–28 km wide (12–15 nm), but narrows to 18 km (10 nm) between Race Rocks and Angeles Point. It then widens again to around 40 km (21 nm) for the next 56 km (30 nm) to the eastern boundary near Whidbey Island. The depth of the Strait decreases gradually inland from around 250 m at mid- channel near its entrance to about 180 m at a distance of 70 km east of Cape Flattery. This eastward shoaling continues to the cross-channel sill that cuts across the Strait to the south of Victoria. Over the sill, depths are relatively shallow at only 55 m. To the east of this submarine ridge, the Strait contains several shallow banks through which the deepest channels lead into Haro Strait. The lesser channels lead into Rosario Strait, Admiralty Inlet, and Deception Passage. For the most part, depths within Juan de Fuca Strait are appreciably less than those in the Strait of Georgia.

The coastline of the Strait is relatively uniform with a low rocky shoreline abutted against cliffs to 20 m high. Centuries of wave action have turned much of the shore into rocky intertidal platforms that are often engulfed in kelp in summer. Though sandy sediments are scarce due to the weather-resistant nature of the rocks, numerous small beaches have developed which, in the eastern portion, consist mainly of pebble–cobble material with logs at the head of the beaches. In the western section there are many small pocket beaches of coarse sediments and a few narrow, sand beaches. Port San Juan near the Pacific entrance is the only major fiordlike opening along the northern shoreline and provides the largest freshwater source, the San Juan River, that flows directly into the Strait. There are also constricted openings into Sooke Inlet and Sequim Bay at the eastern end of the Strait, and a relatively large indentation into Port Discovery. (See Chapter 2 for other coastal features.)

Temperature and Salinity Distributions

With the exception of a few protected bays and partially enclosed regions like Sooke Basin, water temperatures throughout the Juan de Fuca Strait system remain uncomfortably cold year round.

In part, the water of the Strait is kept cool by direct exposure to the Pacific Ocean, which, despite exaggerated tales of a warm "Japan Current" flowing toward the coast,

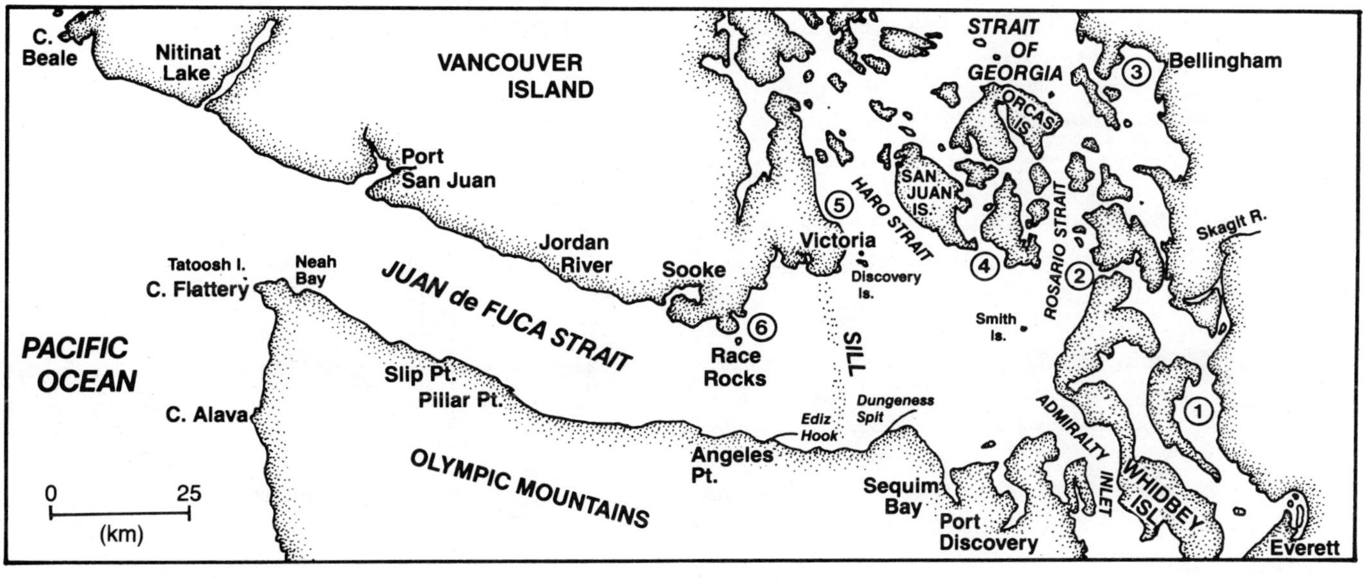

FIG. 11.1. Map of Juan de Fuca Strait and adjoining waters (1) Port Susan, (2) Deception Passage, (3) Bellingham Bay, (4) Middle Channel, (5) Cordova Bay, and (6) Pedder Bay.

coast, usually never warms above 12°C off Vancouver Island below depths of 10 m, even during the warmest summer months. But more importantly, the strong tidal streams through the eastern passes that adjoin the Strait constantly mix the water from top to bottom and make it nearly impossible for the surface layer to retain any appreciable amount of solar heat.

During winter, water temperatures near the surface generally decrease from relatively high values of 8–10°C near the Pacific entrance to comparatively low values of about 8°C in the eastern portion of the Strait (Fig. 11.2a). There is a corresponding reduction in water temperature with depth, though this rarely amounts to more than a few degrees. Nevertheless, it is in winter that near bottom temperatures are warmest in the western half of the Strait. East of the sill, between Victoria and Port Angeles, the water is invariably of uniform temperature from top to bottom because of intense tidal mixing in Haro Strait, Rosario Strait, and Admiralty Inlet.

About March, colder ocean water with temperatures around 6–7°C begins to penetrate up-channel along the bottom of Juan de Fuca Strait, but surface temperatures continue to remain slightly below 10°C (Fig. 11.2b). There is no appreciable change in the latter until the beginning of the spring freshet in the Strait of Georgia, when large volumes of fresh water commence to discharge from the Fraser River. Most of this water, which may warm to over 20°C in the near surface layer of the Strait of Georgia, works its way southward into the eastern basin of Juan de Fuca Strait. Although it has been tidally mixed with colder water enroute, it increases the total heat content of the water in Juan de Fuca Strait and raises average temperatures throughout its entire depth. With local solar heating, patches of surface water in the Strait can attain temperatures of 12–14°C by mid-August but bottom values continue to be cold (Fig. 11.2c). By September, surface water temperatures in the Strait of Georgia have diminished sharply and, in combination with the colder ocean water, enhanced wind activity, and substantially reduced solar radiation, cause surface temperatures along Juan de Fuca Strait to decrease to their eventual winter values of 8–10°C. In contrast, bottom waters begin to increase in temperature as warmer water slowly makes its way inland from the Pacific Ocean. By Christmas, temperatures are nearly uniform again throughout the entire channel.

Salinity of the water from the northern end of Haro Strait to the seaward end of Juan de Fuca Strait generally increases from top to bottom and from east to west (Fig. 11.3a). In simplest terms, the salt distribution can be likened to a "wedge" of saline water that has penetrated up-channel from the Pacific Ocean, but whose furthest point of advance is arrested by the partial barriers created by the series of sills to the east. Under favorable conditions, this "salt wedge" penetrates northward to the Strait of Georgia, where its identity is lost through tidal mixing in the vicinity of Boundary Passage.

In winter, salinities of the surface waters are commonly around 30–31‰, and those of the bottom water near the Strait entrance are relatively high at 33.5‰ (Fig. 11.3a, b). Rainfall and local river runoff can create shallow pools of low-salinity water at the surface at this time, but these are soon dispersed by winds and tides. With the advent of the spring runoff from the Fraser River, salinities in the upper layer can drop to around 26–28‰ in Haro Strait and to 28–30‰ in the eastern portion of Juan de Fuca Strait, as brackish water flows seaward from the Strait of Georgia (Fig. 11.3c). Coincident with this, there can be a slight increase in salinity near the open coast when upwelling (Chapter 5) associated with northwest winds brings cold, salty water onto the continental shelf. Thus, surface salinities in the vicinity of the entrance to Juan de Fuca Strait are generally greatest in summer and least in winter; to a lesser degree this is also true of the deeper water (compare Fig. 11.3a and 11.3c).

Clearly, temperature and salinity distributions in Juan de Fuca Strait are strongly influenced by oceanic conditions, river runoff, and tidal processes. However, seasonal variations in these water properties are quite small and for the most part would go unnoticed by the casual observer.

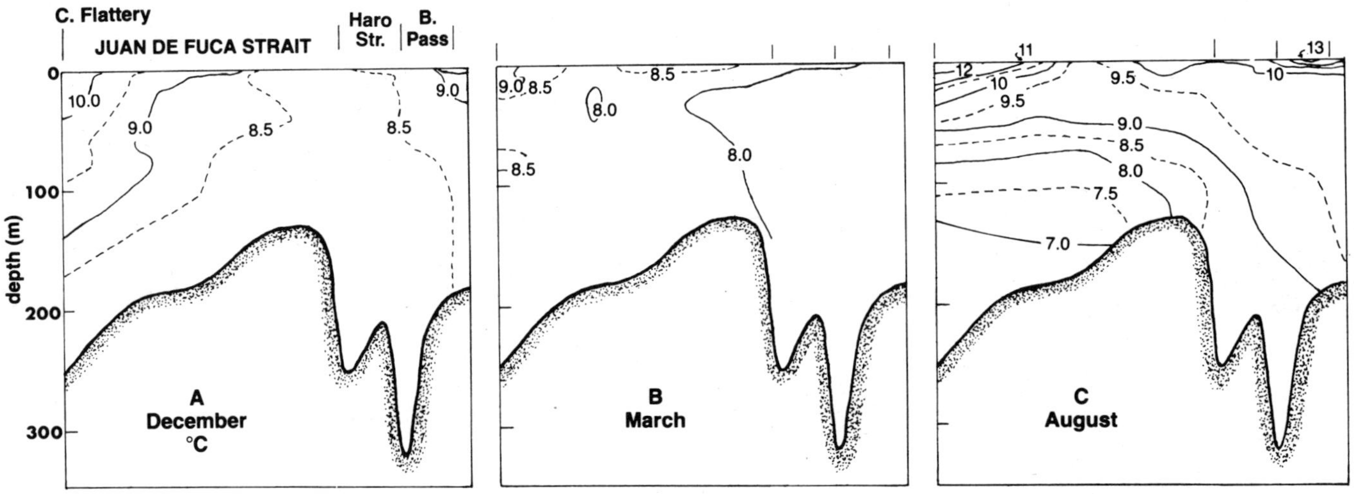

Fig. 11.2. Temperature distribution (°C) from Cape Flattery to Haro Strait, mid-channel; (A) December 1967, (B) March 1968, and (C) August 1968. (Adapted from Crean and Ages 1971)

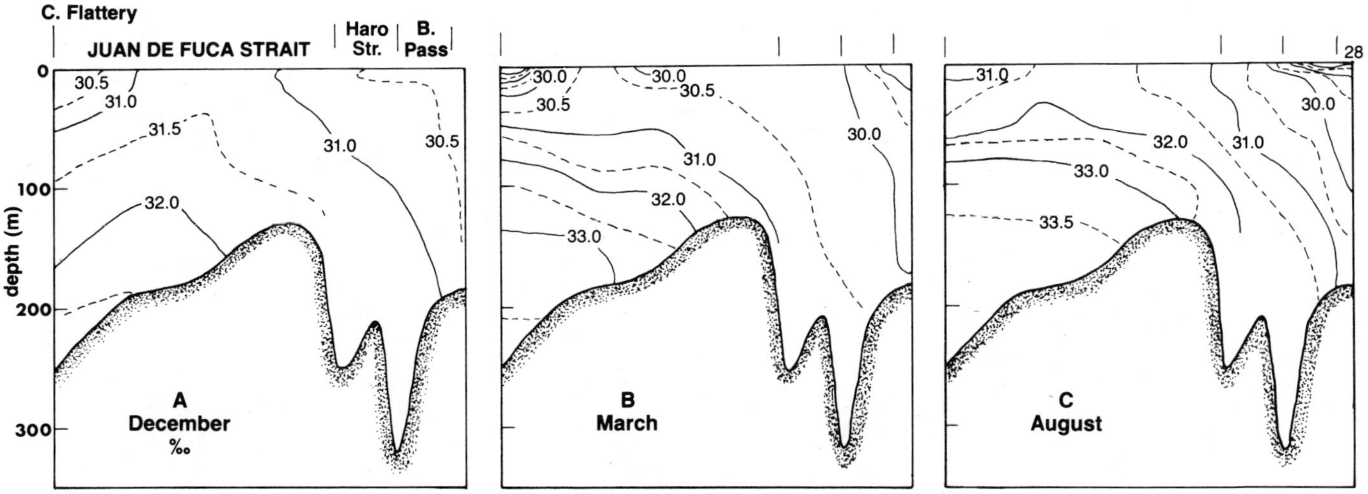

FIG. 11.3. Salinity distribution (‰) from Cape Flattery to Haro Strait, mid-channel; (A) December 1967, (B) March 1968, and (C) August 1968. (Adapted from Crean and Ages 1971)

Wind Patterns

Prevailing oceanic winds off the outer British Columbia–Washington coast are from the southwest in winter and the northwest in summer. Within Juan de Fuca Strait, where the flow of air is strongly influenced by the adjoining mountainous terrain, the corresponding wind directions are easterly in winter and westerly in summer (Fig. 11.4). In most cases, there is an overall increase in wind speed from east to west along the channel, and comparatively weak and variable winds prevail over the eastern sector. On a seasonal basis, winds greater than 15 m/s (30 kn) occur during 10–15 days per month in winter, compared to only 1–2 days per month in summer.

Easterly winds over Juan de Fuca Strait in winter (October–March) are, for the most part, associated with cyclonic oceanic winds of the Aleutian Low (see Fig. 2.14). Because of the funneling effect of the topography, these easterly winds undergo a seaward acceleration along the Strait; at Tatoosh Island off Cape Flattery for example, 38% have speeds in excess of 8.5 m/s (17 kn) whereas at Port Angeles at the eastern end of the Strait only about 5% exceed this value (Fig. 11.5). Average winter winds over the western portion of the channel are 10 m/s (20 kn). Observations taken off the coast from ships indicate that seaward-blowing winds along the Strait generally shift to southeasterlies a few kilometres off Cape Flattery, to suggest a combined influence of the southwesterly oceanic wind pattern and the locally funneled easterly winds of the channel.

Frontal systems that accompany eastward traveling cyclonic disturbances, formed in the central Pacific between the Aleutian Low and the Pacific High, often lead to rapid changes in the wind within the outer reaches of the Strait. As the low moves north of the channel, the prevailing south to southwesterly flow of coastal air backs to west-northwest as the cold front passes. Under certain conditions, moreover, predominant winter winds can be westerly along the Strait rather than easterly. This happened during the winter of 1976, accompanied by southwesterly winds along the outer coast of Washington. Such anomalous wind conditions are presumably linked in part to a greater than normal eastward shift of the midwinter Aleutian Low in the Gulf of Alaska, which in turn intensifies the westerly component of the onshore flow that favors westerly winds along the Strait.

Because of divergence in the airflow in the eastern portion of the Strait, winds to the east of Port Angeles blow predominantly from the west and southwest which, in winter, conforms to the general counterclockwise wind pattern that sets up over the southern half of the Strait of Georgia (see Fig. 10.5). During these months, winds at Victoria are mostly from the north and northeast, although easterlies and southeasterlies occur with regularity (about 20–30% of the time). As a consequence of the paths taken by individual fronts, moreover, winds from the west and southwest also occur. The frequency of such winds is high in October and March (about 30–40%) and low in December and January (about 18%). On the average, these winds are somewhat stronger than those from other quadrants.

Wind patterns in the eastern sector of the Strait are further complicated by the inertia of local air flow. Prior to the passage of a storm with southwesterly winds along the coast, the air is channeled through Puget Sound and around the Olympic Mountains (Fig. 11.6). In the lee of the mountains the winds overshoot, and leave an area of calm airs in the vicinity of Port Angeles. As the storm moves to the northeast, winds in the Strait turn to westerlies and the area of calm airs shifts southward to Puget Sound.

Wind patterns in spring are not appreciably different from winter except for a decrease in frequency of storms and front-related wind shifts.

In summer (June–September), when the air flow is dominated by the North Pacific High off California, winds along the coast are mainly from the northwest. These are redirected to inward-blowing westerlies within the Strait by the steering action of the mountains where

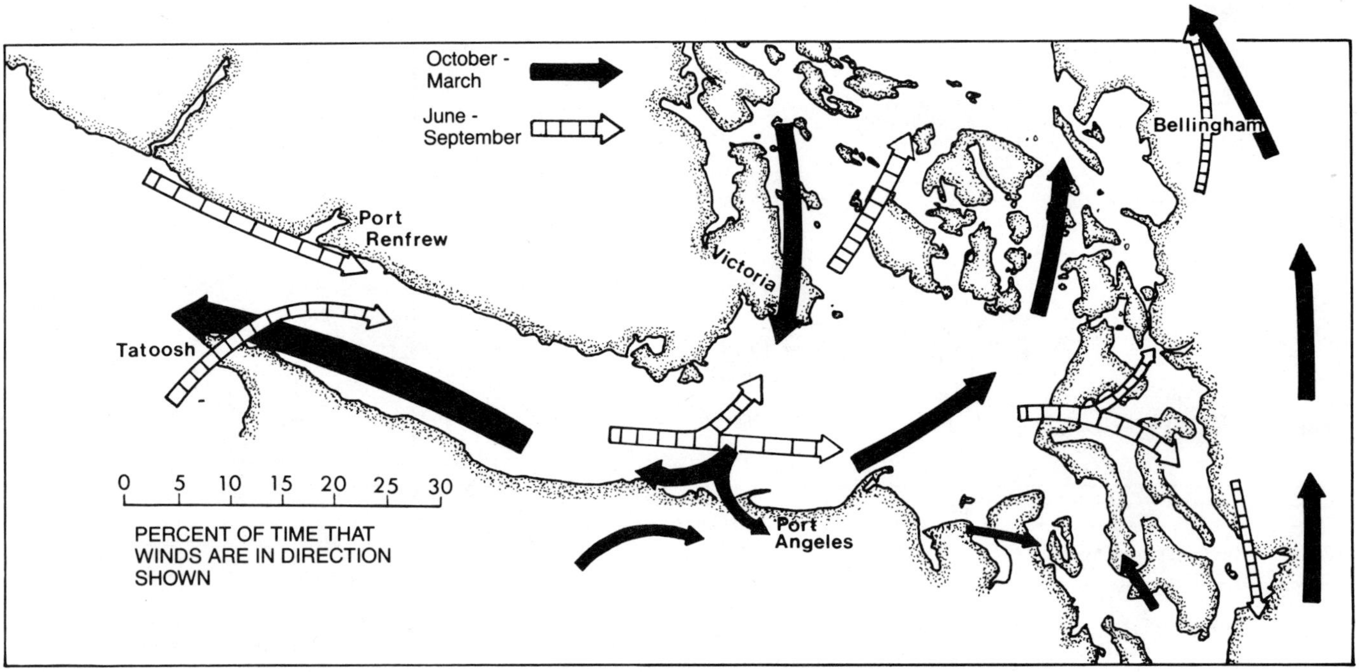

FIG. 11.4. Most frequent pattern of surface winds over Juan de Fuca Strait in winter (solid arrows) and summer (hatched arrows). Thick arrows correspond to wind speeds over 9 m/s (17.5 kn), medium arrows 4.5–9.0 m/s (8.7–17.5 kn), thin arrows less than 4.5 m/s. Comparison of arrow length with scale gives frequency of wind occurrence from given direction. (Compare with Fig. 10.5.) (From Barker 1974)

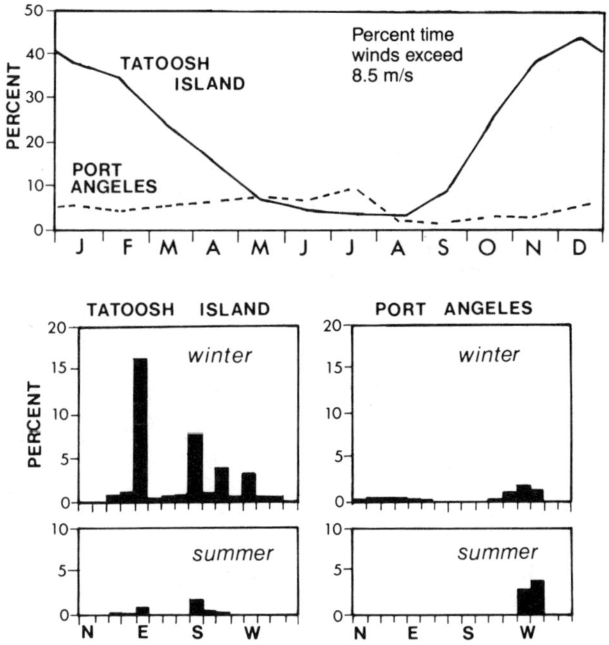

FIG. 11.5. Seasonal distribution of wind speeds greater than 8.5 m/s (16.5 kn) at Tatoosh Island (solid line) off west coast and Port Angeles (broken line). Values in percentages of observations taken 1948–58 for December–February and June–August. (From Holbrook and Halpern 1978)

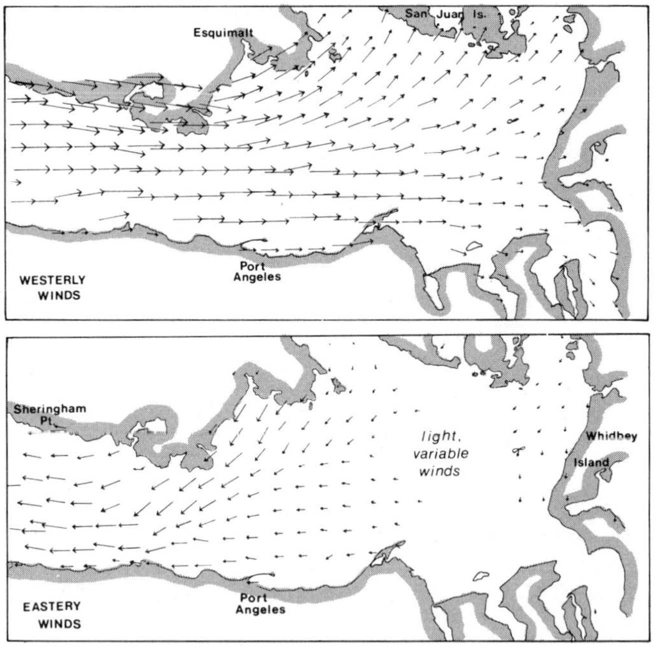

FIG. 11.6. Winds over Juan de Fuca Strait when offshore winds blow from west (*top*) and from east (*bottom*). Wind speeds scaled by relative length of arrows. (From Overland 1978)

they are further enhanced by a strong westerly sea-breeze component. South of Race Rocks the winds branch, part of the air continues eastward into Puget Sound and the remainder blows to the northeast in the direction of the southern Strait of Georgia (Fig. 11.4). Therefore, summer winds at Victoria are predominantly from the southwest at speeds of around 7.5 m/s (15 kn). In fact, from June through August, winds are from the southwest quadrant 80–85% of the time at Victoria and 75% of the time at Port Angeles. Summer winds generally reach maximum speeds around late afternoon, following near zero speeds in the early morning. The frequency of winds from a given direction in the eastern portion of Juan de Fuca Strait is shown in Table 11.1 for Race Rocks. (Unlike many meteorological shore stations, winds measured there are closely representative of conditions over the water, as are winds measured at Tatoosh Island at the seaward entrance to the Strait. Winds recorded at Ediz Hook at Port Angeles are least representative of winds over the water.)

Two important features of the wind patterns in Juan de Fuca Strait not associated with the large-scale oceanic wind regime are the polar outbreaks in winter and the sea breeze in summer. Known locally as gap winds, polar outbreaks frequently occur under clear skies and near-freezing temperatures that accompany the build-up of a high-pressure region over the interior of Washington State. Driven seaward from high to low pressure, the dense arctic air is accelerated from weak winds at Port Angeles to strong easterly winds at the entrance to the Strait. Arctic air also surges southward at this time to produce northerly winds over the Puget Sound–Seattle area.

During summer, a moderately strong westerly sea breeze builds along the Strait as daytime heating of the land draws cooler marine air inward toward the interior of northwestern Washington and southwestern British Columbia. These winds often augment the westerly oceanic winds to cause an up-channel increase in speed, although the west-to-east difference is typically much less pronounced than the down-channel acceleration of the wind in winter.

As with most areas of rough terrain, Juan de Fuca Strait is a difficult region to predict winds. Mariners who have lost faith in "small craft warnings" issued by Canadian and American weather offices should appreciate that such warnings are based on assumptions about the relationship between air-pressure distributions over the region and corresponding surface winds. But, even if the wind-pressure assumptions were 100% reliable, the weatherman must first predict the pressures before he can predict the winds. And second guessing the rapid evolution of pressure systems along the coast is far from straightforward. It is quite common, for instance, for storm fronts approaching the coast to stall, dissipate offshore, and then move suddenly inland in an unpredictable manner. Perhaps high-speed computers combined with a better understanding of atmospheric dynamics will alleviate many of these problems in the future.

Waves

No direct measurements of waves have been made in Juan de Fuca Strait, so empirical wind–wave relationships are relied on to estimate their height distributions. For easterly winds, the maximum attainable seas will clearly be limited by the total along-the-strait fetch of about 160 km (85 nm), and to a lesser extent by the duration and strength of the wind. The same also applies to seas generated by westerlies in the Strait. In the latter case, the associated offshore winds blow parallel to the outer coast rather than parallel to the axis of the channel, whereby the

TABLE 11.1. Percentage occurrence of winds from a given direction, and mean wind speed regardless of direction, for each month at Race Rocks light station (July 1969 to January 1974).

		MONTH											
		J	F	M	A	M	J	J	A	S	O	N	D
N		24	18	14	8	3	2	2	3	9	15	22	26
NE		27	29	17	11	6	3	3	2	13	21	28	29
E		6	12	10	5	4	2	2	2	7	10	12	10
SE		4	7	7	5	7	4	5	3	8	9	8	7
S		2	3	3	3	2	3	2	3	6	5	3	3
SW		6	5	7	7	9	10	10	11	9	9	5	6
W		29	24	40	59	69	75	76	76	46	28	20	17
NW		2	2	2	2	0	1	0	0	1	2	2	2
Calm		0	0	0	0	0	0	0	0	1	1	0	0
Mean speed	(kn)	14.2	10.9	10.6	12.6	13.1	13.4	14.2	13.3	9.3	8.3	10.1	12.2
	(m/s)	7.3	5.6	5.4	6.5	6.7	6.9	7.3	6.8	4.8	4.3	5.2	6.3

seas they generate rarely progress very far inland before they encounter the shore. Moreover, strong winds along the coast are generally linked to rapidly moving frontal systems of limited duration and extent. Sustained polar outbreaks that blow seaward along the Strait for a day or so, and stationary fronts associated with intense low-pressure cells, presumably produce the largest seas.

Table 6.2 shows that fully developed seas in Juan de Fuca Strait can attain significant wave heights of 1.5 m, or most probable maximum heights of 2.7 m, for a 10 m/s (20 kn) wind that blows for at least 10 h over a fetch of 140 km (75 nm). Under these conditions, only about 10% of the waves would be somewhat higher than 3 m. In general, however, the fetch is typically less than 100 km and fully developed seas would not usually be expected to exceed heights of 2 m. Therefore, if Juan de Fuca Strait wasn't directly exposed to the Pacific Ocean, observed waves would be similar to those in the Strait of Georgia. In reality, of course, swells that propagate inland from the open waters appreciably alter the nature of the observed wave field in the Strait by increasing the overall height of the seas and causing them to steepen. Because the continued existence of swell waves does not require energy input from the wind, the longer waves can penetrate the entire length of the Strait regardless of the winds, and slowly decrease in height as they travel along the channel. Based on wave records off Tofino on the west coast of Vancouver Island, maximum probable wave heights near the Strait entrance can be expected to exceed 6 m at least 10% of the time in winter and 3 m about 10% of the time in summer. The period of this swell ranges from about 6 s to over 20 s, but most commonly lies between 9 and 10 s. Due to dispersion, refraction, and dissipation, the height of the swell will continually diminish inward along the Strait. Even exceptionally large, deep-ocean swell will be reduced to a low ground swell by the time it reaches the eastern portion of the channel.

Rips occur off prominent points and over banks, and are especially heavy off Cape Flattery, Race Rocks, New Dungeness, and Point Wilson at the entrance to Admiralty Inlet. There are dangerous rips at the approaches to Haro Strait, Rosario Strait, and in the area from Trial Island to Discovery Island. River discharge from the San Juan River can produce rips in the vicinity of Port San Juan most of the year.

Tides

The oceanic tide that travels northward along the west coast of North America enters Juan de Fuca Strait as a long, progressive wave whose speed and range vary eastward as a result of changes in the depth and geometry of the channel. For practical purposes such as tidal predictions, hydrographers decompose this wave into various constituents (see Chapter 3) whose individual characteristics can readily be determined through shore-based tide gage observations. In this section, the two main constituents, the semidiurnal wave component (M_2 tide) and the diurnal wave component (K_1 tide), will be used to describe the nature of the "real" or combined tide within the Strait.

It takes about 3.5 h for the M_2 tide to propagate from the Pacific entrance of Juan de Fuca Strait to the vicinity of the San Juan Islands, but only 1.5 h for the K_1 tide to cover the same distance. As a consequence, a particular stage of the incoming oceanic tide occurs roughly 1.5–3.5 h later at the eastern end of the channel. There is a further delay of about 1 h as the tides work their way northward through the narrow passes of the San Juan and Gulf Islands into the southern Strait of Georgia.

As elsewhere on the coast, the combination of a semidiurnal tide and a diurnal tide produces mixed-type tides in the Strait. These have two high and two low waters each day with a distinct diurnal inequality in the heights and times between successive high waters and successive low waters, and a consistently greater inequality in low waters in all regions. Unlike other areas, the shift in balance between the M_2 and K_1 constituents causes a change in character of the tide as it moves inland along the channel (Fig. 11.7). Thus, from the entrance to the vicinity of Race Rocks, the tides are mixed, mainly semidiurnal because the M_2 tide outweighs the K_1 tide. From Race Rocks to the southern Strait of Georgia they become mixed, mainly diurnal as the diurnal (K_1) tide gains in importance. This change in the nature of the tide is directly related to the increasing importance of the moon's declination, which causes the K_1 tide in the first place. In fact, the declinational effect grows so strong near Victoria that the tide becomes diurnal with only one pronounced high and low water each day for about 20 days a month (see Fig. 3.5). This effect decreases northward into the Strait of Georgia and southward into Puget Sound where tides are never diurnal. The interplay between the two main components of the tide also produces a change in the sequence of high and low waters part way along the channel; west of Victoria the sequence is Higher High Water, Lower Low Water, Lower High Water, Higher Low Water; but east and north of Victoria the sequence of tides is Higher High Water, Higher Low Water, Lower High Water, Lower Low Water.

As a consequence of the tide's peculiar nature around Victoria, it is imperative not to use it as a reference when attempting to estimate the tides or tidal streams at other localities; Fulford Harbour is much more reliable. The comparison of tides in Victoria Harbour and those in Oak Bay just a few kilometres around the corner to the north is an example of the complexity. Although Higher High

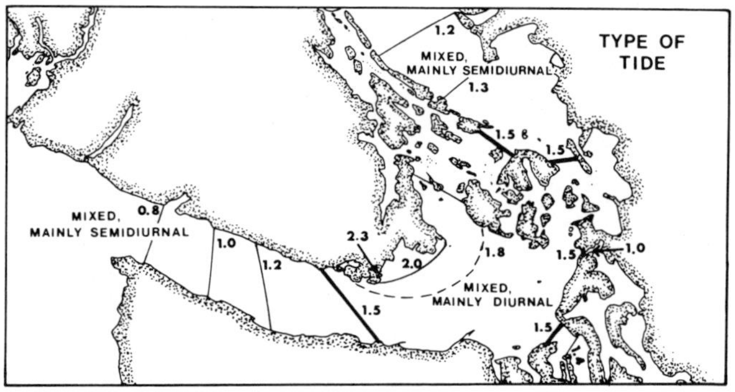

Fig. 11.7. Chart shows change in type of tide along Juan de Fuca Strait. Lines give values of ratio of diurnal tides to semidiurnal tides $(K_1 + O_1)/(M_2 + S_2)$. (See Chapter 3 for further details.) (From Parker 1977)

Water is normally about the same at these two places (2.5 m at Victoria compared to 2.6 m, at Oak Bay) the time of the tide is 58 min later at Oak Bay. Similarly, Lower Low Water is 0.7 m at both locations, but is delayed 22 min longer at Oak Bay.

The decrease in relative importance of the semidiurnal tide along Juan de Fuca Strait is shown in Fig. 11.8a, where its range is plotted as a function of distance. Because the opposing up-channel increase in the range of the diurnal tide (Fig. 11.8b) is not enough to counterbalance this effect, the overall range of the tide also exhibits an eastward decrease toward Victoria. Mean tidal ranges in Fig. 11.9 diminish from 2.4 m off Cape Flattery to a minimum of 1.8 m near Victoria, then increase again to around 2.4 m at the northern end of Haro Strait. A further indication of the complexity of tides near Victoria is the fact that the semidiurnal part of the tide attains its minimum range at Pedder Bay; on the other hand, the diurnal portion of the tide attains its maximum range at this location. Pedder Bay, it seems, is located at a distance from the northern end of the Strait of Georgia where the surviving remnant of one reflected tidal wave that propagates toward the ocean (see Chapter 3, 10) has its maximum cancelling effect on the next tidal wave that propagates inward from the ocean.

The considerable effect of the rightward turning Coriolis force is clearly displayed by the tidal range which, for the K_1, M_2, and total tide, has a marked increase across the Strait from the Canadian to American sides. In the case of the M_2 tide, for instance, the range at Pedder Bay is about 0.7 m; on the directly opposite side of the Strait at Port Angeles the tidal range is 1.0 m. Moreover, there is a tendency of the tide to occur a few minutes earlier on the American side as the tide leans to the righthand side of the channel on its way toward the Strait of Georgia.

Lastly, of course, the tidal range in Juan de Fuca Strait undergoes a biweekly variation due to shifts in the declination of the moon, to changes in the alignment of the sun and moon, and to alteration of the moon's distance from earth. These, together with other periodic changes in the tidal range, were discussed in Chapter 3.

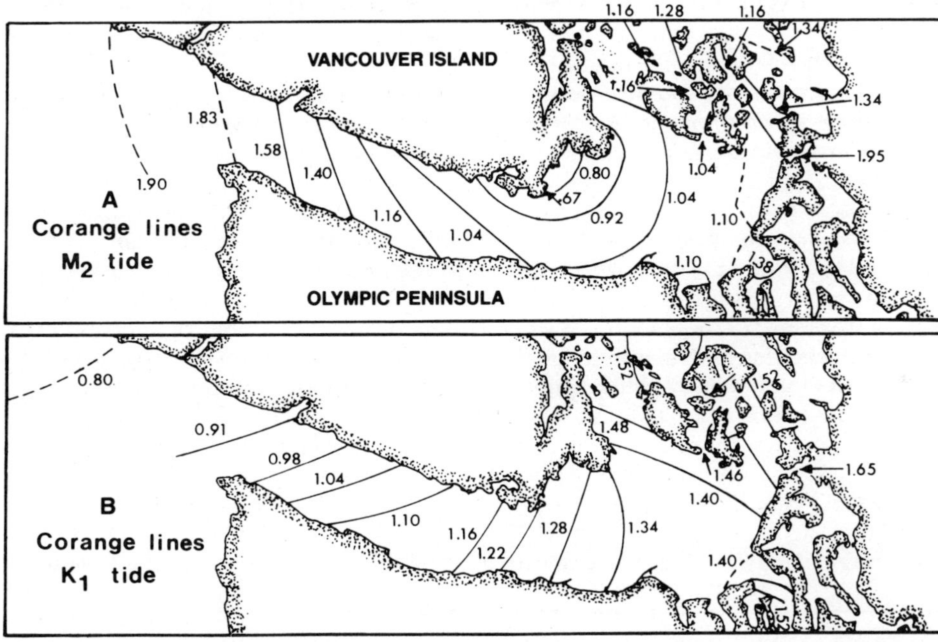

FIG. 11.8. Corange lines (m) in Juan de Fuca Strait. (A) Major semidiurnal constituent, (B) major diurnal constituent. (From Parker 1977)

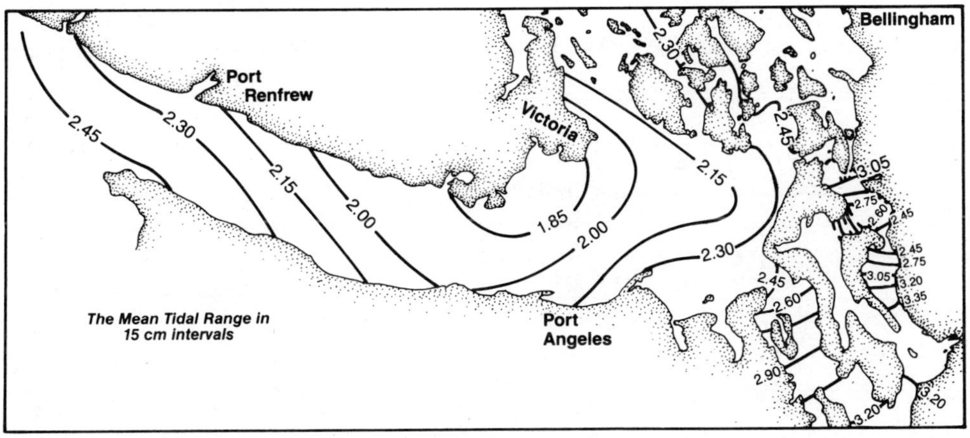

FIG. 11.9. Mean tidal range in Juan de Fuca Strait (m). (From Barker 1974)

Tidal Currents

Major factors that affect currents in Juan de Fuca Strait are tides, freshwater runoff, winds, and along-channel atmospheric pressure differences. The Coriolis force, channel curvature, and bathymetry modify the currents once they become established. Although the relative importance of these factors varies over the length and breadth of the Strait, the combination of a uniform geometry, a fairly regular shoreline, and the lack of direct river discharge tends to produce a comparatively straightforward flow structure. This is certainly true relative to the complex current regimes in the Strait of Georgia. However, Juan de Fuca Strait is not without its own idiosyncrasies. The oceanography of the western sector of the Strait, for example, is influenced strongly by wind and current events along the outer coast; in the eastern sector, "normal" flow patterns are regularly disrupted by strong tidal streams that emanate from connecting channels that open into the main basin. Therefore, what follows is at best a generalization of the flow structure in the Strait, beginning with the tidal streams generated by the oceanic tide.

The tide within Juan de Fuca Strait has the nature of a progressive-type wave that has been measurably modified by the presence of a standing wave component. For a pure progressive tide, flood streams at a particular location are strongest near the time of local high water and ebb streams strongest near the time of local low water. The greater the contribution from a standing wave component, the longer the time of maximum flow lags behind the time of extreme high (or low) water. Therefore, because of the along-channel variation in the importance of the standing wave, the correspondence between the tide height and the tidal streams varies with distance along Juan de Fuca Strait. Near the Pacific entrance, the appreciable standing wave component causes the current (for instance, maximum flood) to lag behind the tide (local high water) by 1–1½ h. In the area between Race Rocks and Angeles Point, however, there is only a small time lag between the tide and the tidal streams, because the tidal wave within the eastern part of the Strait has the nature of a nearly pure progressive wave. East of this region, the roles reverse and the tide lags behind the current as the tidal propagation again begins to depart from that of a progressive wave. This lag increases steadily as far as the Strait of Georgia, where maximum flood streams always precede high water by about 3 h and maximum ebb streams always precede low water by 3 h. The natural sequence of currents in Juan de Fuca Strait always follows the same general pattern of stronger maximum flood, weaker maximum ebb, weaker maximum flood, and stronger maximum ebb (Fig. 11.10).

Flood streams are northward along the Washington coast and turn into Juan de Fuca Strait north of Cape Flattery (Fig. 11.11). In this area, they initially set toward the Vancouver Island shore but further inland they are directed down-channel by the confining geometry of the basin. By maximum flood, the water moves parallel to the axis of the Strait at all locations at speeds of 75–130 cm/s (1.5–2.5 kn) on large spring tides;speeds of 180 cm/s (3.5 kn) can occur on large tides in the eastern portion of the Strait near the approaches to the major channels. Flood currents are generally a little stronger on the U.S. side of the western channel, due in part to the right-turning Coriolis effect. This difference is small, however, and the mariner should not expect to gain any advantage from it. Only when the currents are averaged over many days or weeks, does the influence of the Coriolis force become apparent.

In the vicinity of Race Rocks and Discovery Island off Victoria, flood currents can at times be accelerated to speeds of around 250 cm/s (5 kn) as they are squeezed through narrow channels. Similar conditions prevail at the entrance to Admiralty Inlet and Deception Pass. As in any passage of this type, dangerous rips can form if the currents oppose the propagation of wind waves or ship waves into the area. At Race Rocks, rips will form on the flood if winds are from the northeast or southeast quadrants, a common situation during winter when winds are from these directions 65–80% of the time. Similarly, rips will occur in the vicinity of Discovery and Trial islands when the winds are from the northeast quadrant on the

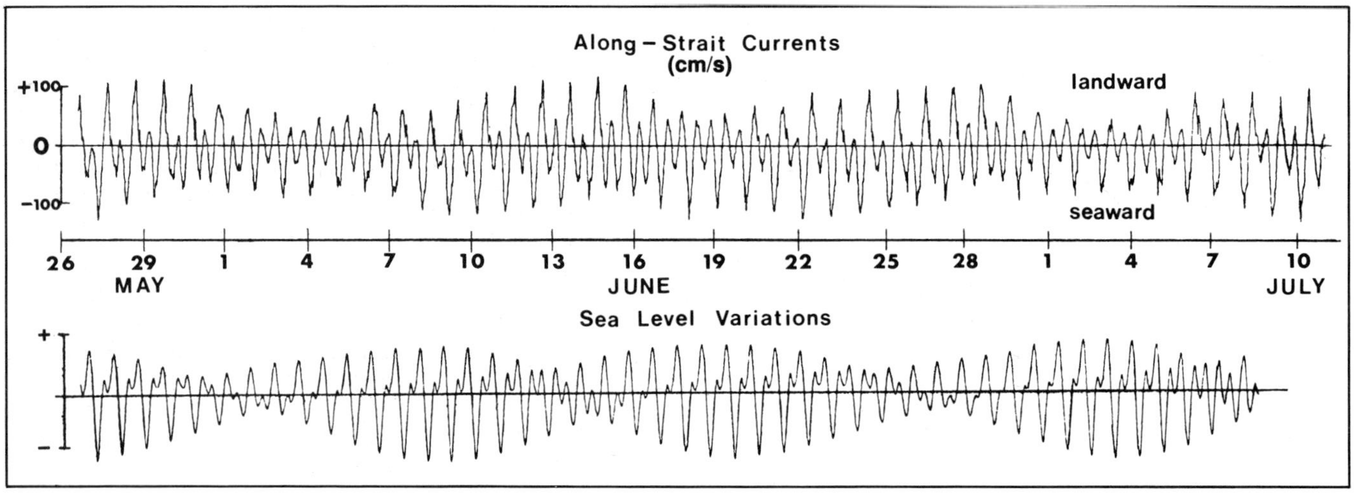

FIG. 11.10. Comparison of ebb–flood sequence and tide height variation in Juan de Fuca Strait. Time in days increases from left to right; currents, cm/s (100 cm/s ~ 2 kn). (Adapted from Fissel and Huggett 1976)

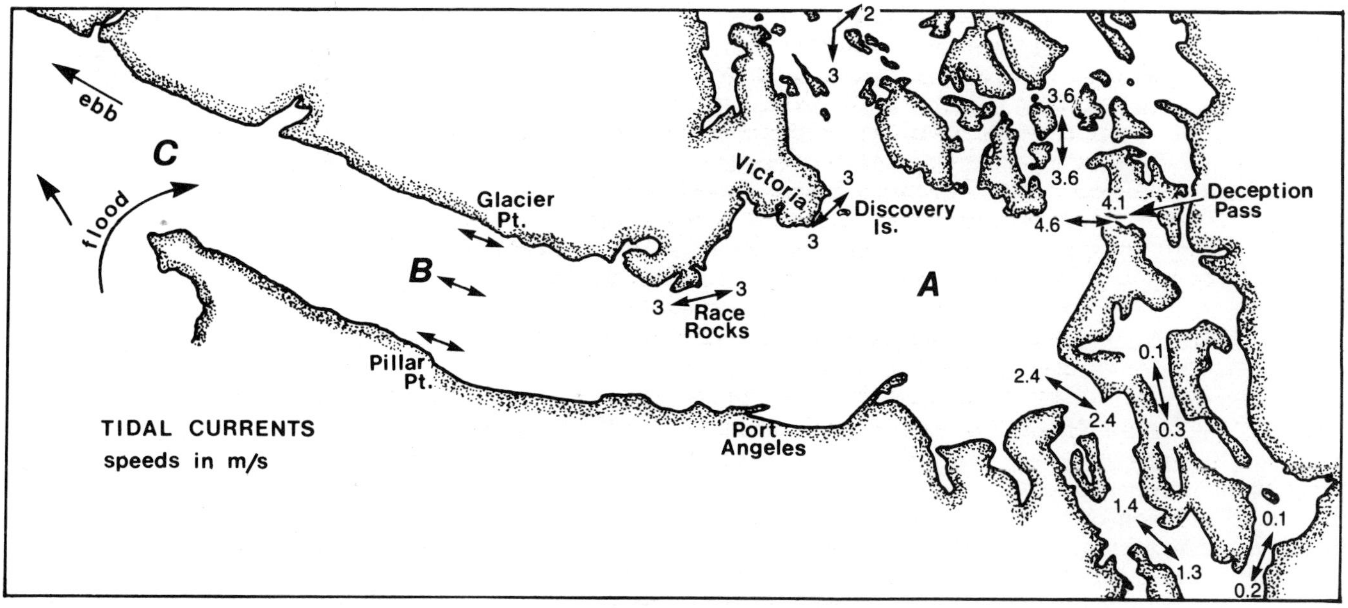

FIG. 11.11. Maximum tidal streams in Juan de Fuca Strait and adjacent passages (1 m/s ~ 2 kn). (A) Maximum ebbs 1.8 m/s, floods 1.5 m/s; (B) maximum streams 1.3 m/s; (C) 0.75 m/s.

flood. In practice, rips will occur anywhere in the Strait where tidal currents are accelerated, such as over banks, shoals, and off prominent points. Because of steep waves, such rips can be dangerous to small boats.

Within the wider eastern section of the Strait, the major portion of the flood sets to the northwest, following the trend of the deeper channels that lead into the Strait of Georgia. Haro Strait accommodates the largest volume of water that moves northward into the Strait of Georgia, followed in turn by Rosario Strait and Middle Channel, while the remainder moves through Admiralty Inlet into Puget Sound. Approximate calculations show that, of the water from Juan de Fuca Strait, 50% goes through Haro Strait, 20% through Rosario Strait, 5% through Middle Channel, and 25% through Admiralty Inlet into Puget Sound. Typical maximum tidal currents for the various passages within, and leading into, Juan de Fuca Strait are shown in Fig. 11.11.

Maximum ebb currents generally flow in the opposite direction to maximum flood currents. Moreover, the change from flood to ebb (and vice versa) at a particular location in the Strait is nearly along a straight line. Without exception, tidal ellipses are flattened and the currents closely rectilinear. Within the narrower portion of the Strait, ebb currents are slightly stronger and of longer duration on the Canadian side than the U.S. side. Near the coastal entrance, tidal currents on the northern side maintain their westerly direction as far as Cape Beale before they merge with the southwesterly setting ebb currents off the B.C. coast. On the Washington side, the streams quickly become part of the southerly tidal flow off Cape Flattery.

As with most channels along the coast, the strength and duration of the tidal streams in the Strait are usually distorted by estuarine-type processes. Consequently, ebb currents are noticeably stronger and of longer duration than flood currents in the top 100 m (Fig. 11.12). Below

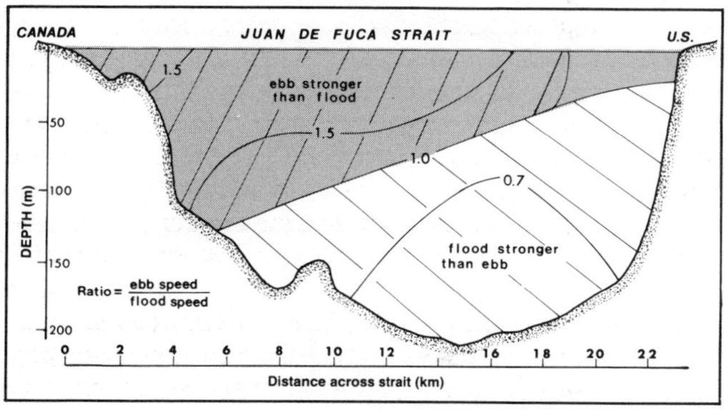

FIG. 11.12. Average ratio of the ebb speed to the flood speed in cross-section from Pillar Point (USA) to Port Renfrew (Canada).

this depth on the other hand, flood currents are normally stronger and of longer duration than ebb currents. This bias toward the ebb in the upper layer is because river runoff into the Strait of Georgia and Puget Sound slowly works its way to the Pacific via Juan de Fuca Strait. By mixing with oceanic water in the passes, the runoff forms a comparatively light surface layer whose general drift seaward somewhat retards the surface flood streams and stengthens the surface ebb streams. Floods are stronger than ebbs below this top layer as oceanic water must drift inward at depth to replace the salt water entrained and carried to the Pacific Ocean within the surface layer.

The "coarse grid" computer model outlined in the previous Chapter gives an overall picture of the depth-averaged tidal streams throughout the main channel of Juan de Fuca Strait. These currents are shown in Fig. 10.14a, b for two stages of a semidiurnal tide. The "fine grid" model provides more detailed information on the flow in the vicinity of the Gulf and San Juan islands. Complete current vector plots for the fine grid are in Fig. 10.15a–c for three stages of the tide.

It is important to remember that these charts do not take into account the presence of the weaker estuarine circulation just discussed. Also, for modeling purposes at the seaward entrance to the Strait, it is necessary to assume that the tidal streams flow parallel to mid-channel, which in reality is not strictly true. A short distance up-channel, the modeled flow should closely resemble actual tidal streams.

A number of especially salient features of the currents are revealed by the computer simulation:

1) During the larger of the two daily floods within Juan de Fuca Strait, a well-defined, counterclockwise backeddy develops to the east of the Race Rocks shoal region. For this to occur, maximum flood streams must attain speeds of around 100 cm/s or more in mid-Strait. Although the details of this flow near Race Rocks are severely limited by the 2-km resolution of the model, it would appear an intense countercurrent is set up, which flows *westward* in the vicinity of Race Passage toward the end of the flood. Moreover, the backeddy continues to expand, while maintaining its strength, until the ensuing ebb. At the time of high-water slack in most of the Strait, the backeddy becomes a strong cross-channel jet that extends a few kilometres from Race Rocks; this disappears once the ebb has been established.

Most informed yachtsmen who have sailed in the Victoria region will attest to the large downstream backeddy on the flood. The existence of a strong counterflow through Race Passage, on the other hand, remains uncertain, although it may account for the apparently anomalous currents Swiftsure racers often encounter in this region.

2) Contrary to the above situation, no backeddy develops east of Race Rocks on the weaker of the two daily floods with maximum flood speeds of around 50 cm/s or less in mid-channel.

3) A strong counterclockwise rotating backeddy develops downstream of Discovery Island on the larger floods (Fig. 11.13); a less intense clockwise eddy forms behind Dungeness Spit on the flood.

4) The turn to ebb begins about 1 h earlier on the U.S. side of Juan de Fuca Strait and comparatively strong currents surge seaward from Admiralty Inlet at the time of high-water slack in the rest of the Strait. At this time, a weak flow extends from Haro Strait across-channel to Port Angeles, where it joins the intensifying ebb along the Washington shore.

5) Flood streams in Haro Strait are strong but confused, especially at the southern approach. Maximum flood speeds are along the western side of San Juan Island and there is a sharp fall-off in speed toward mid-channel; on the Cordova Bay side of the Strait, currents are relatively weak and irregular. Observations in Haro Strait verify the presence of much stronger flood streams along the San Juan Island side of the channel.

6) The flood streams in Rosario Strait split around Cypress Island with nearly equal speeds on either side, but tend to reunite south of Sinclair Island. Tidal streams within Bellingham, Samish, and Padilla bays are always weak despite the large tidal range.

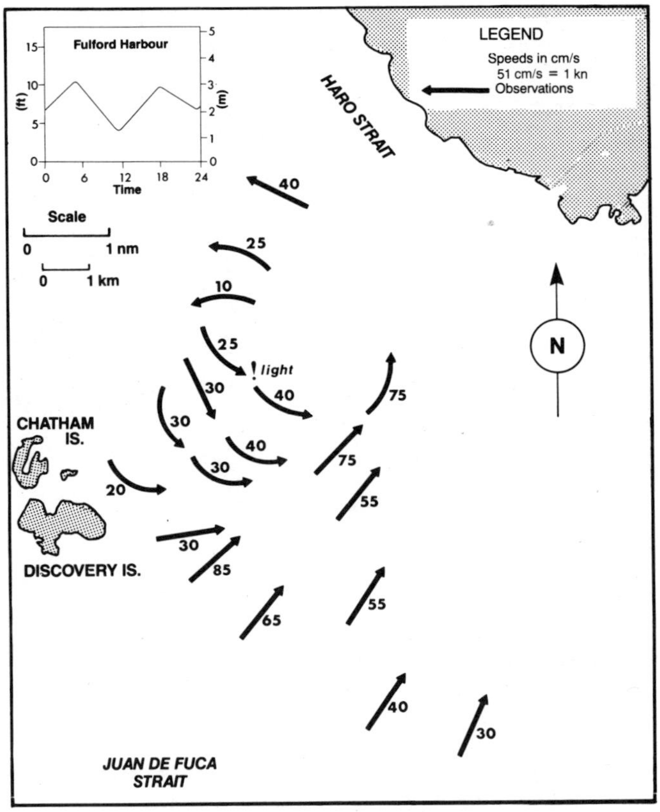

FIG. 11.13. Observed surface flood currents at southern approach Haro Strait. Speeds based on aircraft tracked surface drifters. (Courtesy S. Huggett and P. Crean)

7) After the flood stream rounds the corner at East Point in Haro Strait, it divides, with strong flows in both President Channel and Boundary Passage. These flows then reunite at the Strait of Georgia end of Boundary Passage to produce a jetlike flow that veers to the north in the Strait. At this time, a large counterclockwise backeddy is established north of Saturna and Tumbo islands. As with backeddies in Juan de Fuca Strait, this eddy expands throughout the flood and persists until the ebb has begun. Maximum speeds are off the north shore of those islands and directed to the east.

8) During the ebb, tidal streams in Juan de Fuca Strait are directed seaward everywhere along the channel and no large backeddies are formed, according to the model results. However, there would appear to be a rather weak clockwise backeddy in the lee of Discovery Island.

9) During moderate ebb conditions, there is apparently a weak nearshore countercurrent along the western side of San Juan Island until about maximum ebb. Midway through the ebb this current reverses and joins the normal seaward flow through Haro Strait.

Observed Currents

The computer model of the time-varying currents in Juan de Fuca Strait considers only the contribution from tidal streams generated by tidal forces of the moon and sun. Consequently, actual currents may deviate somewhat from the model results because of the smaller, but far from negligible, contributions by the estuarine current regime, density currents, and wind currents. Eventually, these effects, too, will be included in future models, but until then only direct measurements give a full understanding of the strength and variability of the total flow.

Direct measurements of currents in Juan de Fuca Strait over the last decade by Canadian and American scientists reveal the flow is essentially parallel to the trend of the channel at all stages of the tide. Slight deviations from these directions are possibly due to remnant intrusive motions originating with strong currents from the major tidal passages or to the deflecting influence of points of land, for instance off the coast between Slip point and Pillar Point on the Washington side and east of Otter Point on the British Columbia side (Fig. 11.14).

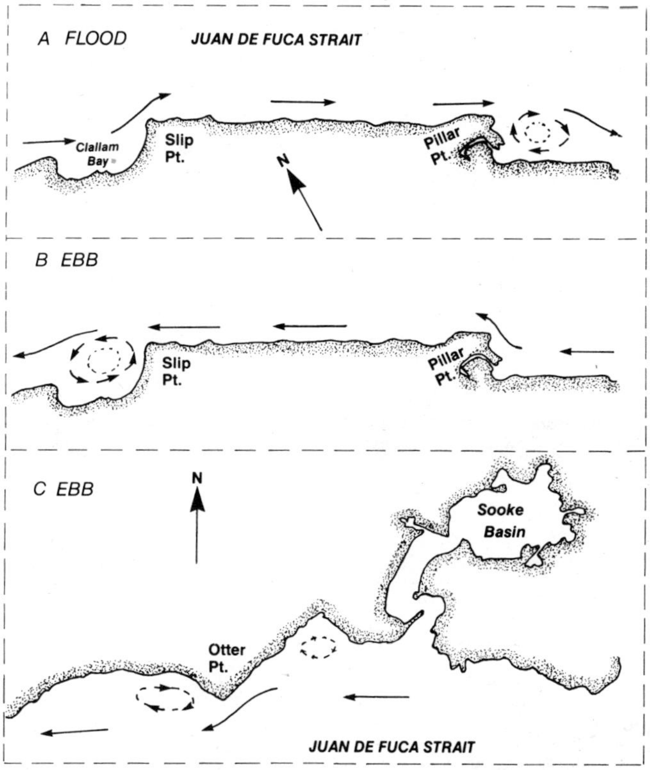

FIG. 11.14. Formation of backeddies in Juan de Fuca Strait. (A) flood, (B) ebb, along southern shore, central Strait, (C) ebb along northern shore, eastern Strait.

Within the broad eastern portion of the Strait, currents are typically east–west, but swing more to a north–south orientation near the approaches to the major channels that lead to the Strait. Due to the confined nature of Juan de Fuca Strait, moreover, the currents tend to be rectilinear (see Chapter 3), whereby the direction of the flood and ebb are diametrically opposite with very little rotary turning. Nevertheless, in mid-channel and off the Pacific entrance, the current vectors have a tendency to rotate clockwise during a tidal cycle to form a somewhat flattened ellipse (Fig. 11.15). The rotary behavior is slightly more developed in the broader, eastern portion of the Strait.

The estuarine circulation (a net seaward flow of relatively low-density water in the upper 100 m of the Strait with a compensating inward flow of more dense water at depth) becomes apparent when the tidal streams are subtracted from the current meter records. Near the surface, this residual flow has typical seaward speeds of 10–20 cm/s (0.2–0.4 kn), becomes as great as 40 cm/s (0.8 kn) in early summer, with maximum values concentrated close to mid-channel in the western portion of the Strait. East-setting residual flow within the lower layer is around 10 cm/s and strongest away from mid-channel (Fig. 11.12). The across-strait tilt of the line separates net seaward flow in the upper layer from net landward flow in the lower layer (Fig. 11.12) and is due to the combined effects of the Coriolis force and channel curvature. In this case, the outflow favors the Canadian side, whereas the inflow favors the American side.

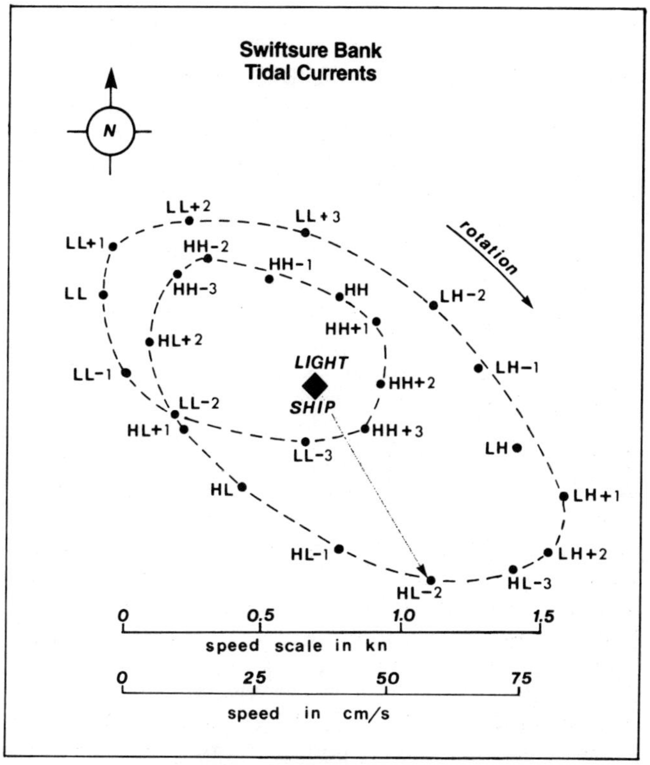

FIG. 11.15. Diagram to determine direction and speed of rotary tidal currents over Swiftsure Bank near Tropic tides (maximum lunar declination). To determine tidal current for particular time look up state of tide at Tofino, B.C., then draw arrow from Light Ship to appropriate time position on dashed curve (e.g. HH–1 means time of Higher High Water less 1 h; LH + 2 Lower High Water plus 2 h, etc.) Orientation of arrow gives direction of current flow, length of arrow its speed as determined from scales. Example shows flow to southeast at 42.6 cm/s at 2 h after Higher Low Water. (Adapted from Marmer 1926)

If the residual currents in Juan de Fuca Strait were steady, or at worst variable in some predictable fashion, very accurate predictions of currents could be obtained by computer simulation models. Unfortunately, there are significant variations in this flow both with time and location due to poorly understood changes in the estuarine processes that produce it. Thus, even though the residual current in the surface layer of the Strait moves seaward on average, it is sometimes observed to "stall" for several days at a time and on occasion reverse direction completely. On a seasonal basis, reversals in the near-surface residual current are least likely in summer, when the driving mechanism for the flow (freshwater discharge into the inshore waters) is at its maximal strength. Prevailing northwest winds at this time also tend to favor a strong, persistent seaward flow. Within the Strait of Georgia, these winds help move the brackish Fraser River runoff into Juan de Fuca Strait, whereas off the coast they drive oceanic surface waters away from the mouth of the Strait, lower sea level, and increase the east–west hydraulic head along the channel.

During fall and winter, the above processes are modified considerably and landward residual currents frequently occur in the near-surface waters with an accompanying seaward residual flow at depth. River discharge is minimal at this time and southerly winds prevail over the coast. Not only do these winds oppose the movement of brackish water into the eastern end of Juan de Fuca Strait but they also tend to raise sea level at the seaward end, which counters the normal east–west hydraulic head associated with the outflow of fresh water. Recent observations indicate that time-averaged, near-surface currents may flow up-channel in winter for 3–10 days, extend as far as Race Rocks 90 km inland, and attain speeds in excess of 50 cm/s (1 kn) for a day or so (Fig. 11.16A). At such times, ebb currents would be weak whereas flood currents would be anomalously strong. Present indications are that such eastward directed residual flows are directly linked to slow (25 km/day) eastward propagation in the Strait of a long internal bore. Generation of this wave is, in turn, apparently linked to southerly storm winds along the outer Washington coast. Figure 11.16B provides further evidence for reversals in surface flow in the Strait during offshore southwesterlies. In this case, warmer oceanic surface water was driven over 100 km along the southern half of the channel before being carried seaward following break-down of the offshore wind system. One thing is certain, no one should expect currents in Juan de Fuca Strait to behave at all times as predicted in the tide tables.

To close this section, a few additional remarks can be made regarding currents in Juan de Fuca Strait.

1) Current predictions for Race Rocks in *Canadian Tide and Current Tables*, Vol. 5 are based on observations taken about 4.5 km south of Race Rocks and not on measurements within Race Passage. Furthermore, the predictions attempt to take into account the residual flow contribution to the ebb and flood speeds, which may possess considerable day to day variability.

2) The times of the turns and maximum tidal streams at locations between Jordan River and Pillar Point on the U.S. side occur 30 min earlier, on the average, than those south of Race Rocks. Variations from this average are at most 30 min but occur irregularly from one tide to the next. Thus, for example, maximum tidal currents half way between Pillar Point and Jordan River may occur between zero and 60 min earlier than at Race Rocks.

3) Current predictions for Juan de Fuca Strait in *Canadian Tide and Current Tables* are based on measurements made in mid-channel near the seaward end early this century and, therefore, should be used with cautious optimism about their accuracy. Variations in the strength and direction of the surface residual flow, together with variable ocean current regimes off the coast, can lead to marked departures of the predictions from actual currents. A comparison of the Race Rocks and Juan de Fuca Strait predictions indicates that times of the turns and maximum tidal streams at the Strait entrance lead those at the eastern

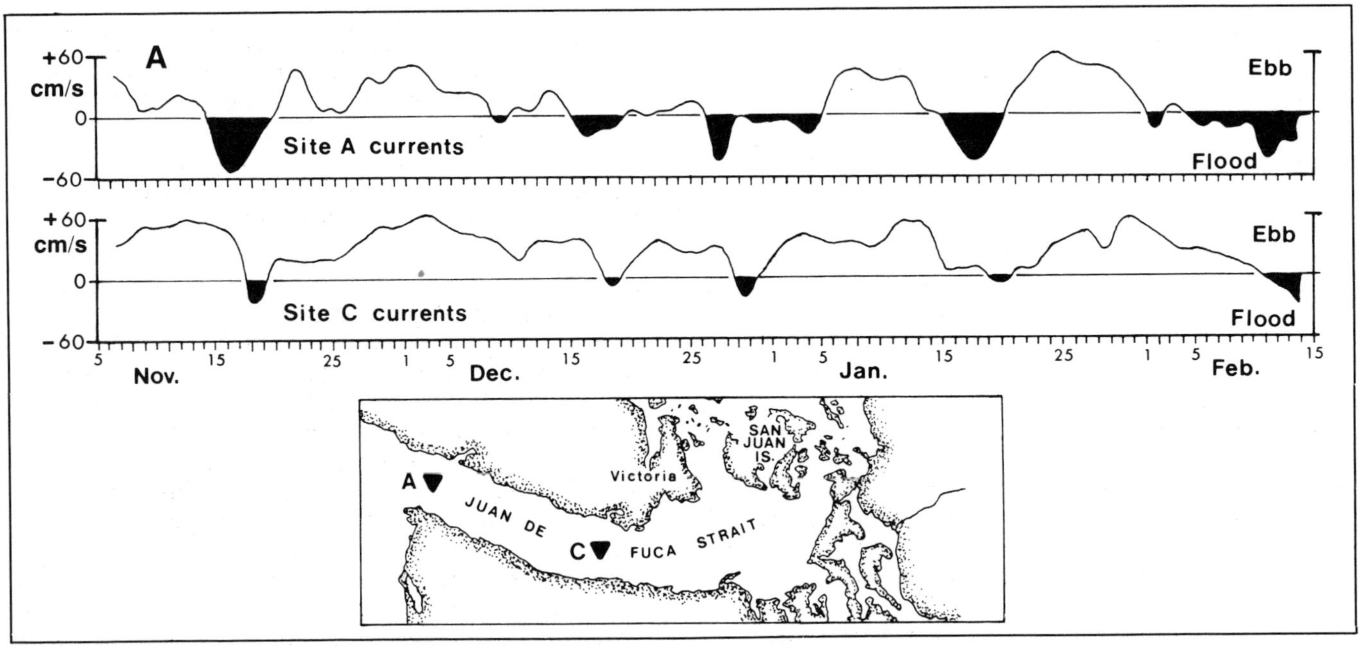

FIG. 11.16. (A) Variation in speed and direction of nontidal, along-channel surface currents in Juan de Fuca Strait. Currents (cm/s) measured at depths of 4 m at A and C from November 1976 to February 1977. Up-channel (flood direction) reversals in mean flow at seaward approach are often few days later at Site C, 60 km inland. (Adapted from Holbrook and Halpern 1978).

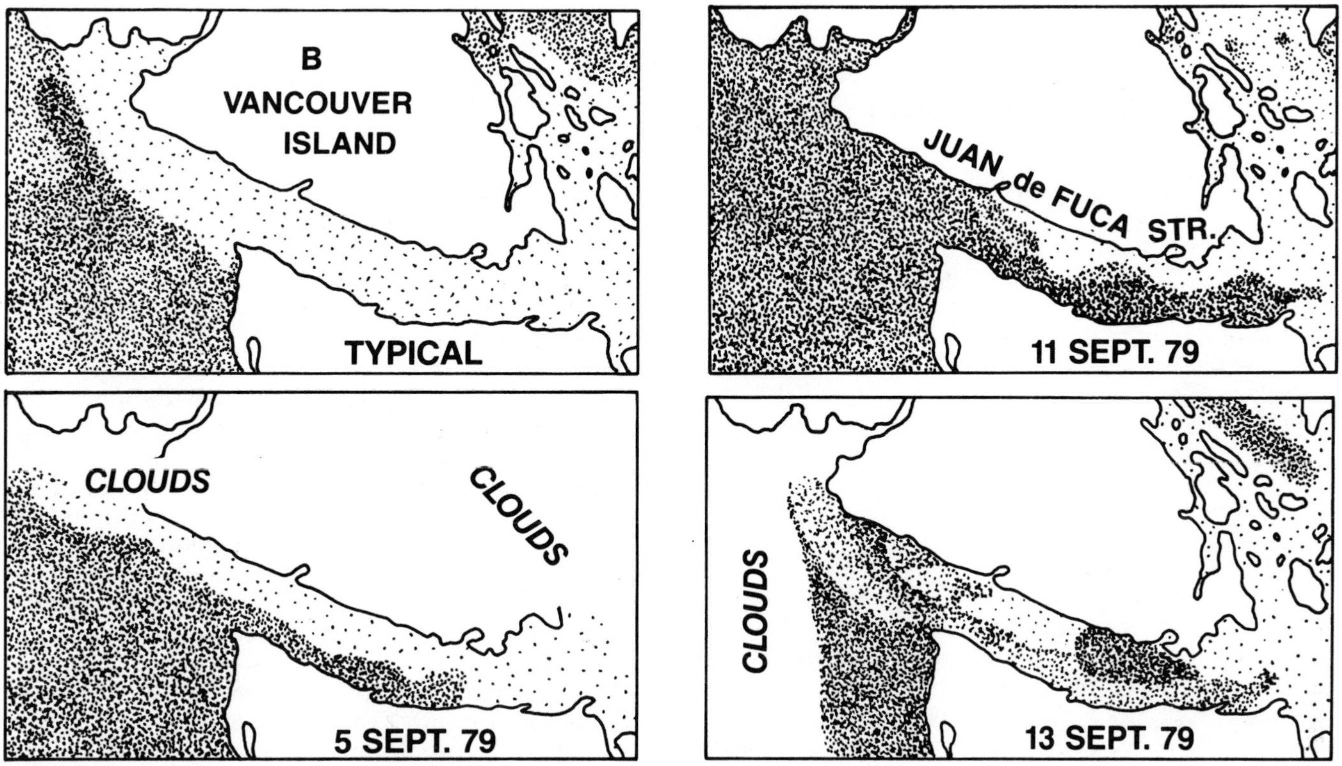

FIG. 11.16. (B) Drawings based on infrared satellite images of sea-surface temperatures, showing sequence of warm-water intrusion (heavy stippling) into Juan de Fuca Strait from the Pacific Ocean in September 1979. Offshore coastal winds were from the southwest and associated with a low-pressure system off the coast. Intrusion was confined mainly to the southern half of the Strait and reached a maximum of 135 km from the entrance. Four days after cessation of the southwest winds, the seaward estuarine circulation was reestablished and the intrusion began to be advected out of the Strait. (Courtesy, J. Holbrook)

end of the channel by 1–1½ h, on the average. The major exception is at the turn of the weaker of the two daily floods that are predicted almost simultaneously along the Strait.

4) Tidal currents over banks and shoals away from the shoreline will generally be stronger than those in adjacent deeper waters. However, due to frictional effects, nearshore tidal streams will be weaker than those in mid-channel and turn to ebb or flood ½ h or so earlier.

5) Nearshore currents downstream of a point of land will form backeddies if there is an accompanying increase in the width of the channel; this occurs to the east of Pillar Point and Race Rocks on larger floods. Nearshore currents will be deflected toward mid-channel as they approach a sudden decrease in channel width as when the ebb approaches Pillar Point from the east. Both deflections and backeddies will be associated with symmetric points like Otter Point near Sooke (Fig. 11.14).

6) Tidal currents in the vicinity of Swiftsure Bank seaward of the entrance to the Strait are strongly rotary, change direction in the clockwise sense about every 12½ h, but only slightly alter their speed as they change direction (Fig. 11.15). Maximum floods are southeast at about 75 cm/s (1.5 kn) and maximum ebbs northwest at around the same rate.

7) Nontidal currents over Swiftsure Bank are primarily the result of coastal wind systems and are predominantly northward in winter and southward in summer. The surface wind currents attain maximum speeds of roughly 100 cm/s (2 kn) with minimum speeds of 25 cm/s. These currents join with the tidal currents to produce a complicated anticlockwise circulation off the entrance to the Strait with westward currents to the north, eastward currents to the south, and a northerly cross-channel current setting toward Vancouver Island from the vicinity of Cape Flattery.

Sooke Inlet

Sooke Inlet is the second of two major bodies of water that open into Juan de Fuca Strait on the Vancouver Island side. Named after the Soke Tribe, which was all but wiped out by a confederation of other local indian tribes around 1848, the inlet actually consists of three main sections (Fig. 11.17): Sooke Basin, the deepest portion with a length of approximately 3 km, a mean depth of 17 m, and greatest depths of about 30 m near its seaward end; Sooke Harbour, basically a 3-km long sill between Billings and Whiffin spits, with an average depth of only 3 m; and Sooke Inlet proper, a short channel that connects the area to the Strait.

The mean tidal range within the inlet is around 2.0 m but reaches 3.2 m on large tides. Freshwater inflow into the area is mostly from the Sooke River which, like other rivers on Vancouver Island, attains maximum discharge in winter and has little runoff in summer. (It is during this

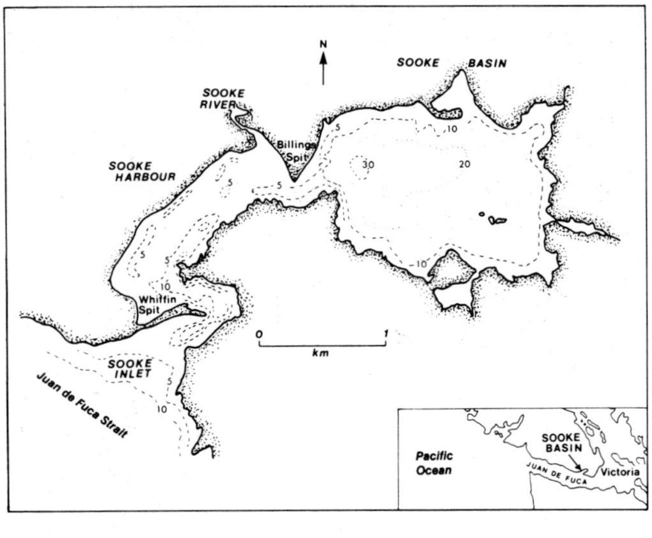

FIG. 11.17. Map of Sooke Basin and adjoining waters. Depths in metres.

low runoff period that the "Sooke Potholes" in Sooke Hills watershed offer some of the most pleasant swimming for residents of the area.)

Water properties throughout Sooke Inlet are largely determined by the combined influence of tidal exchange with Juan de Fuca Strait and freshwater input via the Sooke River. During the winter, salinities are lowest due to river discharge, typically around 20‰ at the suface with a gradual increase to values of 31‰ near the bottom in Sooke Basin as a result of saltwater penetration from the Strait. Water temperatures at this time are comparable to those in the Strait, a chilly 7–10°C, throughout the entire depth of the inlet (Fig. 11.18a). From March onward the water properties begin to change as the river outflow diminishes. This change is marked by a tendency to uniformly high salinities (31–31‰) throughout the entire inlet in summer and an increase in temperatures as solar heat is retained in the upper layer. By August (Fig. 11.18b), the upper few metres of the eastward end of Sooke Basin may be as warm as 20°C depending on weather and wind conditions. Due to tidal mixing, such comfortably warm water will be limited to the Basin; along Sooke Harbour and Sooke Inlet proper there is a marked decrease in temperature toward the ever-cold waters of Juan de Fuca Strait.

The few current observations available for Sooke Inlet show that maximum ebb–flood streams occur off Whiffin and Billings spits in Sooke Harbour, with appreciably lower speeds in other portions of the inlet. Maximum streams are set up roughly midway through a tidal cycle and slack waters are near the times of low and high tide at Sooke. Flood streams along the axis of the inlet during a rise to large high tide can attain speeds of 100 cm/s (2 kn) off Whiffin Spit and 150 cm/s (3 kn) off Billings Spit. Ebb streams during a fall to large low tide, on the other hand, are somewhat weaker because of frictional effects, and attain maximum speeds of approximately 50 cm/s off both prominences. During mean tides, the corresponding speeds for both the ebb and flood are reduced by about a factor of 3. Within Sooke Harbour away from the approaches to the spits, and at the seaward entrance to the inlet, maximum tidal streams are typically less than 15 cm/s, and rise at most to 50 cm/s (1 kn) during large tides.

Tidal streams within Sooke Basin are always weak (less than 20 cm/s; 0.4 kn) and appear to form a clockwise gyre centered over the deeper western portion (Fig. 11.19) due to the fact that the flood keeps mainly to the north side of the basin while the ebb prefers the south side. The confined nature of the weak circulation makes the basin a favorable oyster cultivation region, and accounts for about 25% of British Columbia's commercial oyster production.

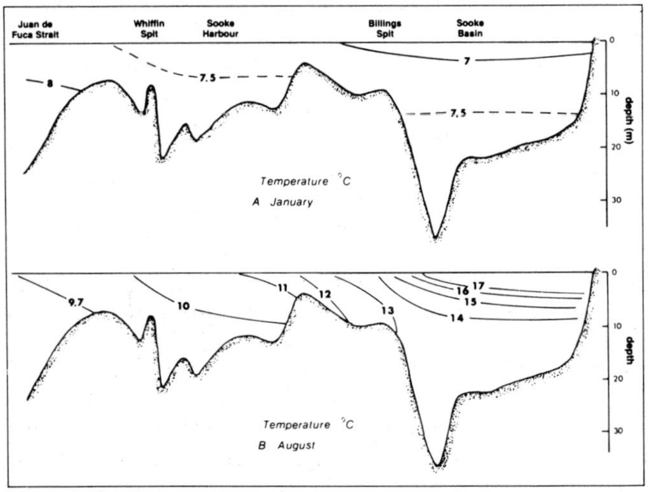

FIG. 11.18. Temperature distribution along mid-channel from Juan de Fuca Strait to head of Sooke Basin (A) January 1966, (B) August 1966. (From Elliott 1969)

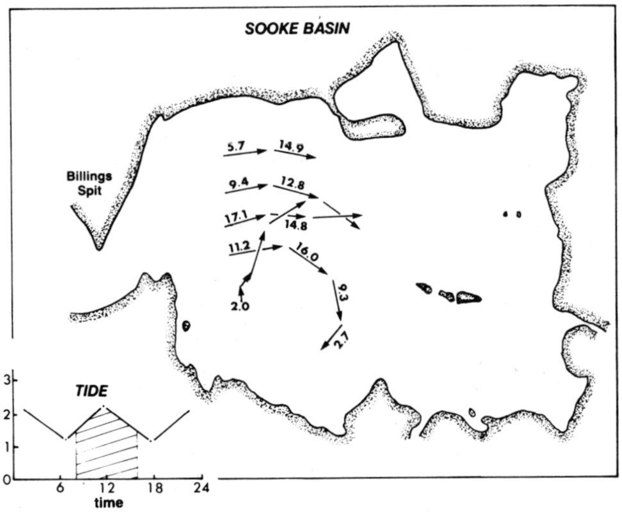

FIG. 11.19. Trajectories of drogues in western sector of Sooke Basin at 2 m, with estimated currents (cm/s). Hatched area lower left shows tide height during time drogues were tracked. (From Elliott 1969)

Chapter 12. Johnstone Strait Region

The complex system of waterways that extends seaward from the northern end of the Strait of Georgia provides access to some of the most spectacular scenery on the British Columbia coast. Snow-capped mountain peaks, cascading waterfalls, steep-sided channels, and surging tidal narrows contribute to a stark ruggedness (Fig. 12.1). Settlements are few and far between, but wildlife is abundant, and it is not uncommon to see a pod of killer whales or a school of porpoises moving easily through the chilly waters. The large numbers of killer whales that frequently gather in Robson Bight, at the western end of Johnstone Strait, create an impressive sight, unequalled in few areas of the coast.

FIG. 12.1. Aerial view of Johnstone Strait near Kelsey Bay, May 1977. View toward mainland side of channel. (Photo by author.)

Background

Formation of the narrow channels along the northeast coast of Vancouver Island dates back to the ice ages, when glaciers carved their way into the arc of igneous rocks that extends across the northern end of the Georgia Depression. Figure 12.2 shows that Johnstone Strait and Discovery Passage presently constitute the main route through the maze of islands that characterize this portion of the coastline. To the northwest of Johnstone Strait, the seaway broadens into Queen Charlotte Strait, a relatively shallow basin that marks the beginning of the coastal lowlands of the Hecate Depression, and then continues through a group of shoal-infested channels into Queen Charlotte Sound.

Queen Charlotte Strait, Johnstone Strait, and Discovery Passage, together with the interconnecting channels, make up a major part of the navigable "inside passage" that separates Vancouver Island from the mainland coasts of British Columbia and Washington State. Johnstone Strait accounts for nearly 20% of the length of this passage, and closely ranks with the Strait of Georgia and Juan de Fuca Strait in importance to the marine environment. Experience has demonstrated, for example, that the channels of the Johnstone Strait region form a key link in the migration route of Pacific salmon. Indicative of this importance is the fact that typically 10–20% of Fraser River sockeye salmon that return to the coast every 4 yr enter the river via Johnstone Strait rather than Juan de Fuca Strait. In certain years of the 4-yr cycle, this "diversion" of salmon through the northern passages is close to half the returning run. In 1978, for example, approximately 57% of the run chose this route rather than Juan de Fuca Strait, and in 1980 there was a record diversion of 70%.

Because the three main channels carry most of the water that flows between the Pacific Ocean and the northern end of the Strait of Georgia, they further influence the physical oceanographic structure of the inside passage and adjoining fiords. The central portion of Johnstone Strait is particularly relevant in this respect, for the simple reason that all such exchanges of water must pass between its shores. The protected nature of the channels has a distinct advantage to marine traffic. Tugboats, bulk carriers, cruise ships, freighters, and pleasure craft are common sights in these waters, and a large percentage of each type of vessel fly the U.S. flag.

Captain George Vancouver was the first European to recognize the importance of these channels to navigation and many modern place names originate with his charting of this region in the summer of 1792. Discovery Passage is named after his command vessel HMS *Discovery* (310 t, 134 crew). Johnstone Strait and Broughton Strait were named personally by Vancouver in honor of James Johnstone and William Broughton, Master and Lieutenant Commander, respectively, of his accompanying consort HMS *Chatham*. Spanish commanders Valdes and Galiano sailed through this region within a few days of Vancouver and, although most of their designated place names are not used on present day charts, their two vessels *Sutil* and *Mexicana* are remembered in the names of Sutil Channel (near Cape Mudge), and in Sutil Point and Mexicana Point on Goletas Channel, which extends seaward from Queen Charlotte Strait. The way in which Helmcken Island, just east of Kelsey Bay, received its name gives an insight into the difficulties early mariners encountered when navigating the rapid and complex currents in Johnstone Strait. Helmcken, who was a medical officer with the Hudson's Bay Company, and later speaker of the Provincial Legislature, spent part of his early career sailing the coast. "In the year 1850 I was on board the company's steamer *Beaver* going to Fort Rupert, and we were passing along Johnstone Strait against a flood tide. As this island was approached, which stands in the middle of the channel, the tide rapidly increased in strength, owing to the island in the way, till the *Beaver* had extremely hard work to make any headway, the vessel sheering about in the swirling current. I asked the Captain the name of the island near which we were struggling along. Captain

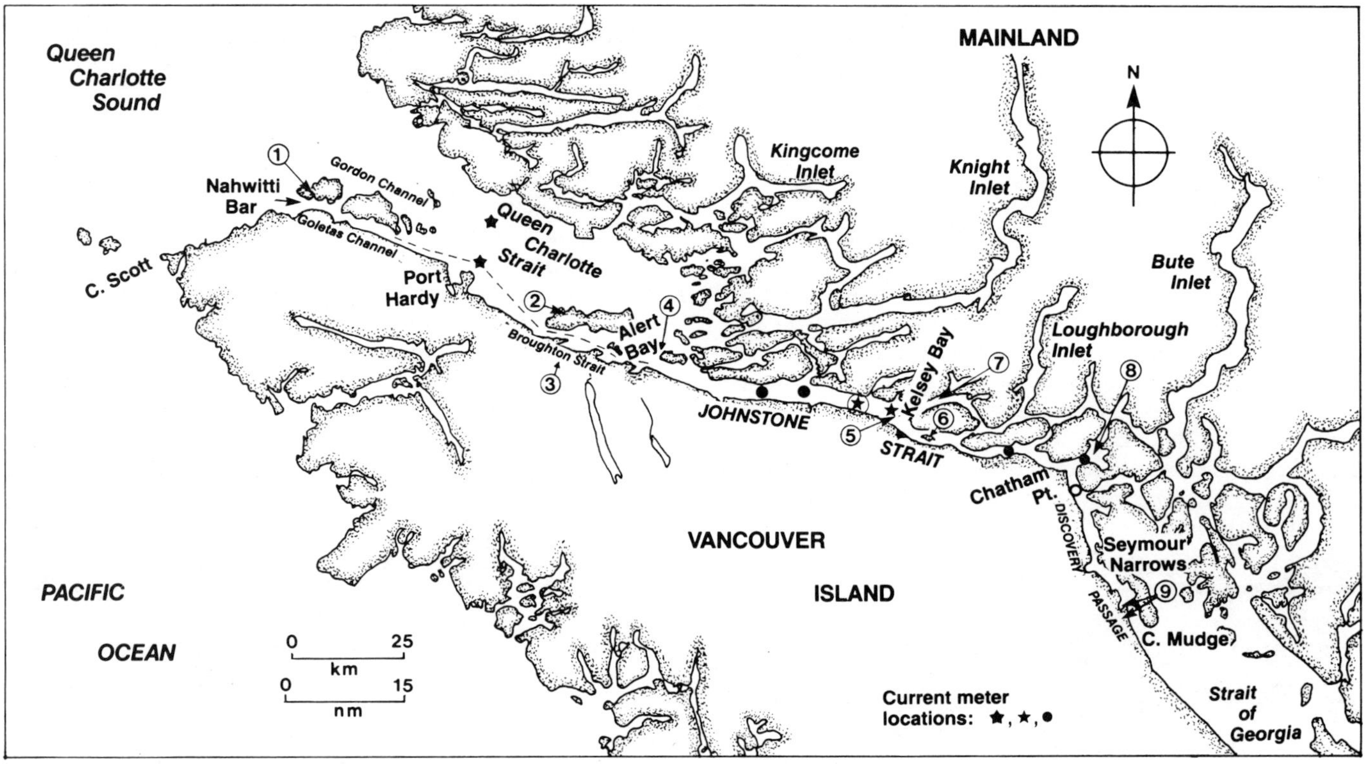

FIG. 12.2. Map of Johnstone Strait region: 1, Bull Harbour; 2, Malcolm Island; 3, Port McNeill; 4, Weynton Passage; 5, Hardwicke and Yorke Islands; 6, Helmcken Island; 7, Sunderland Channel; 8, Nodales Channel; 9, Duncan Bay and Campbell River area. Broken line in Queen Charlotte Strait gives sounding line for bottom profiles, Fig. 12.6, 12.7.

Dodd replied, 'It has no name, but I will call it after you, doctor, for it is like you, always in opposition.' The island has since been known by my name." (Walbran 1971). Dodd's comment on the current in Race Passage has undoubtedly been echoed by many skippers who have tried to stem the 3 m/s (6 kn) currents that surge through the narrows.

Bathymetry and Water Propeties

Although the three main channels of the Johnstone Strait region are part of a continuous seaway, Queen Charlotte Strait has significantly different bathymetry and oceanographic characteristics from Johnstone Strait and Discovery Passage.

Johnstone Strait—Discovery Passage

Johnstone Strait and Discovery Passage are the narrowest of the major channels that make up the navigable inside passage of coastal British Columbia. Between Alert Bay and Kelsey Bay the channel is only about 3.5–4.5 km wide, and from Kelsey Bay to Seymour Narrows its width rarely exceeds 2.5 km. These two channels also contain some of the deepest basins of the inshore waters. Figure 12.3 shows that depths within the western half of Johnstone Strait increase regularly from 70 m over the partial sill off Kelsey Bay to nearly 500 m near the entrance to Broughton Strait. By comparison, the maximum observed depth in the Strait of Georgia is 420 m and in Juan de Fuca Strait a shallow 275 m. Only in some adjoining inlets, such as Knight and Bute, are deeper basins found. Within Discovery Passage and the more constricted eastern half of Johnstone Strait, the bottom is characterized by a highly irregular profile with numerous sills and shoals. Maximum depths in this region are around 250 m.

The large depths and narrow widths are indicative of the steep-sided channels of the region, where safe overnight anchorages are somewhat at a premium. On the other hand, lure fishing can be quite productive, provided the angler takes advantage of the numerous shoals and nearshore underwater shelves and avoids the deeper water that often exists right to the shoreline. Even when doing oceanography at night under strong lights directed onto the surface, there is rarely any sign of life in the dark green mid-channel waters of Johnstone Strait.

As the region is characterized by rapid tidal streams, constricted passages, and numerous shallow sills, the water is in almost constant agitation from top to bottom and never has the oportunity to settle into strongly stratified layers that typify water in inlets and the Strait of Georgia. Therefore, water temperatures increase very slightly inward along Johnstone Strait and Discovery Passage and remain low throughout the year (Fig. 12.3). At the height of summer heating in late July, for example, temperatures of the surface water are usually colder than 10°C, which is appreciably colder than the contemporary values of over 20°C in the central portion of the Strait of Georgia. Even the oceanic surface waters of Queen Charlotte Sound are warmer in summer than those in Johnstone Strait. (This cold surface water is a principal reason for the common occurrence of summer fogs in the area.) Particularly vigorous tidal mixing occurs

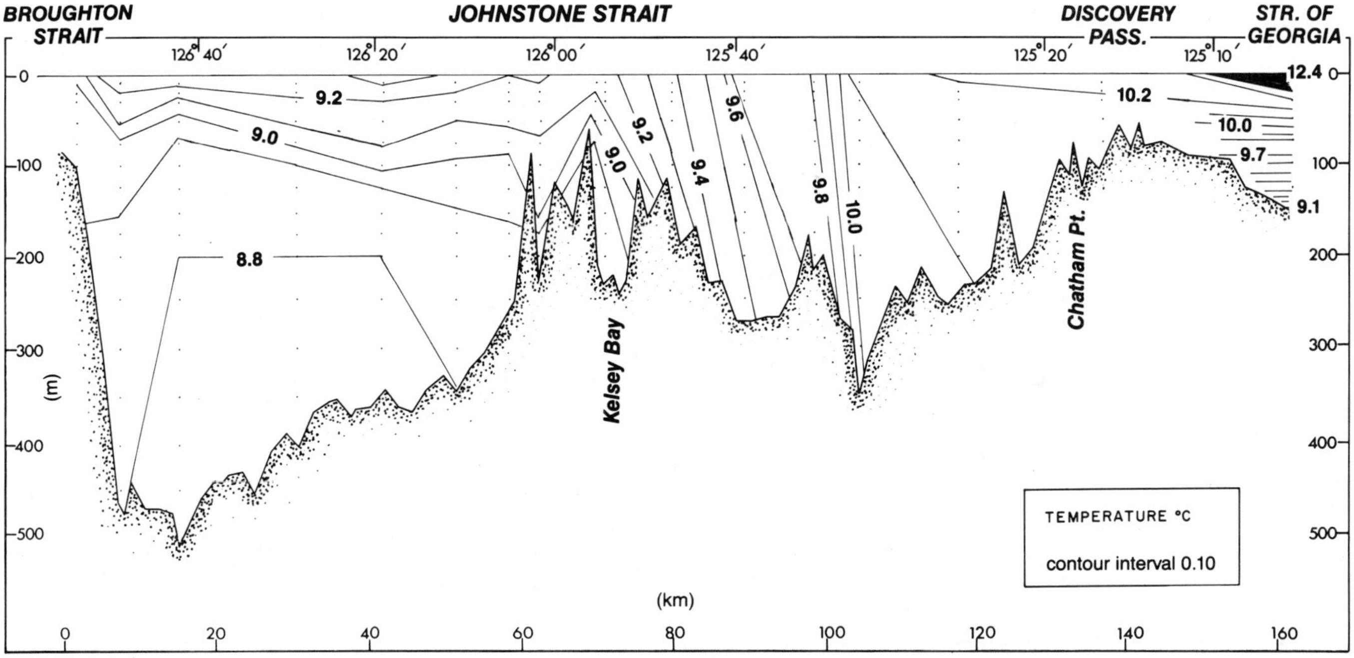

FIG. 12.3. Temperature distribution along mid-channel in Discovery Passage and Johnstone Strait, September 1977. (Note highly structured surface waters in Strait of Georgia in Fig. 12.3–12.5.) (From Thomson et al. 1980)

throughout the year in such areas as Seymour Narrows, Race Passage, and Weynton Passage.

During winter and spring, the waters of Johnstone Strait and Discovery Passage become uniformly cold from top to bottom. Maximum temperatures during these seasons are typically around 7°C and the along-channel variation becomes almost indiscernible. A similar situation exists below 30 m in summer. Just how cold this water can be was made clear to the scientists and crew of a research vessel during a recent May oceanographic survey, when, in the early light of a calm day, they watched helplessly as a large timber wolf died from hypothermia halfway through an attempted swim across Johnstone Strait.

The nearly uniform salinity of the waters within Johnstone Strait and Discovery Passage also reflect the vigorous tidal mixing that takes place over the sills and in the narrow passes (Fig. 12.4). Unlike temperature distribution, however, there is at all times a discernible in-

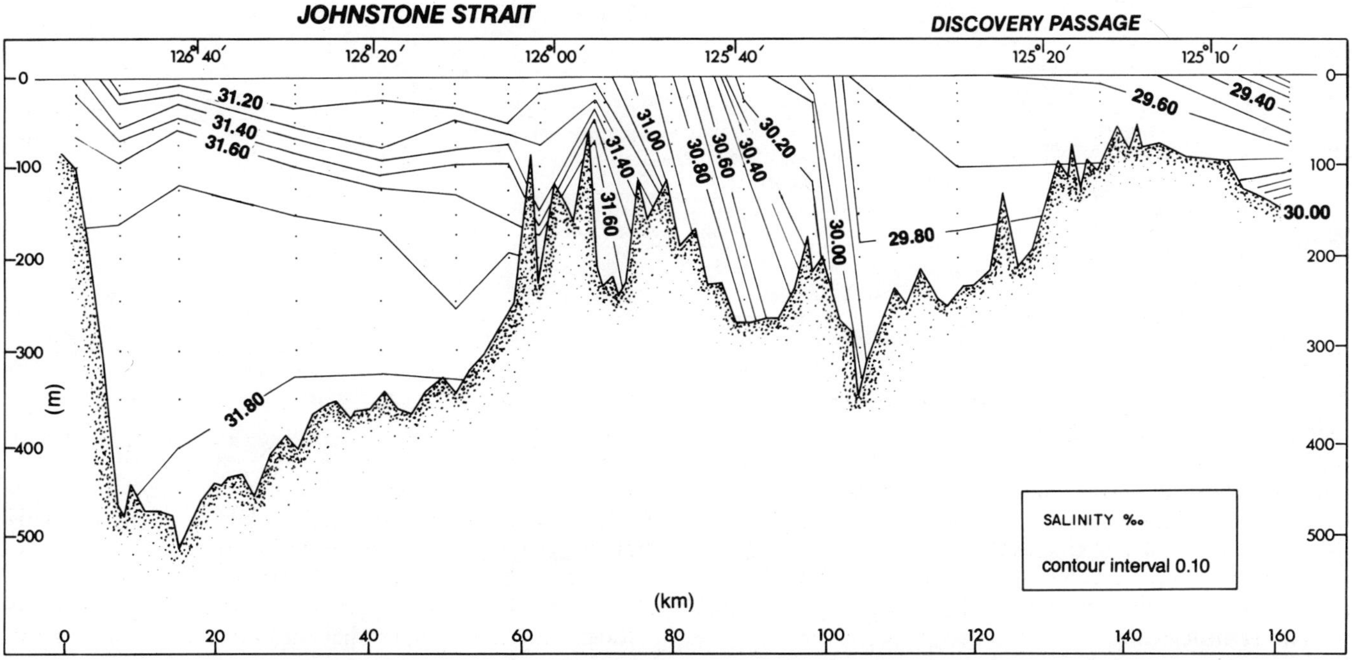

FIG. 12.4. Salinity distributions along mid-channel in Discovery Passage and Johnstone Strait, September 1977. (From Thomson et al. 1980)

crease in saltiness seaward along the channels and a weak but permanent increase in salinity with depth. In Johnstone Strait, salinities increase from a surface minimum of 30‰ to a near bottom maximum of 32‰, and in Discovery Passage the corresponding range is about 26–31‰. Typically, the salinity distributions, and to a lesser extent those of temperature, are similar to other estuarine-type environments on the coast. In the present case, relatively cold, salty Pacific Ocean water continually works its way inward below depths of about 100 m in the two channels while relatively warm, low-salinity water, derived largely from runoff from the Fraser River and the Homathko River in Bute Inlet, makes its way seaward in the top 100 m. Although the fraction of Fraser River runoff that moves seaward via the northern seaway becomes highly diluted by the time it reaches the Pacific Ocean off the northern end of Vancouver Island, it possibly retains enough of its "aroma" that Fraser River salmon, returning southward along the Alaska–British Columbia coast after 4 yr roaming the sea, are able to "smell" their way home to the river mouth.

A striking indication of the well-mixed nature of the waters in Johnstone Strait and Discovery Passage is the uniformity of dissolved oxygen values observed from top to bottom throughout the year (Fig. 12.5). At the seaward metaphorical oceanic wind, the strong bottom currents constantly blow freshly oxygenated water over the floor of the channels to encourage a proliferation of marine life. It was no surprise when, during a recent detailed sampling of the sediments in the western basin of Johnstone Strait, a compacted muddy bottom teeming with worms, brittle stars, crabs, sea urchins, and clams was discovered.

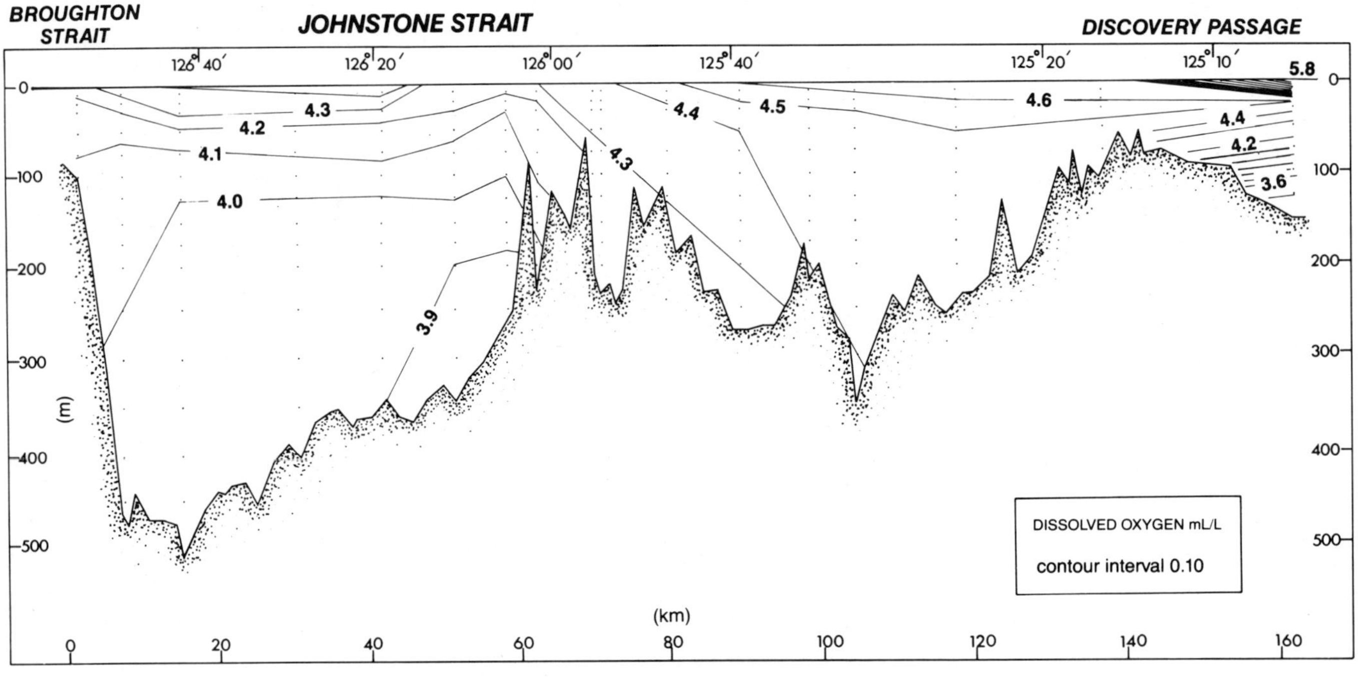

FIG. 12.5. Dissolved oxygen distribution along mid-channel in Discovery Passage and Johnstone Strait, September 1977. (mL/L = millilitres dissolved oxygen per litre of water.) (From Thomson et al. 1980)

end of Johnstone Strait, near-bottom values may even exceed those close to the surface as negatively buoyant flood streams that surge southward from Weynton Passage carry aerated surface waters to the very depths of the basin. Because of this, benthic (bottom) organisms within the two channels never suffer the devastating effects of low oxygen supply that so often plague their counterparts in some of the silled basins and fiords along the coast. Like a

Queen Charlotte Strait

At the seaward end of Johnstone Strait, the floor of the channel rises abruptly to form the 90-km long, shallow, island-strewn basin of Queen Charlotte Strait (Fig. 12.2). The steep-sided walls so characteristic of the narrow seaway to the east are replaced by a broken, shoal-infested coastline adjoining a comparatively low land relief. Shoals and drying rocks are especially numerous within the broad seaward entrance that flanks the mainland shore and within Broughton Strait to the south of Malcolm Island.

The width of the main basin of Queen Charlotte Strait widens from around 13 km at its eastern extremity to more than 26 km at midlength, then narrows to less than 15 km at the seaward approach. Greatest depths within the basin are associated with the narrow, eastward-shoaling trough that extends along Goletas Channel to George Passage north of Malcolm Island. Goletas Channel, with its narrow regular geometry, steep-sided shoreline, depths in excess of 350 m, and shallow entrance sill (Nahwitti Bar), bears a closer physical resemblance to Johnstone Strait than to adjacent Queen Charlotte Strait.

Queen Charlotte Strait is a meeting place for the nearly homogeneous waters that move seaward through the inside passage and the more highly stratified oceanic waters that move inland from Queen Charlotte Sound. Because of this, the temperature and salinity distributions within the basin have slightly more structure than in Johnstone Strait, although the difference really only becomes pronounced in summer. From late fall to early spring, water temperatures generally range around 7–10°C near the surface and about 7–8°C at depth, accompanied by a slight warming trend from east to west (Fig. 12.6). During summer, near surface temperatures

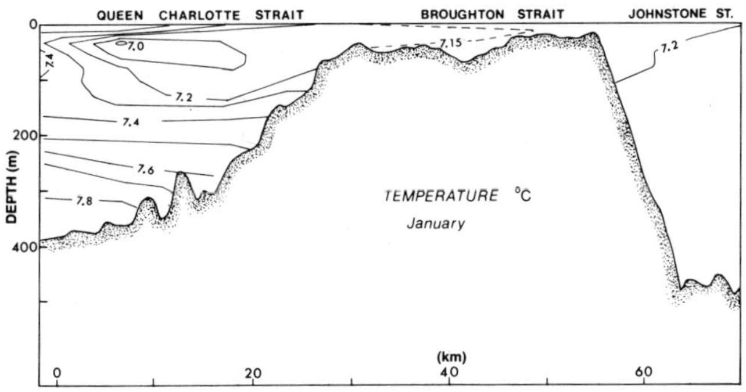

Fig. 12.6. Temperature distribution along southern side Johnstone, Broughton, and Queen Charlotte straits, January 1978. (See broken line Fig. 12.2 for location.) (From Thomson et al. 1980)

rise above 10°C as local runoff covers portions of the Strait with a thin blanket of brackish water capable of retaining a greater proportion of solar heating. Within the more protected embayments along the northern side of the basin, surface temperatures can be expected to exceed 15°C during hot, windless days.

Salinities in Queen Charlotte Strait typically range from 31 to 33‰ throughout the year (Fig. 12.7). As in

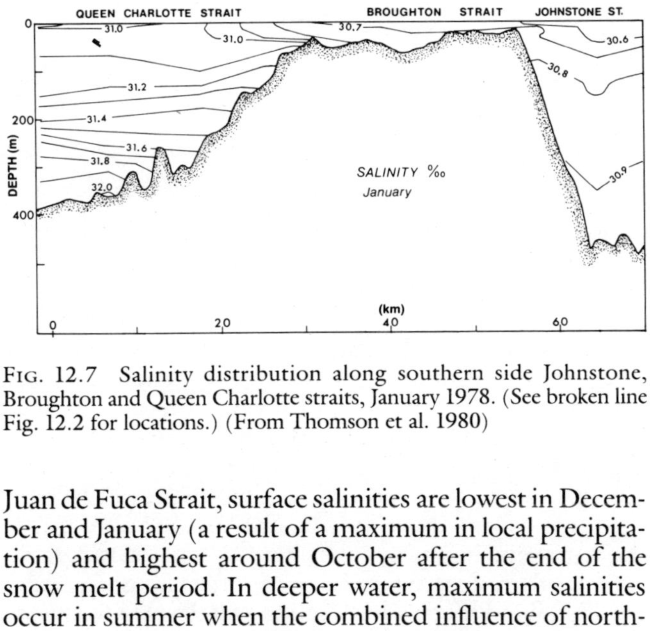

Fig. 12.7 Salinity distribution along southern side Johnstone, Broughton and Queen Charlotte straits, January 1978. (See broken line Fig. 12.2 for locations.) (From Thomson et al. 1980)

Juan de Fuca Strait, surface salinities are lowest in December and January (a result of a maximum in local precipitation) and highest around October after the end of the snow melt period. In deeper water, maximum salinities occur in summer when the combined influence of northwest winds off the coast and a well-established estuarine circulation associated with river runoff induce salty oceanic water outside the entrance sills to spill into the basin. This process all but ceases by late fall with the onset of southeast winds and weakened estuarine flow, and salinities in the deeper portions of the basin gradually decrease to a midwinter low.

Dissolved oxygen values within Queen Charlotte Strait tend to be higher at the surface but lower at the bottom than in Johnstone Strait. There are two main reasons for this: tidal currents are less intense so the more highly oxygenated surface waters are less able to mix with the deeper waters; and dense, low-oxygen oceanic water over the continental shelf can flow to the bottom of the basin. For the most part, however, values remain above 3 mL/L, enough to support most marine animals without any undue hardship.

Winds

The prevailing winds over the waters of the Johnstone Strait region are linked to the large-scale pressure systems of the eastern Pacific Ocean and, guided by the mountainous terrain, blow predominantly in along-channel directions. Only where extensive interconnecting basins from the mainland side open into these passageways do winds sometimes blow from other directions, for example, at the trifurcation area formed by Johnstone Strait, Discovery Passage, and Nodales Channel. The influence of such cross-channel winds dies out quickly a short distance into the main channels; the wind's major effect is to generate rips where it opposes the surface currents.

Within Johnstone and Queen Charlotte straits, prevailing winds are westerly in summer and easterly in winter (Table 12.1). These are funneled into northerly and southerly prevailing winds, respectively, in Discovery Passage. Polar outbreaks from the mainland interior that move down some of the larger fiords such as Knight Inlet, Kingcome Inlet, and Loughborough Inlet can lead to gale force winds over limited areas within the region but, in general, strong winds along the main channels are associated with the passage of frontal systems. During summer, there is a distinct sea-breeze effect in Johnstone and Queen Charlotte straits as rising air over the heated mainland coast draws cooler marine air inland. On clear sunny days, these westerly winds build in strength beginning in the late morning and, combined with the prevailing air flow, can lead to wind speeds of 15 m/s (30 kn) by late afternoon; the western portion of Johnstone Strait appears to be especially susceptible to these winds, the eastern portion less so. The sea breeze dies out just before dusk and is replaced by a considerably weaker land breeze from the east, whose influence is mostly confined to the more exposed waters of Queen Charlotte Strait.

Waves

Wave heights in Johnstone Strait and Discovery Passage are limited by the fetch of the wind and by the narrow, winding nature of the channels. It is unlikely, for example, that waves generated at one end of an along-strait fetch will be able to reach the far end of the channel before running ashore on either side. Even over the longest wind fetch of 65 km in the western portion of

TABLE 12.1. Winds at Chatham Point, Johnstone Strait (*top*). Percentage occurrence of winds from 8 directions for each month, from February 1960 to March 1973. Winds at Bull Harbour, Goletas Channel (*bottom*) from November 1964 to February 1973. (From Sailing Directions British Columbia Coast, Vol. 1, 1976)

		MONTH											
		J	F	M	A	M	J	J	A	S	O	N	D
DIRECTION (Chatham Point)	N	6	3	3	1	1	1	0	1	1	2	4	5
	NE	28	19	20	14	7	5	3	6	9	16	23	26
	E	18	21	18	13	9	7	5	8	11	18	22	19
	SE	18	19	16	21	15	13	10	10	13	22	17	17
	S	1	1	1	1	1	1	1	0	1	1	1	1
	SW	9	12	16	13	20	22	25	27	21	15	11	10
	W	11	14	16	22	32	34	42	36	32	16	12	12
	NW	6	7	8	12	12	15	14	9	9	5	7	8
	Calm	3	4	2	3	3	2	1	2	3	5	3	2
	Mean speed (m/s)	3.8	3.4	3.9	4.1	4.4	4.9	5.5	4.4	3.8	3.4	3.4	3.9
	(kn)	7.4	6.5	7.6	8.0	8.5	9.6	10.8	8.5	7.4	6.6	6.7	7.6
DIRECTION (Bull Harbour)	N	4	2	5	4	5	4	3	4	3	4	3	4
	NE	4	3	3	2	4	3	2	3	3	3	2	3
	E	7	6	4	3	4	3	2	3	6	5	5	5
	SE	47	49	49	35	19	14	9	19	33	50	57	52
	S	10	11	12	13	11	6	8	8	8	8	11	10
	SW	11	10	6	7	7	9	10	9	6	5	5	9
	W	1	2	2	3	9	10	12	10	6	2	1	1
	NW	12	11	16	28	33	34	37	20	16	14	12	13
	Calm	4	6	3	5	8	17	17	24	19	9	4	3
	Mean speed (m/s)	3.8	3.1	3.4	3.1	2.4	1.9	1.6	1.2	1.7	2.8	3.4	3.9
	(kn)	7.4	6.0	6.6	6.0	4.7	3.6	3.0	2.3	3.3	5.5	6.7	7.6

Johnstone Strait, much of the wave energy generated by westerly or easterly winds will be lost to breakers at the shoreline. Because waves produced in the remaining shorter basins will suffer similar fates, long, rolling seas and swell never have the opportunity to develop.

Heights of wind-generated waves within Queen Charlotte Strait are also limited by a fully exposed fetch of only 65 km. For a given wind speed and duration, however, seas will become somewhat higher and have longer periods than those in Johnstone Strait, due to the greater width of the channel. The low, eastward-propagating swells commonly found in Queen Charlotte Strait are for the most part highly attenuated remnants of larger oceanic swell from Queen Charlotte Sound that have negotiated the shoals and passages of Goletas and Gordon channels. When they enter the Strait, these waves undergo a further gradual reduction in height as they move inland. Although no direct measurements have been made of waves in this region, maximum seas will be consistently lower than the 3- to 4-m seas observed at Roberts Bank in the more extensive Strait of Georgia (Chapter 10).

Based solely on the above criteria, it could be assumed that waves in Johnstone Strait and Discovery Passage are insignificant and worthy of little comment. Unfortunately, the presence of strong surface currents in these two channels partially makes up for the short fetches and twisting coastlines, as they contribute to the formation of a short, steep chop more than a metre or two high when winds rise above 10 m/s (20 kn). Because surface ebb currents in these two channels are considerably stronger than surface flood currents, the largest seas for a specified wind speed and duration will develop for winds from the west in Johnstone Strait and from the north in Discovery Passage, in the region north of Seymour Narrows. South of Seymour Narrows, on the other hand, largest wave heights will occur on the flood during periods of strong southeasterly winds in the Strait of Georgia as current-amplified seas and swell propagate into the passage. Especially choppy rip currents are set up near points of land and at entrances to most of the passes that lead into Johnstone Strait, when winds oppose the surface currents. Winds from the east in Johnstone Strait, however, typically will be associated with less choppy seas as they blow in the direction of the average surface flow; in Discovery Passage a similar condition holds for southerly winds.

The short, steep, whitecapping seas that cover the two channels during moderate to strong northwest winds have little effect on larger vessels; onboard the 62.5-m CSS *Parizeau*, for instance, the motion is a gentle roll at worst even during winds in excess of 15 m/s. Although smaller vessels are much more affected by these waves and light displacement pleasure craft pitch and roll awkwardly in the steep seas, heavy displacement tugs or trollers of comparable length cut through the waves with considerably greater ease.

In Queen Charlotte Strait, where surface currents are weaker and the ebb bias less pronounced, amplification of wind-generated waves by opposing currents is not generally an important factor. Only over shoals and in the neighborhood of prominences do rip conditions frequently develop. Nahwitti Bar at the western end of Goletas Channel is particularly well known for its high waves. Sea and swell that cross the 2.5-km wide, 15-m deep sill from Queen Charlotte Sound are amplified by the combined effect of the shoaling bathymetry and the opposing tidal currents. In heavy weather with westerly winds, the sea breaks over the bar and it becomes dangerous to cross in any vessel.

Tides

As in Juan de Fuca Strait, the mixed-type tide that propagates inland from the Pacific Ocean north of Vancouver Island undergoes considerable modification as it makes its way toward the Strait of Georgia. Tides in the vicinity of Alert Bay and Port McNeil, for example, are mostly semidiurnal throughout the month and only take on a truly mixed-type nature around the time of the moon's maximum declination north or south of the equator (Fig. 12.8). In this particular region, therefore, the semidiurnal part of the tide associated with the moon's gravitational pull predominates over the diurnal part associated with changes in the moon's declination. As the tide progresses southeastward, the semidiurnal contribution diminishes in analogy with tides in Juan de Fuca Strait. By the time it reaches Yorke Island (where there are remains of World War II gun emplacements and bunkers) the tide has become mixed, mainly semidiurnal as semidiurnal and diurnal effects begin to become of equal importance.

Near Chatham Point at the eastern end of Johnstone Strait, the contribution from the semidiurnal tide has dropped to 0.7 of the Alert Bay value, whereas that from the diurnal tide has increased by a factor of 1.3. In the vicinity of Duncan Bay, the diurnal contribution becomes sufficiently great that the tide has a diurnal nature and roughly 12 days each month there is only 1 stand of high and 1 stand of low water each day. This state is especially well established when the moon is furthest north or south of the equator, so the moon's declinational effect is most pronounced. South of this area, past Campbell River and into the Strait of Georgia, the tide switches back to mixed, mainly semidiurnal.

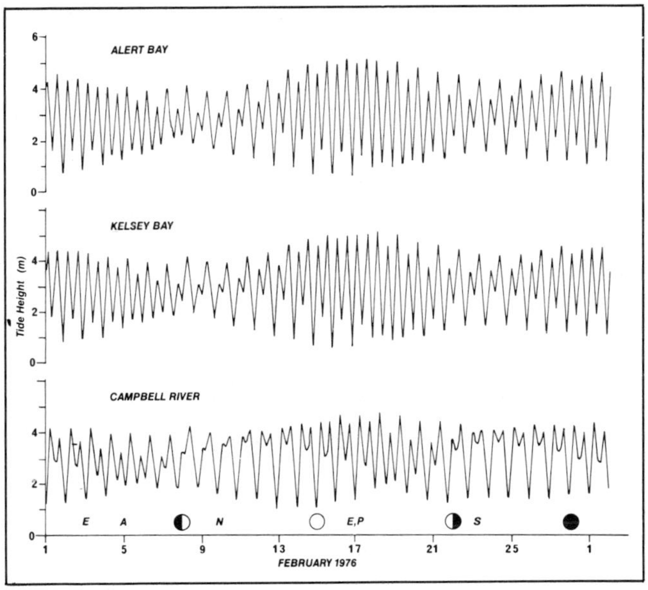

FIG. 12.8. Tide heights over 1 mo at three locations from Broughton Strait to Discovery Passage. Campbell River at southern end Discovery Passage. Times of lunar phases shown; moon on equator (E), maximum north (N) and south (S) declination, and apogee (A) and perigee (P). (Courtesy A. Douglas.)

The change in character of the tide along the channels is further accompanied by a change in tidal sequence. From Cape Scott to Seymour Narrows this sequence is invariably Higher High Water, Lower Low Water, Lower High Water, Higher Low Water. However, south of the Narrows, this sequence is more complicated. For most of the month, the sequence is Higher High Water, Higher Low Water, Lower High Water, Lower Low Water, but for a few days each month just prior to the moon's maximum declination the sequence is identical to that between Cape Scott and the Narrows. Therefore, only a relatively small low tide is expected to follow the highest tide of the day, except just after the moon has crossed the equator, when Higher High Water will be followed by Lower Low Water. The times when this occurs can be determined from the back cover of any *Canadian Tide and Current Tables* for B.C. waters.

The southward propagation of the tides to the Strait of Georgia can be followed by measurements made at various locations throughout the area. The basic information for this purpose is in Table 12.2 where, in addition to the time it takes Higher High Water and Lower Low Water to travel from Cape Scott to a particular location, the mean and extreme tidal ranges for each location have also been listed. For all practical purposes, the along channel propagation for Higher High Water and Lower Low Water is the same and, for illustrative purposes, discussion is confined to Higher High Water alone.

The first thing to notice in Table 12.2 is that the northward propagating tide along the west coast of Van-

TABLE 12.2. Propagation rates for Higher High Water (HHW) and Lower Low Water (LLW) relative to Cape Scott, plus tidal ranges. (Minus sign at Tofino means tides occur earlier than at Cape Scott.) Tidal ranges for extreme spring and extreme neap tides, respectively. South of Seymour Narrows times quoted are for Lower High Water. (Courtesy W. Rapatz)

Location	Travel time from Cape Scott (*min*)		Tidal Range (*m*)	
	HHW	LLW	Spring	Neap
Cape Scott	0	0	4.7	3.0
Tofino	−7	−9	4.1	2.7
Port Hardy	+21	+18	5.6	3.6
Alert Bay	+34	+36	5.5	3.5
Yorke Island	+53	+63	5.3	3.4
Kelsey Bay	+66	+71	5.4	3.4
Chatham Point	+101	+128	4.8	2.9
Seymour Narrows	+149	+270	5.1	3.0
Campbell River	+240	+270	4.6	2.9

couver Island takes about 7 min to cover the 260 km from Tofino to Cape Scott for an average speed of 2230 km/h. Once the tide turns the corner at Cape Scott it requires over 20 min to travel 75 km to Port Hardy in Queen Charlotte Strait, for a speed of 225 km/h, and another 13 min to reach Alert Bay 37 km to the southeast at the reduced speed of 170 km/h. In addition, the tidal range reaches its peak at Port Hardy and then decreases along the channel.

The sluggish nature of the tide after it rounds Cape Scott is, of course, due to the narrowness of the passes and the greatly diminished water depth; tides can move faster in deep water than in shallow water. The latter feature is suggested by the fact that it takes a few minutes longer for Lower Low Water to cover the same distances in Table 12.2 than Higher High Water. After it enters Johnstone Strait, the tide becomes even more sluggish and requires more than 30 min to reach Kelsey Bay, just over 75 km to the east, and an additional 35 min to reach Chatham Point, only 37 km from Kelsey Bay. Respective speeds over these two distances are 150 km/h and 63 km/h. Throughout the journey along Johnstone Strait the tide also decreases in range. This, in part, results from the tide spreading into the various corners of the complex region and from a loss of energy due to friction and mixing processes. At Seymour Narrows, a complicated exchange of water takes place with the result that Higher High Water on the northern side of the Narrows becomes Lower High Water on the southern side of the Narrows. Moreover, both High Water and Low Water are delayed by approximately 2 h at the Narrows. South of Cape Mudge at the northern end of the Strait of Georgia, the diminished southward moving tide then meets the northward moving tide to produce a system of weak and variable currents.

In Fig. 12.8 the tides have been plotted over a month at three locations (Alert Bay, Kelsey Bay, and Campbell River) to show the biweekly variation of the tidal range. Note the reduction in range from one location to the next, the biweekly modification of the diurnal inequality, and the difference in the form of the tidal curves between Campbell River and the more northerly locations.

Currents

Despite its importance to many aspects of the marine environment of southwestern British Columbia, the Johnstone Strait region has only recently received special attention from oceanographers. As a consequence, the assumption must be made that current measurements from a few strategic locations within the major channels can be used to imply the overall flow structure, even though recent measurements have clearly demonstrated the limited applicability of such an assumption, especially where the channel is curved or where shallow sills are present. The symbols in Fig. 12.2 mark the locations of current meter moorings used to study the circulation in the Johnstone Strait region. Current observations from the two starred locations in Johnstone and Queen Charlotte straits will be considered in some detail. In the former case, particular attention will be paid to the circled station because in this region of the channel the mainland coast lies directly opposite the coast of Vancouver Island and, therefore, is the section through which all east–west water exchange must take place.

Johnstone Strait–Discovery Passage

These two channels are characterized by swift and rectilinear tidal currents. In the vicinity of shallow sills and constricted narrows, surface currents are accelerated even further, and take on a turbulent jetlike nature, generally associated with quasi-permanent tidelines that delineate rapid cross-stream changes in speed and direction of the set. Tidelines are especially well defined at the seaward extremity of Johnstone Strait, in the vicinity of Kelsey Bay, and south of Seymour Narrows. Many authors of oceanographic books have referred to Seymour Narrows as an illustration of the maximum strength attainable by tidal currents in the world ocean and to demonstrate their hazard to navigation.

Figure 12.9b–d illustrates the kinds of flow variability at various depths near mid-channel in Johnstone Strait and Discovery Passage; Fig. 12.9a shows the corresponding sea-level variations within the region. Flood currents in Johnstone Strait are toward the east (measured upward in each figure) and ebb currents are toward the west (measured downward in each figure). In Discovery Passage, floods are toward the south and ebbs toward the north. An analysis of these currents indicates they can be separated into three distinct components.

1) Tidal streams associated with the astronomical tide which change speed and direction in a regular manner throughout the day.

2) Estuarine currents driven by freshwater runoff and induced along-channel density gradients which are essentially steady over periods of days.

3) Wind currents confined to the upper layer of the water column which only become important during times of moderate to strong winds.

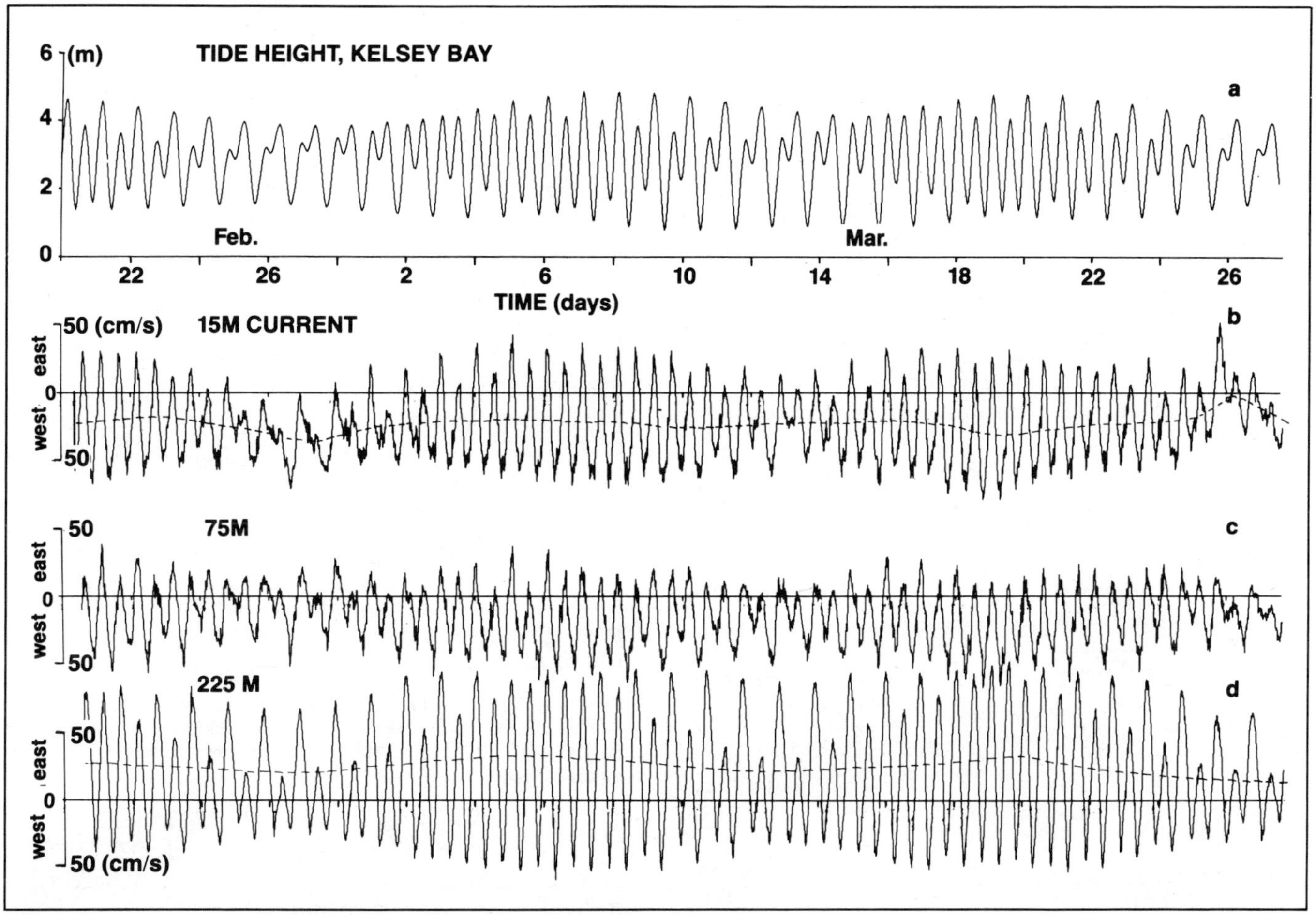

FIG. 12.9 Tides and tidal currents near Kelsey Bay February and March 1973. Tide heights (a) at Kelsey Bay compared with along-channel (east-west) component of currents (b) 15, (c) 75, and (d) 225 m deep, 10 km to west (circled star in Fig. 12.2). Speeds, cm/s; tide heights, m. Lines in (b) and (d) give strength and direction of mean (residual) flow. (From Thomson 1976).

The last two constitute the residual or nontidal current. All three components are in turn affected by the earth's rotation, channel curvature, bottom topography, and the cross-sectional area of the channel. The latter effect, for example, requires that where the width or depth of the channel decreases in the direction of flow, there is an acceleration of the along-channel current to maintain the same volume of transported water and vice-versa.

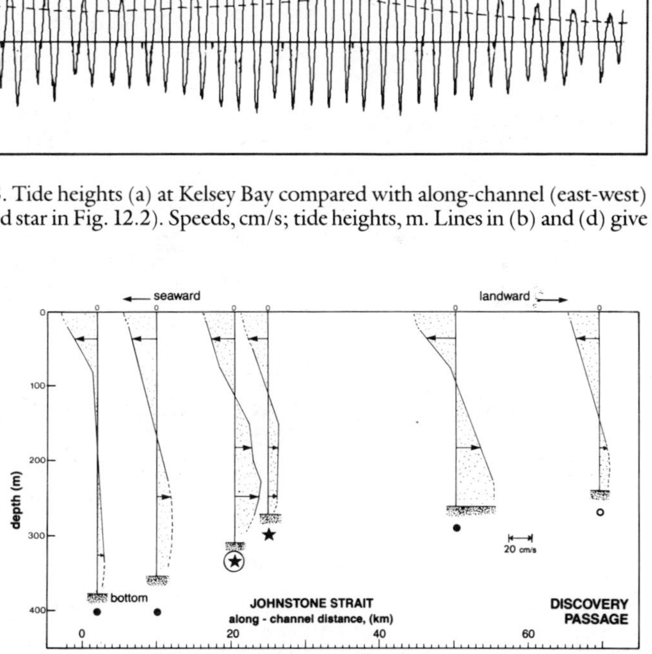

FIG. 12.10. mean flow profiles at various locations in Johnstone Strait and Discovery Passage. Symbols correspond to Fig. 12.2. Resultant currents are westward in upper layer, eastward in lower layer. Speeds obtained by measuring horizontally from vertical axis and comparing to scale.

ESTUARINE CURRENTS

Estuarine currents are a slow, time-varying component of the overall flow that change speed and direction with depth in the channels. In Fig. 12.9b, d, they are represented by the gradually varying line drawn through the more rapidly changing curve for the total current composed of the tidal stream and combined wind-estuarine flow. In the estuarine component, the flow in the upper layer (Fig. 12.9b) is always westward in Johnstone Strait and northward in Discovery Passage; that is, toward the sea. In the lower layer (Fig. 12.9d), the estuarine flow is always to the east in Johnstone Strait and to the south in Discovery Passage; that is, toward the Strait of Georgia. This two-way structure is clearly illustrated in Fig. 12.10, based on current velocity measurements taken at various depths along mid-channel. Above about 100 m the average flow is always seaward whereas below this depth it is invariably landward. Thus, a log on the surface of Johnstone Strait or Discovery Passage would drift progressively toward the open ocean (in the absence of persistent westerly winds) at roughly 20 cm/s (0.4 kn); a powerless submarine near the bottom, by contrast, would gradually drift toward the Strait of Georgia at an average speed of 5–10 cm/s (0.1–0.2 kn). This type of depth-dependent net drift is, of course, common to many inland seaways along the British Columbia–Washington coast (see Juan de Fuca Strait, Chapter 11). As in other coastal regions, moreover, the strength of the estuarine current in the two channels varies over periods of days, weeks, and months in response to changes in the amount of land drainage entering the system, to variations in the large-scale winds over southwestern British Columbia, and to the degree of along-channel tidal mixing but, unlike more exposed channels such as Juan de Fuca Strait, is less susceptible to direct modifications introduced via the open ocean. (Estuarine circulation is discussed more thoroughly in Chapter 2.)

Figures 12.11A, B illustrate the cross-channel structure of the estuarine current at two locations in Johnstone Strait. The flow is probably representative of Johnstone Strait and Discovery Passage as a whole, excluding Seymour Narrows, Race Passage, and Weynton Passage, where tidal jets and fronts are known to disrupt the expected flow patterns. The main feature is that, in the upper seaward flowing layer, the strongest average currents are on the mainland side of the channel whereas in the lower eastward flowing layer, the strongest currents are on the Vancouver Island side. This effect is due in part to the Coriolis effect. Also, because these particular measurements were taken where the channel has a pronounced curvature, the inertia of the water was an important factor in producing the stronger surface currents along the mainland side. The combined effect of the earth's rotation and the curvature of the channel is seen further to produce a downward tilt of the line of zero net velocity, which separates the westward outflow in the upper layer from the eastward inflow in the lower layer. The downward tilt is greater in Fig. 12.11A because of the greater channel curvature involved. Moreover, the cross-sectional area of the upper layer experiences comparatively little along-channel variation except in the narrower passes of the seaway. As a result, near-surface speeds of the estuarine component of the flow are typically around 20 cm/s (0.4 kn) throughout most of Johnstone Strait and Discovery Passage. The cross-sectional area of the lower

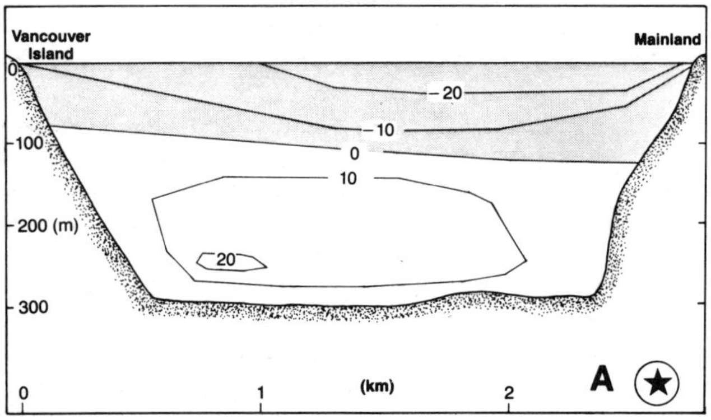

FIG. 12.11. Speed (cm/s) and direction of mean currents at two cross-channel locations in Johnstone Strait. (A) February–June 1973.

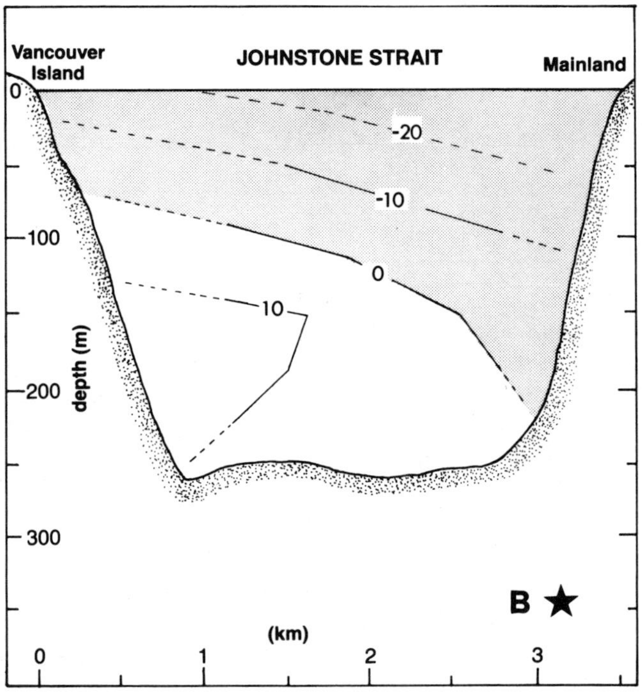

FIG. 12.11. (B) May 1978. Symbols correspond to Fig. 12.2. In shaded upper zone, net flow is westward (negative values); in lower zone, net flow is eastward (positive values). Broken lines represent less certain values of speed.

layer, on the other hand, undergoes large along-channel variations due mainly to changes in depth. This leads to the marked differences in the speed of the nontidal current at depth along the two channels (Fig. 12.10).

Recommendation A slow boat that needs more than a day to cover a particular course in Johnstone Strait or Discovery Passage should keep well to the starboard side of mid-channel. The boat will get maximum assistance from the estuarine flow when heading westward and minimum resistance from this current when heading eastward.

Residual plus Tidal Streams

The tidal stream contribution will now be discussed in relation to the estuarine current. In the absence of any surface wind drift, the sum of these two accounts for nearly all the observed current speed and variability.

Tidal streams are that part of the total current directly associated with the rise and fall of the tide; except where deflected by the bottom topography or coastline they produce no significant net drift over a period of a few tidal cycles, just as the tide produces no net change in the water depth. In reality, of course, those who make tidal predictions lump the pure tidal streams together with the residual currents to get a sort of hybrid tidal flow, a procedure that can be erroneous unless the residual current remains almost constant over many weeks and months. As the residual flow is not generally constant because of variability in the various mechanisms that produce it, current predictions are often incorrect, particularly for times of slack water.

There are a number of things to note about the

combined tidal current and residual current in Fig. 12.9b. For example, the ebbs are much stronger than the floods and the maximum strength of these currents varies in a regular way over a period of 15 days in agreement with the tides. Near mid-channel, the larger of the two ebbs each day is generally over 50 cm/s (1 kn), sometimes reaching 75 cm/s (1.5 kn) during spring tides. On the other hand the larger of the two floods is always weaker than 50 cm/s whereas the weaker flood often produces nothing more than a short period of slack water near high tide. In fact, many days can pass before there is any appreciable surface flood current in Johnstone Strait and Discovery Passage. Within ½ km of either shore the speed of the net current diminishes so that at any given time nearshore speeds are about 20% weaker than those near mid-channel. Moreover, surface currents are usually weaker on the Vancouver Island side than on the mainland side.

As indicated in Fig. 12.9b, the surface ebb currents have a much longer duration than the floods; 8–10 h for ebbs compared with 2–4 h for floods. Nevertheless, near mid-channel the maximum ebbs and maximum floods always occur within ½ h of local low water and local high water, respectively, unless there is a strong wind blowing. Away from mid-channel, the times of maximum ebb or flood become increasingly delayed until at either side they occur roughly 1 h later than at mid-channel. Table 12.3 gives the times of tidal currents at various locations along Johnstone Strait and Discovery Passage based on current observations taken between 1976 and 1978.

TABLE 12.3. Delay times of ebb and flood streams within the Johnstone Strait region. Time (min) for a particular stage of flood (e.g. maximum flood) after its occurrence at western end of Queen Charlotte Strait. Time delay for a particular stage of ebb. As an example, maximum flood occurs at northern end of Discovery Passage about 160 min later than Queen Charlotte Strait.

	Queen Charlotte Strait	Johnstone Strait			Discovery Passage	
	west	west	central	east	north	Seymour Narrows
Flood	0	100	110	140	160	170
Ebb	170	70	60	30	10	0

There is another important facet of the relationship between the tide and the tidal currents in Johnstone Strait and Discovery Passage. The stronger of the two floods each tidal day is not directly associated with the higher of the two high tides, nor is the stronger of the two ebbs directly associated with the lower of the two low tides each day. A rule must be devised, based on the predicted tides, to predict (in a qualitative sense at least) whether to expect strong or weak tidal currents.

Rule for flood currents If an ensuing flood is preceded by Lower Low Water then it will be the stronger of the two floods for a particular 25-h period; if the ensuing flood is preceded by Higher Low Water, it will be the weaker of the two floods for that day. The greater the difference in height between the two low tides the greater the difference in the strength of the two floods.

Rule for ebb currents If an ensuing ebb is preceded by a Higher High Water then it will be the stronger of the two ebbs for a given 25-h period, but if the ensuing ebb is preceded by Higher Low Water then it will be the weaker of the two ebbs for that day. The greater the difference in height between these two high tides the greater the difference between the strength of the two ebbs.

There are of course exceptions to this rule but, generally speaking, the current directions and speeds are closely tied to the height of the tide 6 h earlier. The sequence of the surface currents varies but typically follows the daily pattern larger flood, larger ebb, weaker flood, weaker ebb, although there is usually only a small difference in the strength of consecutive ebbs.

WIND CURRENTS

Winds can generate significant surface currents if the fetch and strength of the wind are sufficiently great. The long, narrow channels provided by Johnstone Strait and Discovery Passage are well suited to this type of situation.

Figure 12.12 illustrates four occasions when the record of the flow was "abnormally" disrupted. The wind data for this region show that these four periods corresponded to times of strong along-channel winds, whose average speeds exceeded 7 m/s (14 kn) for more than 24 h. The most spectacular change in the normally regular pattern of the currents occurred around March 27 when the floods were appreciably enhanced. According to the wind charts, this was the time of the strongest storm of the season and westerly winds blew up-channel at an average speed of over 10 m/s (20 kn) for over 3 days. At its height, this particular storm generated an inward wind-drift current of over 20 cm/s (0.4 kn), and if it is considered that the currents were actually measured at a depth of 15 m, the wind currents right at the surface probably had speeds of around 30 cm/s (0.6 kn) or about 3% of the average speed of the wind. (Unfortunately surface measurements cannot be made with moored instruments in this region because of the possibility of damage by ships.) A somewhat different case occurred around February 27 when the winds were down-channel (easterly). At that time, the ebb currents were strengthened by the wind, and for a period of several days there were no flood currents.

The above findings again bear out some general comments concerning the currents in Johnstone Strait and Discovery Passage. In winter, when the prevailing winds are from the southeast, the wind currents will tend to enhance the ebb currents and, therefore, reduce the strength and duration of the flood currents. This effect will occur less often in summer when the prevailing northwesterlies tend to enhance the floods and, therefore, ensure a regular daily reversal of the flow regime.

ADDITIONAL CURRENT FEATURES

Some general features of the current pattern in these northern channels have been described. Now some particular aspects of these flows will be examined.

One of the more surprising features of the current observations so far is that there appears to be little flow through the entrances to channels that open into

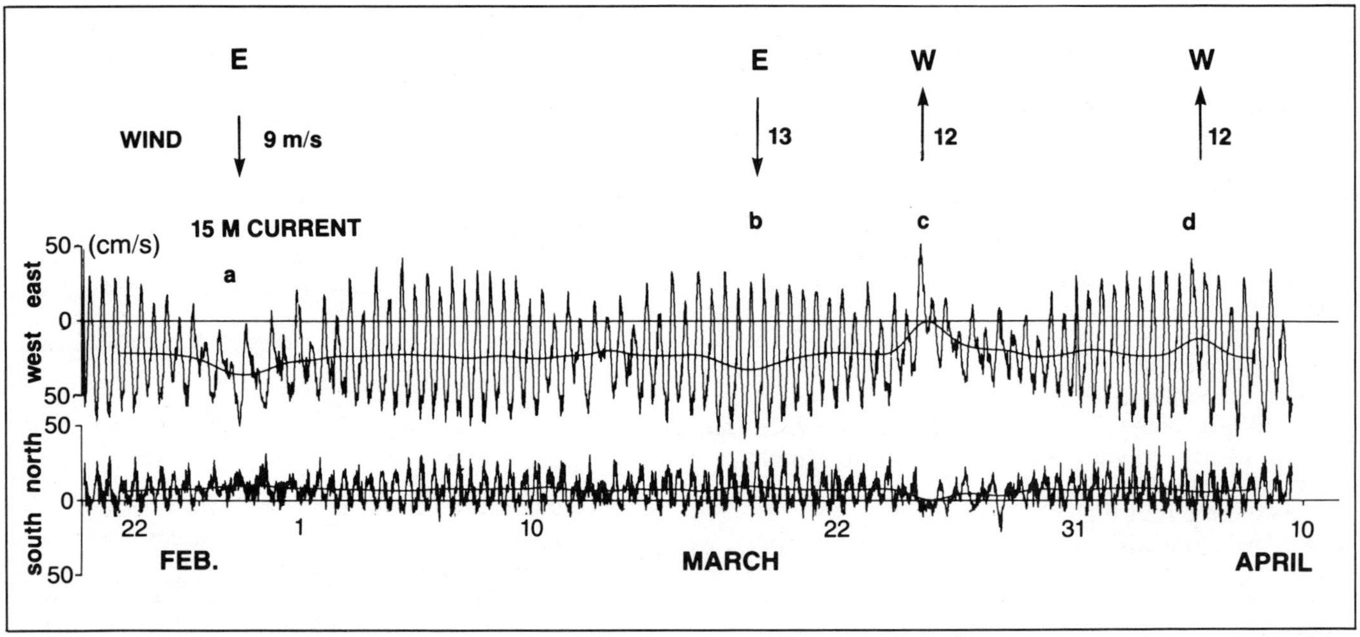

FIG. 12.12. Variations in speed and direction of currents at a depth of 15 m (50 ft) in Johnstone Strait at circled-star in Fig. 12.2. February–April 1973. Top line, along-channel currents; bottom line, weaker cross-channel currents. Flood directions are upward and ebb directions downward. Solid line through more rapidly varying ebb–flood cycles represents residual flow persistent to west and north (i.e. like persistent ebb). Residual current disrupted on four occasions (a–d) by strong winds; numbers next to arrows give wind speed (m/s). Downward arrows correspond to easterly winds, upward arrows to westerly winds.

Johnstone Strait or Discovery Passage from the mainland side. Therefore, strong currents are mostly confined to the main basins except in the very narrow tidal passes such as Yuculta and Dent rapids further inland. At the entrance to Sunderland Channel north of Kelsey Bay, for example, surface currents were almost nonexistent at a time when currents in Johnstone Strait, less than ½ km to the south on the other side of Yorke Island (Fig. 12.2), had speeds of nearly 100 cm/s (2 kn). Measurements from Nodales Channel that adjoins the northern end of Discovery Passage also reveal weak currents compared to those along the main channels to the west. Therefore, it would appear that tidal streams along Johnstone Strait and Discovery Passage prefer the larger channels and little branching of the flow takes place.

The brief current velocity measurements obtained in the 500-m deep, western end of Johnstone Strait indicate that tidal currents from Weynton Passage penetrate to the very bottom of the basin during the flood. The result is an eastward-advancing wall of water with a strong bottom jet that sweeps up-channel at maximum speeds in excess of 1.5 m/s (3 kn). A sharply defined tideline usually marks the leading edge of the flood waters that push their way into the Strait from the adjoining passages. Formation of the strong penetrative jet in this case is analogous to that of Rupert Inlet described in Chapter 3, with negatively buoyant flood streams that sink as they enter the less dense water of the Strait.

At the opposite end of the deep western basin of Johnstone Strait, it is over the shallow 70-m deep sill on the Kelsey Bay side of Race and Current Passages that some of the most vigorous tidal mixing on the coast occurs (see Fig. 12.3, 4, 5). Within the confined passages themselves, the swirling ebb and flood streams can flow at more than 3 m/s (6 kn) and there is little maneuvering room for larger vessels or tugs with a tow. These factors led to establishment of a maritime separation scheme in the vicinity of Helmcken Island, that requires traffic to proceed westbound via Current Passage and eastbound via Race Passage (Chart No. 3523). Mariners should note that the times of slack water in Race Passage are nearly simultaneous with those at Seymour Narrows, despite the 65-km separation and, therefore, are obtainable from tide tables. This correspondence also holds true for low-water slack in Current Passage but not, surprisingly, for high-water slack. For some reason (possibly the differences in inertia of the flood streams in each channel), the turn to ebb in Current Passage occurs some 75 min earlier than in Race Passage! It may be this peculiarity that accounts for the sharply defined tideline that lies in a north-west–southeast direction just west of Hardwicke Island on the ebb; waters on the northern side of the demarcation line can be ebbing westward at the same time those to the south flood eastward or flow westward at a somewhat slower rate. On the change to flood, the slow westward advance of the tideline is arrested, reverses direction, and approaches Kelsey Bay along the Vancouver Island shore. The waters behind the line are commonly distinguished by their turbulent churning motions and by an accompanying choppy sea, which breaks at the leading edge of the advancing convergence zone. Eventually, the tideline enters the two passages where it is swept away during the change to flood.

Perhaps the most interesting feature of the currents is the presence of a strong, westward-propagating internal tide in the deep western basin of the Strait. An analysis of current meter records from this basin (Thomson and Huggett 1980) indicates these long internal waves of tidal period are continually generated by the ebb–flood motions of the astronomical tide over the shallow sill immediately west of Kelsey Bay. Within 10 km of the sill, these internal tides produce depth-variable, along-channel currents with upper layer speeds of about 20 cm/s (0.4 kn), which accounts for the marked difference in the strength of the tidal flows between the top and bottom layers. Yet, despite the magnitude of these internal motions, less than 0.3% of the total energy carried eastward by the inward-propagating, astronomical tidal wave goes into their generation at the sill. The internal tides themselves undergo rapid frictional damping as they travel toward Broughton Strait, with current speeds attenuating by a factor of ⅓ within a distance of a single wavelength (25 km) from their place of origin. Half way along the 100-km channel any trace of the internal tides has all but disappeared.

Queen Charlotte Strait

Present knowledge of the flow structure in this region is based on 4 mo of velocity measurements taken at the western end of the main basin (Fig. 12.2). Essentially, the current patterns have a similar synthesis to those in Johnstone Strait — tidal streams, estuarine currents, plus wind-drift currents. Because of the considerably greater cross-sectional area of Queen Charlotte Strait, however, all but the wind-generated surface currents have appreciably lower speeds compared to Johnstone Strait.

At the shallow, more northerly location, the estuarine component of the flow was seaward to a depth of roughly 100 m, with maximum near-surface speeds of about 15 cm/s (0.3 kn). Extrapolation of the data implies that the deeper waters to the bottom at 135 m were inland but at a much slower speed, consistent with the normal kind of estuarine flow structure found throughout the inland seaways (Fig. 12.13a). The estuarine flow at the deep, more southerly location was quite different (Fig. 12.13b). In this case, the average drift was negligible to a depth of 15 m, then gave way to a comparatively strong inflow to a depth of about 250 m. The estuarine component was again weak from this depth to the bottom at 330 m. If the measurements were characteristic of Queen Charlotte Strait, it means that the lower-density surface waters that flow out of Johnstone Strait eventually make their way seaward via the northern half of the basin. In contrast, the higher density oceanic waters that eventually plunge to the deeper portions of Johnstone Strait gradually work their way eastward at middepths on the southern side of Queen Charlotte Strait. Much of this oceanic water appears to move through Goletas Channel and, based on the bathymetry of the Strait, most likely prefers to continue eastward, following the deep trough that terminates north of Malcolm Island. During the height of the snowmelt period in early summer, there is probably a strengthening of the estuarine current in the surface layer and a more definitive seaward flow over the entire breadth of the Strait.

Tidal streams in the Strait are mixed, mainly semidiurnal so the two floods and two ebbs each lunar day are generally of unequal strength. The larger daily flood and ebb typically have speeds of around 20–25 cm/s (0.4–0.5 kn) but increase to as much as 30 cm/s during large spring tides. (Stronger flows will, of course, arise over shoals and

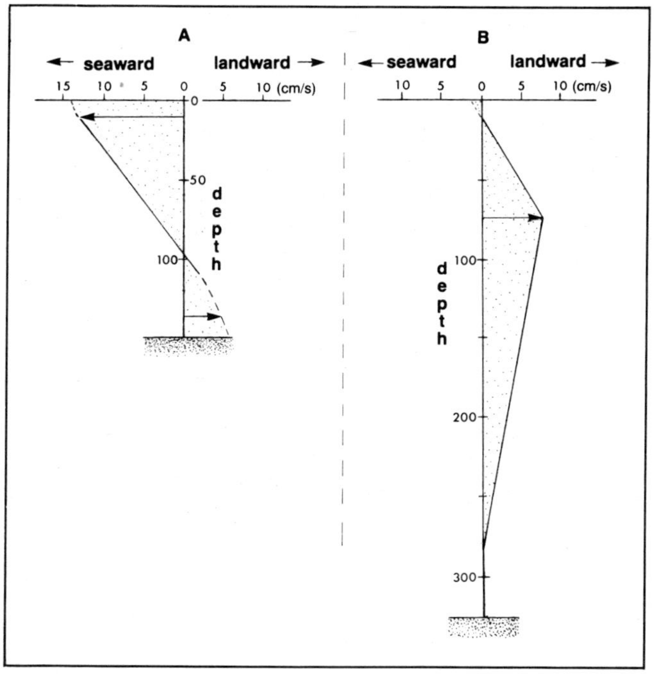

FIG. 12.13. Speed and direction of mean currents at two locations, western Queen Charlotte Strait, January–May 1977. (see Fig. 12.2 for locations.) (A) northern side; (B) southern side. Net flow generally seaward (westward) on northern side and landward (eastward) on southern side.

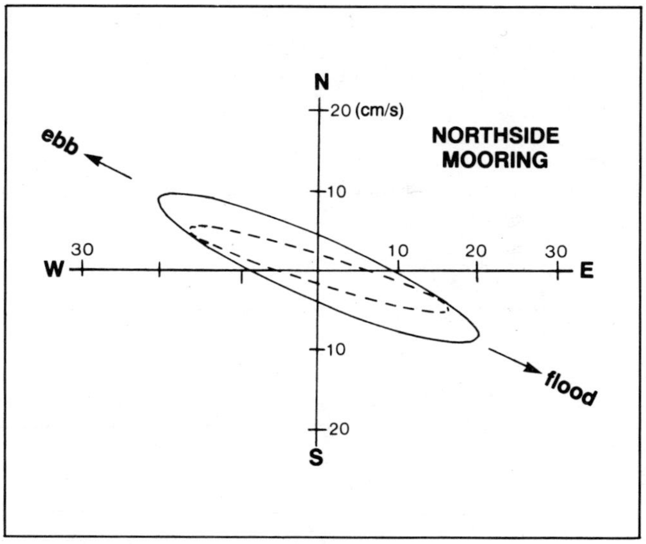

FIG. 12.14. Ellipses for semidiurnal tidal currents at two depths, northside mooring location, Queen Charlotte Strait. Solid line for 15 m depth; broken line for 75 m depth. Speeds cm/s, and currents rotate clockwise around ellipses (see Rotary Currents, Chapter 3).

in constricted passes between islands.) Tidal ellipses are flat and oriented parallel to the axis of the basin at all depths (Fig. 12.14). Except near the shore or in the vicinity of shoals, therefore, tidal streams flood to the east-southeast and ebb to the west-northwest and there is negligible cross-channel set. When the estuarine circulation is added to these pure tidal streams, the flow structure over the northern portion of the basin resembles a somewhat weakened version of Johnstone Strait, with relatively strong ebb currents and weak flood currents in the upper

100 m and the reverse situation at depth (Fig. 12.15a). Except during summer months, on the other hand, surface currents over the southern half of the basin possess little of this ebb bias (Fig. 12.15b).

Wind-generated surface drift in open oceanic basins like Queen Charlotte Strait was discussed in Chapter 4, 10. The only additional comment to make here is that the prevailing northwesterlies in summer tend to produce drift currents that partially counter the net seaward flow in the upper layer. During moderate to strong winds, flood currents become of comparable strength and duration to the ebb currents. The opposite effect develops in winter under the prevailing southeasterlies and, as in Johnstone Strait, there may be periods of a day or so when floods are weak or nonexistent. Further details on the currents in Queen Charlotte Strait await future oceanographic studies.

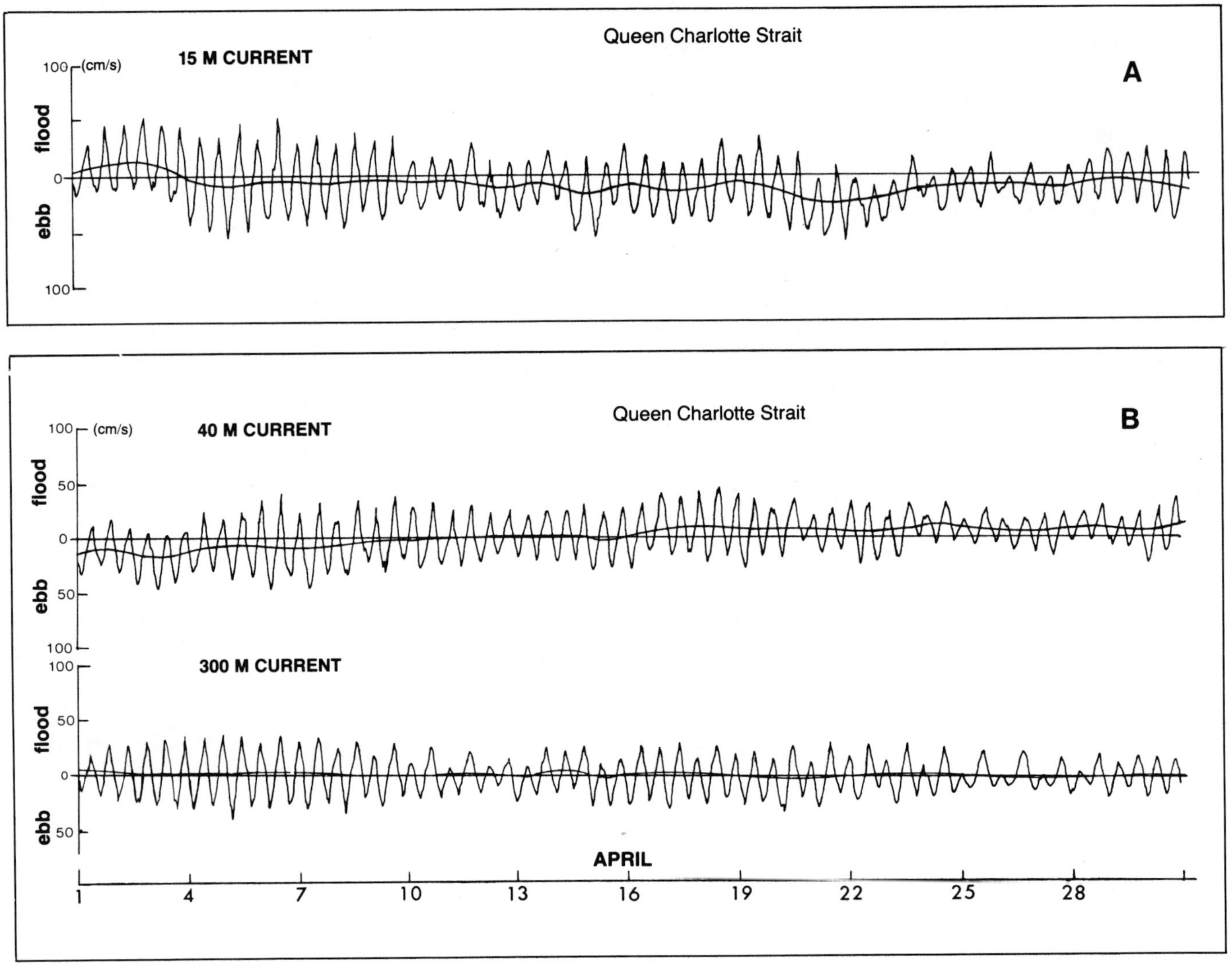

FIG. 12.15. Along-channel currents at western end Queen Charlotte Strait, April 1977 (A) northside, at depth of 15 m, (B) southside, at 40 and 300 m. Smooth line through more rapidly varying tidal currents represents residual flow.

PART V OCEANOGRAPHY OF OFFSHORE WATERS

Chapter 13. Deep-Sea Region

To this point attention has been mainly on the oceanography of the protected coastal waters. Discussion will now center on the exposed waters seaward of the British Columbia–Washington coast (Fig. 13.1) where the ocean is characterized by active seafloor crustal movements, rotary tidal streams, ever-present swell, wide meandering drift currents, and distinct water mass domains .

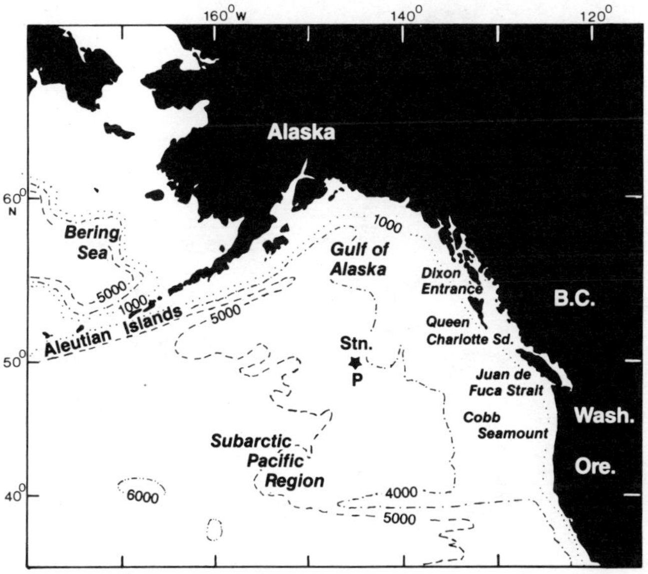

FIG. 13.1. Map of the subarctic Pacific region of North Pacific Ocean (eastern sector). Depth contours in metres.

Plate Tectonics

The continental margin of British Columbia and Washington is a region of active tectonism, a result of interactive stresses between the Pacific, North American, Explorer, and Juan de Fuca crustal plates (see Fig. 1.12). These plates (Chapter 1) are like giant flagstones that abutt one another as they drift above the supporting asthenosphere. The boundaries between the plates consist of a major fault, a seafloor spreading ridge system, and a subduction zone that mutually intersect to form a "triple junction" off northwestern Vancouver Island. To the northwest of this point, oceanic crustal rocks generated with the 150-km length of the Explorer Ridge system are moving northward relative to the North American Plate parallel to the Queen Charlotte Fault. This fault is one of the world's most important zones of crustal dislocation and is the locus of numerous large earthquakes. To the south of the triple junction, oceanic rocks of the Juan de Fuca plate and its subplate, the Explorer plate, which are generated within the ridge system (Explorer Ridge and its southwestward continuation, the Juan de Fuca Ridge) are moving eastward relative to the North American plate at a rate of 4 cm/yr. At the region of convergence, approximately beneath the continental slope, the oceanic crust is descending beneath the continental crust within a subduction zone. The resorbed crustal material ultimately melts, resulting in intrusion into the overlying continental crust. Active volcanoes such as mounts St. Helens and Baker have been formed where this material has been extruded to the surface.

Physiography

To the east, the Pacific west coast is bounded by the rugged mainland Coast Mountain Range that rises over 2000 m within 70 km of the ocean and is dissected by numerous fiords, sounds, and straits. Seaward of these mountains are the segmented mountain chain of the Alaska Panhandle, the Queen Charlottes, Vancouver Island, and the Olympic Peninsula. The continuity of the mountain chain is disrupted by three major low-lying depressions—Dixon Entrance, Queen Charlotte Sound, and Juan de Fuca Strait. Mountains of the Queen Charlottes, Vancouver Island, and the Olympics reach maximum elevations of 1200 m, 2200 m, and 2400 m, respectively.

The shallow continental shelf that borders the coast varies considerably in width. It is almost nonexistent off the Queen Charlotte Islands where the continental slope drops precipitously to depths of over 2500 m within only 30 km of the shore. Over millions of years, sediments derived from erosion of the adjacent land that moved down-slope in this region have filled the Queen Charlotte Trough at the base of the slope with over 1000 m of flat-lying strata (Fig. 13.2). To the west of Vancouver Island the shelf widens, from 20 km in the north to 80 km in the south, with a further increase to 100 km off the coast of Washington. The shelf is a large sediment trap which, in places, is underlain by 3000 m of land-derived materials carried seaward during the past 60 million yr by currents and glaciation. Cuts through the shelf are either due to past glacial action when sea level was much lower than at present or to fracturing along faults in the earth's crust.

The north-to-south widening of the shelf is generally accompanied by a decrease in the angle of the continental slope. The slope is by no means featureless. Many troughs and ridges trend parallel to the coast and numerous canyons cut across the slope. There are about 30 of these steep-walled canyons between Cape St. James and Cape Flattery, or an average of one every 20 km.

The gentle continental rise that typically lies at the base of appreciably steeper continental slopes is for the most part absent on the west coast. The main exceptions are the Nitinat Deep Sea Fan at the seaward end of canyons that lead from Juan de Fuca Strait and the smaller fans that lie west of the canyons emanating from Queen Charlotte Sound. Water depths gradually increase west-

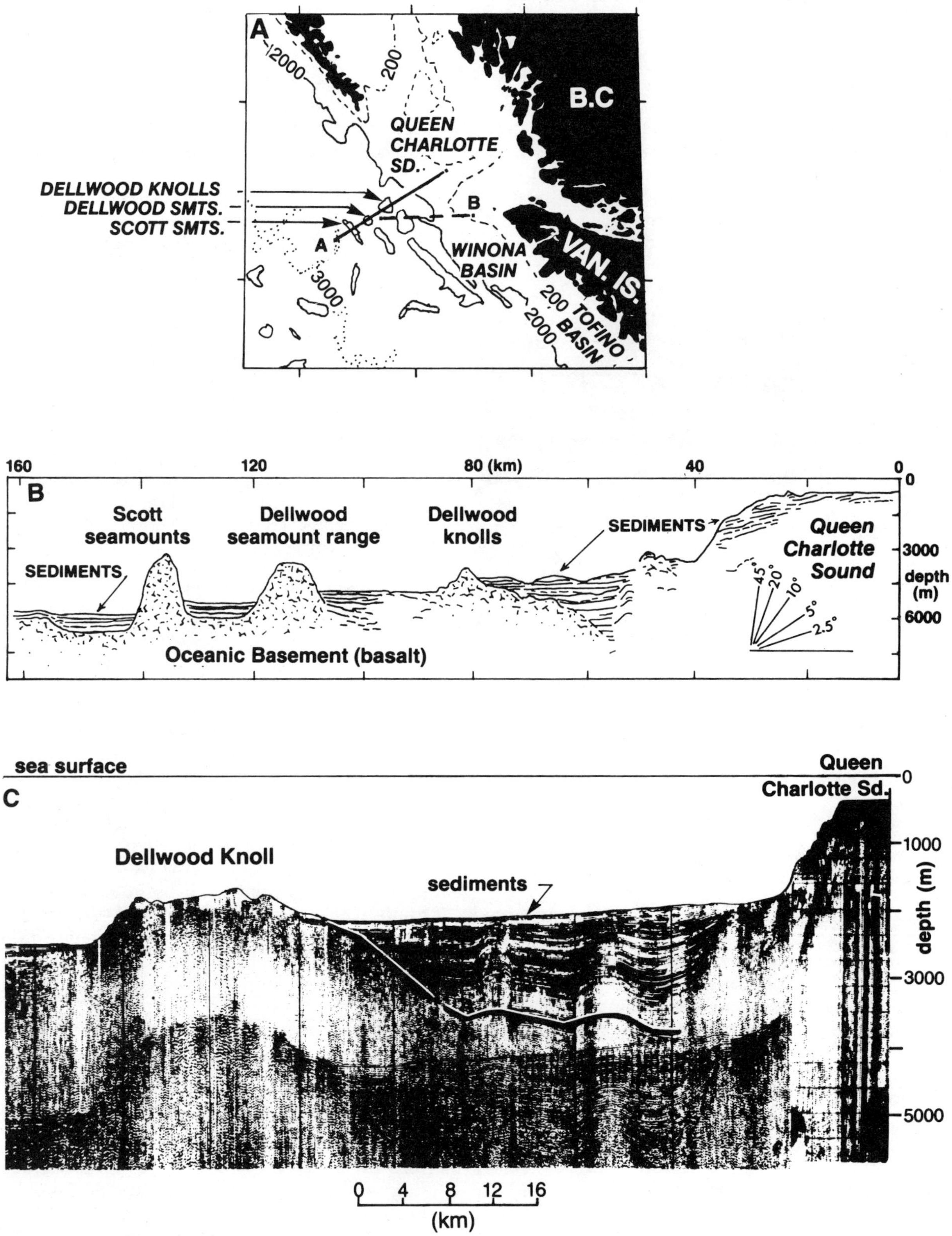

FIG. 13.2. (A) Seismic lines taken over northern continental margin seaward from Queen Charlotte Sound (B) Seismic lines "A". This shows seaward dipping sediments under the continental shelf, disrupted strata at the base of the continental slope, and a damming of sediment by the Dellwood Knolls and seamount ranges. Lower right scale gives true slope angles (vertical scale in diagram is exaggerated). (C) Seismic line "B". This shows deep sediments of Winona Basin at the base of the slope. Thick solid line indicates probable depth of basement rock beneath sediments. (Courtesy D.L. Tiffin)

ward and attain values in excess of 4000 m over the broad abyssal plain more than 1500 km offshore.

Water Properties

The density of seawater in the Northeast Pacific is determined primarily by salt content (salinity) rather than temperature. As a result, oceanographers have been able to define a number of oceanic domains based on salinity structure which, as shown in Fig. 13.3, include: the dilute domain, where freshwater discharge from the coast has diluted the upper layer of the ocean; the upwelling domain adjacent to the coast from California to British Columbia, where northwest winds cause deeper cold water to rise to the surface; the transition domain that marks the gradual change in the water properties between the subtropical and subarctic regions; and the ridge domain, where the density domes upward in a core region in the Gulf of Alaska about which the currents flow in a counterclockwise direction.

The 32.6‰ isohaline is used to delineate the maximum westward penetration of coastal runoff influence and normally lies many hundreds of kilometres offshore. Because west coast rivers discharge considerably more water in warm months than cold, this line is much further westward of the coast in summer and fall than in winter and spring. The salinity of the sea surface also tends to be patchy with large "puddles" of brackish-type water surrounded by water of considerably higher salinity of non-coastal origin. Infrared (heat sensitive) photographs taken from satellites (Fig. 13.4) often show the movement of cold, brackish water from the protected coastal areas to the continental shelf and slope region of the British Columbia–Washington coast.

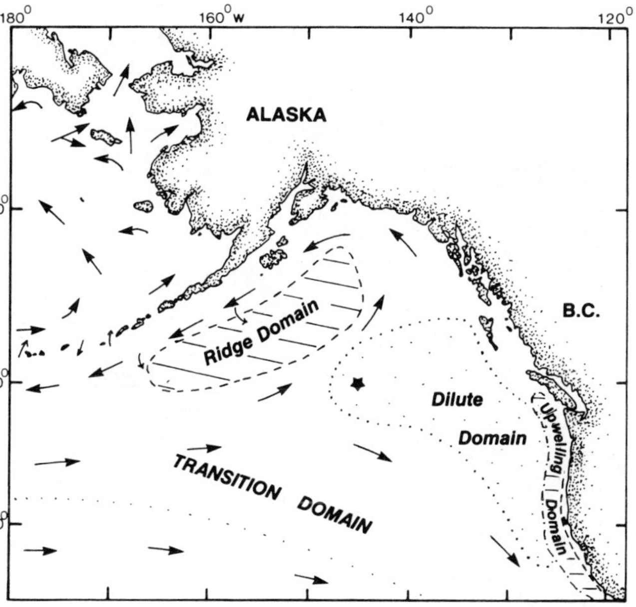

FIG. 13.3. Approximate extents of oceanic domains and prevailing current directions in subarctic Pacific region. Star is Ocean Station P. (Modified after Favorite et al. 1976)

Within the upwelling domain that extends from Vancouver Island to southern California, mean surface temperatures in summer can be more than 5°C lower, and mean salinities 0.1–0.3‰ higher, than in immediately adjacent offshore areas. At Cape St. James, north of the upwelling region, surface salinities are maximum between October and December, and minimum between July and September. By contrast, the surface waters adjacent to Vancouver Island are most saline from July to August, when upwelling is prevalent along the coast, and least saline from November to February, when river discharge from Vancouver Island valleys is at a peak.

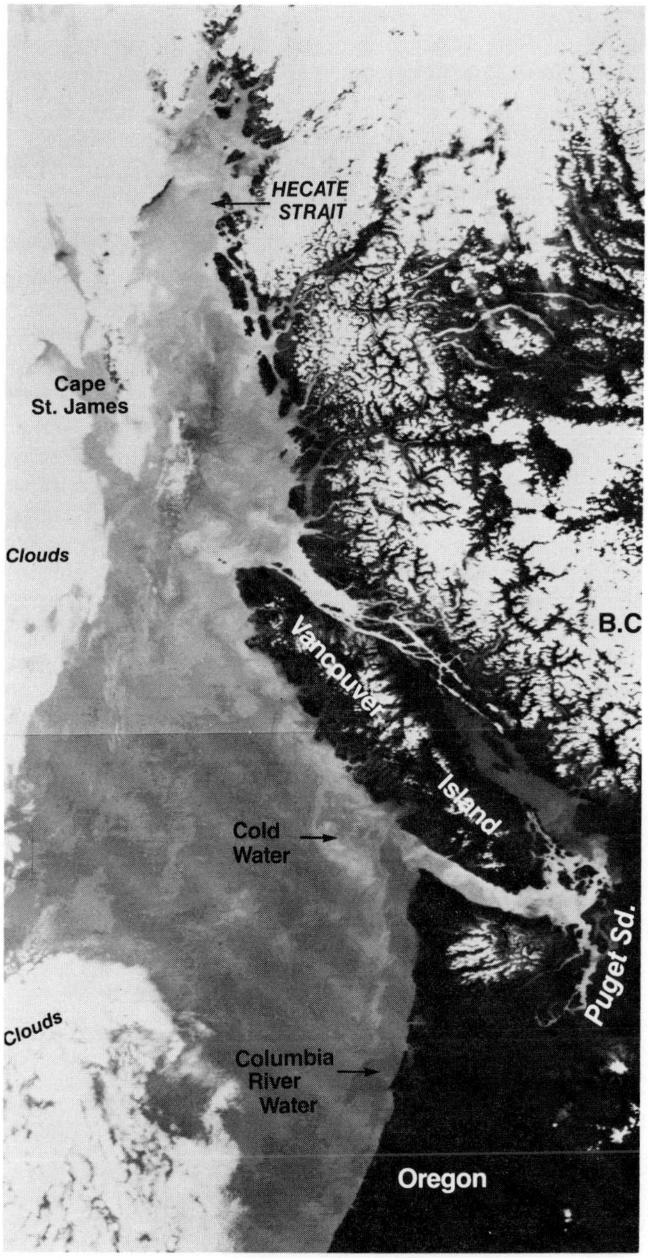

FIG. 13.4. Offshore movement of cold surface waters off Vancouver Island (NOAA infrared satellite photograph, Sept. 9, 1975). Relatively cold coastal waters appear whitish off B.C. coast; warm waters appear dark. Light areas off Oregon are probably cold upwelled water. (Courtesy S. Tabata)

The ridge domain represents the core of the counterclockwise Alaska Gyre. The bowing-up of the density (i.e. salinity) surfaces here is due to winds of the Aleutian Low that cause waters in the upper few hundred metres to diverge and deeper waters to slowly rise upward to take their place, a process not unrelated to upwelling. It is these upward motions in conjunction with the earth's rotation that set up the horizontal currents of the gyre—the winds are said to "spin-up" the ocean.

In addition to the water mass domains, the salinity structure in the Northeast Pacific Ocean is characterized by three distinct vertical layers or zones (see Fig. 2.12). The upper zone extends to a depth of approximately 100 m, and is the region of major seasonal variability with respect to almost all oceanographic properties including temperature, salinity, dissolved oxygen content, plankton distribution, and nutrients. Beneath this, to around 200 m, is the permanent halocline, a stable layer of water with a comparatively large (for the ocean) salinity increase of 1‰. In the lower layer, salinity and temperature change more gradually to the bottom and undergo very minor alteration over periods of years to centuries.

Surface salinities in this sector of the Pacific Ocean range from highs of around 33‰ in midocean to about 29–32‰ in coastal areas in winter, with extreme lows in the range of 25–30‰ off the mouth of the Columbia River and the coast of Vancouver Island in late spring. At the very abyss of the region at 4000 m, the salt content remains essentially unchanging at about 34.8‰.

From November to March average sea-surface temperatures range from about 10°C off Washington to around 7°C just north of the Queen Charlotte Islands, plus or minus a few degrees. By midsummer surface water temperatures outside of the upwelling zone are about 15–16°C off Washington and 12–14°C off British Columbia (see Fig. 2.10). Temperatures diminish rapidly with depth beneath the surface, and below 150–200 m are generally colder than 5°C near midocean and 7°C near the continental shelf.

Seasonal Heating and Cooling

The way the upper ocean stores and releases solar energy through the year is a good illustration of the close interaction between the atmosphere and the ocean. When the vertical structure of the water off the coast in February or March is measured for example, there is very little difference in temperature from the surface to a depth of roughly 100 m. Below this isothermal layer, between 100–150 m, a decrease of a degree or so occurs (the permanent thermocline) followed by a very gradual temperature decrease to the bottom (Fig. 2.12). The same measurement in summer reveals a thin top layer of relatively warm water (12–15°C) below which the temperature drops off rapidly by 5–10C° within 50 m (the seasonal thermocline), followed by a more gradual decrease down to the weak permanent thermocline starting at around 100 m.

In March, the upper zone of much of the Northeast Pacific is of uniform temperature. With the advent of spring, solar energy absorbed by the surface water during daylight hours is not completely radiated back into space at night so there is a net gain of heat. This warmer water is mixed downward by the mechanical stirring action of wind-generated waves and currents until it reaches a depth of enhanced density stability or a depth at which the stirring becomes insignificant (Fig. 13.5a). By summer, rapid warming of the surface waters has begun but, because of light winds, the additional heat becomes mixed to only shallow depths of a few tens of metres before it encounters a region of stability. The outcome is that, by early August, the upper 100 m of the ocean is arranged into layers with an isothermal upper mixed layer 10–20 m thick (depending on wind conditions) that overlies a number of layers of rapid temperature decrease, the seasonal thermoclines. Below these gradient regions, temperatures decrease more gradually until the top of the lower zone. During periods of windless summer weather, temperatures in the top few centimetres of localized areas of

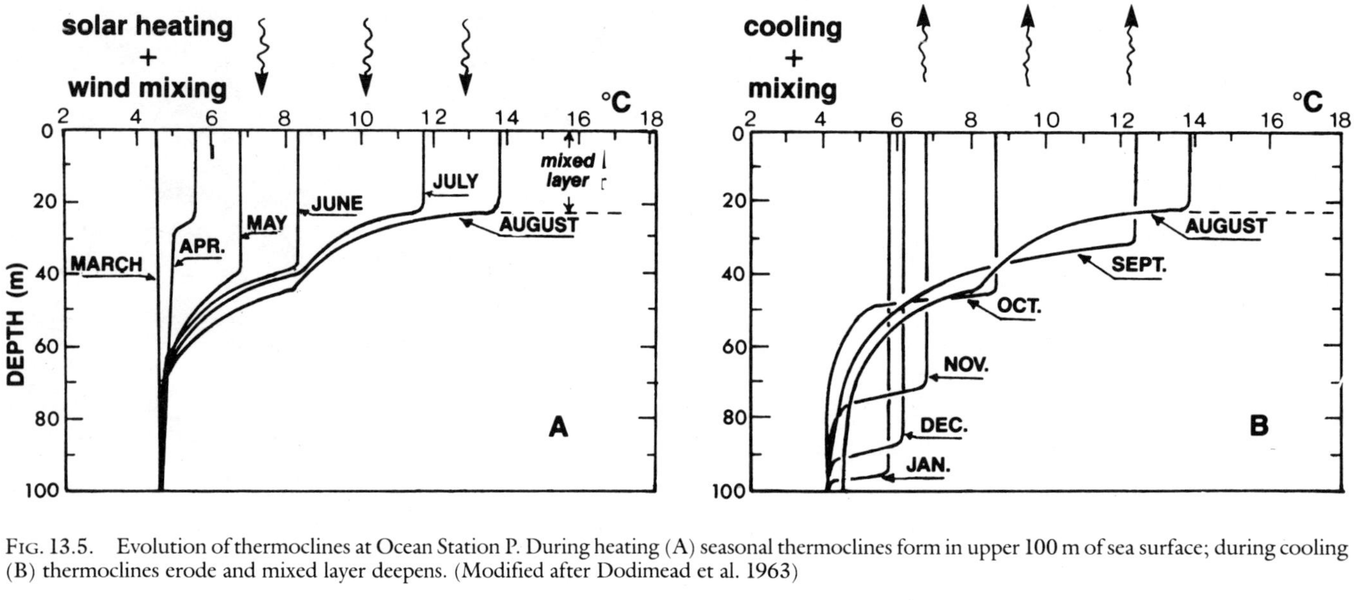

FIG. 13.5. Evolution of thermoclines at Ocean Station P. During heating (A) seasonal thermoclines form in upper 100 m of sea surface; during cooling (B) thermoclines erode and mixed layer deepens. (Modified after Dodimead et al. 1963)

the Northeast Pacific may rise above 20°C during daytime, though such pools of warm water tend to be quickly destroyed by external conditions such as winds and air cooling.

After August, the amount of solar energy available to the upper portion of the Northeast Pacific tapers off and the ocean begins to release stored heat to the atmosphere and helps fuel the ensuing storms. Throughout the fall and winter, the winds encourage this process by generating waves and currents that churn up the waters of the upper zone, continually bringing deeper water in contact with the cooler air, so a deepening wind-mixed layer is formed (Fig. 13.5b). The heat loss in the upper zone is further accelerated by the presence of cold air over the ocean that rapidly cools the surface water, which then becomes heavier than the warmer water below and sinks. Known as convective overturning, this mechanism allows colder water to penetrate below the depth of direct wind influence to erode the deeper temperature gradients. Around January each year an isothermal upper zone extends to the top of the halocline where it is prevented from going any deeper by the salt-controlled stability of the water column. At this point, overturning and wind stirring progressively remove the last vestiges of summer heating in the upper 100 m or so of the ocean until, by March, the process is ready to be repeated. However, there are slight year-to-year variations. During especially cold winters the temperature of the upper zone may fall below the temperature at the top of the permanent thermocline and, as illustrated by the inversion in Fig. 2.12, creates a layer of relatively warm water at the base of the upper zone by the next summer known as a dicothermal layer. Such features are common to the open ocean and also, it seems, to inlets and sounds in B.C. coastal waters. In the case of inlets, seasonally variable intrusions of relatively warm or cool oceanic water are also responsible for formation of these layers.

Before closing this section, there is a characteristic difference between the vertical structure of the open ocean and that of the inshore waters. Within the open ocean, strong thermoclines are transient seasonal features always situated above the well-established main halocline, whereas in protected coastal waters dominated by estuarine discharge, the major halocline and thermocline are always coincident.

Climatology

As outlined in Chapter 2, the prevailing winds in the northeast sector of the Pacific Ocean are determined by two major large-scale pressure cells, the Aleutian Low and the North Pacific High. Variations in the relative intensity and location of these two cells affect the pronounced seasonal cycle in the strength and direction of the prevailing offshore winds as well as their long-term climatic variability. The large pressure systems also guide the movement of synoptic-scale wind patterns, with lifetimes of a day to several weeks, associated with smaller-scale cyclonic and anticyclonic pressure cells (lows and highs) and their frontal systems. Storm tracks are commonly directed northeastward between the centers of the two major pressure cells.

Prevailing Winds

The Aleutian Low builds in intensity from about August to January, then tapers off until July when it is no longer evident in the monthly average surface pressure maps. The Pacific High on the other hand reaches maximum intensity between June and August when it dominates most of the Northeast Pacific. As a result, the strong cyclonic (counterclockwise) winds that prevail over the oceanic area from late fall to early spring give way to weaker anticyclonic winds in summer. These features are clearly revealed in the monthly averaged wind speed distributions plotted for every second month in Fig. 13.6. (A more detailed analysis of the average winds adjacent to the British Columbia and Washington coasts is in Table 13.1) From October to March, when cyclonic winds are most prevalent over the open ocean, prevailing coastal winds are from the southern quadrant (S, SE, and SW) about 40–50% of the time. During these months, northwest winds occur only about 10–20% of the time. The arrival of summer and anticyclonic oceanic winds is heralded by the onset of predominantly west to northwest winds along the outer coast (40–50%), together with a reduction in average wind speed. Along the coasts of Oregon and California, summer winds are almost invariably from the north whereas in winter both northerly and southerly winds predominate.

Synoptic Winds

Seasonal modifications in the prevailing oceanic wind patterns are accompanied by changes in the number and intensity of traveling cyclonic disturbances. For instance, storms occur more often and with greater strength in fall and winter than in spring and summer, a fact clearly demonstrated by a spectral analysis of 10 yr of wind data at Ocean Station P (Fissel 1975). According to this analysis, there is a doubling of peak energy associated with synoptic-scale oceanic wind systems between summer and fall as a result of enhanced storm activity. (Energy here is a measure of the wind speed squared. Peak energy corresponds to that cyclic component of wind speed with the most energy.) This is followed by a slight 10% reduction in peak wind energy between fall and winter and by a further reduction of 50% from winter to spring. These findings are consistent with monthly averaged wind speeds at P, which vary from about 6 m/s (12 kn) in summer to over 10 m/s (20 kn) in winter. The rather abrupt intensification of winds that takes place in fall is presumably linked to favorable conditions for cyclonic development that arise in the oceanic area between the strengthening Aleutian Low and the weakening North Pacific High.

A further finding of Fissel's analysis is that the most probable time span between maximal wind conditions decreases from about 4½ days in summer to around 2½ days in fall and winter. Moreover, the average range in wind speed over these periods varies from 6.5 m/s (13 kn) in summer to more than 9 m/s (18 kn) in winter. Typically, then, the mariner's respite between peak winds is a few

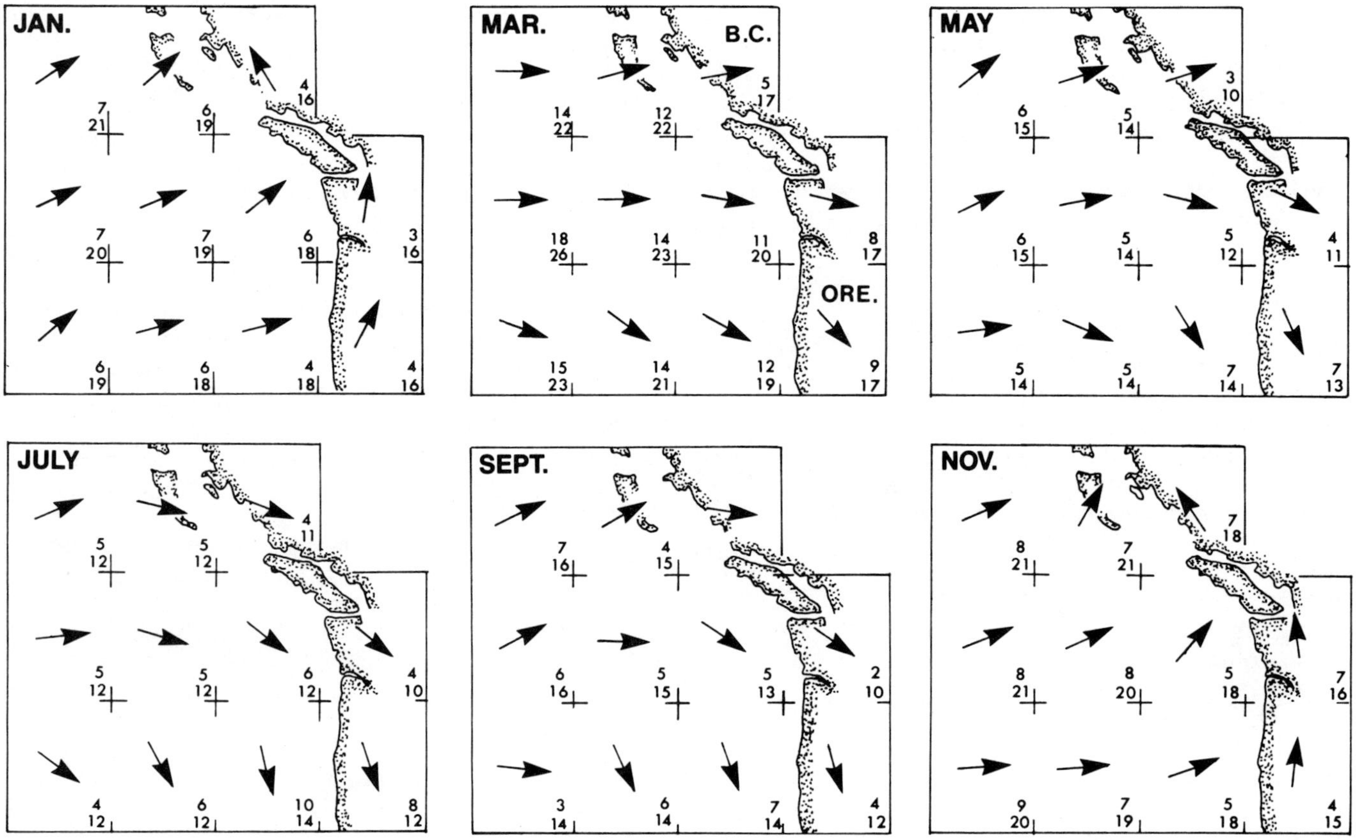

FIG. 13.6. Wind directions and speeds off B.C.–Washington coast. Arrows show net (resultant) direction of wind for month, averaged over a 5 × 5° oceanic region. Upper number in each region gives speed (kn) of resultant wind, lower number gives wind speed regardless of direction (kn). (Adapted from National Marine Fisheries Service)

TABLE 13.1. Coastal winds at selected stations. Cape St. James and Estevan Point give the prevailinig wind direction and time (%) winds are from that direction; direction with maximum average speed (m/s); average wind speed regardless of direction (m/s and kn). Last three stations give prevailing wind direction and wind speed regardless of direction. (Adapted from Anon 1971, and British Columbia Sailing Directions Vol. 1, 1976; Vol. 2, 1977)

		Jan.	Feb.	Mar.	Apr.	May	June	July	Aug.	Sept.	Oct.	Nov.	Dec.
Cape St. James, B.C.		NE(19%) SE(14.8)	SE(19%) SE(13.4)	SE(19%) SE(12.3)	NW(26%) SE(12.8)	NW(33%) SE(9.4)	NW(39%) SE(9.2)	NW(50%) NW(9.3)	NW(36%) SE(9.2)	NW(29%) SE(9.8)	SE(19%) SE(13.4)	SE(20%) SE(14.3)	SE(20%) SE(15.5)
	(m/s)	11.3	11.1	10.1	9.8	8.1	7.9	8.0	7.2	8.3	10.2	10.9	11.8
	(kn)	21.9	21.6	19.6	19.0	15.8	15.3	15.5	13.9	15.5	19.8	21.1	22.9
Estevan Point, B.C.		SE(38%) NW(7.9)	SE(33%) NW(7.6)	SE(30%) NW(7.2)	SE(25%) NW(7.3)	NW(31%) NW(7.3)	NW(35%) NW(6.6)	NW(36%) NW(6.3)	NW(29%) NW(5.4)	SE(27%) NW(6.0)	SE(44%) NW(6.7)	SE(41%) NW(7.3)	SE(42%) NW(7.3)
	(m/s)	5.2	5.4	5.5	5.7	5.3	5.1	4.8	4.2	4.3	5.0	5.4	5.9
	(kn)	10.1	10.4	10.8	11.2	10.3	10.0	9.5	8.2	8.4	9.7	10.4	11.5
Umatilla Reef, Wash.		S	S	S	SW	W	W	W	SW	S	S	S	S
	(m/s)	9.3	8.2	7.2	7.7	6.7	6.2	5.1	3.6	4.1	6.7	8.2	8.8
	(kn)	18.0	16.0	14.0	15.0	13.0	12.0	10.0	7.0	8.0	13.0	16.0	17.0
North Head, Wash.		E	SE	SE	NW	NW	NW	N	N	N	SE	SE	E
	(m/s)	7.1	6.5	6.3	6.2	5.9	5.7	5.4	5.0	5.2	5.7	6.9	7.2
	(kn)	13.8	12.7	12.2	12.0	11.5	11.1	10.4	9.7	10.2	11.1	13.4	14.1
North Bend, Ore.		SE	SE	SE	NW	NW	NW	NW	NW	NW	SE	SE	SE
	(m/s)	4.2	3.8	4.0	4.1	4.5	4.3	5.2	4.4	3.4	3.0	3.2	3.7
	(kn)	8.2	7.3	7.8	8.0	8.7	8.4	10.2	8.5	6.7	5.9	6.3	7.2

days shorter in winter than in summer. Over the year, the mariner can expect a period of 1.5–10.0 days between the passage of successive high or lows, with a most likely separation of 3.1 days.

Guided by the prevailing westerly flow of the upper atmosphere, synoptic disturbances that originate over the North Pacific Ocean consistently move from west to east, conforming to the age-old dictum that "weather drifts to the east." Storms over the Northeast Pacific follow two main routes regardless of season. Both are directed into the Gulf of Alaska (Fig. 13.7). One track is to the northeast from the vicinity of the Aleutian Islands, the other to the northeast from the midocean area west of Station P. During their lifetimes, storms along these tracks cross 1500–2000 km of ocean.

With the advent of winter, lows begin to follow paths toward the coasts of British Columbia and Washington (Fig. 13.7). For example, storms formed over the open ocean at the southern edge of the Aleutian Low between November and March travel in the northeasterly direction until they eventually lash the coasts of southwest British Columbia and northwest Washington. Also at this time, major storm systems from the Gulf of Alaska are directed to the southeast along the outer coast and bring with them strong winds and rain. The occasional storm hits the coast in April and May, although usually with much less fury than in winter, and by summer all storms pass into the Gulf of Alaska well north of the Canadian coast.

As the previous discussion intimates, the likelihood of encountering gale force winds (Force 8) in the Northeast Pacific during a given month is greatest in the Gulf of Alaska, but lessens toward the B.C.–Washington coast as part of a general southward trend. (Gales are, of course, rare in the subtropics irrespective of season.) Table 13.2 shows that, on the average, gale force winds in areas A and B in Fig. 13.7c are encountered with greatest frequency from November through February along the B.C. coast, with a maximum of nearly 10% of the time in January. From May until September, the frequency of these winds is 1%, so that early August is generally the most pleasant time to venture offshore.

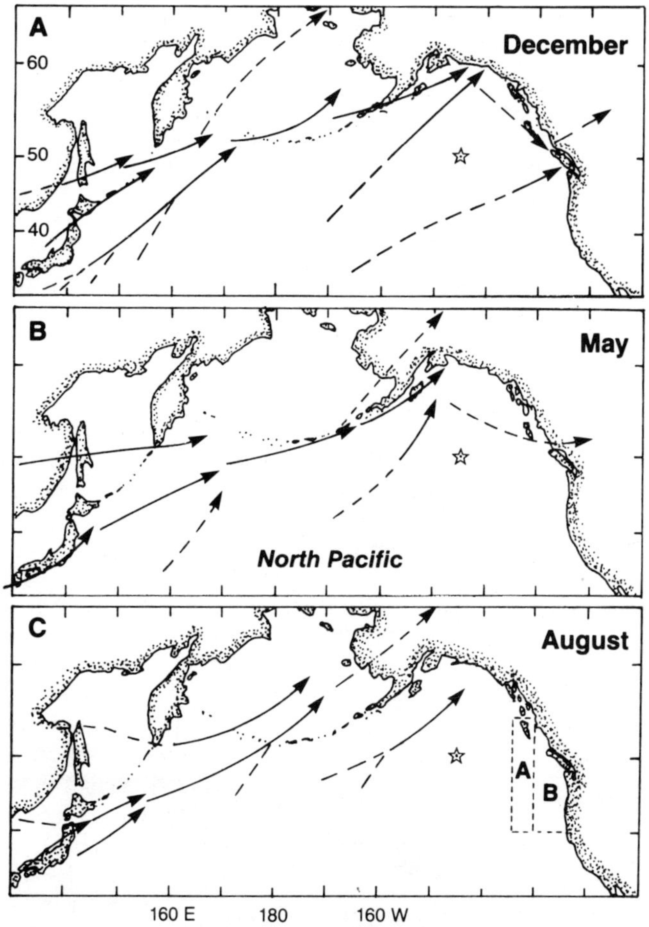

Fig. 13.7. Typical tracks followed by storms over North Pacific Ocean for three seasons. Solid lines denote primary tracks; broken lines, secondary tracks. Arrow heads terminate in areas of maximum storm frequency. Star gives location of Station P. Gale frequencies in (A) and (B) are in Table 13.2. (Adapted from U.S. Pilot Coastal Charts 1978).

Table 13.2. Frequency of gales for each month. Average percentage of ship reports from (A), (B) Fig. 13.7 when winds at least Force 8 (gale force) were recorded for the month. Where 0 is given, gales may have been recorded but too infrequently to give a percentage value. (From U.S. Pilot Coastal Charts)

Area	Jan.	Feb.	Mar.	Apr.	May	June	July	Aug.	Sept.	Oct.	Nov.	Dec.
A(%)	10	7	5	2	1	0	1	0	1	4	6	10
B(%)	10	6	5	3	1	3	3	1	3	4	8	9

Air Temperature

Due to the moderating effect of marine air, the west coast has a temperate climate with cool summers and mild, wet winters. The mean daily temperatures presented in Table 13.3 for Tofino (Vancouver Island coast), Tatoosh Island (Washington coast), and Ocean Station P (midocean) are characteristic of exposed regions where temperatures range from about 5°C in January to about 15°C in August. Tofino and Tatoosh Island naturally register greater temperature extremes than at midocean because of the influence of local topography, an effect that increases

Table 13.3. Average daily temperatures (°C) for each month. (from British Columbia Sailing Directions, Vol. 1, 1976; Anon 1976)

	Jan.	Feb.	Mar.	Apr.	May	June	July	Aug.	Sept.	Oct.	Nov.	Dec.
Ocean Station P,	5.0	5.0	4.4	5.6	7.2	8.9	11.1	13.3	13.3	10.6	8.3	6.1
Tofino, B.C.	4.4	5.6	5.6	7.2	10.2	12.7	14.3	14.6	13.1	10.3	6.9	5.1
Tatoosh Island, Wash.	5.6	6.2	7.1	8.6	10.5	12.2	13.1	13.2	12.5	11.1	8.6	6.7
Kamloops, B.C.	−5.6	−0.6	4.4	9.4	14.4	18.3	21.1	20.0	15.0	8.9	2.2	−2.2

markedly inland from the coast (Kamloops, Table 13.3). In summer, for instance, heating the land will lead to balmy coastal air temperatures but will have little effect a short distance out to sea. Similarly, summer upwelling of cold subsurface water by prolonged northwest winds within a narrow strip off Vancouver Island and Washington can lead to lower than normal air temperatures at the coast. In winter, Arctic winds from the northeast may cause subnormal air temperatures at the coast. With the obvious exception of the region immediately adjacent to the coast, air temperatures over the ocean are within a few degrees of sea-surface temperatures.

Cloud Cover

A sometimes discouraging aspect of the open Northeast Pacific Ocean is the lack of truly sunny weather. Even on the calmest summer days the sky seems to be invariably obscured by low cloud. Decades of observations bear this out. North of about latitude 40°N, the average cloud cover per month always exceeds 75%, whereas in subtropical regions like Hawaii it is around 50% most of the year. Over the North Pacific as a whole, the percentage cloud cover for each month has a nearly zonal trend with maximum coverage along a broad east–west band seaward of the B.C. coast (Fig. 13.8). Coastal cloud cover

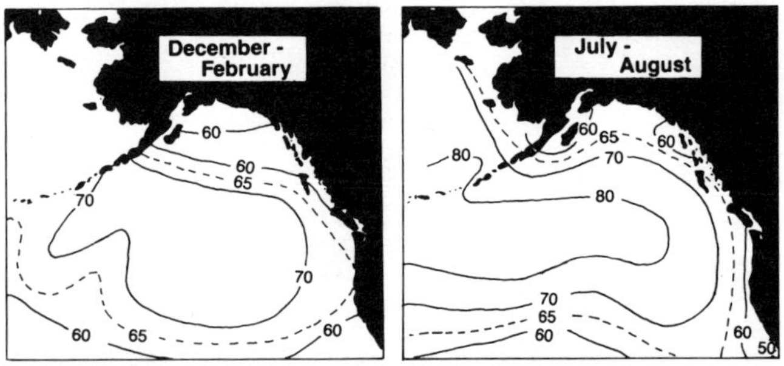

FIG. 13.8. Average percentage cloud cover over Northeast Pacific Ocean, winter and summer. (From Dodimead et al. 1963)

tends to be slightly lower than its deep-sea counterpart at a given latitude, especially off California where values are commonly around 60%. Off the B.C.–Washington coast, low cloud cover increases from a midwinter value of 65% to a maximum of over 70% in July.

Fog

Regular visitors to the west coast will notice the greater incidence of foggy days in summer compared to winter. This is not a local phenomenon but extends from Vancouver Island to northern California and is especially well established along the rugged coast of Oregon. Even on the hottest summer days the fog may persist, sometimes retreating a little seaward of the shore then returning again to dampen the enthusiasm of the sunbathers (Fig. 13.9).

Climatological data from both Tatoosh Island and Tofino show that August is the peak month with an average of 15 days of fog compared with only a few days a month from November through May. The foggy period lasts from about June to October and corresponds to the

FIG. 13.9. Sea fog, Kalaloch Beach, Olympic National Park, September 1979. Temperatures a few kilometres inland at the time were around 25°C and northwest winds were blowing along British Columbia–Washington Coast. (Photo by author.)

time of prevailing northwest winds. In this case, fog forms when the warm moist air carried by these winds contacts the colder coastal water whose temperature, in many instances, is well below normal due to upwelling of colder water from below.

Waves

Historically, most information on the seasonal distribution of wave heights and directions in the ocean has been derived from decades of visual estimates reported by officers of deep-sea vessels. Although wave statistics obtained in this way appear to be fairly reliable, the method has a number of drawbacks. It provides somewhat dubious statistics on waves away from the major shipping lanes, typically underestimates wave heights, and is biased toward good weather conditions. Recent comparisons between visual and measured wave heights reveal that mariners consistently underestimate the lower wave heights, but are generally accurate for high waves (Fig. 13.10). At Station P, most visual estimates were 50% lower than measured for significant waves of less than 3 m,

FIG. 13.10. High seas generated by gale force winds in Northeast Pacific Ocean, early May 1972. (Photo by author)

whereas for wave heights of over 5 m agreement between the two was good. These factors should be borne in mind when using wave frequency diagrams such as those published in *U.S. Coastal Pilot* or *Sailing Directions, British Columbia Coast.*

Heights

As expected, wave heights off the coast vary in accordance with oceanic winds. Both sea and swell are highest in fall and winter when the strength, duration, and fetch of winds are greatest, and least in summer when these wind factors are minimal. During the roughest months from October to March, seas exceed 1.5 m about 30% of the time off the southwest coast and 60% of the time off the more inhospitable coast of the Queen Charlotte Islands. The occurrence of swell heights in excess of 1.5 m for these months increases from approximately 25% off Vancouver Island to 50% off the Queen Charlottes, with fall conditions slightly worse than winter conditions. In summer, the occurrence of rough-to-high seas (heights in excess of 1.5 m) in these two areas diminishes to about 15 and 40% of the time, respectively, while moderate-to-high swell conditions (over 1.5 m) occur about 10 and 20% of the time. Hence, with regard to sea conditions as well as winds, summer is the most favorable season to set a southward course along the west coast.

Direction

Swell, by its very nature, is little affected by local winds so height and direction of travel is effectively determined by the prevailing wind direction. For this reason, and because of the short fetches for northerly to easterly winds, swell is directed toward the coast for a considerable distance offshore. From October to March the most favored directions are from the southwest and west for regions north of Vancouver Island, and from the southeast around to the northwest off Vancouver Island and Washington. By summer, swell is most likely to be from the west and northwest off the latter coast and from the southwest to northwest off the Queen Charlottes.

Seas, on the average, are spread over a greater directional range than swell and undergo more rapid shifts in direction over periods of time. This is especially true in fall and winter when storms and their inherent pattern of cyclonic winds are directed toward the coast. Off Vancouver Island, however, seas are from the southeast and northwest slightly more often that other directions, because of the tendency for southeast storm winds to veer to the northwest with the passage of a front. Waves in this area are from these two directions nearly 25% of the time from October to March, yet a few hundred kilometres further seaward they are more likely to be from the southwest quadrant. In spring and summer, seas are more aligned with the prevailing winds, i.e. from the northwest, although seas off the Queen Charlottes continue to arrive from a wide range of directions. During July, August, and September, sea waves are from the northwest an average of 20% of the time along the entire outer coast and calms account for another 7%.

Measured Waves

Direct measurements of wave heights have been made at three locations seaward of the B.C–Washington coast: at Cobb Seamount, 500 km southwest of Cape Flattery, via a self-contained, bottom-mounted pressure recorder; at Ocean Station P, 1500 km west of the coast, by Tucker shipborne wave recorders; and at a location 5.5 km southwest of Long Beach on Vancouver Island by a tethered waverider accelerometer buoy.

Of the three locations, only Station P is representative of a truly deep-sea environment. There, the distribution of wave directions is almost identical to the corresponding distribution of wind directions; both winds and waves for example are from the southwest quadrant roughly 70% of the time. This quadrant is also the direction of maximum possible fetch and subsequently the direction from which largest waves are expected to arrive. As Fig. 13.11a illustrates, the mean significant wave height, $H_{1/3}$, for all seasons is a maximum of around 5.5 m (18 ft) for waves from the southwest to northwest and a minimum of about 2.5 m (8 ft) for waves from the northeast quadrant, the direction of minimal wind fetch. The observed relationship between wind speed and significant wave height at station P is in Fig. 13.11b, together with the theoretical relationship.

In addition to the dependence of average wave height on direction, there is a pronounced variation in height with season. Average significant wave heights range from 4.9 m to 5.5 m (16–18 ft) between October and April, diminish to 1.2–2.1 m (4–7 ft) in June through August, then abruptly increase again to around 5 m some time between August and September, coinciding with a doubling of peak wind energy in the region (see previous section). When lumped together over an entire year, the wave height distribution in the region of Station P has the form of Fig. 13.12. This shows, for example, that on an

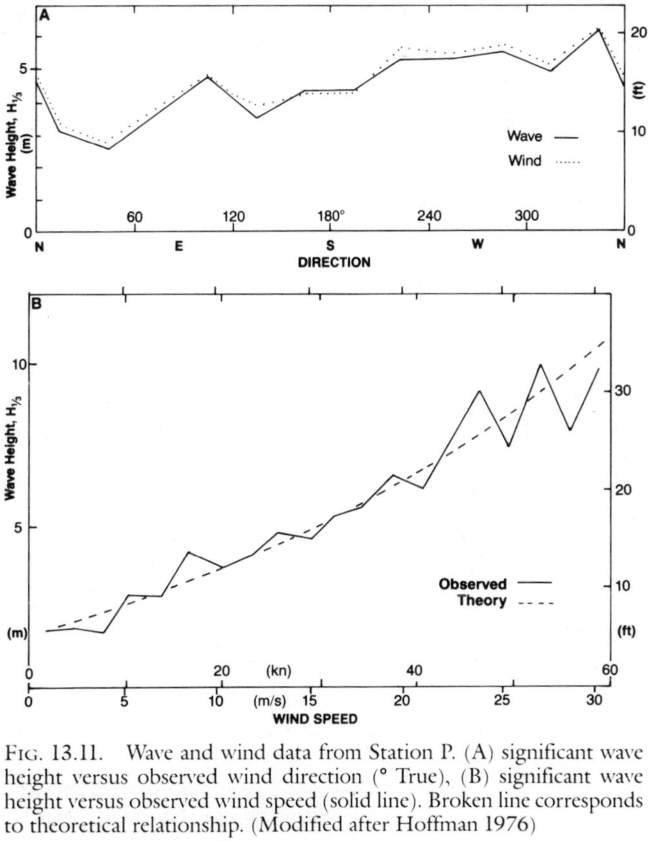

FIG. 13.11. Wave and wind data from Station P. (A) significant wave height versus observed wind direction (° True), (B) significant wave height versus observed wind speed (solid line). Broken line corresponds to theoretical relationship. (Modified after Hoffman 1976)

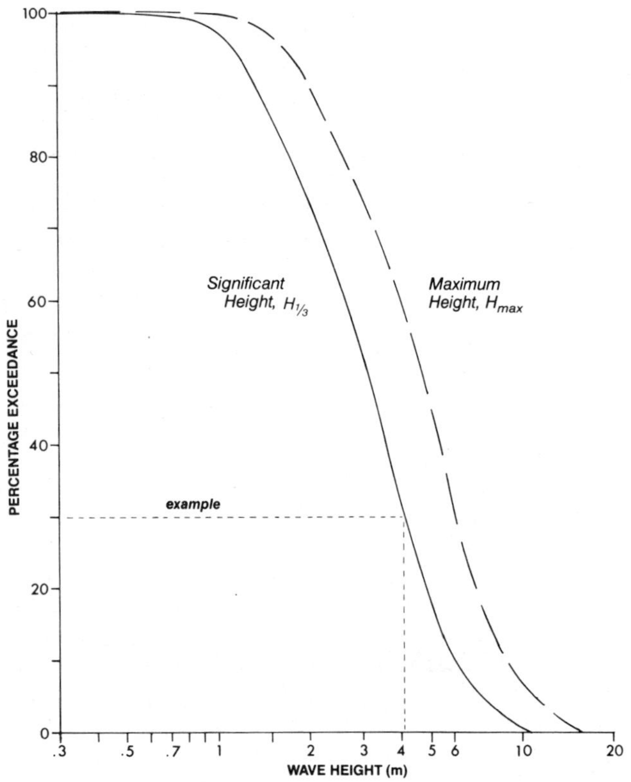

FIG. 13.12. Percentage of time waves at Station P exceed a particular height. Curves based on shipborne wave-recorder data from CCGS *Quadra* Jan. 1–Dec. 31, 1978. Significant wave height exceeds 4.1 m about 30% of time, but rarely exceeds 10 m. (Adapted from Marine Environmental Data Service 1979)

annual basis significant waves exceed a height of 4 m roughly 30% of the time, but only rarely exceed 10 m. The peak period of these deep-sea waves is 9–10 s, and such waves occur about 18% of the time; very long swell with periods of 17–20 s and wavelengths of hundreds of metres occur approximately 6% of the time. Figure 13.13a, b exemplifies the appreciable daily variation that significant wave heights undergo in this area of the North Pacific Ocean, regardless of season.

Wave energies at Cobb Seamount are similar to those in midocean with one important difference, the peak energy at the seamount occurs at a significantly longer period; 12 s at Cobb versus 9 s at Station P. Why swell waves that propagate over the seamount should have greater periods is still not explained. Among the possibilities are errors in data acquisition and analysis, influence of coastal currents, shoaling effects over the seamount, and meteorological differences associated with winds in the two regions.

Wave data have been collected off Vancouver Island since 1970 as part of a Canadian Wave Climate Study and provide the longest running survey of its kind on the coast. As noted earlier, records from the wave rider buoy permit estimation of the significant wave height over a 20-min interval every 3 h barring malfunctions. According to these measurements, significant wave heights averaged over all seasons are less than 3 m (10 ft) about 86% of the time and less than 6 m more than 99% of the time, and 1.2 – 2.1 m over 55% of the time (Fig. 13.14). The most common wave period, the period a mariner would expect to observe in a moderate sea, is 10 s (18% of the time), while maximum periods of 20 s corresponding to very

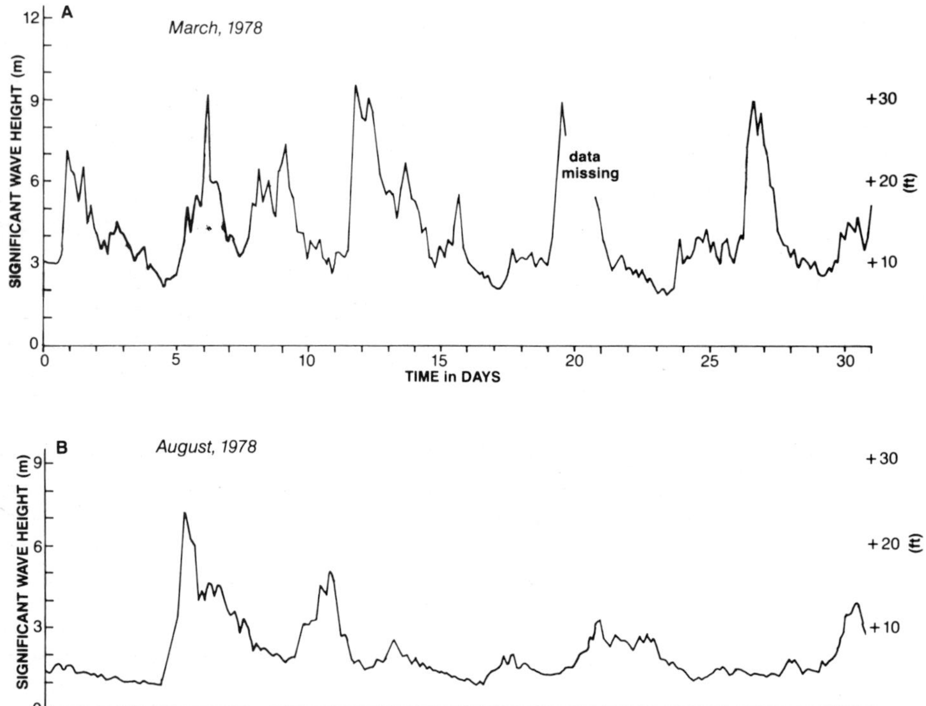

FIG. 13.13. Variation in the significant wave height (A) March and (B) August 1978 at Station P. Data based on wave-recorder measurements collected every 3 h from CCGS *Vancouver*. (Adapted from Marine Environmental Data Service 1979)

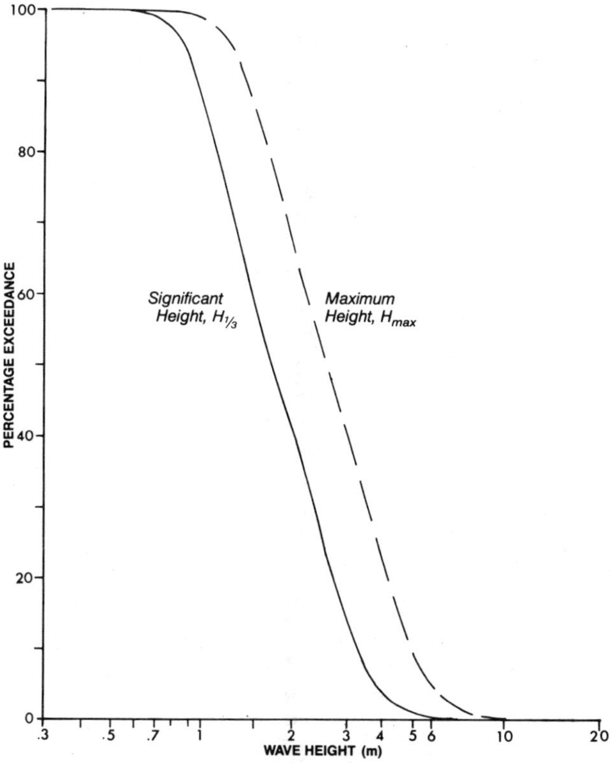

FIG. 13.14. Percentage of time waves off southwest coast of Vancouver Island exceed particular height. Data collected from wave-rider accelerometer buoy in 40.2 m of water, 5.6 km southwest of Long Beach.Compare with Fig. 13.12 (Adapted from Marine Environmental Data Service 1977).

long swells occur around 5% of the time. Only 6% of all waves have periods shorter than 6 s, whereas 50% have periods of 10 s or more.

The kind of day-to-day variability to be expected for waves off the coast is similar to Fig. 13.13 a, b for Station P, albeit with generally lower heights. A feature worth noting here is that, even during months of relatively low "average" waves, there can be periods of exceptionally high waves, and vice versa.

Tides

Present knowledge about the range and propagation of deep-sea tides is interpolated from coastal installations, including midocean islands such as Hawaii. However, because coastal tides are subject to modification by local topography, onshore winds, atmospheric pressure fluctuations, water density variations, and shallow-water effects, they may deviate appreciably from tides offshore. To obtain a global picture of the tide in the North Pacific Ocean it is necessary to resort to computer-simulated models, which can be verified where possible by coastal observations.

The general features of most tidal models are similar, although they often differ markedly in details. In the North Pacific, each model predicts the presence of an amphidromic point (or "pivot point") in the main semidiurnal tidal constituent in the vicinity of the Hawaiian Islands (Fig. 13.15). (Because this is the largest constituent, it is considered "the" tide.) As the exact location of this pivot point has yet to be found by observation, the theoretical location varies from model to model, but not enough to affect the present discussion. First, the tidal wave propagates counterclockwise around the amphidromic point so the cotidal lines (lines joining oceanic regions where high tide occurs simultaneously) radiate outward from the center like warped spokes from an axle. These lines intersect the coast of North America at right angles, or nearly so, and indicate that the tide propagates northward parallel to the coastline. Because the semidiurnal tide moves at a speed of 29°/h around the amphidromic point, or one complete circuit of 360° every 12½ h it takes but 3 h for a particular stage of the tide to travel from southern California to the northern end of Graham Island in the Queen Charlottes. Five hours later the same stage of the tide will be well along the Aleutian Islands. This rapid speed means that the tide covers the 1000 km from Washington to Graham Island in less than 1 h. Closer to the amphidromic point, the speed of the tidal wave decreases but so does the circumference of the circle it must travel (by analogy with a wheel) so that an observer would still find the same cyclic rise and fall of sea level over a lunar day.

Lines of equal tidal range (the corange lines, Fig. 13.15) spread radially outward from the center of the amphidromic point and attain a nearly uniform value of

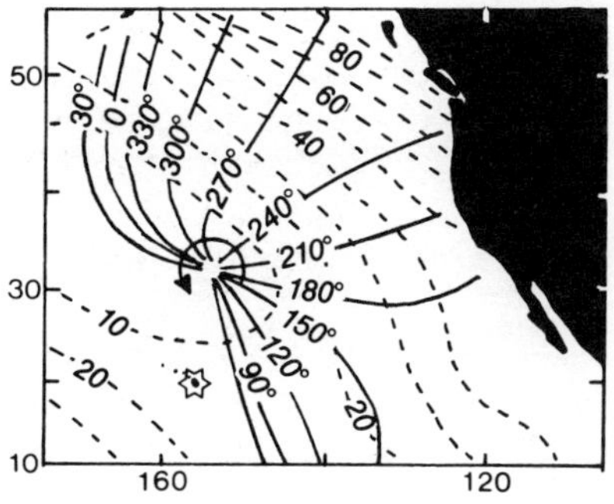

FIG. 13.15. Computer-generated cotidal and corange chart for the semidiurnal (M_2) tide off Pacific coast, North America. Solid lines (spokes) connect points with given stage of tide at same time (i.e. same phase). A difference of 29° in phase equals a difference of 1 h in time. Tide propagates counterclockwise around amphidromic point located north of Hawaii (star). Broken lines give the range of semidiurnal tide (cm).(From Bogdanov et al. 1964)

just over 2 m along the outer B.C. coast. If a further 2 m contribution is added from the diurnal tide and other smaller tidal components, a maximum average range of approximately 4 m is obtained for the full offshore tide. Off California, the full tidal range is half that off the B.C. coast while at Hawaii the maximum range is only 45 cm (1.5 ft) due to its close proximity to the amphidromic point. Persons accustomed to the large tidal ranges of the B.C. coast may find the almost nonexistent tides on the Hawaiian beaches a little strange at first, and may need to adjust to the fact that one's beach towel doesn't need constant rescuing from an encroaching tide.

Recently, Canadian and American scientists successfully measured offshore tidal variations by taking advantage of the relatively shallow depths over some of the seamounts in the Pacific Ocean. Pressure-recording devices accurate to 0.01% of the water depth (e.g. 1 cm in 100 m of water) were placed on the pinnacles of Cobb, Union, Bowie, Surveyor, and Patton seamounts (see Fig. 2.3), all with depths of less than 300 m. Tidal records from these locations, in conjunction with coastal stations and offshore islands off Alaska, were used to plot the cotidal and corange charts in Fig. 13.16, which, incidently, agree quite closely with the computer modeled results in Fig. 13.15. Plans are now underway for a more detailed look at the offshore tides using seamounts and shore stations in the North Pacific.

cific Ocean is general and based on ship logs, measurements of the density structure, and drift bottle experiments, all of which have their limitations. In contrast to this, the circulation over the continental margins is reasonably well understood because of extensive current metering programs undertaken in the last decade by oceanographers of Oregon State University, the University of Washington, the Pacific Marine Environmental Laboratory (Seattle), and the Institute of Ocean Sciences (Sidney).

For convenience, the current at a given depth and location may be separated into two basic components: (1) tidal streams generated by the astronomical tidal forces, and (2) nontidal currents produced by all other processes. The latter category includes the large-scale current systems driven by the prevailing surface winds and by horizontal density gradients set up by differential heating of the global ocean. Tidal streams are the more regular of the two. They are periodic in that they essentially repeat themselves once every 12½ h (for semidiurnal currents) or every 25 h (for diurnal or mixed currents) and, with the exception of nearshore or shallow-water regions produce no net flow when averaged over a lunar day. Their speeds and directions change little with depth and they tend to be

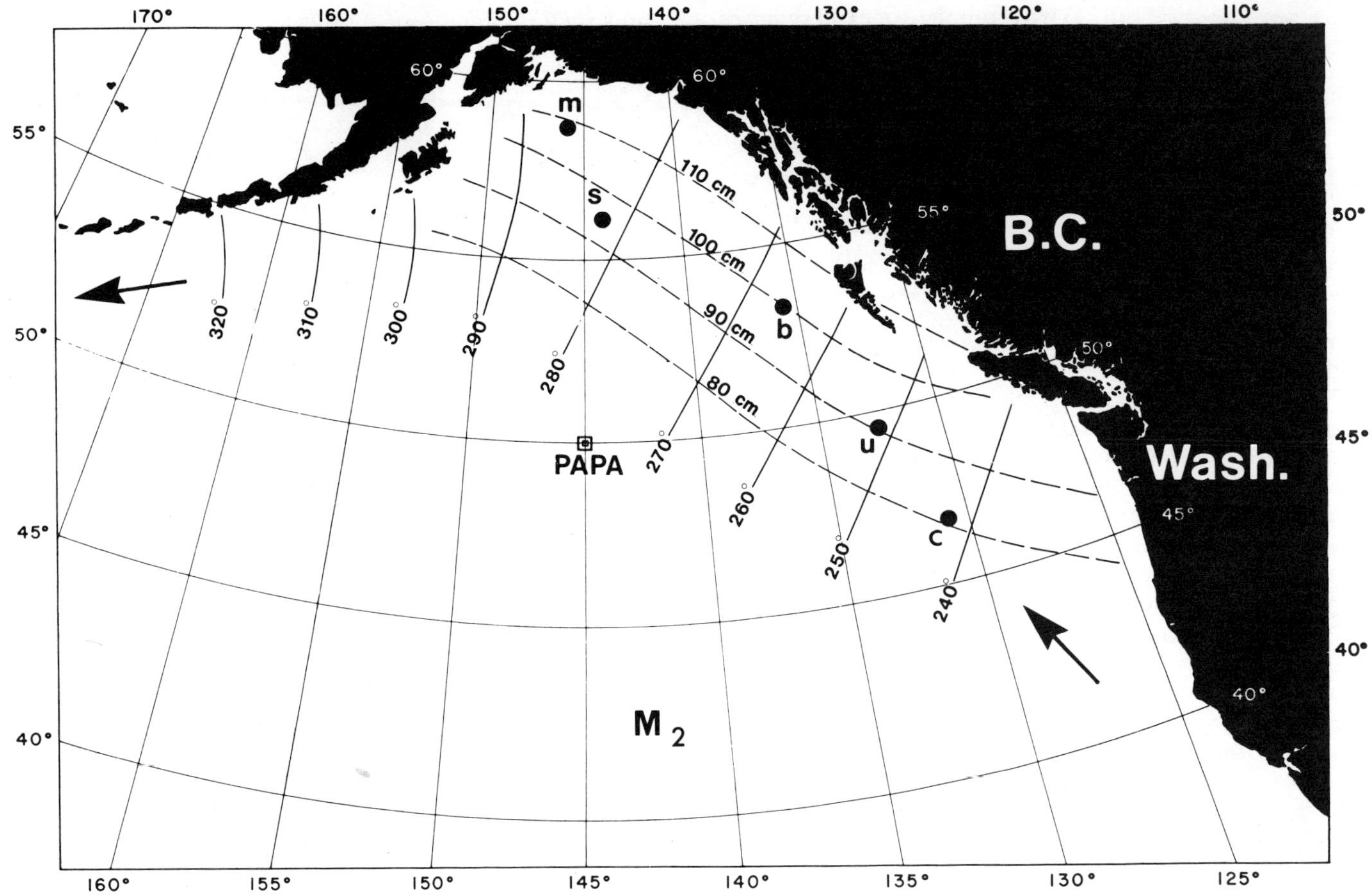

Fig. 13.16. Cotidal and corange chart of semidiurnal tide for northeast Pacific based on measurements at Cobb (c), Union (u), Bowie (b) and Surveyor (s) seamounts and near Middleton Island (m). Arrows denote direction of tidal propagation. (From Crawford, Rapatz and Huggett, 1981.)

Currents

In relation to the inshore waters of the B.C. coast, present knowledge of speed, direction, and variability of the offshore circulation is poor. Part of the reason for this is that the open ocean is less restricted in its response to the range of natural forces and is consequently more complex. Equally to blame is the deficiency of current measurements in an often hostile offshore environment. Most knowledge of the deep-sea currents in the Northeast Pa-rotary (see Chapter 3). Nontidal currents on the other hand are generally steady for periods of days to months and may possess considerable vertical structure in both their speed and direction, with higher velocities typically in the upper layer of the ocean. Nontidal currents usually predominate in the open ocean seaward of the continental margin; however, tidal streams become of equal importance near the coast due to their acceleration over the relatively shoal continental slope and shelf.

Tidal Streams

The horizontal flow at a particular depth and location can be represented by a vector, an "arrow" that points in the direction of flow, whose length is proportional to the speed. A few kilometres from shore, a tidal stream vector will gradually alter its orientation with time, and take a semidiurnal period of 12½ h to complete one full swing of 360°. This rotation may either be clockwise or counterclockwise, depending on location, and is accompanied by a cyclic variation in speed. Because of the diurnal inequality of the mixed tides along the B.C.–Washington coast, the speed of the stream will usually not return to the value it had 12½ h earlier but will require 2 full cycles (i.e. 25 h) to repeat itself.

Because the tide progresses northward along the west coast of North America, flood streams are in the northerly direction and ebb streams in the southerly direction. Over the continental shelf, maximum flood occurs within a few hours of high water at the coast and maximum ebb within a few hours of low water, and both attain maximum speeds of around 50 cm/s (1 kn) during spring tides. It is imperative that the times of high and low water be derived from tide tables for exposed coast locations such as Tofino, rather than from protected locations such as Astoria. In the latter case, the tide gage is located inland of a bar at the mouth of the Columbia River that delays the tide about 1 h from that at the coast. Observations have shown that the current vectors usually rotate clockwise off the coast, so the tidal streams are directed to the west midway through a falling tide, and to the east midway through a rising tide. Speeds in the offshore–onshore direction diminish to zero within a few kilometres of shore as tidal streams are confined more and more to flow parallel to the coast. Also, tides are shifted by about 30 min over the shallow continental shelf, so deepwater streams reach maximum velocities later than their nearshore counterparts.

The strongest tidal streams likely to be encountered on the west coast occur near points of land such as Brooks Peninsula and around offshore islands. Tidal streams in the vicinity of the Scott and Triangle islands off northern Vancouver Island, for instance, are observed to set to the southeast (ebb) and to the northwest (flood) at speeds of 1.0–1.5 m/s (2–3 kn). The intense tidal mixing that occurs in the vicinity of these offshore islands is responsible for the formation of pronounced tidelines and, perhaps indirectly, for the proliferation of sea life in the area (Pl. 13).

Long-period measurements by current meter arrays moored at distances to 100 km offshore reveal interesting details of the tidal streams off the coasts of Washington and British Columbia. Seaward of the continental slope, tidal streams are predominantly semidiurnal with pronounced biweekly variations in speed, are elliptical rotary, attain maximum speeds of 1–5 cm/s, and undergo little attenuation with depth. At the time of their cycle, when they are directed parallel to the bottom contours or roughly parallel to the coastline, these tidal streams tend to be slightly stronger than when they are across bottom contours. Moreover, for periods of days to weeks when the nontidal component of the circulation dominantes the tidal streams, the direction of the offshore current may remain relatively steady although its strength varies in accordance with the changing directions of the superimposed tidal streams. This effect is similar to the superposition of tidal streams on the estuarine circulations of Juan de Fuca Strait and Johnstone Strait (Chapter 11, 12).

From the deep-sea region to the continental slope and shelf, a significant modification of the tidal stream pattern takes place. Off the west coast of Vancouver Island, the tidal streams become increasingly more diurnal in nature, more aligned with the direction of the coastline and appreciably stronger, with speeds in excess of 50 cm/s. Whereas the relatively weak deep-sea streams are essentially repeated twice per lunar day, the enhanced tidal streams over the continental slope and shelf are repeated once per lunar day. Tidal streams over the slope and shelf regions of Washington and Oregon show a similar acceleration in speed and rectilinearization of direction, but retain the semidiurnal variability of their deep-sea counterparts. The change from semidiurnal tidal flow to diurnal tidal flow off Vancouver Island is noteworthy, especially when it is considered that the tidal heights at the coastal tide stations are predominantly semidiurnal (e.g. Fig. 3.5, Tofino). The reader may at this point suspect a paradox. How can the rise and fall of the sea level have a cyclic variation over half a lunar day when the coastal tidal currents have a cyclic variation of one lunar day? A possible explanation is that mechanisms are at work that affect the tidal currents but have little effect on tidal heights. The tide is due mainly to astronomical forces that cause sea level to rise and fall a few metres over a semidiurnal period, with accompanying to and fro motions of the water. However, superimposed on these tidal flows can be significant wind- or tide-induced currents of tidal period associated with wavelike circulation patterns called continental shelf waves or bathymetrically induced internal tides (e.g. Chapter 6; Fig. 6.22) both of which produce sea level variations of only a few centimetres at the coast. Continental shelf waves, whose restoring mechanisms are a sloping seafloor and the Coriolis force, can be of diurnal period but not semidiurnal period poleward of 30° latitude and are necessarily confined to continental margins. If superimposed on the astronomically generated tidal streams, these may produce enhanced diurnal currents over the continental slope and shelf but not in deep water. Internal tides generated at the edge of the continental shelf, on the other hand, are generally of semidiurnal period but, by canceling the semidiurnal component of the astronomically produced tidal streams, could lead to the same effect, namely a dominance of the diurnal tidal streams over the semidiurnal tidal streams at the continental margin of Vancouver Island. Present evidence clearly reveals that the anomalously strong diurnal tidal currents over the shelf are due to continental shelf wave motions. In fact, the measurements off Vancouver Island provide the best proof yet for the existence of such motions. Generated indirectly by the diurnal component of the astronomical tidal stream, the presence of these waves means that the time difference between the tide at the coast and the tidal flow over the shelf increases by roughly one hour for every 15–20 km along Vancouver Island north of Juan de Fuca Strait. The tide versus current relationship in this region is obviously not a simple one.

Despite the considerably higher speeds of the tidal streams on the shelf compared to the deep-sea region, they are not always sufficient to reverse the direction of flow associated with the nontidal component of the circulation, which is also stronger over the shelf. Therefore, except during spring tides and periods of weak coastal winds, the direction of flow may be dominated by the mean currents although the speed of the net current will undergo marked daily variations due to the tidal streams. A further discussion of nontidal currents is presented in the section on deep-sea currents.

Before concluding this section, the inertial currents discussed in Chapter 4 can be important factors that affect short-term availability in the offshore circulation. Mainly, they may cause a disruption in the normal ebb–flood pattern of the tidal streams, not only by altering the speed and direction of the overall flow for a few days, but by changing the duration of the apparent ebbs and floods. A common observation of officers on research vessels who try to maintain fixed offshore LORAN positions for periods of hours is that the rhythm of tidal currents is "strange." Inertial currents could well account for this effect. Such motions are especially vigorous for about the first day following passage of a frontal system.

Deep-Sea Currents

For present purposes, the nontidal component of the deep-sea currents can be associated with the quasi-steady oceanic gyres that are primarily driven by the prevailing winds. Perturbations to this large-scale circulation caused by local winds or some other effect result in temporary departure of the current from the norm over periods of a few days (e.g. inertial type currents), but these are unpredictable, so concentration will be on the average current regimes.

The prevailing surface circulation in the North Pacific is outlined schematically in Fig. 13.17. Like the winds the two main trans-Pacific currents in this region, the North Pacific Current and Subarctic Current are to the east. Both originate in the strong boundary currents off the east coast of Asia. The largest and most important of these boundary flows is the Kuroshio (or Japan Current), a well-defined, warm-water stream about 150 km wide that flows in a northeasterly direction from the equatorial Pacific. With core speeds to 250 cm/s (5 kn), the Kuroshio hugs the continental slope until, at 35–40°N, it separates from the Japanese coast. The seaward extension of this current rapidly widens and decelerates to eventually become the North Pacific Current, a broad, slow, easterly drift with speeds around 5–10 cm/s (or 4–8 km/day). East of about 150°W the current turns southeasterly and enters the subtropics as part of the general clockwise circulation of the Central Pacific Gyre.

The northern edge of the North Pacific Current is delineated by a rather remarkable feature known as the Subarctic Boundary. Essentially an oceanic "front," it is consistently within a few degrees of 40°N across most of the ocean, but begins to dip toward 35°N about 1500 km off the California coast and eventually disappears. The main characteristic of this boundary is that the 34.0‰

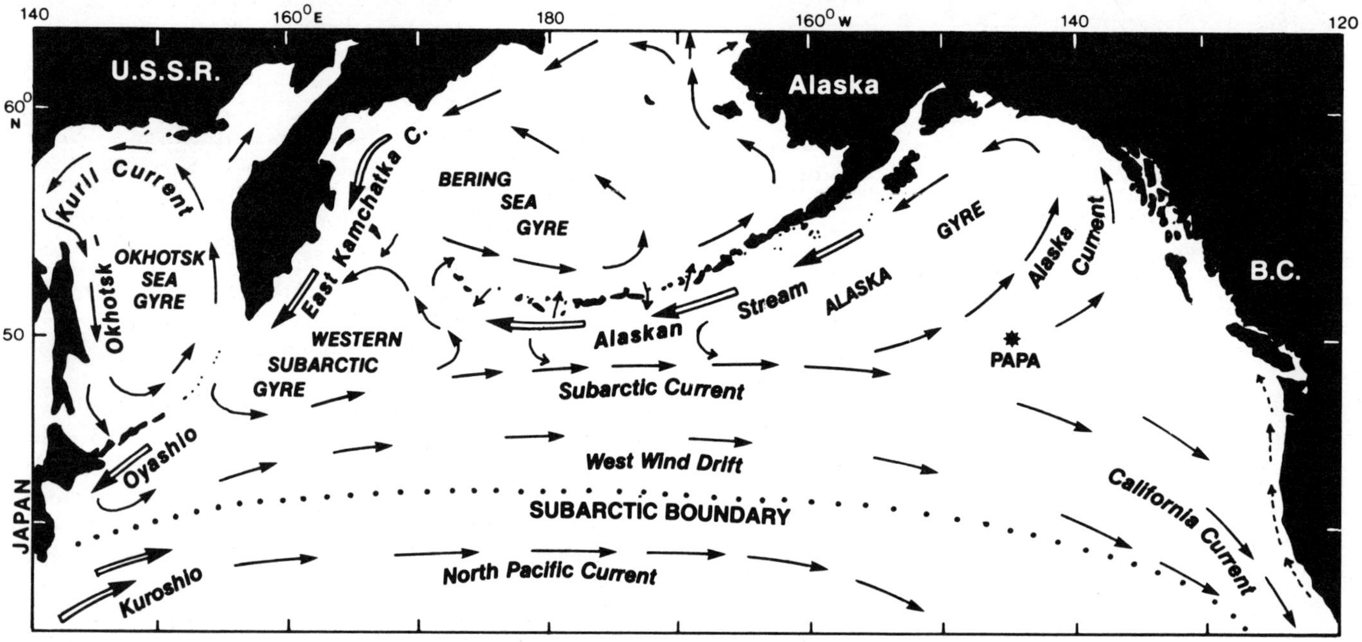

FIG. 13.17. Schematic diagram of prevailing surface currents in North Pacific Ocean. Double arrows are intense boundary currents speeds, typically 1–2 m/s (2–4 kn); over most of region speeds are less than 0.25 m/s (0.5 kn). Subarctic Boundary separates subarctic Pacific Region to north from subtropic Pacific Region to south. Broken arrows correspond to winter Davidson Current off Oregon–Washington coast.

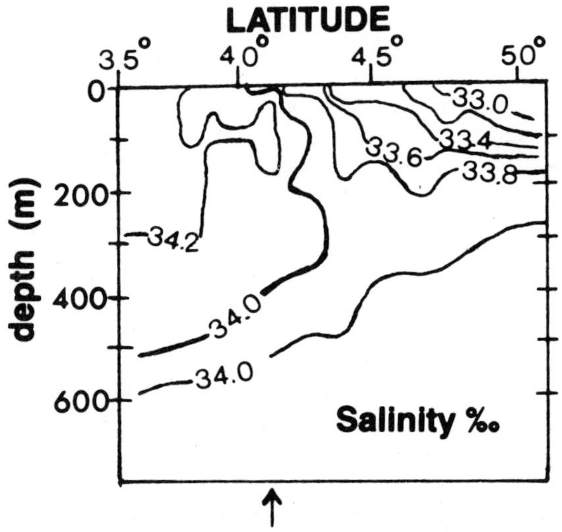

FIG. 13.18. Salinity distribution in north–south section along 160°W, summer 1958. Subarctic Boundary (arrow) is where the 34.00‰ isohaline rises to surface (see Fig. 13.17).

isohaline rises abruptly to the surface from a depth of 200 m (Fig. 13.18). In addition, there is a marked northward increase in the pH of surface water across the boundary, and north of the line precipitation exceeds evaporation, whereas to the south evaporation exceeds precipitation. Some scientists consider the Subarctic Boundary to be the southern limit of salmon migration.

At the Asian coast, the second most important boundary current is the Oyashio, a narrow but intense stream that transports cold water to the southwest along the continental slope. Fed by waters from off the Kamchatka Peninsula and the Sea of Okhotsk, the Oyashio swings to sea just north of 40°N, to eventually form the Subarctic Current, a broad, slow easterly drift with speeds of around 5–10 cm/s (0.1–0.2 kn). Part of the Oyashio also mixes with the Kuroshio off the Japanese Coast to create water for the so-called Transition Domain, which separates the Subarctic and North Pacific Currents (Fig. 13.18). The surface waters in this domain have drift speeds comparable to the latter two currents and may be said to constitute the West Wind Drift, a name that has recently lost favor with some oceanographers as it is essentially part of the Subarctic Current system.

Near the coast of North America a divergence in the prevailing wind pattern causes the Subarctic Current to divide into two branches; a northern branch curves to the northeast into the Gulf of Alaska as the Alaska Current, and a southern branch turns to the southeast as the California Current. The formation of these two branches is by no means abrupt. Instead, the Subarctic Current bifurcates in a rather confused manner within the region between 45 and 50°N and 130–150°W (Fig. 13.19a, b). Here currents are characterized by numerous eddies and meanders of variable spatial extent that range from tens to hundreds of kilometres imbedded in an overall eastward drift. Although the latter typically will be slower than 10 cm/s (0.2 kn), speeds associated with the eddying currents may be in excess of 25 cm/s for brief periods of time. On a seasonal basis, division of the Subarctic Current is expected to be most abrupt during winter when the Aleutian Low and North Pacific High form a well-defined pattern of strong counterclockwise winds to the north of about 45°N and clockwise winds to the south. Division of the flow during this time will be mostly confined to the southeast portion of the region (Fig. 13.19a), whereas in summer, when wind patterns are less clearly established, the current will split over the region as a whole (Fig. 13.19b). There is also evidence in this region of a weak westward flowing Subarctic Countercurrent. According to a study based on U.S. and Japanese ship-drift reports, the countercurrent is a midwinter feature confined to a belt between 48 and 51°N that attains maximum westward speeds of around 10 cm/s (0.2 kn) in January. The location of this current appears to coincide with a change in curvature of the cyclonic prevailing winds in winter and occurs most frequently east of 160°W.

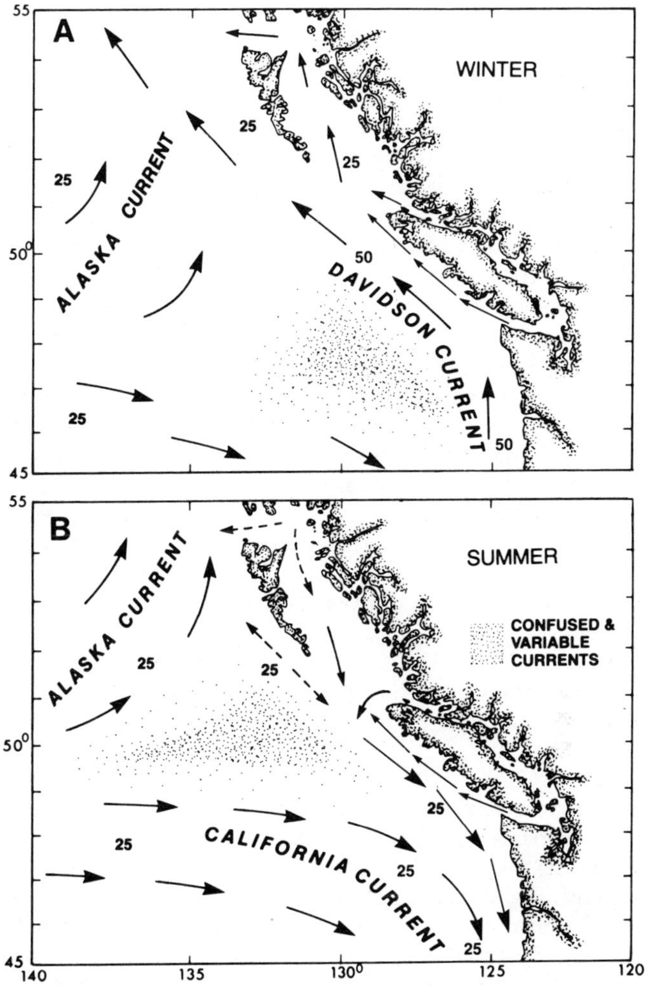

FIG. 13.19. Prevailing surface circulation off British Columbia–Washington coast in winter and summer. Broken arrows indicate uncertain currents. Numbers give speeds (cm/s).

Coastal Currents

The Alaska Current flows northward off the continental shelf of the Queen Charlotte Islands and southwest Alaska as a broad, slow drift. Speeds range from an average of about 25 cm/s (0.5 kn) in summer to around 35 cm/s (0.7 kn) in winter, but may be accelerated to 75 cm/s (1.5 kn) by strong southeast winds in winter. Moreover, as the Aleutian Low begins to intensify in early fall, so too will the circulation round the Alaska Gyre. This, combined with a coastward shift of the axis of the Alaska Current, leads to stronger nearshore currents in winter than in summer. Occasionally, in fact, strong northwest winds in summer will reverse the nearshore portion of the Alaska Current which, coupled with a weakening of the Alaska Gyre, creates a system of eddies and irregular meanders along its coastal boundary.

Once they attain the northernmost reaches of the Gulf of Alaska, the waters of the Alaska Current are constrained to turn southwestward along the coast of the Alaska Peninsula and Aleutian Islands as a strong current known as the Alaskan Stream. As with other currents on the western sides of oceanic regions, such as the Gulf Stream, Kuroshio, and East Australian Current, the Alaskan Stream is a narrow boundary current with mean speeds in excess of 1 m/s (2 kn). The seaward edge of the flow is often marked by eddies, and branches of the current frequently separate to reenter the Subarctic Current, or to flow through the narrow passes between the Aleutian Islands into the Bering Sea. A portion of the flow also reaches Asia where it turns south to rejoin the Oyashio.

The California Current is a poorly defined and variable equatorward extension of the Subarctic Current. It is most persistent off California, where it sets southward about 50% of the time each month at average speeds of around 20 cm/s (0.4 kn); it is least persistent seaward of the Oregon–Washington coast, where it sets to the south only 20% of the time. Because it is a southerly drift, the California Current gains importance in summer months during extended periods of northwest winds. Within 100 km of the Washington–Oregon–California coast, southerly currents are observed 30–50% of the time between March and September, with average speeds of 25 cm/s and maximum speeds to 100 cm/s (2 kn). Current velocity observations from northern California to Vancouver Island indicate that the reversal from prevailing northward flow to prevailing southward flow over the shelf occurs within a period of a few days in March or April. Known as the "spring transition," this rather abrupt reversal generally begins first off California then works its way northward off Oregon and Washington to coastal Vancouver Island where it occurs a week or so later.

Starting in late autumn or early winter, the California Current is shifted offshore by the Davidson Current, a seasonal current observed to set northward from 32°N to the coast of Vancouver Island. Northward flow persists until early spring (March), when the California Current again moves inshore. As suggested by Table 13.4, the onset of the Davidson Current during the "fall transition" coincides with a reversal of the prevailing coastal winds from northwest to southeast, and disappearance of this current coincides with a return to northwest winds. During October, when the California and Davidson currents are competing for coastal prominence, flow is equal to the south and north 23% of the time but becomes mainly to the north until March, and is particularly well established in January. Further evidence for the summer–winter reversal in the nearshore currents is in the offshore distribution of Columbia River water. During summer, the river plume extends 300–400 km seaward and over 500 km to the southwest, whereas in winter, river-diluted seawater extends 50–100 km seaward and from 100 km south of the river mouth to Juan de Fuca Strait. Winter observations also have shown evidence of Columbia River water well along the American side of the Strait.

Another coastal current worth mentioning at this point is the California Undercurrent, a sometimes strong northward-setting subsurface flow that originates off California and hugs the continental slope at depths below 200 m. This undercurrent appears to exist as far as northern Vancouver Island — though not necessarily as one

TABLE 13.4. Speeds and directions of prevailing and secondary currents within a zone 20 km wide along the Oregon–Washington coast (40–48°N). Northerly winter flow represents the Davidson Current system, southerly flow from March through September the California Current system. (From Boisvert 1969)

Month	Prevailing Current				Secondary Currents			
	direction	%	range of speeds		direction	%	range of speeds	
			(cm/s)	*(kn)*			*(cm/s)*	*(kn)*
Jan.	N	31	30–100	0.6–1.9	S	16	25–55	0.5–1.1
Feb.	N	29	30–80	0.6–1.6	E	19	25–65	0.5–1.3
					S	17	20–45	0.4–0.9
Mar.	S	40	30–115	0.6–2.2	N	19	25–105	0.5–2.0
Apr.	S	39	35–95	0.7–1.8	SE	15	30–95	0.6–1.8
May	S	48	30–105	0.6–2.0	N	12	25–80	0.5–1.6
June	S	31	30–80	0.6–1.6	SE	17	30–55	0.6–1.1
July	S	48	35–80	0.7–1.6	SE	14	25–80	0.5–1.6
Aug.	S	39	35–80	0.6–1.6	N	13	25–40	0.5–0.8
Sept.	S	31	30–80	0.6–1.6	E	20	20–30	0.4–0.6
Oct.	S	23	25–75	0.5–1.5	E	13	25–50	0.5–1.0
	N	23	25–60	0.5–1.2	W	10	20–30	0.4–0.6
Nov.	N	38	25–85	0.5–1.7	S	23	20–55	0.4–1.1
Dec.	N	28	30–70	0.6–1.4	S	19	20–45	0.4–0.9

continuous well-defined flow structure — and to reach maximum speeds in excess of 10 cm/s within a core centered at 200–300 m depth over the inner portion of the continental slope. From September to April, the undercurrent off Vancouver Island is often disrupted for several weeks at a time by the passage of 100-km wide offshore eddies. These oceanic "storms" drift more or less parallel to the coast and, in the case of the more intense eddies, carry with them circulating currents that attain speeds of more than 25 cm/s (0.5 kn) over depths of many hundreds of metres. From May to August, the eddies seem to be more diffuse and produce only minor short-lived disruptions of the undercurrent.

Shelf Currents

As indicated in the previous section, coastal winds are a major factor in determining the nontidal circulation over the continental shelf and slope. Current measurements obtained off Oregon, Washington, and British Columbia within the last decade provide conclusive evidence of a direct correlation between wind and current. In particular, alteration in the along-shore flow over periods of days and longer can be linked to changes in the strength and direction of the prevailing coastal winds. For example, approximately one inertial period (about 15 h) after the onset of a moderate to strong north to northwest wind, the nontidal component of the current over the shelf at all depths is typically directed to the south-southeast, parallel to the coastline, at speeds approaching a few percent of the wind speed (Fig. 13.20). This one-to-one correspondence of wind and current diminishes over the slope and essentially disappears by the time deep water is reached and offshore influences prevail. With a shift to south to southeast winds, the along-shore currents will again reverse direction. (An important exception to this is described in the next subsection.) Moreover, as mentioned previously the nontidal mean flow frequently masks the tidal streams so the shelf currents may remain nearly unidirectional for periods of days or weeks, but with variations in speeds produced by the superimposed tidal motions. Because of the great spatial extent of the wind systems, these reversals in the nontidal currents that flow over the shelf occur almost simultaneously over distances of hundreds of kilometres.

Observations further show there is a direct correlation between the along-shore component of the coast wind and nontidal variations in sea level of a few centimetres at the coast. A northwest wind will tend to move water slightly offshore resulting in a small lowering of sea level; a southeast wind will move water onshore and cause sea level to rise. In theory then it may be possible to use sea level measurements to deduce the speed and direction of the mean along-shore currents without measuring them directly. An additional feature of this weak onshore–offshore movement of the shelf waters is the pronounced seasonal effect on bottom water replacement in inlets, sounds, and other basins adjoining the outer coast. It has been found, for example, that during the summer, bottom waters on the continental shelf off Vancouver Island typically have low dissolved oxygen content. This water has possibly upwelled onto the shelf from greater offshore depths to compensate for the seaward Ekman flow near the surface induced by the prevailing northwest winds. When conditions are right, the shelf bottom water may then move inland and spill into the deeper portions of the coastal basins, as in Alberni Inlet and Nootka Sound. Therefore, the deep water in these basins is replaced with low oxygen water contrary to the type of processes described in previous sections. During the winter, on the other hand, the bottom waters on the shelf have relatively high dissolved oxygen content due to an offshore flow at depth associated with predominantly southeast winds and nearshore downwelling. Water that eventually works its way into the coastal basins in winter is of relatively high oxygen content.

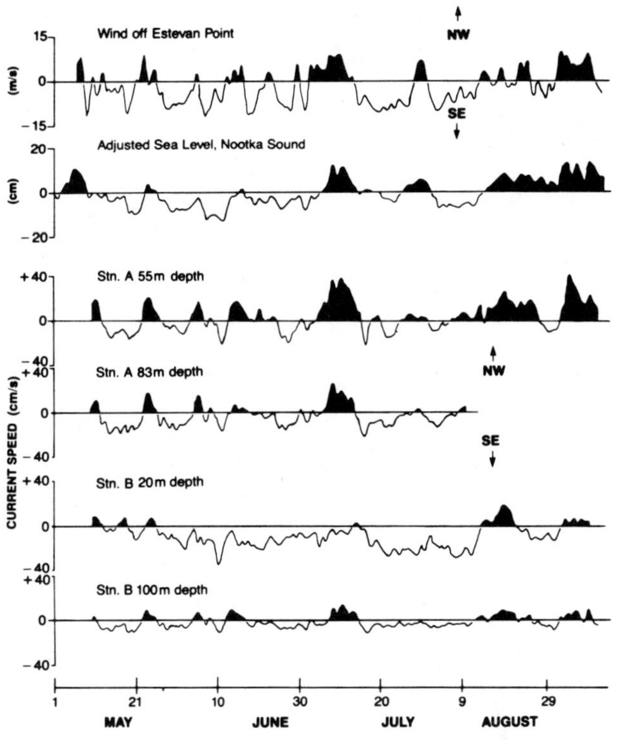

FIG. 13.20. Daily variations in along-shore components of wind and currents over continental shelf seaward of Estevan Point, west coast Vancouver Island, summer 1979. Also nontidal variation in sea level in Nootka Sound with values (cm) above or below mean sea level. Solid (or positive) portions represent winds and currents directed to northwest; open (or negative) portions represent flow to southeast. Depths (m) where currents were measured. Station A, approximately 15 km southwest of Estevan Point in 100 m of water; Station B, about 30 km southwest Estevan Point in 120 m of water (Courtesy W. R. Crawford)

Vancouver Island Coastal Current

Canadian fishermen working the west coast of Vancouver Island have long known about the existence of a persistent nearshore 'counter-current' that flows to the northwest against the prevailing winds during the summer fishing season. Recent current measurements confirm the existence of such a net nontidal flow over the inner portion of the continental shelf region (Fig. 13.21). In summer, this coastal current appears to be confined to

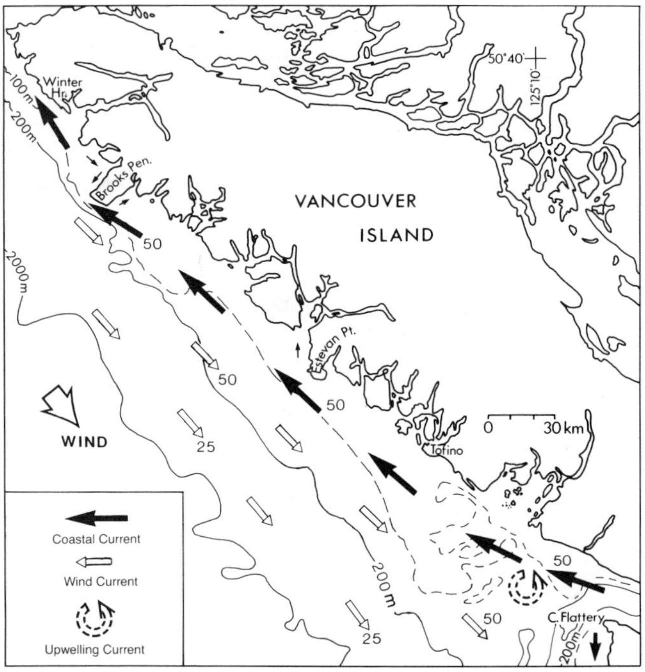

FIG. 13.21. Prevailing surface currents during summer off Vancouver Island. The coastal current is confined to the inner portion of the continental shelf and appears to attain its greatest speed at a distance of approximately 15 to 20 km from the coast. The wind-driven current attains maximum speed near the edge of shelf as delineated by the 200 m depth contour. In winter, the directions of the wind and wind current are reversed. (Numbers give typical maximum speeds in cm/s in summer.)

within 15 to 20 km of the shore and to extend more than 200 km northward from the entrance to Juan de Fuca Strait to beyond Brooks Peninsula. Maximum current speeds are about 50 cm/s (1 kn) near the surface and 15 cm/s near the bottom. The speed of the coastal current varies in accordance with the speed and direction of the prevailing winds although only northwesterlies of more than 15 to 20m/s (30 to 40 kn) appear to be sufficiently strong to cause a reversal of the current. Once the wind strength diminishes, the process generating the coastal current again begins to predominate and the flow sets northward. In contrast, the nontidal surface current over the seaward portion of the shelf is almost invariably to the southeast in summer, in direct response to the winds. During winter, the coastal flow merges with the northward wind-driven current and is not readily distinguishable as a separate identity. Current speeds at this time of the year are frequently greater than 75 cm/s (1.5 kn).

Present conjecture is that the coastal current is a continuation of the relatively low density, nearsurface flow that emanates from the Canadian side of Juan de Fuca Strait. The flow is driven northward by the alongshore pressure gradient created by the volume of water exiting the Strait. Northwesterly winds oppose the pressure gradient and weaken or reverse the coastal current; southeasterly winds augment the pressure gradient and strengthen the flow. The main balance of forces that keeps the current from spreading northward is between that part of the pressure gradient directed offshore and the Coriolis force directed onshore. Prevailing summer northwesterlies may also help confine the coastal current by establishing upwelling frontal zones over the mid-shelf region. In winter, runoff from coastal Vancouver Island possibly combines with the southeast winds to intensify the flow especially near the coast. However, regardless of its origin, the coastal current is clearly important to the physics and biology of the nearshore region and is a prime candidate for future investigation.

Exploding a Myth

A popular and often promoted notion is that the climate of the west coast is moderated by a warm Japan Current, envisioned by many as a sort of "ribbon of warm water" that stretches from the Orient to the shores of British Columbia. The concept is pleasantly appealing but totally fictitious. To begin with, the true Japan Current (the Kuroshio) is confined to the western side of the Pacific and its eastward extension, the North Pacific Current, flows well south of the latitude of British Columbia. Moreover, the current that approaches the west coast, the Subarctic Current, is a dead-slow, ill-defined drift that originates from a cold body of water not a warm one. Perhaps even more revealing is the realization that it takes some 2–5 yr for a parcel of water to cross the expanse of the North Pacific Ocean at the laborious speeds of the Subarctic Current, and during that time the temperature of the top 100 m or so of the drift current is almost completely controlled by exchanges of heat with the atmosphere and not by the origins of the current.

Put very simply, the temperate climate of the west coast is not due to a warm current but due to a prevailing onshore flow of marine air whose temperature is regulated by heat exchange with the upper ocean. This regulation is brought about by the fact that the ocean has a much larger heat capacity than the atmosphere and can store vast amounts of heat with only small changes of temperature below the sea surface. (A summer-to-winter range of only 10C° is common for the open ocean at a depth of 10 m or so, while at Kamloops, for example, the corresponding range in air temperature is 25C°; Table 13.3). Therefore, in summer the ocean will cool the marine air and in winter it will warm it. Also, because of the locations of the North Pacific High and Aleutian Low, relatively cool marine air from the northwest tends to cover the coast in summer, whereas relatively warm air is transported onshore from the southwest in winter. At this point, one might ask why the Atlantic coast has a more severe climate than the Pacific coast? The answer again lies in the eastward movement of weather systems, and the fact that the Maritimes are affected by continental air masses with their much greater seasonal variability in temperature and precipitation.

Chapter 14. Northern Shelf Region

Queen Charlotte Sound, Hecate Strait, and Dixon Entrance form a continuous coastal seaway over the continental shelf of the Canadian west coast (Fig. 14.1). Except for the broad lowlands along the northwest side of Hecate Strait, the region is typified by a highly broken shoreline of islands, isolated shoals, and countless embayments which, during the last ice age, were covered by glaciers that spread seaward from the mountainous terrain of the mainland coast and the Queen Charlotte Islands. The irregular countenance of the seaway is mirrored by its bathymetry as re-entrant troughs cut landward between shallow banks and broad shoals and extend into Hecate Strait from northern Graham Island. From an oceanographic point of view it is a hybrid region, similar in many respects to the offshore waters but considerably modified by estuarine processes characteristic of the protected inland coastal waters. Deep-sea processes, tides, winds, and river discharge are all important factors in the establishment of the currents and water structure in this semiexposed marine environment.

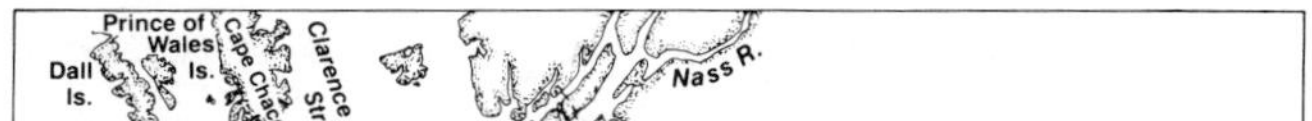

FIG. 14.1. Map of northwest Pacific coast. Broken lines show extent of coastal lowlands; wave measurements taken at star near Prince Rupert.

A Brief History

Long before the arrival of the "discoverers," the north coast Indians had evolved one of the most distinctive and sophisticated cultures in the Americas. The Haida, or "people" of the Queen Charlotte Islands, were especially skillful seamen and fishermen, whose sleek war canoes were almost as long as the ships of the early Spanish and British explorers. The Haida also were gifted carvers and produced a volume of art work which, like that of the mainland tribes of the Kwakiutl and Tsimshian, is only now becoming appreciated by the general public.

The first Europeans to sail the west coast of British Columbia were Spaniards. Under the command of Juan Perez they reached the vicinity of the Queen Charlotte Islands in 1774 before returning to a landfall at Nootka Sound on Vancouver Island. Quadra followed in 1775, but it was not until after Cook's voyage of 1778 with the *Resolution* and *Discovery* that the white man, or "Yets-haida" (iron men) as the Haida called them, began to explore in earnest the northern coastal waters. During his sojourn at Nootka that year Cook had received a number of soft, luxuriant sea otter furs which, after his death in Hawaii, members of his crew sold for high prices on arrival in China in 1779. News of his ships' windfall caused trading ships of several nations to sail to the coast to obtain sea otter furs for the China market.

It was the search for furs that brought about most of the early exploration of the coast. Dixon Entrance, for example, was named after Captain George Dixon, master of the 200-t *Queen Charlotte*, that passed the mouth of the channel in 1787 during a trading expedition for sea otter furs. (The channel was named in 1788 by Sir Joseph Banks who had been botanist and naturalist on Cook's round-the-world voyage in the *Endeavour* from 1768 to 1771 and in whose honor Banks Island in Hecate Strait is named.) "As he sailed southward Dixon met with a large island or islands, where he purchased a large number of sea otter skins. Rounding the southern termination of the land which he named Cape St. James, he sailed northward along the eastern shore until he recognized ahead the high mountains seen some days before, northward of a large opening in the coastline. Dixon thus ascertained the land he had been trading along was a large island or islands, and gave them the name of his vessel, Queen Charlotte isles. In 1789 Captain Gray, of the American trading sloop *Washington*, visited these islands, and named them after his vessel. As the ownership of the islands was established in Great Britain . . . the group has long been known by their present name." (Walbran 1971)

Hecate Strait received its name in 1861 after H. M. surveying vessel *Hecate*, a paddle-wheel sloop, 780 t, 5 guns, and brigantine rigged that worked the British Columbia coast from December 1860 to December 1862. The name for Queen Charlotte Sound was given in 1786 by the commander of the *Experiment* after the wife of George III of England, and adopted by Vancouver during his charting of the coast between 1792 and 1794.

Shoreline Features

The coastal seaway lies within the Hecate Depression, a segment of a continuous, low-lying area that extends from Puget Sound to Alaska (Fig. 1.11, 14.1). This depression includes the narrow, 20–50 km wide plain along the island-strewn mainland coast (the Hecate Lowland), the low relief northeast corner of Graham Island (the Queen Charlotte Lowland), and the rocky flatlands of northern Vancouver Island (the Nahwitti Lowland). To the east lie the Coast Mountains, to the west the open Pacific Ocean and the insular mountains of the Queen Charlotte Islands. The latter are especially rugged and plunge steeply into the sea off western Moresby Island. Nearly vertical sea cliffs over 200 m high are common and the short gravelly beaches that have managed to form are mostly associated with the numerous fiords that indent the outer coastline. The contrast between this area and the inner coast is striking.

The shoreline of the Nahwitti Lowland consists of shallow, open embayments, rocky intertidal zones, long stretches of gravelly beaches, and numerous small river deltas. Within the Hecate Lowland, the coastline is characterized by discontinuous sand, gravel, and boulder beaches between low rocky headlands. The Skeena River which enters these lowlands some 20 km south of Prince Rupert, has the second largest delta in British Columbia, and extends about 30 km to the west into Chatham Sound and adjoining channels. Unlike most deltas in British Columbia, it is neither at the head of an inlet nor directly exposed to a large body of open water, and exhibits no extensive tidal flats. However, the river sediments have effectively broadened the delta by "cementing" together the various islands situated seaward of the river mouth. Also, currents at the mouth of the river, which may attain speeds in excess of 1.5 m/s (3 kn), have formed extensive bars. The bars have megarippled surfaces with heights to 0.5 m and spacings over 10 m.

The Queen Charlotte Lowland has some of the most fascinating beach forms in British Columbia. Along Virago Sound, large beaches have been created through wave erosion and onshore current-induced movement of offshore bottom sediments. Between Massett and Rose Point to the east, similar processes have led to the formation of a continuous sand beach with a 200-m wide foreshore that slopes gently into Dixon Entrance (Fig. 14.2) The backshore is characterized by a series of dunes and ancient beach forms to 10 m above present sea level. In addition to widening the beach, the vast nearshore supply of sediment is transported eastward in the littoral zone, where it eventually contributes to the growth of Rose Spit that projects more than 12 km to the northeast of the trees on Rose Point. The beach-type environment continues some 65 km south of Rose Spit, but with a foreshore that is steeper,

Fig. 14.2. Broad sandy beach, McIntyre Bay, facing Dixon Entrance, Graham Island, Queen Charlotte Islands. Beaches on northeast Graham Island are unlike most on British Columbia coast. Sediment is plentiful, beaches wide, with sandy and extensive backshore dunes. (Courtesy J. Harper.)

narrower, and of coarser sediment than along the north coast. Sea cliffs in this region range from 15 to 60 m high and dunes are present 100–200 m inland of the backshore (see Fig. 8.11). In contrast to the north coast, beach sediments are actively being removed by littoral processes, which then carry them northward to Rose Spit.

Another pronounced depositional feature within the confines of the Queen Charlotte Lowland is Sand Spit which curves for nearly 5.5 km half way across the entrance to Skidegate Inlet. As with the gradually eroding beaches to the north, it suggests the presence of a net northward surface drift along the western shores of Hecate Strait.

Bathymetry

Dixon Entrance is an east–west depression in the continental shelf bounded over most of the 170-km length by the mountainous Dall and Prince of Wales islands to the north and by Graham Island to the south. At the seaward end it is split into two 400-m deep channels by Learmonth Bank, a shallow ridge that rises to within 35 m of the surface (Fig. 14.3). To the east, these channels recombine to form a single depression that gradually shoals to 270 m over a sill south of Cape Chacon, followed by more rapid shoaling over the submarine ridge that separates Dixon Entrance from Chatham Sound. The main axis of the depression continues into Clarence Strait, a 400-m deep trench leading 200 km seaward through the Alaska Panhandle.

Hecate Strait is the shallowest of the major channels that make up the Hecate Depression. It is also the least exposed and has the most regular bathymetry. The axis of the Strait is a narrow, 220 km long submarine valley that hugs the mainland flank, with depths that diminish from around 300 m in the south to about 50 m in the north. The northwest side of the Strait is a broad platform of glacial sands and gravels less than 100 m deep, adjacent to the flat coastal plain or strandflat of east Graham Island.

The bathymetry of Queen Charlotte Sound is considerably more complex than either Hecate Strait or Dixon Entrance because of shallow banks and three broad troughs that slice inland across the continental shelf from depths of 350–400 m (Fig. 14.3). Of these, the northern trough is the most irregular and most extensive and trends 270 km northward along faulted, ice-deepened depressions to form the deep channel of Hecate Strait. The central and southern troughs, on the other hand, shoal rather smoothly toward the mainland coast; the southern trough divides about 60 km from the shelf edge to send a branch into Queen Charlotte Strait to the southeast. All three troughs have low sills at the seaward entrances and are separated by wide shallow banks that are heavily fished by draggers. These sills and banks appear to have glacial origins and are made up in part by recessional moraines and stream outwash deposits left by retreating ice sheets. Exploratory oil wells drilled on the banks in the 1960s often encountered 3 km of marine sediment.

Due to the seaward funneling of land-derived sediments from the coastal mountains, the continental slope off Queen Charlotte Sound is comparatively gentle. There is also some evidence to suggest that the Queen Charlotte Islands once abutted Vancouver Island and effectively sealed off the Sound from any direct exposure to the Pacific Ocean. The subsequent northward movement of the Queen Charlottes along the Queen Charlotte fault created the present-day configuration.

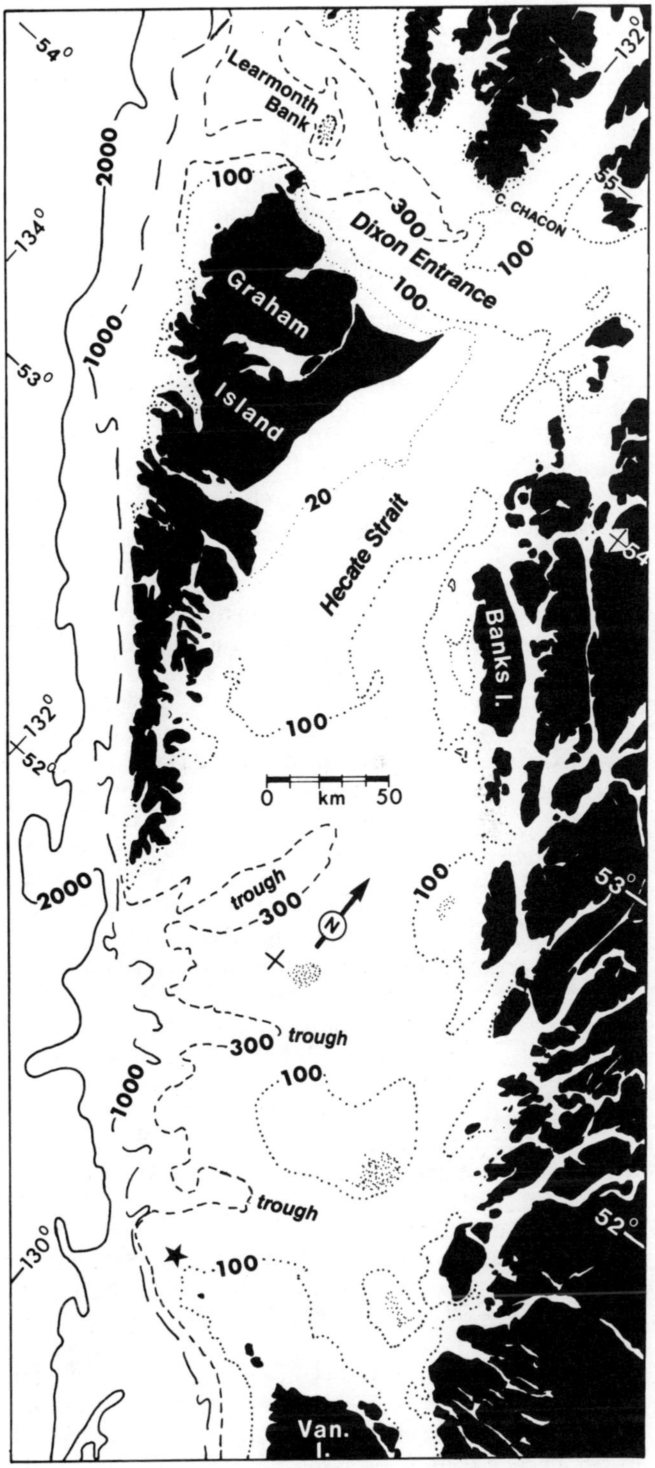

FIG. 14.3. Bathymetry of northwest coastal waters (m). Speckled areas denote shallow portions of banks. (Adapted from Chase et al. 1975)

Glaciation

On numerous occasions over the last million years, the coastal seaway has been inundated with vast ice sheets that spread from the mountainous terrain at the times of world-wide climatic cooling. During the nonglacial intervals the thick layers of debris deposited by the advancing and retreating glaciers were reworked by rivers, waves, and tidal currents to help produce the diverse littoral sediments, landforms, and bathymetry of the region.

The present-day structure of the seaway was well established by the end of the last glacial period, but because of the isostatic depression of the earth's crust by the weight of the ice, the Hecate Depression and adjoining fiords remained submerged as the glaciers withdrew from the coast. The marine and glaciomarine muds deposited over the region at this time were then raised far above present sea level as the crust eventually rebounded upward to its new equilibrium. In general, the maximum elevation to which marine deposits are now found on land decreases from the inner coast toward the outer coast, in correspondence with the seaward decrease in past ice thickness. At Kitimat, for example, marine deposits are at elevations in excess of 200 m; on the outer coast of the Queen Charlotte Islands they are less than 20 m above sea level.

The position of sea level relative to the land has undergone several major variations since the last glaciation. In the Terrace–Kitimat area, for instance, the coast attained maximum submergence about 11,000 yr ago (compared to about 13,000 yr ago in the Strait of Georgia region), whereas the outer coast of the Queen Charlotte Islands around this time was raised some 30 m above present sea level. According to marine geologists, the latter condition stems from the fact that the removal of water from the oceans to create the world glaciers (eustatic effect) was of greater importance in this area than the depression of the crust (isostatic effect) by a regionally thin ice sheet. The difference in vertical movement between the inner and outer coasts as a result of the interaction of crustal plates also has contributed to their different sea-level histories. Aside from the outer coast, deglaciation was accompanied and succeeded by isostatic rebound, and with it a relative fall in sea level. The land–sea boundary on the inner coast reached a position similar to that of today about 9,000–10,000 yr ago, but continued to fall a further 10 m over the next few thousand years. Except for a few minor fluctuations, sea level has been on the rise ever since, and will continue to rise with the worldwide shrinkage of glaciers. An exception to this trend is the north coast of Graham Island where prograding beaches give evidence of tectonic uplift and an accompanying seaward advance of the shoreline. This evidence is supported by nearly 40 yr of tide-gage records from Sitka and Juneau in the Alaska Panhandle where sea levels are observed to be falling at average rates of 2.3 mm/yr and 11.5 mm/yr, respectively. However, at Prince Rupert, much further to the east, sea level is rising at about 0.5 mm/yr. These rates may be compared with observed falls in sea level at Tofino and Neah Bay of approximately 0.8 mm/yr and 1.3 mm/yr, or with rises in sea level at Victoria and Seattle of 0.6 mm/yr and 1.9 mm/yr, respectively.

Temperature and Salinity

Temperature and salinity of the surface waters within the coastal seaway vary seasonally in accordance with the amount of incoming solar energy and the input of freshwater runoff. Over most of the region the annual range of surface salinity is from 28 to 32‰, with maximal values in winter and minimal values in summer. Water on the western side of the seaway tends to be saltier than on the mainland side, due to drainage from the mainland mountains. This effect is most pronounced in Chatham Sound and along the northern side of Dixon Entrance, which receive most of the silty outflow of the Nass and Skeena rivers. The combined discharge of these rivers of around 5700 m^3/s (200,000 ft^3/s) during early June rivals that of the Fraser River.

Average sea surface temperatures vary from about 6°C in April to around 14°C in August; waters of Dixon Entrance are consistently colder than those of Hecate Strait and Queen Charlotte Sound. The temperature of surface waters can depart conspicuously from the average for a given month (Table 14.1). And, unlike the situation in midocean, maximum sea-surface temperatures in the coastal seaway do not occur at the end of the heating season in mid-August, nor minimum temperatures at the end of the cooling season in March, because wind-driven currents bring in relatively cold water in summer and relatively warm water in winter to partially counteract the effect of solar heating. There is also a reverse seasonal variation in the deep waters of the seaway, which are colder and more saline in summer than in winter. Two processes can account for this reversal in trends; first, northwest winds along the outer coast in summer move surface waters seaward and thereby cause deeper oceanic water to move onto the continental shelf, and second, the increased outflow of brackish water from the seaway in summer strengthens the deep inward flow of the estuarine circulation.

TABLE 14.1. Ranges of monthly averaged sea-surface temperature, wind speed, and air temperature for the northwest coast waters (50–55°N). Minimum values of water temperature are typically at northern end, maximum values at southern end of seaway for each month. Maximum speeds are along outer coast, minimum speeds along mainland coast of the seaway each month. Minimum values of air temperature are typically at northern end, maximum values at southern end of the seaway. (Adapted from U.S. Department of Commerce 1978)

Month	Sea Surface (°C)	Wind Speed (*m/s*)	Air (°C)
Jan.	3–11	6–9	3–6
Feb.	3–11	9	4–6
Mar.	2–12	6–9	4–6
Apr.	3–12	6–8	6–8
May	5–14	5–6	8–10
June	7–16	4–6	11–12
July	9–18	4–6	13–15
Aug.	10–20	4–6	14–16
Sept.	9–18	4–7	13–15
Oct.	7–17	5–9	10–14
Nov.	5–14	7–10	6–10
Dec.	4–13	6–10	4–8

The influence of freshwater discharge is most strongly felt in Dixon Entrance in late spring and summer when runoff from the Nass and Skeena rivers tends to flow seaward within a comparatively warm brackish layer 10–20 m thick that hugs the northern side of the channel. In contrast, cooler, saltier oceanic water generally prevails over the southern half of the channel at this time.

Winds

Prevailing wind conditions over the inland waterway are associated with the same semipermanent, large-scale pressure systems that govern the deep-sea wind regimes, namely the Aleutian Low of winter and the North Pacific High of summer. Major differences in the winds over the two regions exist, nonetheless. Winds over the seaway, for example, commonly blow parallel to the coast because of the channeling effect of the bordering mountains. And there is a marked drop in average wind speeds toward the mainland shore, due perhaps to dissipation of wind systems as they cross the coast and to enhanced frictional drag caused by the surrounding terrain. A further generalization is that the frequency and intensity of southeasterlies is greater over the northern sector of the seaway than over the southern sector, whereas the frequency and intensity of northwesterlies is greater in the south than in the north.

The most pleasant months to cruise the coastal seaway are June, July, and August. Mean winds at this time range from 5.5 m/s (11 kn) at the seaward approaches to 4.0 m/s (8 kn) near the mainland coast, and speeds in excess of Force 8 (34 kn) occur less than 5% of the time (Table 14.1). Prevailing winds are from the west in Dixon Entrance and from the northwest in Queen Charlotte Sound. Conditions usually remain favorable until early September, but by October deteriorate rapidly with the onset of storm activity. The prevailing winds shift to the southeast and sharply intensify. Average wind speeds in October range from around 9.5 m/s (19 kn) along the outer coast to 6 m/s (12 kn) along the eastern side of the seaway and exceed Force 8 about 5–10% of the time. From November to February winds blow at Force 8 or greater more than 10% of the time. Throughout most of this period average wind speeds vary from 10 m/s (20 kn) at the seaward approaches to Dixon Entrance and Queen Charlotte Sound to about 7.5 m/s (15 kn) along the mainland side. Although southeast winds are most frequent (20% occurrence), winds from the south, southwest, and west occur with almost equal regularity. From April to June there is a gradual abatement in the strength of the wind and a return to more frequent westerlies.

In addition to gale force winds associated with migrating depressions and their attendant frontal systems, strong Squamish winds periodically blow seaward along the mainland inlets of the seaway during periods of clear winter weather. Usually triggered by a fall in pressure offshore, these outflows of cold interior air may attain speeds in excess of 25 m/s (50 kn) and penetrate many kilometres beyond the mouth of the inlet before their momentum is diffused by lateral spreading. Williwaws (a somewhat similar phenomenon that originates from the seaward drainage of cold air from more local areas of high ground) are violent, squally winds of short duration that frequently blow down the narrow inlets and valleys of the west coast of the Queen Charlotte Islands.

Fog

Table 14.2 gives the seasonally averaged days of fog at four locations over the north coast waters. In concert with the outer coasts of Vancouver Island and western Washington, the percentage occurrence of sea fog over the coastal seaway is greater during summer (June through September) than during autumn or winter (October through February). Fog tends to be more prevalent over the seaward approaches to Queen Charlotte Sound and Dixon Entrance than over the mainland side or over Hecate Strait, although due to the absence of intense upwelling in these regions there is nothing that resembles the persistent summer fogs of the exposed coast to the south (Table 14.2).

TABLE 14.2. Number of days of fog during 3-mo intervals at four locations on northern coast. (From British Columbia Sailing Directions, Vol. II, 1977)

Location	Jan.–Mar.	Apr.–June	July–Sept.	Oct.–Dec.
Prince Rupert	1	4	10	1
Cape St. James	5	7	11	7
Langara	1	3	6	1
Sandspit	2	2	2	2

The foggiest months are August and September at 10–15% of the time. Fog occurs less than 5% of the time from October through February, about 5% of the time in March and April, and around 10% of the time during May, June, and July. Fresh winds that sometimes accompany foggy conditions in summer may effectively lift the moisture to a very low deck of stratus clouds with a slight improvement of visibility but, in general, near zero visibilities are likely to persist for many days over wide areas of the seaway.

Waves

Because of the limited extent and partially sheltered nature of the seaway, wave conditions are typically less severe than in the open ocean to the west. There is also a gradation within the seaway itself, with average wave heights in Dixon Entrance somewhat lower than those in Queen Charlotte Sound but higher than in Hecate Strait.

Dixon Entrance is protected from ocean waves that arrive from the southern quadrant, and Queen Charlotte Sound from waves that arrive from the northern quadrant. Swell entering the inland seas from the west, on the contrary, travel a considerable distance before the height is noticeably reduced by refraction and breaking in the gradually shoaling waters; over the shallow banks in Queen Charlotte Sound and Learmonth Bank in Dixon Entrance swell may in fact be amplified by the focussing effect of the bottom contours (see Chapter 6). Learmonth Bank, in particular, is noted for its sizable rips formed when such amplified waves move against an accelerated ebb current over the bank.

Queen Charlotte Sound and the southern end of Hecate Strait are especially vulnerable to deep-sea waves from the southwest which, when aided by favoring winds, undergo little attenuation as they propagate northward. In addition, frontal systems are known to rapidly generate steep mountainous seas over the broad shoal area adjacent to the east coast of the Queen Charlotte Islands and leave the hapless mariner no choice but to ride them out.

On a seasonal basis, wave conditions are most severe in autumn and winter and mildest in spring and summer. Hazardous wave heights in excess of 3.5 m occur on the average 20–30% of the time along the outer reaches of the seaway from October through January, but diminish in frequency to around 10% along the mainland side. From February until March the frequency of seas over 3.5 m reduces to 15% on the outer coast and 5% on the mainland coast. By late spring, the chance of encountering hazardous seas anywhere in the region is less than 5%, a condition that prevails until late September when storms again begin to lash the coast with renewed vigor. The "sudden" jump in wind strength and wave activity that transpires between September and October is, of course, characteristic of the entire coast.

The suddenness at which extreme autumn wave conditions can arise in the coastal seaway was made abundantly clear to workers on the Shell Drill Rig (SEDCO 135F) in 137 m of water off Cape St. James in Queen Charlotte Sound in 1968. Beginning in mid-October the rig was battered by heavy seas for 16 days, as one storm after another generated waves of 9–15 m. Between storms, seas never fell below 3 m. "On October 22, a rapidly moving storm with gusting winds of 80 kn produced the worst conditions experienced. Sea and swell from the southeast and southwest combined to form significant waves exceeding 65 ft (20 m) in height. On at least one occasion, a wave close to 100 ft in height smashed against the rig. The height of this monster was reliably observed since the wave trough exposed a lower support form of the rig while the wave crests passed just below a control room extending below the main deck. The standby ship *Gulf Jean* reported solid walls of spray over the main deck with passage of the larger waves.

"The most dangerous aspect of these storms is not the magnitude of the waves generated but the rapidity with which operating conditions change from difficult to dangerous or impossible. At 1500 on October 22 the seas were reported 10 ft in height. Within 8 hr the waves were 60 ft in height and were extremely dangerous for all ship operations. Shortly after this, there was an even more rapid decrease in wave height." (James 1969). Figure 14.4 shows the wave heights reported by personnel of Southeastern Drilling and Shell of Canada, aboard the rig during the storm.

The cause of the rapid onset of extreme waves in this case was twofold. First, the storm was advancing eastward as quickly as the larger waves it was generating, and thereby caused a build-up of wave energy at the leading edge. Second, arrival of the waves coincided with an opposing ebb current out of Queen Charlotte Sound. The 30.5-m "monster" was likely a freak wave created for that single brief moment when the larger waves were in step in the vicinity of the rig.

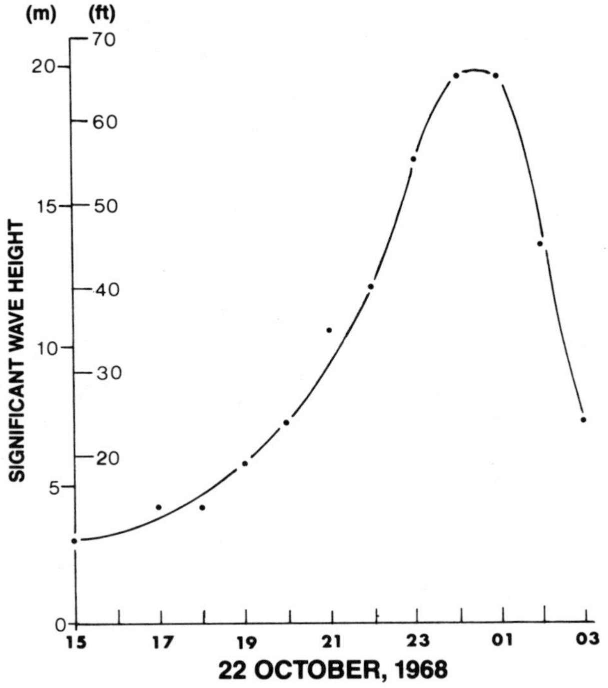

FIG. 14.4. Wave heights reported from SEDCO 135F drilling rig anchored off Cape St. James in 137 m water. (From James 1969).

Chatham Sound

From September 1972 to June 1973 wave measurements were obtained from a waverider buoy moored in 90 m of water approximately 2 km west of Prince Rupert Harbour. It represents one of the few attempts to determine wave height distributions within the north coast waters, although, because of the sheltered nature of the region, results are not representative of the seaway as a whole.

Most waves were found to be lower than 1 m in height and to have short periods of 2–5 s, typical of locally generated seas in Chatham Sound. Low swells that entered the Sound after propagating the 150 km length of Dixon Entrance were, nevertheless, common, and swell periods in the range of 8–10 s accounted for 20% of the wave observations. By the time such swells reached the inner portion of the coast, however, the effects of refraction and dissipation had considerably reduced the heights, so that only 10% were higher than 1 m. During the observation period, the significant wave heights remained below 3 m. In winter, conditions of eastward propagating swell often persisted for 2–3 days, but with wave heights in excess of 2 m for less than a day each time.

Tides

Tides within this region are mixed, predominantly semidiurnal, and co-oscillate with tides in the adjoining North Pacific Ocean. Shoaling effects cause the range of the semidiurnal tide to increase from about 2.4 m across the mouth of Queen Charlotte Sound to around 3.0 m midway along Hecate Strait, and from 2.4 m at the mouth of Dixon Entrance to 3.7 m along its eastern shore including Chatham Sound (Fig. 14.5). Corresponding values for the mean tide are from 3.0 m at the entrance to Queen Charlotte Sound to 4.8 m midway along Hecate Strait, and from about 3.5 m at the entrance to Dixon Entrance to 5.0 m at Prince Rupert. Tidal ranges over 7 m are encountered on large tides within Skidegate Channel that separates Graham and Morseby islands and in the vicinity of Prince Rupert. There is an amplification of the mean range to as much as 4.9 m as it moves up the deep narrow inlets of the mainland coast.

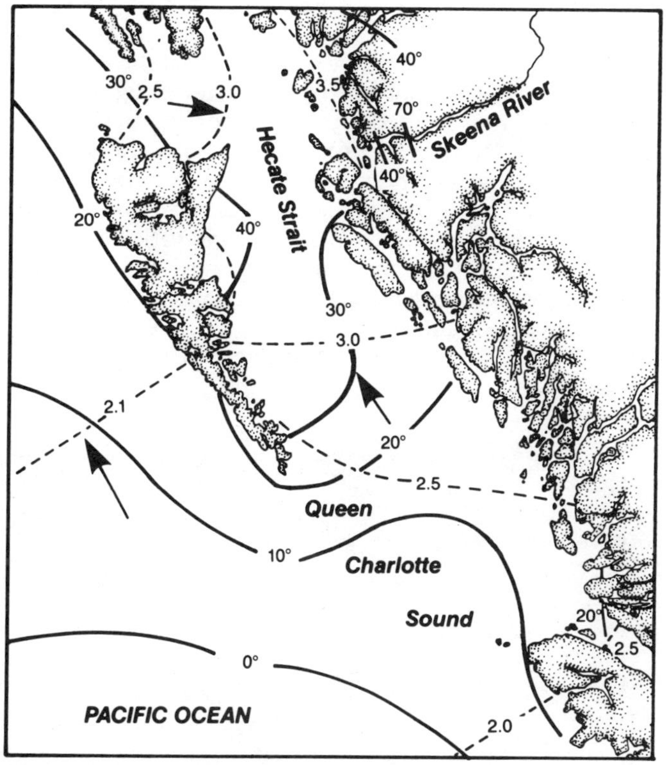

FIG. 14.5. Corange and cotidal values for semidiurnal tide. Tidal range (broken line) in metres; tidal phase (solid line) in degrees. Difference of 29° corresponds to time difference of 1 h. Arrows give direction of tide propagation.

In connection with the Strait of Georgia tides that propagate from the north and south meet at the northern end of the Strait. A similar situation occurs in the northern sector of Hecate Strait, where the north-going flood in Queen Charlotte Sound meets the south-going flood from Dixon Entrance. This confluence of the tides takes place over a somewhat ill-defined region, but appears to occur some 45–55 km further north in winter than in summer. Because the tidal waves will be partly reflected in this region, tides in Hecate Strait and Dixon Entrance are expected to behave like modified standing waves.

It takes about 30 min for the tide to reach the northern end of Hecate Strait from either of the two oceanic approaches, then the combined tide swings westward across the Strait to arrive at the shores of Graham Island some 15 min later. Consequently, a particular stage of the tide within the entire seaway differs by at most 1 h, not too unlike the 30 min difference that characterizes the Strait of Georgia tides. This fact, combined with recent current observations, supports the notion that the tides within the open seaway closely resemble standing waves.

Circulation

Surface water motions within the seaway are dominated by semidiurnal tidal streams that consistently undergo modification by the effects of wind, runoff, bathymetry, and shoreline configuration. During certain times of the year, wind-generated currents and seaward-moving brackish water may be strong enough to temporarily mask the normal ebb and flood of the surface streams, and result in persistent directional sets for durations of days. As expected, such conditions are particularly well established in many adjoining inlets and passages during spring and summer, when river discharges create a persistent ebb bias in the surface layer and a weaker flood bias at depth.

Tidal Streams

In the absence of vigorous winds or runoff, surface currents within Hecate Strait and Queen Charlotte Sound consist of clockwise-rotary tidal streams that alter direction and speed in a regular manner over a cycle of about 12½ h. The rotary nature of the tidal flow is best developed in the outer reaches of Queen Charlotte Sound, where horizontal motions are comparatively unrestricted by the land. Principal floods are to the northeast and principal ebbs to the southwest in this area (Fig. 14.6). During spring tides, maximum speeds are of the order of

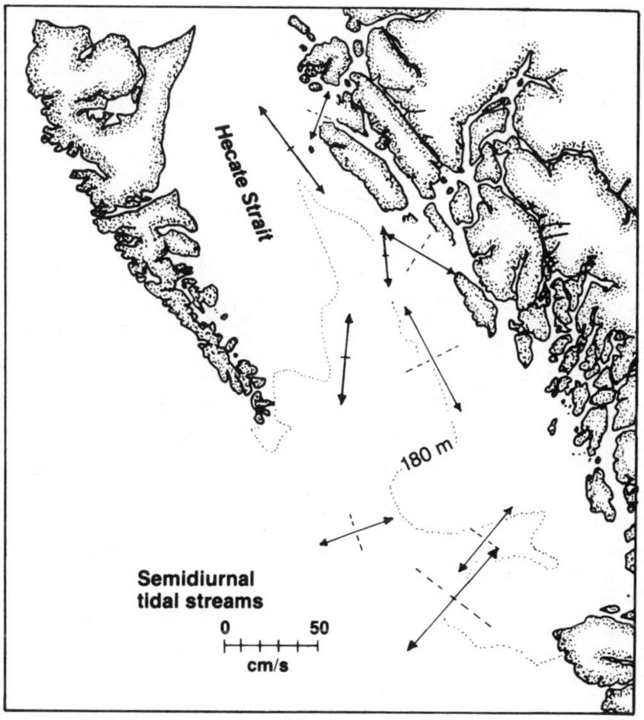

FIG. 14.6. Observed semidiurnal tidal streams at a depth of 15 m in Queen Charlotte Sound and Hecate Strait. Solid lines and arrows give orientations of major ebb and flood directions; broken lines at right angles give minor flow directions for rotary tidal currents. Scale measures speed relative to midpoint of each axis. (Courtesy S. Huggett).

50 cm/s (1 kn) but decrease to around half this value during neap tides. Slack water is nonexistent, as the effect of the earth's rotation causes the current vectors to cyclically alter direction with little change in speed. Nearer the shore, the tidal ellipses generally become increasingly rectilinear with major axes aligned parallel to the trend of entrant channels. In many cases, the flow is accelerated as it negotiates the constricted mouths of these inland leading waterways. As always, there are exceptions. It appears that tidal streams in the central portions of some broader openings that adjoin Queen Charlotte Sound retain their rotary nature because of the curvature of the shoreline and the splitting of the streams into still further channels. (Readers wishing more detailed information on currents within the inland channels are referred to *Sailing Directions British Columbia Coast (North Portion)* Vol. II, 8th ed. 1981.)

Tidal streams in Hecate Strait are basically rectilinear due to restrictions on cross-strait flow by the valleylike bathymetry (Fig. 14.6). Therefore, floods are directed to the north and ebbs to the south at maximum rates of roughly 50 cm/s. Again, there is considerable distortion of surface tidal streams near the broken shorelines of the mainland and Moresby Island coasts. The times of a particular stage of the current along Hecate Strait (for example, maximum ebb or flood) are progressively delayed from those at the seaward entrance to Queen Charlotte Sound, and reach a maximum delay of just over 2 h at the latitude of Banks Island (Fig. 14.7). The fact that the tide itself is delayed by less than 1 h over the same distance would suggest it gradually departs from a standing-type wave as it travels the length of the Strait. This is ineed what happens as the northward propagating tide in Hecate Strait merges with the portion of the tide that advances along Dixon Entrance. From a practical point of view, the times of maximum flood or ebb in northern Hecate Strait are 2–3 h later than the times of local high or low water.

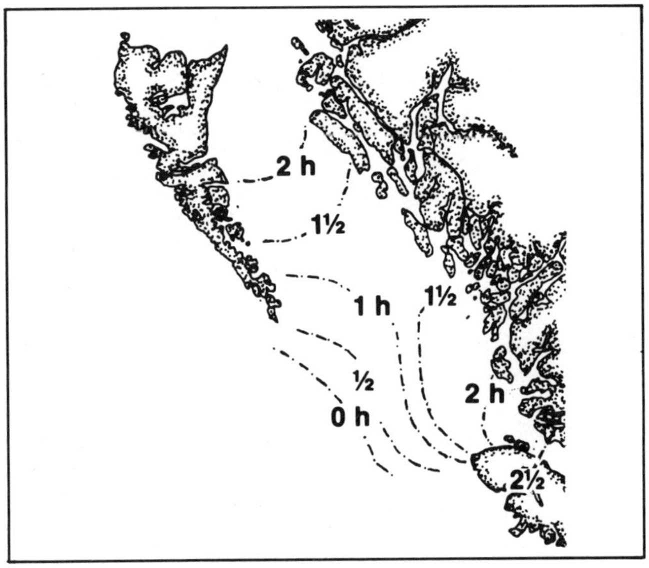

FIG. 14.7. Times of particular stage of tidal streams, relative to open coast. For example, off Banks Island maximum flood occurs 2 h after maximum flood at entrance to Queen Charlotte Sound. (Courtesy S. Huggett)

To the north of the meeting area of the tides, there is a reversal in the ebb and flood directions as principal ebbs are directed to the north and principal floods to the southeast. The confluence of tides in combination with the right-angle geometry of the coastline further complicates the tidal streams within the central portion of Dixon Entrance. Instead of near uniform strength, flood streams are decidedly stronger on the southern side of Dixon Entrance and ebb streams stronger on the northern side (Fig. 14.8). As a consequence, a net counterclockwise vortex is established over a tidal cycle with its center about half way between Cape Chacon and Rose Spit. Currents within this vortex attain greatest speeds (50–100 cm/s) around the outer circumference approximately midway between high and low water, as measured at the shore, and diminish inward to the vortex core. An intense westward setting stream is formed south of Cape Chacon where the ebb from Clarence Strait converges with the ebb that originates from northern Hecate Strait. Similarly, on the flood there is comparatively strong eastward flow north of Graham Island at a time when weak and variable currents exist off Cape Chacon. Further toward the mouth of Dixon Entrance, tidal streams run parallel to the trend of the channel axis and eventually merge with the northward setting flood and southward setting ebb along the outer coast.

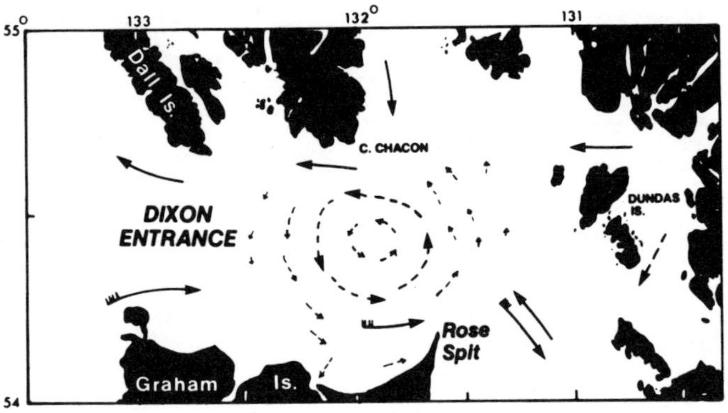

FIG. 14.8. Tidal streams in Dixon Entrance. Flood streams strongest on south side of channel, ebb streams strongest on north side. Broken arrows show net (resultant) cyclonic circulation created by asymetric tidal current pattern. (Adapted from Crean 1967)

Near-bottom tidal motions within the seaway are weaker than their surface counterpart and differ slightly in direction because of the more immediate constraints of the bathymetry. Speeds of 15–25 cm/s appear to be characteristic of the region.

Nontidal Currents

Until now, the important influences of wind and runoff have been ignored. When added to the tidal streams (which must be done to obtain a truly representative picture of the circulation) they may radically alter the characteristics of the tidally induced currents and even suppress them completely. The difficulty with these nontidal components of the circulation is they are often subject to rapid spatial and temporal variations associated with transient fluctuations in wind and runoff patterns. On a short period basis their predictability is almost nil and attention must be confined to prevailing seasonal conditions. As noted in Chapter 4, inertial currents have an important, albeit intermittent, influence on the currents in Queen Charlotte Sound and southern Hecate Strait. Whether they are a factor in Dixon Entrance has yet to be determined. A brief description of inertial currents in the Queen Charlotte Sound–Hecate Strait region is presented in the next subsection.

During November through February, the low freshwater drainage from the land has but a minor influence on currents in the seaway. The reverse is true of the winds, which commonly blow with considerable vigor from the southeast over the entire north coast. To appreciate fully the importance of these winds to the long-term, nontidal component of the flow within the seaway, the study should begin along the outer coast. In this region, southeast winds initially generate a downwind surface drift which is then deflected to the right by the earth's rotation. This, in turn, leads to a net onshore Ekman transport within the top 100 m or so of the water column. Where this transport is blocked by the coast there is an onshore accumulation of surface waters and a depression of the nearshore isopycnals, with only a partially compensating offshore transport at depth. Resulting pressure gradients are then balanced by the establishment of northward coastal currents in the upper layer. If the winds die, the piled-up surface waters collapse seaward and the current disappears. If the winds reverse to the northwest, the antithesis occurs; the surface Ekman transport is offshore, isopycnals are raised (upwelling), and the ensuing coastal current is southward.

At the entrance to Queen Charlotte Sound there is no coastal barrier and surface waters that would normally converge over the continental shelf under the influence of persistent southeast winds are directed northward into the Sound. Aided by the direct action of southeast winds over the seaway, they then flow northward into Hecate Strait as a well-defined, relatively warm surface current (see Fig. 13.19a). Typical speeds of this current are presumably of the order of 3% of the wind speed averaged over several days, but can be expected to vary markedly over periods of half a day or longer and be slightly faster on the eastern side of the channel. If the winds are weak-to-moderately strong, the wind-induced flow will bear northward across the eastern portion of Dixon Entrance into Clarence Strait. The remaining portion of Dixon Entrance will continue to be dominated by the tidal vortex. If, on the other hand, the winds are strong over the coast, the northward transport of surface water through Hecate Strait will create a concentrated seaward flow around Cape Chacon and along the northern side of Dixon Entrance in addition to the northerly set through Clarence Strait. At the mouth of Dixon Entrance the westward progressing surface current veers sharply to the northwest as it merges with the prevailing currents along the outer coast of Alaska.

By early to late spring (March–May), the strength of the southeast winds begins to moderate. This is accompanied by a reduction in the onshore convergence of oceanic surface water and a weakening of the northward wind-generated current that moves through Hecate Strait into Clarence Strait. During the summer (June–August) southeast winds are minimal and there is negligible wind-induced flow within the coastal seaway. Regeneration of the northerly set follows the reestablishment of intense southeast winds some time between September and October and the cycle is completed.

In contradistinction to the winds, freshwater runoff can be expected to modify the circulation of the northern seaway most effectively from late spring to early summer, when large volumes of snow melt are discharged from the multitude of coastal estuaries. The fresh water spreads laterally over the denser salt water where it reaches the seaway and soon becomes mixed to near uniform salinity by winds and tidal currents. This results in the formation of an upper zone of brackish water that is forced to move seaward because of the hydraulic pressure head set up by the accumulation of fresh water along the mainland shores of the seaway. As it progresses, the upper zone becomes increasingly more saline through entrainment of oceanic water from below. To compensate for this loss of salt, sea water intrudes inward at depth, but at a considerably slower rate than the surface outflow.

Therefore, the effect of the runoff is to superimpose a measurable, albeit highly variable, estuarine-type circulation on the tide and average wind-generated circulations. Among other factors, the strength of the brackish outflow over the seaway will, in the absence of winds, vary in accordance with the amount of runoff from contiguous waterways and the degree of lateral spreading. For these reasons the most well-defined estuarine circulation is within Chatham Sound and the fairly confined eastern sector of Dixon Entrance into which pour the voluminous discharges of the Skeena and Nass rivers. Throughout the rest of the seaway, significant estuarine flow is generally limited to the immediate vicinity of mainland shores. Runoff has only a marginal influence on the surface currents in Queen Charlotte Sound and Hecate Strait, which are at all times dominated by tidal streams and wind drift. Attention will be focused on the Dixon Entrance region.

Beginning in late April, there is a rapid rise in the discharge from the Nass and Skeena rivers until respective peaks of 3000 and 4200 m^3/s (105,000 and 150,000 ft^3/s) are attained about the beginning of June, a 10-fold increase over midwinter values (Fig. 14.9). The freshet is then followed by a gradual reduction in outflow until October when a secondary peak associated with winter precipitation occurs. After leaving Chatham Sound the freshwater discharge advances in a general northwesterly direction into the southern reaches of Clarence Strait and the adjoining passages. In summer the flow divides in this region, one branch continues northward through Clarence Strait, the other turns southward around Cape Chacon into Dixon Entrance (Fig. 14.10). The latter branch then spreads into the west central portion of the channel as a result of the indigenous tidal vortex before eventually making its way seaward via the deep channel north of Learmonth Bank. Because the seaward discharge of fresh water acts in concert with the last stages of the westward wind-driven currents in this region, there is a major flushing of brackish water into the Pacific Ocean, and a strong intrusion of cold, saline oceanic water at

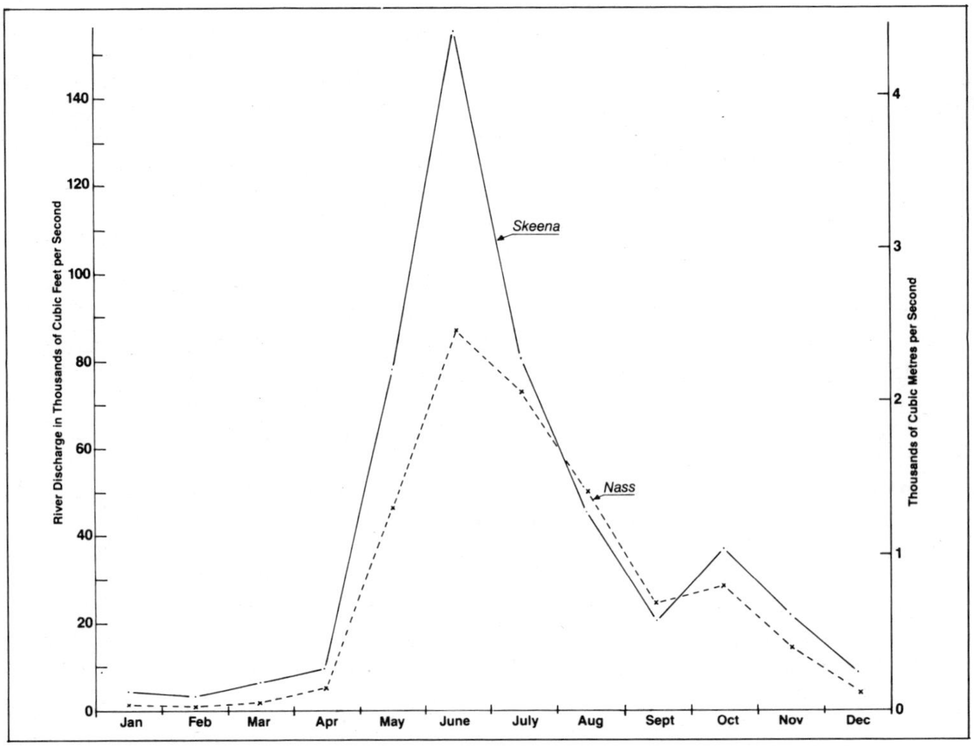

FIG. 14.9. Average monthly volume discharges for Skeena and Nass rivers, 1972. Skeena measurements were taken at Usk, 145 km upstream, and Nass River at Aiyansh, 70 km upstream. Significant volume is added to Skeena downstream of Usk, so values shown underestimate amount of fresh water entering Chatham Sound. (From Water Survey of Canada 1974)

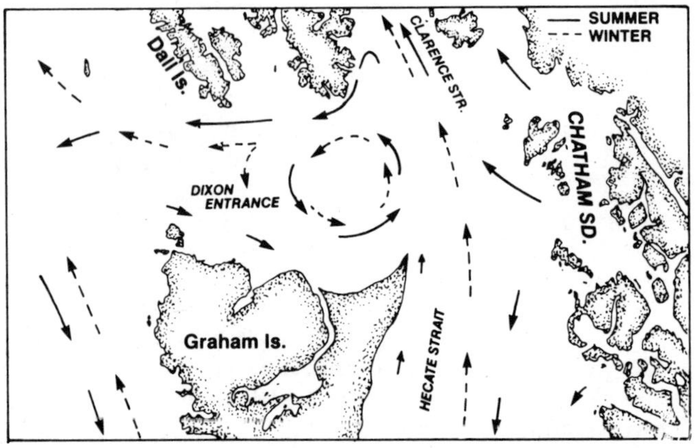

FIG. 14.10. Nontidal surface currents in summer (solid arrows) and winter (broken arrows) for northwest coastal waters. Summer conditions correspond to northwest winds and large river runoff; winter conditions to southeast winds and low runoff. (Adapted from Crean 1967)

depth. "Such flushing is particularly important economically in the case of Dixon Entrance where strong seaward movements of this type may seriously affect the extensive ground fishery in northern Hecate Strait, particularly during the crucial period when the eggs or larvae are free floating." (Crean 1967)

With a subsequent fall in discharge from these rivers by late summer, the outward flow of brackish water becomes mainly confined to Clarence Strait and to the northern shores of Dixon Entrance. On the other hand, westerly winds at this time probably enhance the net inflow of high-salinity surface water along the southern side of Dixon Entrance. By late autumn the nontidal component of the circulation becomes dominated by winds, and what remains of the brackish upper zone is flushed seaward through Clarence Strait. This condition prevails throughout the winter and early spring until the return of warmer weather and melting of snow in the high elevations of the coastal mountains.

Inertial Currents

Current meter observations from the nine mooring sites shown in Fig. 14.6 reveal that intermittent, wind-generated inertial currents frequently dominate the surface circulation within Queen Charlotte Sound and Hecate Strait. Weak remnants of these motions also occur near the bottom to depths of 260 m and within the entrances to the larger channels (e.g. Browning Entrance and Caamaño Sound) that open into the seaway on the mainland side.

Generation of inertial currents in this region is principally by moderate-to-strong, southeast frontal winds that accompany passage of extratropical cyclones ("storms") across the north coast (Fig. 14.11). The current vector necessarily rotates in the clockwise direction (looking downward) and its tip traces out a circular path approximately every 15½ h (1 inertial period at 51° latitude). The latter feature can be verified by noticing that the north–south velocity component of the inertial current in Fig. 14.11 leads the east–west component by 90° (i.e. by $90°/360° \times 15.5 = 3.9$ h) and that the strengths of the two components are always about equal. Because of this rotation, especially strong inertial currents result if the southeast winds veer to southwesterly at about20–30°/h following passage of the front so they remain

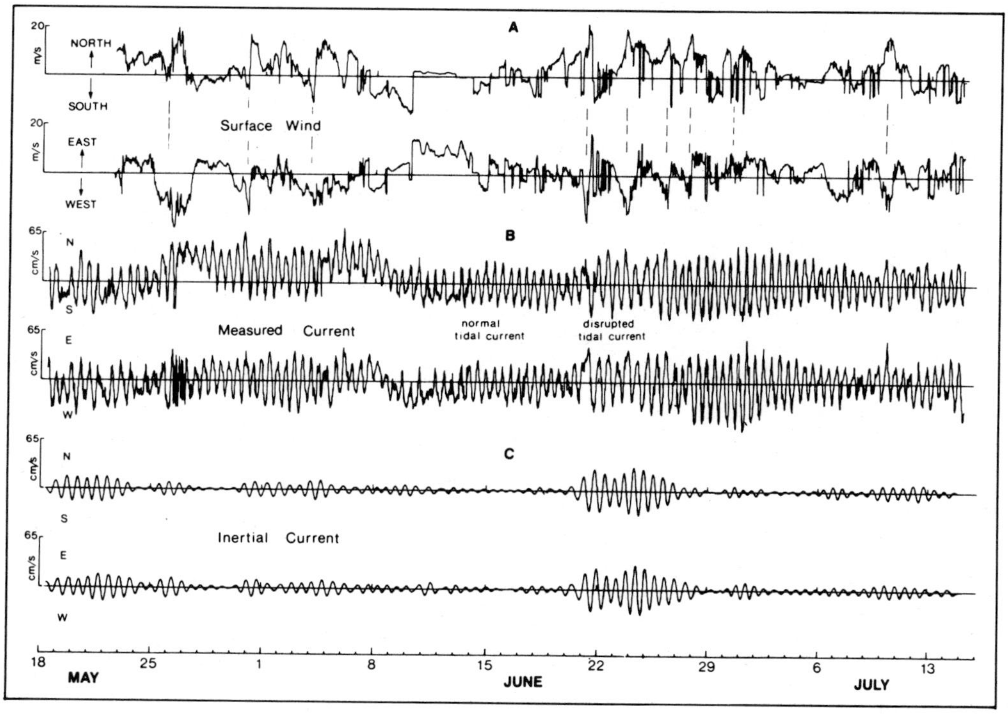

FIG. 14.11. Winds and currents in Queen Charlotte Sound from May to July, 1977. (A) North–south and east–west components of the wind according to oceanographic convention, i.e. direction to which wind is blowing. For example, on May 26 winds were *toward* the northwest or *from* the southeast at over 20 m/s (40 kn). (B) North–south and east–west components of the actual current measured at a depth of about 15 m (50 ft) at a location 70 km west-northwest of Cape Scott (star in Fig. 14.3). Winds were measured from a mooring 100 km northwest of current meter location. (C) North–south and east–west components of inertial currents, extracted from the observed records in (B) using digital filtering methods. (In the observed curret records, inertial currents were masked by tidal and estuarine currents and had to be isolated using computational techniques.) Note that each onset of strong inertial currents was roughly coincident with passage of a frontal system marked by vertical dashed lines in (A).

aligned with the current during its inception. The near-surface speed of the current is highest roughly 15–20 h after onset and may exceed 50 cm/s (1 kn) within the exposed western half of Queen Charlotte Sound; a speed of 75 cm/s was recorded on June 22, 1977, at an exposed mooring site 130 km northwest of Cape Scott. Through a combination of turbulence, mixing processes, and dispersion, the downward propagated kinetic energy of these motions is rapidly attenuated with depth and wind-generated inertial currents are mainly confined to the upper 50 m or so.

The speeds of near-surface inertial motions are greatest near the entrance to Queen Charlotte Sound and decrease in a generally northeastward direction over the seaway. Typical maximum speeds of 50 cm/s at the Pacific entrance diminish to less than 20 cm/s off Banks Island in northern Hecate Strait and 30 cm/s in the eastern sector of the Sound. Moreover, because the storms that generate the currents travel to the northeast, there is a corresponding delay in the time of onset of the current toward the northeast. The inertial currents produced by a particular storm usually persist locally for about 2½ days and represent transient oceanic responses to rapid variations in the wind. However, if the time between successive storms is a multiple of the inertial period and less than 3 days, the gradually decelerating currents generated by one storm will be given an accelerating boost by the following storm in the sequence, and the current will persist. This was the case beginning on June 21 in Fig. 14.11, where four successive storms, each with peak winds about four inertial periods (2½ days) apart, kept the inertial currents going for more than a week; the much stronger motions about 50 km northwest of this location lasted nearly 12 days. (Instances where successive storms are "out of sync" lead to rapid deceleration of established inertial flows by adverse winds.)

In the previous chapter it was noted that the most common time delay between successive winter storms in the Northeast Pacific is 2½ days. This, in conjunction with the greater strength of winter winds, suggests that extended periods of intense inertial oscillations are a common feature of the Queen Charlotte Sound–Hecate Strait region from late autumn to early spring.

As demonstrated by the observed currents in Fig. 14.11, inertial currents in excess of 10 cm/s effectively disrupt the normal rhythm of the rotary tidal streams. Current-table predictions would be useless at such times as both the strength and direction of surface flows are modified by the wind-generated currents. A less obvious, but equally important, facet of inertial motions is that they are a prime mechanism for transferring wind energy downward into the water column and, therefore, are a major source of energy for mixing processes within the upper layer of the ocean.

GLOSSARY

GLOSSARY

Abyssal — deep portions of the ocean, typically below 2000 fa (3700 m).

Abyssal hills — small elevations (often extinct volcanoes) distributed over large tracts of the seafloor.

Abyssal plain — a flat or gently sloping, almost featureless region of the deep ocean overlain with sediments.

Acceleration of gravity (g) — the rate of change with time of the velocity of an object caused by the earth's gravitational force.

Advection — the transport of a property (e.g. heat or salt) by fluid motions such as currents or winds.

Aerobic — conditions where free oxygen is present; requires free oxygen for organic existence and growth.

Amphidromic point — a point at which there is no tide and around which the cotidal lines rotate over a tidal cycle. Tidal amplitudes (corange lines) increase outward from this point to cover an area called the amphidromic region.

Amplitude — the magnitude or half-cycle displacement of a wavelike fluctuation as measured from its mean or equilibrium value. The amplitude of an ocean wave is half its trough-to-crest height, or the height from the still water level to the wave crest; in the case of the tide, amplitude equals half the range.

Anaerobic — lacking in free oxygen. Certain bacteria can live in such conditions.

Angular momentum — the momentum of a particle or parcel of water arising from its rotation about an axis. Its value is proportional to the particle's mass, linear velocity, and distance from the rotation axis.

Anomalistic month — the average period of the moon's orbit around the earth with respect to the lunar perigee; equal to about 27.55 days.

Anticyclonic — a sense of rotation opposite to that of the earth's rotation; clockwise in the northern hemisphere and counterclockwise in the southern hemisphere.

Antinode (antinodal line, antinodal point) — the position (or line) of maximum displacement in an oscillating region of water. The region midway between two nodal lines in a standing wave pattern.

Apogee — position in the moon's orbit that is furthest from the earth (opposite to *perigee*).

Asthenosphere (or *mantle*) — the region of the earth with plastic-flow characteristics between the crust (lithosphere) and the core.

Attenuation — a reduction in wave amplitude.

Backing — a change in the wind or current direction in a counterclockwise sense (opposite to *Veering*).

Bar — an elongate ridge of sand, gravel, or other unconsolidated material built in shallow water by the action of waves and currents, especially at the mouth of a river or estuary.

Basalt — a fine-grained, dark-coloured igneous rock. The oceanic crust and granitic continents are thought to be underlain by basaltic material.

Bed load (traction load) — quantity of bottom sediment transported by a current through traction, that is by saltation, rolling, and sliding.

Benthic — pertaining to the bottom terrain of oceanic regions.

Bloom — a rapid growth of phytoplankton in spring and fall that causes large temporary concentrations of the organism. Can result in discolored or toxic water known as red tide. (See *Plankton*).

Bore (hydraulic jump) — an almost vertical, solitary wave front caused when the water speed associated with the crest begins to exceed the translation speed of the wave. Waves on beaches commonly deform into a bore prior to breaking. The water flow associated with a bore changes abruptly from a region of low depth and high speed to one of high depth and low speed with a series of undulations behind the leading crest (see *Tidal bore*).

Bottom friction — the retarding force on water moving over the bottom. It increases with increased roughness of the bottom, current speed, and degree of turbulence in the water.

Brackish water — water having a salinity of 0.50–17‰, common to nearsurface estuarine environments from late spring to early autumn.

Buoyancy — the net upward force exerted on an object or parcel of fluid due to the difference in density between it and the surrounding fluid; quantitatively equal to the weight of the water displaced by the object.

Cat's paw — a puff of wind; a light breeze that temporarily ruffles patches of the water surface.

Chop — short-crested waves that spring up following onset of a moderate breeze, and break readily at the crest.

Coastal plain — a plain composed of gently sloping or flat strata of sedimentary material fronting the coast and generally derived from a recently emerged portion of the sea bottom.

Cobble — a water-worn rock fragment between 64 and 256 mm in diameter, larger than a pebble but smaller than a boulder.

Constituent, tidal — a harmonic (sinusoidal) contribution in a mathematical expression for the tide-generating force used to specify a selected cycle in the tide or tidal current. Each constituent has a fixed period but an amplitude that is determined by observation or through a model.

Continental margin — the portion of the seafloor adjacent to the continent and separating it from the deep sea. The continental margin includes the continental rise, continental slope, and continental shelf.

Convective overturn — the process by which the surface waters, cooled sufficiently by the air to make them more dense than underlying waters, sink and are replaced by an upward motion of less dense water. In the subarctic Pacific region of the North Pacific Ocean, overturning takes place in winter but is limited to the upper 100 m of water due to the presence of the permanent halocline between 100 and 200 m.

Corange line — a line joining regions of equal tide range.

Coriolis force (Coriolis effect) — an apparent force due to the earth's rotation that acts at right angles on a moving object or fluid. The force is proportional to the speed and latitude of the object, is zero at the equator and a maximum at either pole for a fixed speed.

Cotidal line — a line joining regions with simultaneous occurrence of high water. Synonymous with cophase line.

Countercurrent — a surface current parallel to, but setting in the opposite direction to, the main current.

Crust — see *Lithosphere.*

Cusp — one of a series of low mounds of beach material separated by crescent-shaped troughs spaced at roughly equal intervals along a beach.

Cyclonic — a sense of rotation in the same sense as that of the earth's; counterclockwise in the northern hemisphere and clockwise in the southern hemisphere.

Day — the time for one rotation of the earth. There are different types of day depending on whether the sun, moon, or a star is used as the point of reference for the rotation; solar day (24 h), lunar day (24.84 h), and sidereal day (23.93 h), respectively.

Dead water — the condition whereby a vessel expends considerable propulsive power in the generation of an internal wake as it travels in a region with a shallow brackish layer overlying a deeper more saline layer.

Declination — the angle that the sun, moon, or other celestial object subtents with the plane of the earth's equator.

Density — the ratio of the mass of any substance to the volume it occupies. Typical ocean water has densities in the approximate range of 1.020–1.028 g/cm^3.

Diatom — a single-celled plant (phytoplankton) covered with two overlapping porous shells of silica. Diatoms are one of the most abundant marine organisms and a primary source of food in the sea.

Dispersion — the separation of a group (or packet) of waves into component parts due to differences in propagation speeds of the waves.

Diurnal — meaning daily and pertaining to motions which complete a cyclic pattern every solar or lunar day.

Drift current — a broad, slow ocean current principally driven by the large-scale winds.

Drogue — a current measuring device with sufficient subsurface area and depth penetration to be mainly carried by the oceanic current rather than by the effects of winds and waves.

Earth tide — cyclic displacement of the earth's crust caused by the tide-generating forces of the sun and moon.

Eddy — a quasi-circular movement of water, of relatively small area, formed in the lee of obstructions or along the edge of two regions with different current speeds or directions.

Edge wave — a near-shore (shore edge) wave with an accompanying pattern of horizontal currents that progresses parallel to the coast with crests normal to the coastline. Wave heights diminish rapidly seaward of the coast.

Embayment — a shoreline indentation that forms an open bay.

Entrainment — the process by which relatively high-density fluid is incorporated into an overlying layer of less dense fluid as a result of breaking wavelike disturbances or turbulent motions at the interface between the fluids. In an estuary, entrainment produces at net transport of salt from the saline ocean water to the overlying brackish layer.

Epicenter — the point on the earth's surface directly above the point of origin (focus) of an earthquake.

Equilibrium theory — a model of the tide that assumes a continuous ocean of uniform depth over the earth in which the water responds instantly to the tide-generating forces of the sun and moon. Land masses, bottom friction, and inertia are ignored in the model.

Equilibrium tide — the hypothetical tide generated by the tide-generating forces in the equilibrium theory of the tide. Also called the gravitational tide or astronomical tide.

Fan — a relatively smooth, fan-shaped feature that normally slopes away from the lower termination of a canyon or canyon system. Also, deep-sea fan and submarine fan.

Fault — a fracture or fracture zone in rock along which relative movement of opposing sides has occurred.

Fetch — the area of the ocean surface where seas are generated by a wind with nearly constant direction and speed. Also, the spatial extent of this area measured in the wind direction.

Flocculation — the process by which fine particles of clay suspended in fresh water aggregate into lumps on contact with salt water and settle out of suspension; a depositional mechanism common to estuarine regions.

Fracture zone — an extensive linear zone of irregular seafloor topography characterized by steep-sided or asymmetrical ridges, troughs, or escarpments.

Frequency — a measure of the number of oscillations or cycles per unit time; the reciprocal of the time duration (period) of an oscillation. (A wall outlet in North America for example, has a voltage oscillation frequency of 60 cycles/s.)

Frictional drag — the retarding force exerted on a fluid moving relative to a solid surface or to another region of fluid as a result of momentum exchange processes. The momentum exchange can arise through molecular motions or through eddylike motions. (See *Laminar flow, Turbulent flow*)

Front — a comparatively sharp horizontal transition between two fluids of different density generally accompanied by a change in flow velocity. Atmospheric fronts are delineated by changes in temperature, oceanic fronts by changes in temperature and/or salinity.

Geostrophic flow — currents or winds on length scales that exceed approximately 100 km where the horizontal pressure gradient is exactly balanced by the Coriolis force.

Glaciomarine — pertaining to unsorted and nonlayered marine sediments, generally unconsolidated, deposited by or underneath a glacier.

Granite — a coarse-grained igneous rock of mainly feldspar and quartz comprising a significant portion of the upper continental crust.

Gyre — a large oceanic region of closed or nearly closed horizontal circulation. In the northern hemisphere cyclonic gyres exhibit a net divergence, anticyclonic gyres a net convergence of surface waters.

Halocline — a layer characterized by a sharp increase in salinity with depth. In the North Pacific Ocean, there is a permanent halocline at 100–200 m.

Heat capacity — the ratio of the heat absorbed (or released) by a fluid to the corresponding temperature rise (or fall). For example, 1 cal is required to raise the temperature of 1 g of liquid water by 1C° so the heat capacity is 1.0 cal/g C°; for ice the value is 0.55 cal/g C°.

Homogeneous fluid — a fluid with a spatially uniform density.

Hydraulic current — a current produced in a channel by a difference in water elevation (hydraulic head) at the ends. May be due to a difference in the range or time of the tide in the water bodies that the channel connects.

Hydraulic jump — see *Bore*.

Hydrocarbon — organic compounds composed of hydrogen and carbon; natural gas and petroleum are common examples.

Igneous rock — rock formed at the earth's surface when molten material cools. (see *Magma*)

Interface — a surface separating two fluids across which there is an abrupt change in temperature, salinity, or density.

Internal tide — a long subsurface wave of mainly semidiurnal period generated within the body of an oceanic region or contiguous channel through the interaction of the astronomical tide with the bottom topography.

Iso- — prefix meaning alike, equal, or same; from the Greek word isos: equal.

Isobar — A line that connects points of equal pressure.

Isohaline — a line that connects points of equal salinity.

Isotherm — a line that connects points of equal temperature.

Japan Current — a popular, but not recommended, name for the Kuroshio that flows northeastward over the continental slope off Japan. Also an incorrectly used name for the eastward setting Subarctic Current.

Jetty — a structure that extends into a body of water designed to direct and confine the current and to prevent shoaling of a channel due to littoral drift. A jetty projecting seaward to retard coastal erosion is called a groyne; a jetty breaking the wave force is called a breakwater. Other jetties include sea walls and training walls.

Kinetic energy — the energy of an object or parcel of fluid by virtue of its motion. Kinetic energy is proportional to mass and the square of the speed.

Knoll — a relatively small, isolated elevation of a rounded shape on the seafloor.

Laminar flow — a flow in which the fluid moves smoothly and in parallel layers. Laminar flow over solid surfaces conforms to the general topography. (Opposite to *Turbulent flow*)

Land breeze — a generally light wind that blows seaward due to greater cooling of the land than water.

Larva (plural *larvae*) — the young, self-sustaining and independent stage of an animal's life cycle prior to assuming the characteristics of the parents.

Lee wave — a type of internal wave formed downstream (i.e. in the lee) of a topographic feature such as a sill or, in the case of the atmosphere, a hill or mountain range.

Lithosphere — the solid outer portion of the earth's surface above the asthenosphere (mantle). Oceanic lithospheres have an average thickness of 5–7 km, continental lithospheres an average thickness of 35 km.

Littoral (intertidal) — of or pertaining to the seashore. Commonly, that part of the benthic zone between the high- and low-water levels.

Littoral current — any current in the littoral zone that flows parallel to shore and usually driven by breaking waves. Synonymous here with longshore current.

Littoral drift — the material transported in the littoral zone by a littoral current.

Littoral zone — according to beach terminology, the indefinite zone that extends from the shoreline to just beyond the breaker zone. In biological oceanography it is part of the benthic region from the high-water line to 200 m depth.

Longshore current — (see *Littoral current*)

Magma — molten rock under the earth's surface from which igneous rock is derived by solidification. Magma extruded onto the earth's surface is called lava.

Magnetic anomaly — a distortion of the normal configuration of the earth's magnetic field resulting from local concentrations of ferromagnetic materials.

Magnetic domain — a small region in a ferromagnetic material where, because of molecular interactions, the individual molecular magnets are all aligned parallel to one another. The directions of magnetization in different domains are not necessarily the same as neighboring domains. In an unmagnetized specimen, domains are randomly oriented and the resultant magnetization is zero.

Mantle — (see *Asthenosphere*).

Maritime climate — a regional climate under the predominant influence of the ocean, and characterized by small diurnal and annual ranges in air temperature.

Marsh (salt marsh) — a flat, vegetated area at or above the high-water mark that is flooded by spring tides or during storm surges.

Meteorological tide — the variation in water level due to meteorological effects such as winds and atmospheric pressure.

Mixed layer — the surface layer of a body of water mixed to near homogeneity by wind-wave action or by convective overturning. Mixed layers can also be created above the bottom through the action of turbulent motions.

Mollusk — a soft-bodied, unsegmented animal with gills and commonly protected by a calcareous shell. Includes snails, squid, octopuses, and clams.

Momentum — property of a moving fluid that is proportional to the product of its mass times velocity. A force is required to change the momentum of a fluid; similarly, a fluid will produce a force if it imparts momentum to an object or another parcel of fluid.

Month — period of the moon's orbit around the earth. Types of month include sidereal, anomalistic, and synodical depending on whether the period of revolution is measured relative to a fixed star, perigee, or the sun. The calendar month is a rough approximation to the synodical month.

Mud flat — a muddy or sandy coastal strip often submerged at high tide.

Natural frequency (resonant frequency) — the frequency, or set of discrete frequencies, of oscillation in a body of water determined by the physical characteristics of the basin such as its depth, shape, dimensions, and density structure. Seiches in semienclosed embayments or harbors typically attain maximum amplitudes at an integer number of frequencies that are naturally tuned to the frequency of the external mechanism that cause the displacements.

Noble gas (inert gas) — any group of rare gases that include helium, neon, argon, krypton, and xenon and exhibit great stability and extremely low chemical reaction rates.

Node (also *nodal line, nodal point*) — the position or line in an oscillating region of water where the motions are a minimum or nonexistent. For a standing wave, vertical displacements are least and horizontal current greatest at a nodal point. (see *Amphidromic point*)

Nontidal current — the contribution to the net current at a particular location not directly caused by the tide-generating forces of the sun and moon (Synonymous with *Residual current*)

Nutrient — a chemical compound needed for the growth of marine plants.

Ocean (weather) station — a specifically located area of the ocean surface roughly 200 nm on a side. Ocean weather ships are equipped to obtain comprehensive meteorological and oceanographic observations at these stations.

Orographic — of or pertaining to mountains. Orographic effects include modifications of winds and rainfall by the presence of mountain ranges.

Osmosis — the movement of molecules or ions across a selectively semi-permeable membrane.

Osmotic pressure — the pressure exerted on a membrane as a result of the difference in chemical concentration in the fluids on either side.

Overtopping — the flow of water over a natural or man-made structure due to wave run-up or a surge.

Oxygen, dissolved — oxygen gas that has dissolved in water. The concentration of dissolved oxygen in the sea at normal temperatures and pressures typically lies between 0 and 8 mL/L; the concentration at saturation decreases with increasing temperature and salinity. In the Northeast Pacific Ocean, oxygen levels decrease rapidly with depth beneath the permanent halocline to a minimum of 1 mL/L or less around 1500 m then gradually increase with depth.

Paleomagnetism — the remnant of the earth's ancient magnetism retained during the initial formation of a material.

Paleontology — the study of ancient life by fossil evidence.

Percolation — the processes by which wave motions force water into the spaces between bottom sediments.

Perigee — the position in the moon's orbit that is nearest the earth. (opposite to *Apogee*)

pH — a measure of the alkalinity or acidity of a solution based on the concentration of the hydrogen ion. A pH value of 7 indicates a neutral solution, less than 7 is acidic, and greater than 7 is alkaline.

Physiography — the description of the features and phenomena of nature; the science that deals with the nature and origin of the earth's topographic features.

Pinnacle — any high tower or spire-shaped pillar of rock, or coral, alone or cresting a summit.

Plankton — drifting marine organisms with limited swimming capability; include microscopic phytoplankton and zooplankton as well as certain macroscopic organisms, such as jellyfish.

Plastic material — a material that can be deformed continuously and permanently in any direction without rupture through application of pressure.

Plume — the sediment suspension carried by fresh water that flows out over the surface of a region of saltwater.

Polar outbreak — the seaward flow of cold, dry, stable air along inlets subsequent to its formation over high altitude continental land masses.

Potential energy — a stored form of energy associated with the displacement of a fluid from its equilibrium or undisturbed level.

Prevailing — pertaining to the wind or current most commonly observed during a specified period, such as a month or season.

Primordial — of or related to the beginning or initial times of the earth's history.

Promontory — a high point of land that projects into a body of water; a headland.

Propagation of waves — the transmission of waves through a fluid.

Province — a region identifiable by a group of similar physiographic features whose characteristics differ markedly with surrounding areas; example, seamount province.

Pycnocline — a zone of comparatively rapid change in density with depth.

Rectilinear current (reversing current) — a time varying current such as a tidal current, which flows alternatively in roughly opposite directions with a slack water at each reversal in direction. Found in confined channels, straits, or passes.

Reef — rocks near or at the surface that may constitute a hazard to navigation.

Residual current — (see *Nontidal current*)

Resonance — a condition of wave amplification that arises when the oscillation frequency of the applied force on a system matches the frequency of the particular wave or basin oscillation on which the force is acting.

Ridge — a long, narrow elevation with steep sides that often separates ocean basins as part of the major oceanic mountain systems with global extent.

Rip — a turbulent agitation of the water produced by opposing waves and currents or by rapid flow over an irregular bottom. Often called tide rips in regions where currents are predominantly tidal in nature.

Rocky intertidal platform — a flat or gently sloping wave-eroded platform in the lower foreshore zone on a rocky coast.

Rotary current — a current that flows continually, partly or wholly unrestricted by coastal boundaries, and which changes direction through all points of the compass during a single cycle.

Salinity — the quantity of dissolved salts in seawater in parts per thousand (‰) by weight.

Saltation — a form of sand transport in a fluid when individual particles bounce off the bed and are carried some distance downstream before again sinking to the bottom.

Sea breeze — a light to moderate wind that blows toward the land caused by greater heating of land than water.

Sea fog — an advection-type fog produced when relatively moist air is transported from a region of warm surface waters to one of cooler surface waters, leading to condensation of the lower layer of air.

Sea stack — a tall, columnar rock separated from the coast by differential wave erosion.

Semidiurnal — pertaining to motions that undergo a variation of one cycle once every half lunar day.

Set-up — elevation of the water surface above normal level due to an onshore transport of water by wave action (wave set-up) or directly by wind-induced flow (wind set-up).

Shingle — well rounded, often flat, waterworn pebbles larger than about 16 mm.

Shore — the narrow strip of land immediately adjacent to and contacting the sea, including the tidal zone. A beach is a shore formed of unconsolidated material.

Sidereal month — the average period of the moon's orbit around the earth with respect to a fixed star. One sidereal month equals 27.32 mean solar days.

Solitary wave — an isolated wave crest in the form of a traveling mound of water, whose amplitude may be comparable to the water depth, and not accompanied by a preceding or trailing trough. Also, a wave of translation in which the water parcels move only in the direction of wave advance.

Stability — the resistance to overturn or mixing in a column of water. Stability increases with increased stratification and with decreased vertical change in current velocity with depth.

Stratification — the condition in which a column of water is layered, continuously or in abrupt steps, according to density with lighter water overlying heavier water.

Synodical month — the average period of the moon's orbit around the earth with respect to the sun, or the time between identical lunar phases. One synodical month equals 29.53 mean solar days.

Terrace — a level or nearly level topographic feature on a steeper slope e.g. a relict beach terrace on the side of a seamount.

Thermocline — a layer that marks a sharp change in temperature with depth. In the ocean, there are short-lived, transient thermoclines, seasonal (temporary) thermoclines, and permanent thermoclines.

Tidal bore — a tidal wave that advances up a relatively shallow and sloping estuary in the form of a solitary wave. (see *Bore*)

Tidal current — the cyclic horizontal flow of water associated with the combination of the tidal stream and more slowly varying residual (or nontidal) currents.

Tidal flats — an area of marshy, muddy, or sandy land that is inundated by the tide.

Tidal stream — the cyclic horizontal flow of water produced solely by tide-generating forces of the moon and sun. The horizontal component associated with the rise and fall of the tide. (Called tidal current in the United States).

Tidal wave — a long, propagating shallow-water wave of tidal period generated by tide-generating forces and modified by the Coriolis force, bottom friction, and seafloor topography. A misnomer for tsunami and storm surge.

Turbidity current — a dense, fast flowing current of sediment-laden water that moves down a submarine slope.

Turbulence — a state of fluid motion with irregular and random fluctuations that is describable only in terms of its statistics. Turbulent motions provide a more intense mechanism for the redistribution of momentum, energy, and suspended material than the slow molecular diffusion processes.

Turbulent flow — a flow of significant irregular and random motions, usually superimposed on a more organized flow pattern.

Undercurrent — a subsurface current typically with a

different speed and direction from the prevailing surface current.

Undertow — a seaward current near the bottom on a sloping inshore zone, the result of the return of water carried onshore by waves; a common misnomer for rip current.

Up-rush, (swash, run-up) — the rush of a breaking wave onto a beach.

Vector — a quantity with both magnitude and orientation relative to a given coordinate system or grid. A current vector (current velocity) requires both speed and direction for complete description. A current vector can be represented by an arrow with the length of the arrow proportional to the speed, and the tip of the arrow pointing in the direction of the flow.

Veering — a change in wind or current direction in a clockwise sense. (opposite to *Backing*)

Velocity — a vector quantity that gives both the speed and direction. The rate of change in time of an object's position.

Viscous damping — the attenuation of a fluid oscillation due to the fluid's internal friction or viscosity.

Water mass — a body of water characterized by a particular (limited) range of temperatures and salinities that allow it to be distinguishable from surrounding oceanic regions.

Wind drift — a large-scale ocean current where the Coriolis force and frictional forces are dominant. Also used here to describe a localized, weak, near-surface current generated by the wind.

Wind waves — waves formed and amplified by the wind.

Zooplankton — the animal forms of plankton; include jellyfish, worms, mollusks, and a wide variety of other marine animals. The primary grazers of phytoplankton and, in turn, the principal food for many larger marine animals, such as squid and baleen whales.

REFERENCES

REFERENCES

(* denotes general interest publication)

AGES, A. 1979. The salinity intrusion in the Fraser River: salinity, temperature and current observations, 1976, 1977. Pac. Mar. Sci. Rep. 79-14: 193 p.

AGES, A., and A. WOOLLARD. 1976. The tides in the Fraser estuary. Pac. Mar. Sci. Rep. 76-5: 100 p.

ANON. 1971. Oceanography of the nearshore coastal waters of the Pacific relating to possible pollution. Environ. Prot. Agency Water Qual. Off. Vol. 1: 188 p.

1973. The Burrard Inlet-Howe Sound area, preliminary description of existing environmental conditions. Unpublished manuscript, Environment Canada, 69 p.

1976. Tables of temperature, precipitation and sunshine. B.C. Min. Agric. 82 p.

ASHLEY, G. M. 1977. Sedimentology of a freshwater tidal system, Pitt River–Pitt Lake, British Columbia. Ph.D. Thesis. Univ. British Columbia. 234 p.

BARAZANGI, M., and J. DORMAN. 1969. World seismicity maps compiled from ESSA, coast and geodetic survey, epicenter data 1961–1967. Bull. Seismol. Soc. Am. 59: 369–380.

BARKER, M. L. 1974.* Water resources and related land uses Strait of Georgia–Puget Sound basin. Dep. Environ. Geogr. Pap. 56: 55 p.

BASCOM, W. 1964.* Waves and beaches. Doubleday and Co., New York. 267 p.

BLANCHARD, D. C. 1963. Electrification of the atmosphere by particles from bubbles in the sea, p. 71–202. ***In*** Progress in oceanography, Vol. 1. Pergamon Press.

BOGDANOV, K. T., K. V. KIM, and V. A. MAGARIK. 1964. Numerical solution of tide hydrodynamic equations by means of BESM-2 electronic computer for the Pacific area, p. 161–235. ***In*** Munk et al. [ed.] Tides off-shore: transition from California coastal to deep-sea waters. Geophys. Fluid Dyn.

BOISVERT, W. E. 1969. Major currents off the west coasts of North and South America. Nav. Oceanogr. Off. Tech. Rep. 221: 34 p.

BRETSCHNEIDER, C. L. 1952. The generation and decay of wind waves in deep water. Trans. Am. Geophys. Union 33: 381 – 389.

1958. Revisions in wave forecasting: deep and shallow water. Proc. 6th Conf. Coast. Eng. p. 30 – 67.

BUCKLEY, J. R. 1977. The currents, winds and tides of northern Howe Sound. Ph.D. Thesis. Univ. British Columbia. 228 p.

BUDINGER, T. F. 1967. Cobb seamount. Deep Sea Res. 14: 191–201.

CAMPBELL, N. J. 1954. A study of lateral circulation in an inlet. Ph.D. Thesis. Univ. British Columbia. 56 p.

CANADIAN HYDROGRAPHIC SERVICE. 1976. Sailing Directions British Columbia Coast (south portion), Vol. 1, 10th ed. 399 p.

1977. Sailing Directions British Columbia Coast (north portion), Vol. 2, 7th ed. 413 p.

1979. Small craft guide. Vol. 1, 4th ed. 238 p.

CHASE, R. L., D. L. TIFFIN, and J. W. MURRAY. 1975. The western Canadian continental margin, p. 701–721. ***In*** Canada's continental margins and offshore petroleum exploration. Can. Soc. Pet. Geol. Mem. 4.

CLAGUE, J. J., and B. D. BORNHOLD. 1980. Morphology and littoral processes of the Pacific coast of Canada, p. 339–380. ***In*** S. B. McCann [ed.] Coastlines of Canada. Geol. Surv. Can.

CRAWFORD, W. R., W. J. RAPATZ and W. S. HUGGETT, 1981. Pressure and temperature measurements on seamounts in the North Pacific, *Marine Geodesy*, *5*, 43–54.

CRAWFORD, W. R. and R. E. THOMSON, 1982. Continental shelf waves of diurnal period along Vancouver Island, *J. Geophys. Res.*, *87*: 9516–9522.

CREAN, P. B. 1967. Physical oceanography of Dixon Entrance British Columbia. Bull. Fish. Res. Board Can. 156: 66 p.

CREAN, P. B., and A. AGES. 1971. Oceanographic records from twelve cruises in the Strait of Georgia and Juan de Fuca Strait, 1968. Dep. Energy, Mines Resour. Vol. 1: 55 p.

DAWSON, W. B. 1920. The tides and tidal streams, with illustrative examples from Canadian waters. Dep. Nav. Sci. Ottawa. 43 p.

DIETZ, R. S., and J. C. HOLDEN. 1970. Reconstruction of Pangea: breakup and dispersion of continents, Permian to present. J. Geophys. Res. 75: 4939–4956.

DODIMEAD, A. J., F. FAVORITE, and T. HIRANO. 1963. Salmon of the North Pacific Ocean — Part II. Review of oceanography of the subarctic Pacific region. Int. North Pac. Fish. Comm. Bull. 13: 195 p.

DOHLER, G. 1964. Tides in Canadian waters. Can. Hydrogr. Serv. 14 p.

EKMAN, V. W. 1905. On the influence of the earth's rotation on ocean currents. Ark. Mat. Astron. Fys. 2: 1–52.

ELLIOTT, J. A. 1969. A synoptic study of Sooke Basin. Inst. Oceanog. Univ. British Columbia Rep. 22: 32 p.

EMSLIE, J. H. 1979. Winds at Jericho, p. 51, 84. ***In*** Pacific Yachting. Interpress Publications, Vancouver.

FAVORITE, F., A. J. DODIMEAD, and K. NASU. 1976. Oceanography of the subarctic Pacific region, 1960–71. Int. North Pac. Fish. Comm. Bull. 33: 187 p.

FARMER, D., and J. D. SMITH. 1978. Nonlinear internal waves in a fjord, p. 465–493. ***In*** Hydrodynamics of estuaries and fjords. Proc. 9th Liege Colloq. Ocean Hydrodyn. Elsevier, Amsterdam.

FISSEL, D. B. 1975. A frequency analysis of ten years of surface atmospheric data at Ocean Weathership 'PAPA' (50°N, 145°W). M.Sc. Thesis. Univ. British Columbia. 136 p.

FISSEL, D. B., and W. S. HUGGETT. 1976. Observations of currents, bottom pressures and densities through a cross-section of Juan de Fuca Strait. Pac. Mar. Sci. Rep. 76–6: 68 p.

GALVIN, C. J. 1972. Waves breaking in shallow water, p. 413–456. ***In*** R. E. Meyer [ed.] Waves on beaches and resulting sediment transport. Academic Press, New York.

GIOVANDO, L. F., and S. TABATA. 1970. Measurements of surface flow in the Strait of Georgia by means of free-floating current followers. Fish. Res. Board Can. Tech. Rep. 163: 69 p.

GOYETTE, D., and H. NELSON. 1977. Marine environmental assessment of mine waste disposal into Rupert Inlet, British Columbia. Fish. Environ. Can. Rep. EPS 5PR-77-11: 93 p.

HENRY, R. F., and T. S. MURTY. 1972. Three-dimensional circulation in a stratified bay under variable wind-stress, p. 125–140. ***In*** J. Nihoul [ed.] Mem. Soc. R. Sci. Liege.

HERLINVEAUX, R. H. 1962. Oceanography of Saanich Inlet in Vancouver Island British Columbia. J. Fish. Res. Board Can. 19: 1–37.

HOFFMAN, D. 1976. Analysis of wave spectra at Station "PAPA". Webb Inst. Nav. Archit. New York. 55 p.

HOLBROOK, J. R., and D. HALPERN. 1978. Variability of near surface currents and winds in the western Strait of Juan de Fuca. Pac. Mar. Environ. Lab. (NOAA).

HOOS, L. M., and G. A. PACKMAN. 1974. The Fraser River estuary, status of environmental knowledge to 1974. Spec. Estuar. Ser. 1: 518 p.

HOUGHTON, D. 1969.* Acapulco '68. Weather 24: 2–18.

HUGHES, B. A. 1969. Characteristics of some internal waves in Georgia Strait. Def. Res. Establ. Pac. Tech. Mem. 69–2: 9 p.

HUGHES, B. A., and H. L. GRANT. 1978. The effect of internal waves on surface wind waves. 1. Experimental measurements. J. Geophys. Res. 83: 443–454.

IVERSEN, H. W. 1952. Laboratory study of breakers. U.S Natl. Bur. Stand. Circ. 521: 9–32.

JAMES, R. W. 1969. Abnormal changes in wave heights. Mariners Weather Log. 13: 252–255.

JEFFREYS, H. 1925. On the formation of waves by wind. Proc. R. Soc. London, Ser. A. 107: 189–206.

KING, C. A. M. 1966.* Beaches and coasts. Edwards Arnold Ltd., London. 403 p.

KINSMAN, B. 1965. Wind waves. Prentice-Hall, N. J. 676 p.

KOMAR, P. P. 1976.* Beach processes and sedimentation. Prentice-Hall, N. J. 429 p.

LACOMBE, H., and P. TCHERNIA. 1972. Caractères hydrologiques et circulation des eaux en Méditerranée, p. 25–36. *In* D. J. Stanley [ed.] The Mediterranean Sea: a natural sedimentation laboratory. Dowden, Hutchinson and Ross, Stroudsburg.

LEBLOND, P. H. 1972. The theory of tides. Dep. Oceanog. Univ. British Columbia. 98 p.

LEBLOND, P. H., and L. A. MYSAK. 1978. Waves in the ocean. Elsevier Scientific Publishing Company, Amsterdam. 602 p.

LEVINGS, C. D., and N. G. MCDANIEL. 1980. Data report on effects of dissolved oxygen deficiency in bottom waters of Howe Sound: trawl data January 1978 to March 1979 and oceanographic data August 1977 to March 1979. Can. Data Rep. Fish. Aquat. Sci. 217: 87 p.

LUM, K. 1975. Erosion of the Point Grey cliffs. BSc. Thesis. Univ. British Columbia. 40 p.

MCLEAN, D. G. 1975. Marine erosion at Towers Beach. BSc. Thesis. Univ. British Columbia. 127 p.

MALLORY, J. K. 1974. Abnormal waves on the south east coast of South Africa. Int. Hydrogr. Rev. 51: 99–118.

MARMER, H. A. 1926. Coastal currents along the Pacific coast of the United States. U.S. Coast Geod. Surv. Spec. Publ. 121: 80 p.

MARINE ENVIRONMENTAL DATA SERVICE, Department of Fisheries and Oceans. (unpublished manuscripts)

Waves recorded off Roberts Bank, B.C., Station 106, February 7, 1974, to April 3, 1976. File 108-5.

Waves recorded off Sturgeon Bank, Station 102, February 7, 1974 to April 1, 1976. File 102-5

Waves recorded off Station 'PAPA', C.C.G.S. Quadra/96, C.C.G.S. Vancouver/100, February 2, 1974 to January 10, 1979. File 100-4 M.

Waves recorded off Tofino, B.C., Stn. 103, June 26, 1970 to August 21, 1976. File 103-3.

Waves recorded off West Vancouver, B.C., Stn. 106, December 19, 1972 to May 18, 1974. File 106-1.

Waves recorded off Prince Rupert, B.C., September 28, 1972 to June 13, 1973. File 104-1.

MILES, J. W. 1957. On the generation of surface waves by shear flows. J. Fluid Mech. 3: 185–204.

MURTY, T. S. 1977. Seismic sea waves — tsunamis. Bull. Fish. Res. Board Can. 198: 337 p.

MYSAK, L. A., 1980. Recent advances in shelf wave dynamics, *Rev. Geophys. Space Phys.*, 18: 211–241.

NATIONAL MARINE FISHERIES SERVICE (NOAA). 1977a. Sea surface temperature charts, for July, August, and December 1977 in fishing information supplements. U.S. Dep. Comm.

1977b. Fishing information. J. A. Renner [ed.] U.S. Dep. Comm. No. 1, 3, 5, 7, 9, 11.

NEWTON, I. 1687. Principia Mathematica. 680 p. (Revised translation by F. Cajori, Univ. California Press, Berkeley (1946)).

OVERLAND, J. E. 1978. Oil-spill trajectory modeling, p. 36–41. *In* G. A. Cannon [ed.] Circulation in the Strait of Juan de Fuca: some recent observations. U.S. Natl. Ocean. Atmos. Adm. Tech. Rep. ERL 399-PMEL29.

PARKER, B. B. 1977. Tidal hydrodynamics in the Strait of Juan de Fuca–Strait of Georgia. NOAA Tech. Rep. 69: 56 p.

PARKS CANADA. 1974.* Be oceanwise . . . otherwise . . . . Dep. Indian North. Aff. 1 p.

PHILLIPS, O. M. 1957. On the generation of waves by turbulent wind, J. Fluid Mech. 2: 417-445.

PICKARD, G. L., and K. RODGERS. 1959. Current measurements in Knight Inlet, British Columbia. J. Fish. Res. Board Can. 16: 635–684.

PIERSON, W. J., G. NEUMANN, and R. W. JAMES. 1971. Observing and forecasting ocean waves. U.S. Nav. Oceanog. Off. 284 p.

PORE, N. A. 1964. The relation of wind and pressure to extratropical storm surges at Atlantic City. J. Appl. Meteorol. 3: 155–163.

RAFF, A. D., and R. G. MASON. 1961. Magnetic survey off the west coast of North America, 40°N to 52°N latitude. Geol. Soc. Am. Bull. 72: 1267–1270.

RIDDIHOUGH, R. P., and R. D. HYNDMAN. 1976. Canada's active western margin — the case for subduction. Geosci. Can. 4: 269–278.

SANDILANDS, R. W. 1971/72. Hydrographic charting and oceanography on the west coast of Canada from the eighteenth century to present day. Proc. R. Soc. Edinburgh Sect. B 73: 75–83.

SCHUMACHER, A. 1939. Stereophotogrammetrische Wellenaufrahmen. (Stereophotogrammetric wave photos.) Wiss. Ergebnisse Dtsch. Atl. Exped. Forsch. Vermessungsschiff Meteor, 1925–1927 Vol. 7: 86 p.

SCRINGER, J. A., and W. HALLIDAY. 1971. Bathymetry of Pacific seamounts Bowie and Union. Deep Sea Res. 18: 123–126.

SENGUPTA, S. 1966. Studies on orientation and imbrication of pebbles with respect to cross-stratification. J. Sediment. Petrol. 36: 362–369.

SNODGRASS, D., G. W. GROVES, K. F. HASSELMANN, G. R. MILLER, W. H. MUNK, and W. H. POWERS. 1966. Propagation of ocean swell across the Pacific, Philos. Trans. R. Soc. London Ser. A 259: 431–497.

SPAETH, M. G., and S. C. BERKMAN. 1967. The tsunami of March 28, 1964, as recorded at tide stations. U.S. Coast. Geod. Surv. Tech. Bull. 33: 86 p.

SVERDRUP, H. U., and W. H. MUNK. 1947. Wind, sea and swell theory of relationships in forecasting. U.S. Dep. Navy Hydrogr. Off. Publ. 601: 44 p.

TABATA, S. 1972. The movement of Fraser River-influenced surface water in the Strait of Georgia as deduced from a series of aerial photographs. Pac. Mar. Sci. Rep. 72—6: 69p.

THOMSON, R. E. 1976. Tidal currents and estuarine-type circulation in Johnstone Strait, British Columbia. J. Fish. Res. Board Can. 33: 2242–2264.

THOMSON, R. E., and W. S. HUGGETT. 1980. M_2 baroclinic tides in Johnstone Strait, British Columbia. J. Phys. Oeanogr. 10: 1509–1539.

THOMSON, R. E., W. S. HUGGETT, and L. S. C. KUWAHARA. 1980. Data record of current observations Volume VIII, Discovery Passage, Johnstone Strait, and Queen Charlotte Strait, 1976, 1977, 1978, (1979). Inst. Ocean Sci. Sidney, B.C. 262 p.

TYNER, R. V. 1951. Paths taken by the cold air in polar outbreaks in British Columbia. Dep. Transp. Meteorol. Div. Tech. Rep. 106: 13 p.

URSELL, F. 1956. Wave generation by wind. p. 216–249. *In* G. K. Batchelor and R. M. Davies [ed.] Surveys in mechanics. Cambridge Univ. Press.

U.S. ARMY COASTAL ENGINEERING RESEARCH CENTER. 1977. Shore protection manual. Vol. 1: 188 p.

U.S. ARMY CORPS of ENGINEERS. 1976. Ediz Hook beach erosion control. Gen. Des. Memo. 21: 14 p.

U.S. DEPARTMENT of COMMERCE (NOAA). 1978. Climatic Atlas of the outer continental shelf waters and coastal regions of Alaska, Vol. 1, Gulf of Alaska. Univ. Alaska, Anchorage. 439 p.

VAN DORN, W. G. 1974.* Oceanography and seamanship. Dodd, Mead and Co., New York. 481 p.

VINE, F. J., and D. H. MATHEWS. 1963. Magnetic anomalies over ocean ridges. Nature 199: 947–949.

WALBRAN, J. T. 1971.* British Columbia coast names, 1592–1906. J. J. Douglas Ltd., Vancouver. 546 p.

WALDICHUK, M. 1957. Physical oceanography of the Strait of Georgia, British Columbia. J. Fish. Res. Board Can. 14: 321–486.

1958. Drift bottle observations in the Strait of Georgia. J. Fish. Res. Board Can. 15: 1065–1102.

WATER SURVEY of CANADA. 1974. Historical streamflow summary. Dep. Environ. Water Resour. Br. 694 p.

WICKETT, P. W. 1973. An unusually strong current in Hecate Strait, September, 1968. Fish. Res. Board. Can. Tech. Rep. 395: 23 p.

WIGEN, S. O., and W. R. WHITE. 1964. Tsunami of March 27–29, 1964, west coast of Canada. Dep. Mines Tech. Surv. 6 p.

WILSON, B. W. 1964. Generation and dispersion characteristics of tsunamis, p. 413–444. *In* Studies on oceanography.

WILSON, E. E. 1976. Hypothermia and cold water survival. Mar. Weather Log 20(3): 136–138.

WOLFERSTAN, W. 1976. *Pacific yachting's cruising guide to the Gulf Islands and Vancouver Island from Sooke to Courtenay. Interpress Publications Ltd., Vancouver, B.C. 190 p.

WYLLIE, P.J. 1976.* The way the earth works. John Wiley and Sons, Inc., New York. 296 p.

YORATH, C.J., B.D. BORNHOLD, AND R.E. THOMSON, 1979. Oscillation ripples on the northeast Pacific continental shelf. Mar. Geol. 31: 45–58.

APPENDICES

Appendix A. Metric – English Units and Equivalents

Length

1 kilometre (km) = 1000 metres = 0.6214 statute miles = 0.5396 nautical miles

1 metre (m) = 100 centimetres = 3.281 feet (ft) = 0.5468 fathoms (fa)

1 centimetre (cm) = 10 millimetres (mm) = 0.3937 inches (in.)

1 nautical mile (nm) = 6080 ft = 1.152 statute miles = 1.853 km

1 statute mile = 5280 ft = 0.8684 nm = 1.609 km

1 fathom (fa) = 6 ft = 1.829 m

1 foot (ft) = 0.3048 m

1 inch (in.) = 2.540 cm

1° latitude = 60 nm = 111.18 km

1° longitude (at 50° latitude) = 38.5 nm = 71.3 km

Area

1 square km (sq km *or* km^2) = 0.3861 sq miles

1 hectare (ha) = 10,000 m^2 = 2.47 acres

Volume

1 cubic km (cu km *or* km^3) = 0.240 cu miles

1 cubic m (cu m *or* m^3) = 35.3 cu ft = 264.2 U.S. gallons (gal.) = 220.0 Imperial gal.

1 cubic cm (cm^3) = 0.0610 cu in.

1 litre (L) = 1000 cm^3 = 0.264 U.S. gal. = 0.220 Imperial gal.

Speed

1 km/h = 27.78 cm/s = 0.6214 miles per hour (mph) = 0.5396 kn

1 m/s = 3.60 km/h = 2.237 mph = 1.943 knots (kn)

1 knot (nm/h) = 1.151 mph = 1.853 km/h = 51.44 cm/s

1 mph = 0.8684 kn = 1.6093 km/h

1 ft/s = 30.48 cm/s = 0.6214 mph = 0.5396 kn

The term "knot" originates with the Dutch who used a piece of wood or log attached to a knotted line to measure ship motion. The log was thrown overboard and the line payed out. The number of equally spaced knots that passed through the sailor's hands in a certain period of time was a measure of ship's speed relative to the water. There were 47 ft 3 in. (14.4 m) between knots and time was measured via a 28-s sandglass.

Mass

1 kilogram (kg) = 1000 grams = 2.205 pounds (lb)

1 gram (g) = 0.0353 ounces (oz)

1 metric ton (t) = 1000 kg = 1.102 short tons = 2205 lb

Pressure

1 atmosphere = 1013.25 millibars = 760 millimetres mercury (mm Hg) = 29.92 in. Hg = 14.7 lb/sq in.

1 millibar (mb) = 0.750 mm Hg = 0.0145 lb/sq in. = 100 pascals (pa)

In the ocean, the ambient pressure increases by one atmosphere for every 10 m (or 33 ft) of depth. An increase of 1 mb in atmospheric pressure depresses sea level by about 1 cm.

Temperature Conversion

°C = (°F − 32°) × 5/9 °F = (9/5 × °C) + 32

At sea level fresh water freezes at 0°C = 32°F and boils at 100°C = 212°F.

Constants

Gravitational acceleration at the earth's surface (g) = 980 cm/s^2 = 32 ft/s^2

Equatorial radius of earth = 6378 km

Polar radius of earth = 6357 km

Mean earth–moon distance = 384,393 km

Mean earth–sun distance = 149,450,000 km

Average depth of oceans = 3729 m

Total volume of oceans = 1350 million km^3

Area and mean depth of

Pacific Ocean = 181 million km^2; 3940 m

Atlantic Ocean = 94 million km^2; 3575 m

Indian Ocean = 74 million km^2; 3840 m

Arctic Ocean = 12 million km^2; 1117 m

Appendix B.
Marine Research Institutions of the Northwest Pacific Coast

Department of Fisheries and Oceans Pacific Environment Institute West Vancouver, B.C.	Marine ecology, biological and chemical oceanography
Department of Energy, Mines and Resources Pacific Geoscience Centre Sidney, B.C.	Marine geology and geophysics
Department of Fisheries and Oceans Institute of Ocean Sciences Sidney, B.C.	Chemical and physical oceanography, ocean engineering, hydrography, marine ecology
Department of Fisheries and Oceans Pacific Biological Station Nanaimo, B.C.	Biological and physical oceanography, fisheries management and research, marine ecology
Department of National Defence Defense Research Establishment Pacific Esquimalt, B.C.	Ocean acoustics, ocean engineering, marine physics

Department of the Navy Applied Physics Laboratory Seattle, Washington	Ocean engineering, marine physics
International Pacific Salmon Fisheries Commission New Westminster, B.C.	Management of Fraser River pink and sockeye salmon stocks
National Oceanic and Atmospheric Administration National Ocean Survey Seattle, Washington	Physical oceanography, ocean engineering, hydrography
National Oceanic and Atmospheric Administration Northwest Fisheries Center Seattle, Washington	Deep-sea fisheries research, physical oceanography
National Oceanic and Atmospheric Administration Pacific Marine Environmental Laboratory Seattle, Washington	Physical, biological, chemical, and geological oceanography
Oceanographic Institute of Washington Seattle, Washington	Ocean engineering (operates the Northwest Regional Calibration Center for testing and calibrating oceanographic instruments)
Oregon State University School of Oceanography Corvallis, Oregon	Biological, chemical, geological, and and physical oceanography; ocean engineering and fisheries (operates the Marine Science Center at Newport, Oregon)
Department of National Defense Royal Roads Military College Esquimalt, B.C.	Physical oceanography
Simon Fraser University Department of Biological Sciences Burnaby, B.C.	Marine ecology, biological oceanography
University of Alaska Institute of Marine Sciences Fairbanks, Alaska	Biological, chemical, geological, and physical oceanography; ocean engineering
University of British Columbia Department of Oceanography Vancouver, B.C.	Biological, chemical, geological, and physical oceanography
University of Victoria Department of Biology Victoria, B.C.	Marine ecology and biological oceanography (the Bamfield Marine Station is affiliated with the Universities of British Columbia and Victoria)
University of Washington Department of Oceanography College of Fisheries and College of Engineering. Seattle, Washington	Biological, chemical, geological, and physical oceanography; ocean engineering and fisheries (operates the Friday Harbor Marine Biological Station on San Juan Island)
Walla Walla College Marine Biological Station Anacortes, Washington	Marine ecology
Western Washington State College Aquatic Studies Program Bellingham, Washington	Aquatic sciences (operates Shannon Point Marine Center, Anacortes)

Appendix C. Immersion Hypothermia

Immersion hypothermia is the subnormal temperature of a body that has been immersed in water. Except in tropical waters warmer than 20–25°C, it is the primary threat to life during prolonged immersion. Death from cardiac arrest usually follows once the normal rectal temperature of around 37.6°C drops below 25.9°C.

The effect of the cold water is to lower body temperature. This leads to a slowing of the heartbeat, a reduction in the metabolic rate, and an increase in the amount of carbon dioxide in the blood. The victim's mental capacity is impaired and becomes an important factor in his chances of survival. Numerous reports from shipwrecks and accidents in cold waters show that people may become confused or delirious, to further increase the possibility of death by hypothermia.

The length of time a person can survive in the ocean depends on the temperature of the surface water, the type of garment worn and, to a lesser extent, on the person's behavior. Approximate human survival times are presented in the following table. These can vary with body type. Thin people, for example, become hypothermic before fat people. Extremely fat people may last almost indefinitely in water near 0°C, provided they are warmly clothed.

Water temperature	Exhaustion or unconsciousness	Expected time of survival
0°C	15 min	15–45 min
0– 5	15–30 min	30–90 min
5–10	30–60 min	1– 3 h
10–15	1– 2 h	1– 6 h
15–20	2– 7 h	2–40 h
20–25	3–12 h	3–indefinite h
25	indefinite	indefinite

The rate of hypothermia can be reduced by proper behavior and gear. This has been conclusively demonstrated by researchers at the University of Victoria who closely monitored more than 500 immersions in oceanic waters with temperatures between 4 and 16°C (Wilson 1976). It was found that if the critical heat loss areas could be protected, then survival time would increase. The Heat Escape Lessening Posture (HELP) was developed for lone persons and the "Huddle" for small groups. Both methods require life preservers. HELP shows how to hold the upper arms firmly against the sides of the chest, keep the thighs together, and raise the knees to protect the groin area. In the Huddle, people face one another and keep as close together as possible. These positions improve survival time in 9°C water (common to much of the B.C. coast in winter) to 4 h, about twice that of a swimmer and one and one-half times that of a person in a passive position. Scientists at the University of Victoria have also been instrumental in the design of floater coats and the development of methods to treat hypothermia victims. It is found, for example, that warming the outside of a person who suffers from hypothermia may hasten his death because further strain is put on the heart as the outer blood vessels expand. Blankets alone serve little purpose because the person's outer skin does not emit heat. The best treatment is to have the person inhale temperature regulated "steam", which then distributes the heat where it is needed, to the inner organs. (Further information on hypothermia can be found in *British Columbia Small Craft Guide* Vol. 1, 1979 p. 25–26.)

Appendix D. Lunar Versus Solar Tide-Generating forces

The approximate magnitude of the tide-generating force for the earth–moon pair has the relationship $f \sim m/d^3$, and that for the earth–sun pair the relationship $F \sim M/D^3$, for the following quantities:

$m = 7.347 \times 10^{25}$ g is the mass of the moon
$M = 1.971 \times 10^{33}$ g is the mass of the sun
$d = 3.844 \times 10^5$ km is the average distance from earth to the moon
$D = 1.495 \times 10^8$ km is the average distance from earth to the sun

The ratio of these two forces is then $f/F = (m/M)(D/d)^3 = 2.193$ so that the moon's tide-generating force is more than twice that of the sun's. Clearly, the moon's closer proximity to the earth more than makes up for its much smaller mass.

Color Plates

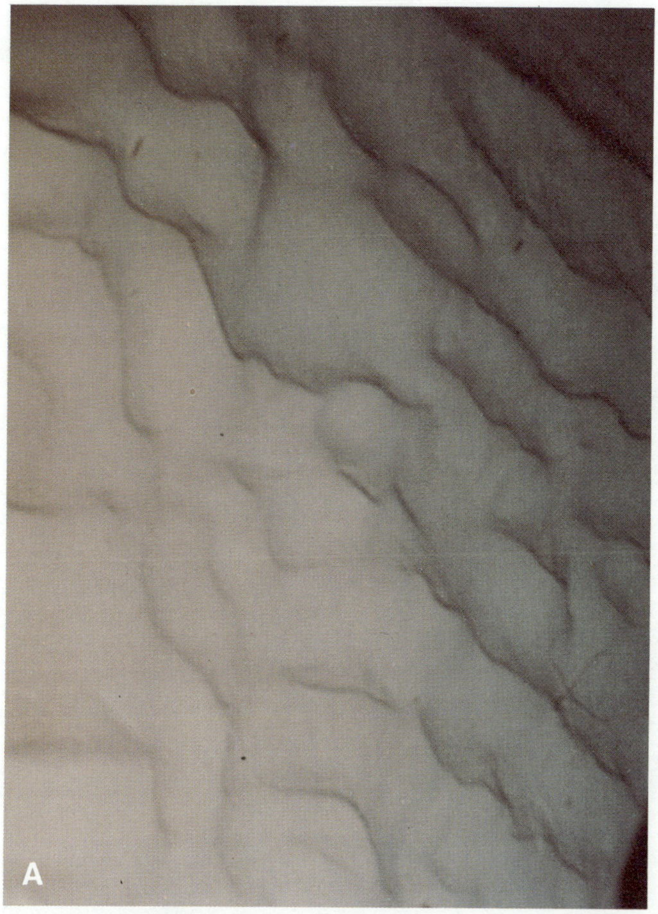

PLATE 1. A relict beach terrace at 100 m depth on Cobb Seamount photographed from *Pisces IV* submersible. (A) Current-induced sand ripples. (Courtesy S. Huggett)

PLATE 1(B). Rounded pebbles. (Courtesy S. Huggett)

PLATE 2. Howe Sound, August 6, 1972, near Lower Low Water. Light colored regions are silt-laden surface brackish waters originating from Squamish River at the top. Compare with Fig. 10.39. (Government of Canada Air Photo 1972)

PLATE 3. Southerly end of Long Beach, Wickaninnish Bay between Tofino and Ucluelet on west coast Vancouver Island. For location see Fig. 8.12. (Government of British Columbia Air Photo 1978)

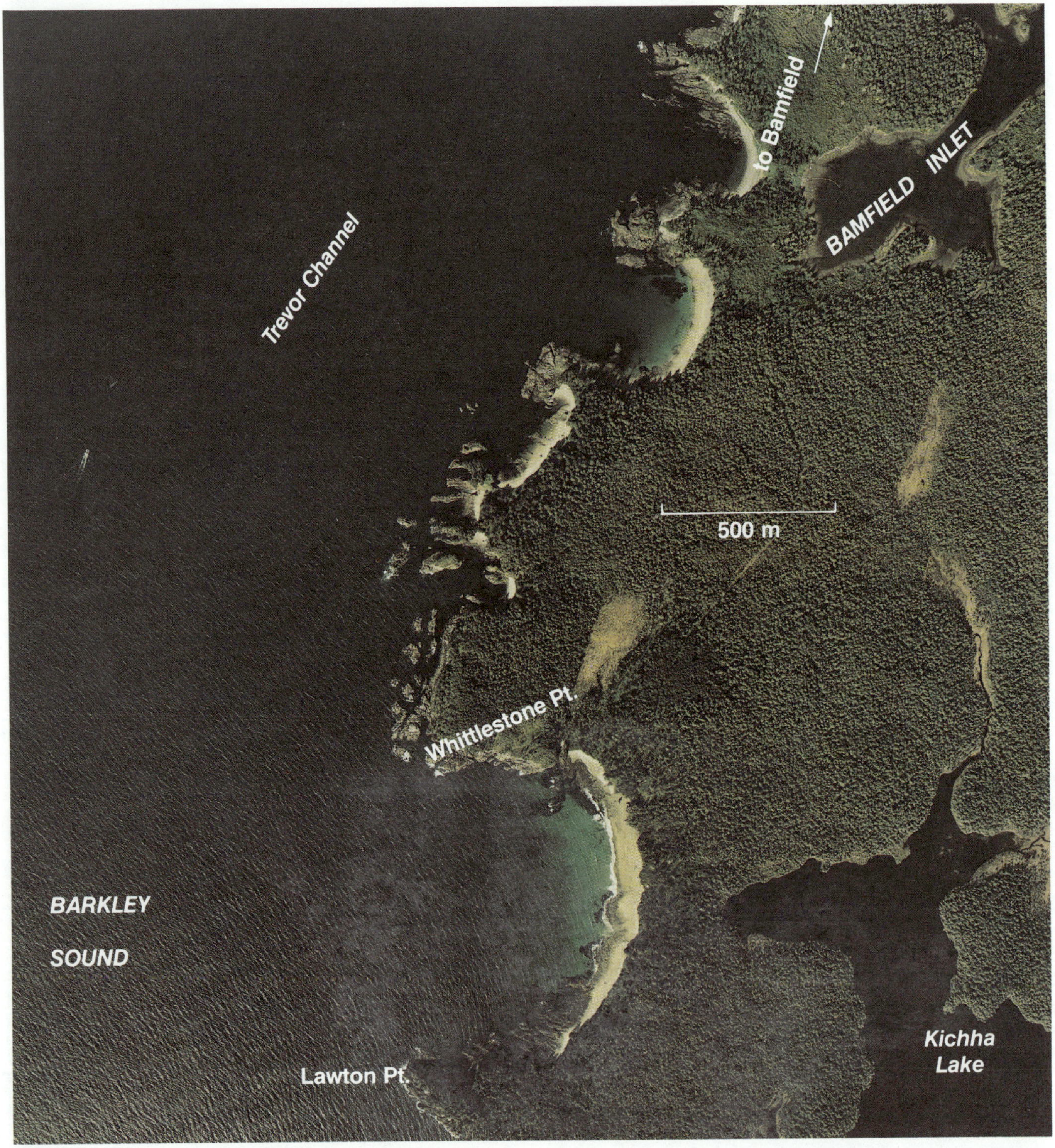

PLATE 4. Pocket beaches on Trevor Channel, Barkley Sound, west coast Vancouver Island, 5 km southwest of Bamfield. Sands and other unconsolidated sediments accumulate between more weather resistant headlands. (Government of British Columbia Air Photo 1978)

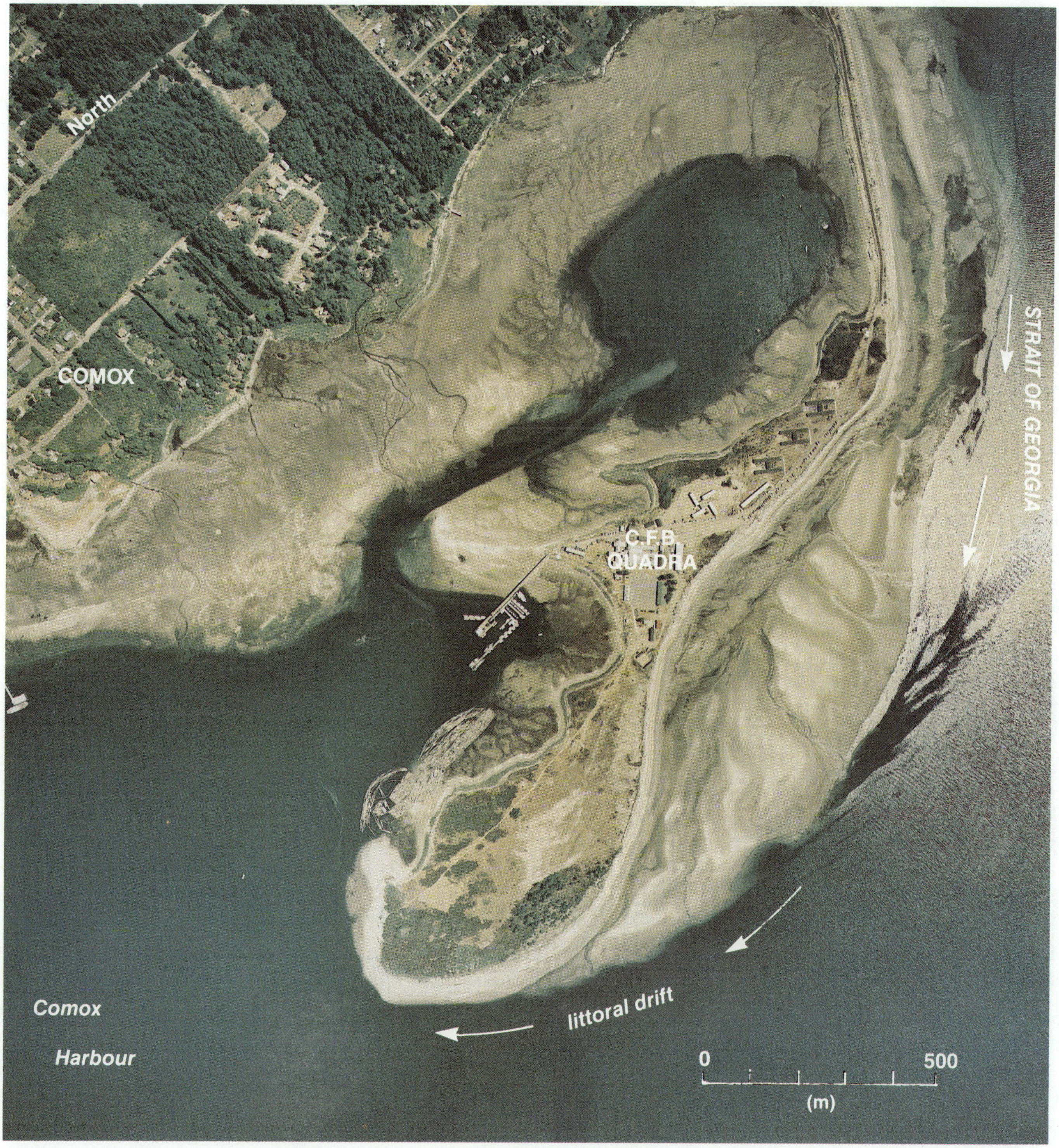

PLATE 5. Goose Spit, near Comox, Strait of Georgia. Large sand ripples exposed during low tide. (Government of Canada Air Photo 1978)

PLATE 6. Sea arch; Hole-in-the-Wall, west coast Vancouver Island. (Courtesy R. Macdonald)

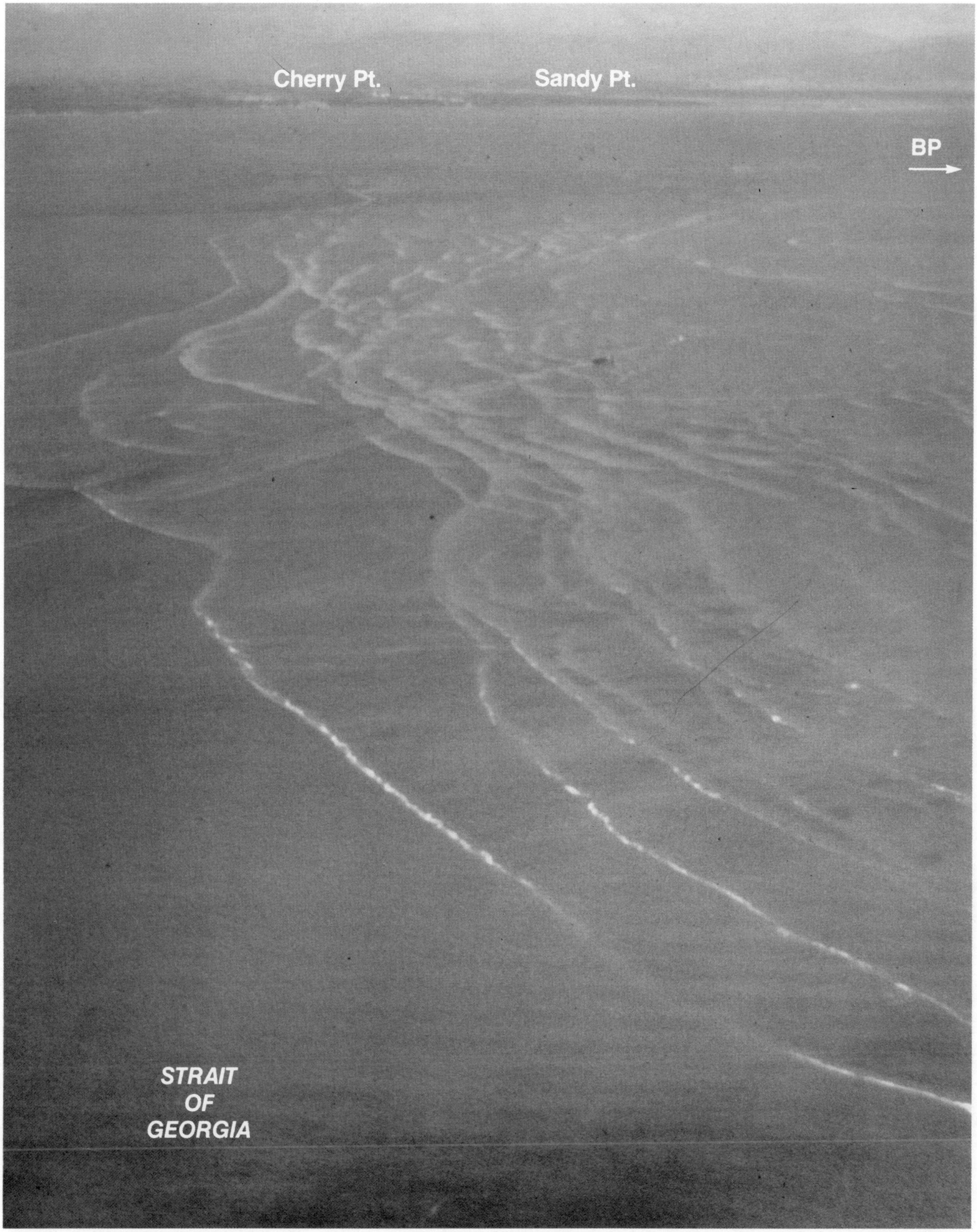

PLATE 7. Surface slicks generated by internal gravity waves propagating on shallow pycnocline (sharp density interface) in Strait of Georgia late May 1972. Waves propagate northwestward (to left) from vicinity of Boundary Passage (BP). View to east toward U.S. mainland 5 km south of Birch Bay. Barge in upper central is about 70 m long. (From LeBlond and Mysak 1978)

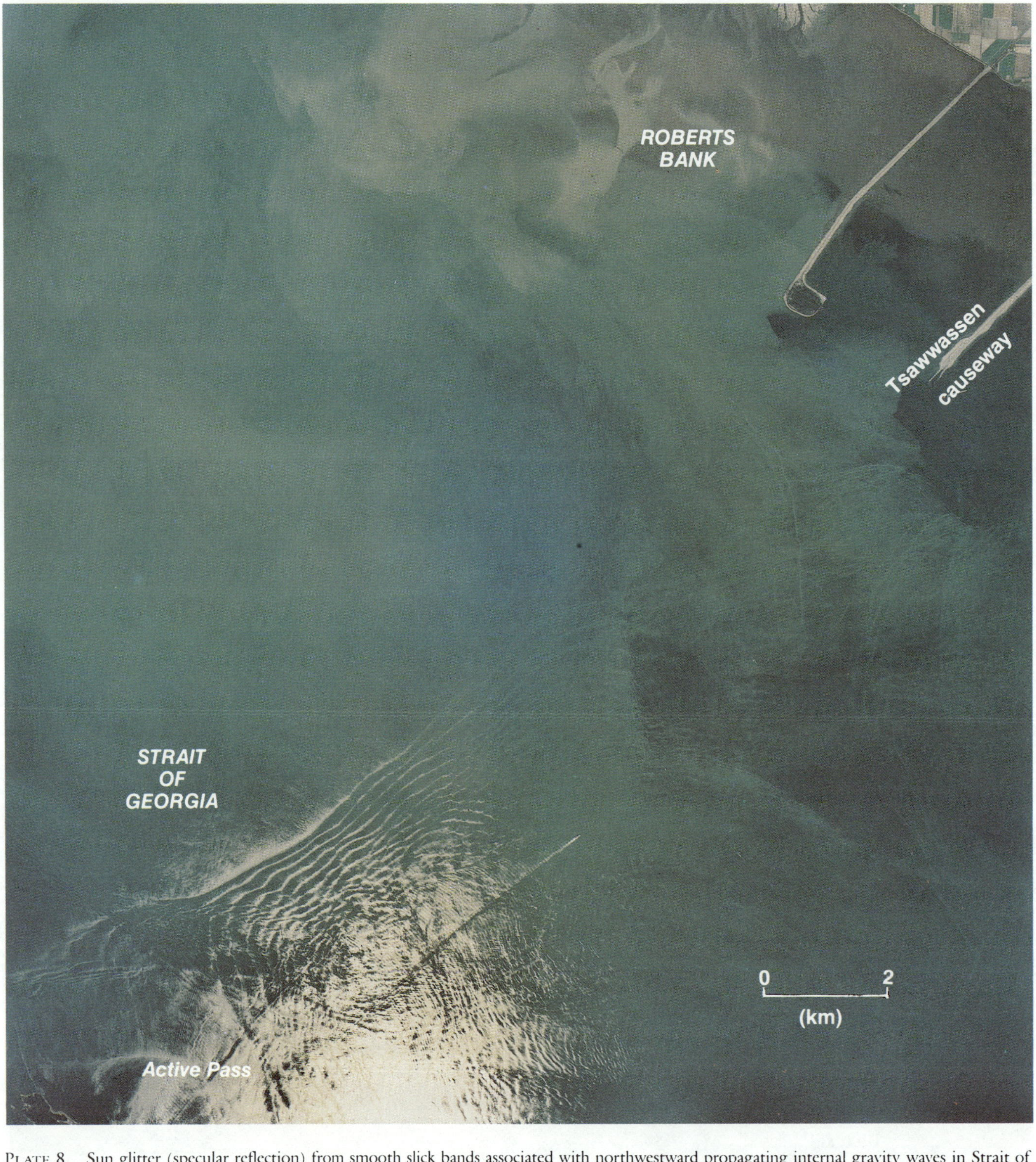

PLATE 8. Sun glitter (specular reflection) from smooth slick bands associated with northwestward propagating internal gravity waves in Strait of Georgia, August 1972. Fraser River delta, Roberts Bank Superport (Westshore Terminals), and Tsawwassen Causeway, upper right; Active Pass, lower left. See Fig. 10.1 for location. (Government of British Columbia Air Photo 1972)

PLATE 9.A. *Noctiluca* red water (red tide) at head of East Sound, Orcas Island, San Juan Islands, July 1976. Eel grass at bottom grows at shoreline. (Courtesy F. J. R. T. Taylor)

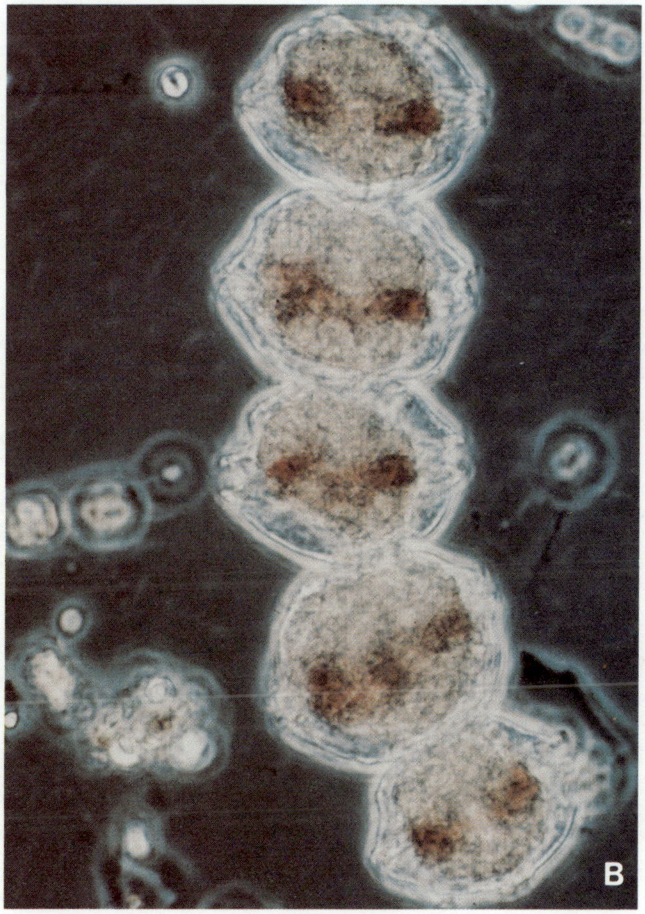

PLATE 9.B. Phytoplankton *Gonyaulax Catenella*, best known of three organisms responsible for paralytic shellfish poisoning in B.C. waters; cells approximately 40 μ (0.04 mm, 0.0016 in.) across. (Courtesy F. J. R. T. Taylor)

PLATE 10. (A) Extension of turbid surface waters from northern end of Quatsino Narrows during early flood, May 8, 1974. (See Fig. 4.5, 4.7 for locations.) (B) Surface turbidity around northern end of Quatsino Narrows and mouth of Rupert Inlet, Aug. 3, 1973. (Courtesy D. Goyette)

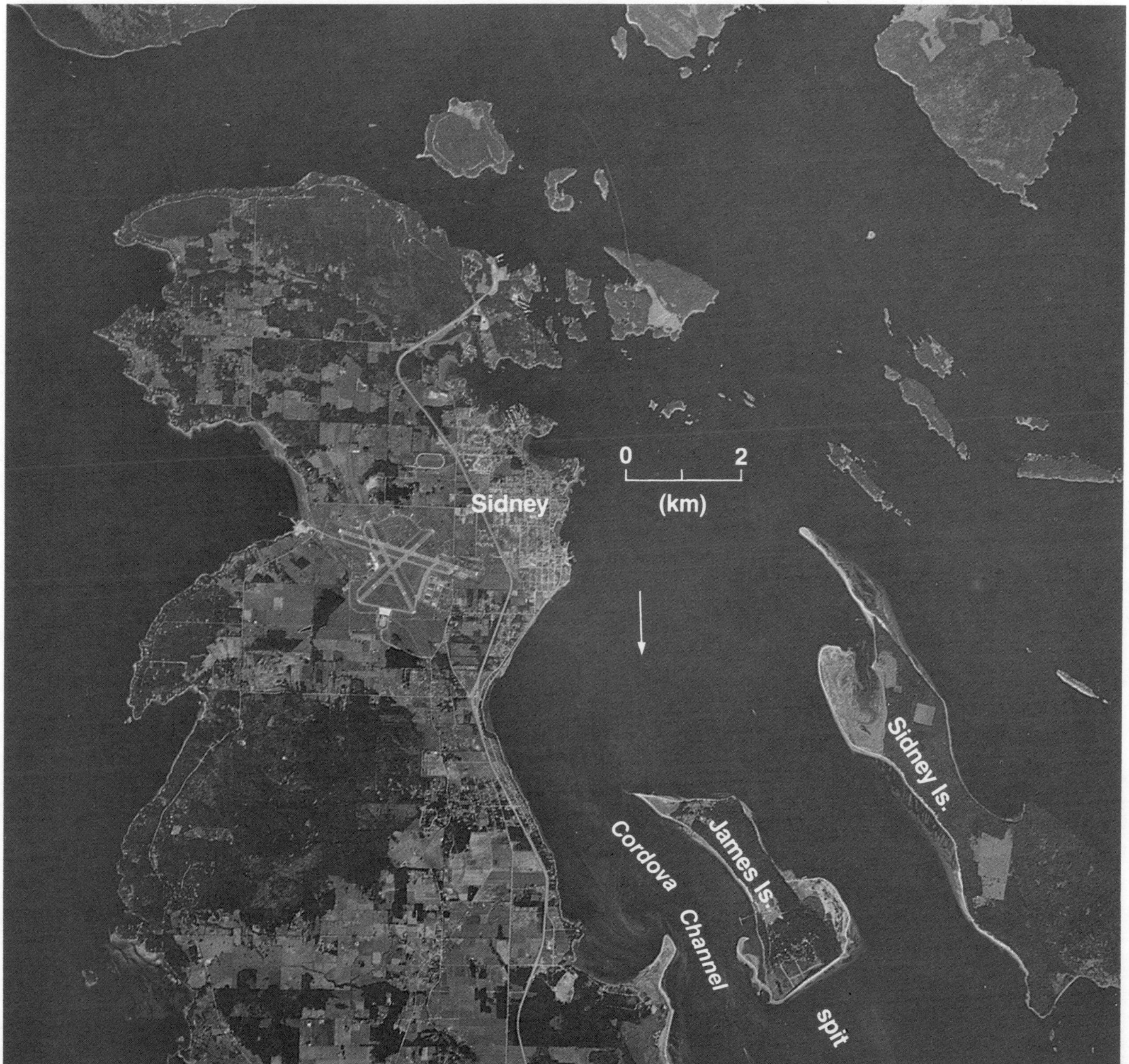

PLATE 11. James Island, southeast of Sidney, Vancouver Island. Two light-colored, wakelike trails (presumably suspended bottom sediments) appear to have formed in the lee of the island at the sides of James Spit as the southward flowing ebb current separates from the end of the island. (Photo also shows Sidney Spit at the northern end of Sidney Island; see Fig. 2.26d.). (Government of British Columbia Air Photo, August 1972.)

PLATE 12. Area between the Middle Arm of Fraser River (A) and Point Atkinson (B) approximately 3 h after low water, August 1972. Sediment-laden water originates mainly from South Arm of the river (below the photo) and moves northward off delta front to combine with silty outflow from North Arm (C). A portion of water has entered outer basin of Burrard Inlet and is being advected through First Narrows (D). Some dispersed sediments reached Passage Island (E); others formed counterclockwise eddy west of Point Grey (F). Ships anchored in English Bay are aligned in flow direction (arrows). Note sand ripples on exposed sand flats and clear wake behind power boat (p). For location map see Fig. 10.1, 10.33. (Adapted from Government of British Columbia Air Photo 1972)

PLATE 13. Colony of sea lions on a rocky outcrop immediately west of Sartine Island, Scott Island Chain, northwest tip of Vancouver Island, June 1979. (Courtesy B. Minkley)

Index

Index

Roman type is used for location in text, italic decimal numbers are used for figure numbers, numbers preceded by the letter t are Table numbers, italic numbers preceded by the letter *P* are Plate numbers.